Waves occur widely in nature and have innumerable commerc responsible for the sound of speech, bow waves created by me earth's atmosphere, ultrasonic waves are used for medical im used for the synthesis of new materials.

Although much has been written about these linear and nonlinear waves, most books approach the topic at an advanced level. This book provides a thorough, modern introduction to the subject.

Beginning with fundamental concepts of motion, the book goes on to discuss linear and nonlinear mechanical waves, thermodynamics, and constitutive models. In contrast to many classic works, which were limited to nonlinear waves in gases, this text also includes liquids and solids as integral parts of the subject. Among the important areas of research and application that will benefit from this text are impact analysis, shock wave research, explosive detonation, nonlinear acoustics, and hypersonic aerodynamics.

Students at an advanced undergraduate/graduate level will find this text a clear and comprehensive introduction to the study of wave phenomena, and it will also be a valuable professional reference for engineers and applied physicists.

Introduction to Wave Propagation in Nonlinear Fluids and Solids

Introduction to Wave Propagation in Nonlinear Fluids and Solids

D. S. DRUMHELLER
Sandia National Laboratories

CAMBRIDGE
UNIVERSITY PRESS

PUBLISHED BY THE PRESS SYNDICATE OF THE UNIVERSITY OF CAMBRIDGE
The Pitt Building, Trumpington Street, Cambridge, CB2 1RP, United Kingdom

CAMBRIDGE UNIVERSITY PRESS
The Edinburgh Building, Cambridge CB2 2RU, United Kingdom
40 West 20th Street, New York, NY 10011-4211, USA
10 Stamford Road, Oakleigh, Melbourne 3166, Australia

First published 1998

Printed in the United States of America

Typeset in Times Roman

Library of Congress Cataloging-in-Publication Data

Drumheller, D. S.
Introduction to wave propagation in nonlinear fluids and solids /
D. S. Drumheller.
p. cm.
Includes bibliographical references (p. –) and index.
ISBN 0-521-58313-6 (hardback). – ISBN 0-521-58746-8 (paperback)
1. Wave motion. Theory of. 2. Nonlinear waves. I. Title.
QA927.D77 1998
531′.1133 – dc21 97-3266
CIP

A catalogue record for this book is available from the British Library

ISBN 0 521 58313 6 hardback
ISBN 0 521 58746 8 paperback

To
Phylis,
my love, my friend,
who gave me MeLinda and Paul
one Monday morning

Contents

Preface

. . . one day, well off the Patagonian coast, while the sloop was reaching under short sail, a tremendous wave, the culmination, it seemed, of many waves, rolled down upon her in a storm, roaring as it came. I had only a moment to get all sail down and myself up on the peak halyards, out of danger when I saw the mighty crest towering masthead-high above me. The mountain of water submerged my vessel. She shook in every timber and reeled under the weight of the sea, but rose quickly out of it, and rode grandly over the rollers that followed. It may have been a minute that from my hold in the rigging I could see no part of the *Spray's* hull. Perhaps it was even less time than that, but it seemed a long while, for under great excitement one lives fast, and in a few seconds one may think a great deal of one's past life.

—Captain Joshua Slocum
Sailing Alone Around the World, 1895

During July 1994 while I was writing this book, fragments of the comet Shoemaker–Levy 9 crashed into Jupiter. As each fragment entered the Jovian atmosphere, it formed a shock wave, similar to the bow wave in front of a boat. Heat generated within these waves ignited the atmosphere of Jupiter creating fireballs the size of the earth. Before this event, a friend, Tim Trucano, and several other physicists at Sandia National Laboratories were immersed in a theoretical effort to predict the effects of these impacts. Tim told me that the energy released by the impacts might actually be observable from earth by telescope. At that time, few people believed this would be possible. Actually, Tim's comment proved to be quite an understatement. The impacts produced a spectacle viewed around the world.

In essence, this book describes the theory these people used to make a most successful prediction of the comet impacts. Other people have used the same theory to describe many different events such as volcanic eruptions, Mach waves in front of supersonic aircraft, explosive forming of metals, impact absorption of armor plate, and the treatment of human kidney stones by medical lithotripsy. In fact, nonlinear waves occur widely in nature and have innumerable commercial uses. They are also an experimental tool providing scientists easy access to pressures and temperatures that are unattainable by most other laboratory techniques. The scientific and commercial demands for accurate predictions of nonlinear wave phenomena are growing, and powerful computers and commercial software make these types of computations accessible to a wide range of people. In addition, there is an imposing volume of research literature and several treatises devoted exclusively to the theory of nonlinear waves. What is missing is a comprehensive text written at an introductory level.

I have designed this book to fill the need for an introductory text. In writing it, I have been greatly influenced by two classical works: *Supersonic Flow and Shock Waves* by R. Courant and K. O. Friedrichs (Springer-Verlag, 1948) and *Physics of Shock Waves and*

High-Temperature Hydrodynamic Phenomena by Ya. B. Zel'dovich and Yu. P. Raizer (Academic Press, 1966). Indeed, these works are truly impressive. However, upon opening them, we quickly discover that an in-depth knowledge of continuum mechanics and thermodynamics is a prerequisite. Moreover, the primary focus of these works is on nonlinear waves in gases.

In contrast, here I only require you to have a basic understanding of differential equations. The fundamental concepts of motion, deformation, strain, and stress are presented in Chapter 1. In Chapter 2, the elastic material is introduced. This is followed by analysis of both linear and nonlinear mechanical waves. Chapter 3 contains a complete development of thermodynamics with a particular emphasis on nonlinear wave processes. Chapter 4 is devoted to a detailed development of constitutive models for a variety of gases, liquids, and solids. In keeping with the introductory nature of this book, in large, I restrict the analysis to a single spatial dimension. Clearly, I hope this book will prepare you to read the classical texts mentioned above. However, I do go beyond these works in several important ways. I include not only gases but also liquids and solids as an integral part of the development; I use modern constitutive theory, which has evolved over the past 30 years; and I place special emphasis on presenting the results in both spatial and material coordinates. This emphasis is driven not only by the needs of the constitutive formulations for solids but also by the development of modern numerical methods. As I explain in the appendix, numerical integration of the equations of wave theory are often performed using either *Lagrangian algorithms*, which operate in material coordinates, or *Eulerian algorithms*, which operate in spatial coordinates. Sometimes both methods are employed in parallel in the same computation.

This book is suitable for a one- or two-semester university course taught at either the advanced undergraduate or graduate level. When taught as a one-semester course, the instructor might limit the topics to Chapters 1 and 2, which cover the fundamentals of continum mechanics and the theory of mechanical waves. The analysis in these chapters includes the linear wave equation, simple nonlinear waves, shock waves, and wave–wave interactions. During a two-semester course, the instructor could then proceed into the development of thermodynamics and the analysis of steady waves. Topics in the second semester might include ideal gases, Mie–Grüneisen solids, elastic-plastic solids, fluid-saturated porous solids, detonations, and phase transitions. In large part, I have developed the solutions in this text using analytic integration and graphical representation of nonlinear algebraic systems. Occasionally I employ routine numerical integration of ordinary differential equations, but I avoid numerical integration of partial differential equations. However, I do include sufficient information to allow you to write an elementary computer code and experiment with the numerical integration of the partial differential equations for both linear and nonlinear waves.

Although this book has been written with the student in mind, I believe it also serves as a reference for the advanced professional. For example, Chapter 1 contains detailed descriptions of stress and strain for uniaxial and triaxial symmetries that are often encountered in routine analysis tasks. I have also included diagrams of the characteristic solutions for all of the elementary wave–wave interactions in Chapter 2. Indeed, these diagrams are unique to this book. Chapter 3 contains a convenient summary of the relationships between the thermodynamic coefficients and the four fundamental thermodynamic potentials. And Chapter 4 covers a wide range of constitutive models, which before this work could only be found in the research literature.

The genesis of this book can be found in my years of research while a member of the shock physics group at Sandia National Laboratories. I owe much to my past association with many members of that group. I am particularly indebted to one of those people, Dr. Timothy Trucano. Tim had the courage to read the manuscript three times and to sit through innumerable technical discussions during our years of lunches together. Through his thoughtful advice he has had great influence on this work. Dr. Rebecca Brannon and Hari Simha reviewed the manuscript in detail, and I am very grateful for their comments. James Cazamais, Dr. J. Erni Dunn, Sandy Powell, and Dr. Jack Wise also provided me with invaluable comments and material. In addition, I owe much to my long collaboration with Professor A. Bedford of the University of Texas. "Marc" Bedford has always been a close friend, and I have always had exceptional respect for his professional advice.

Perhaps the real reason behind my writing this book can be found in another of my pastimes. Climbing rock walls has been one of the passions of my life, and it has not escaped my notice that it is an obvious metaphor for learning. Every climber knows that some climbs are not worth the effort while others are gems to be climbed over and over again. My favorite climb is *Belle Fourche Buttress* on Devils Tower in Wyoming. It is a thin, elegant crack up a seemly impossible, vertical wall. It is not the hardest climb I have done – it is just the most enjoyable. Every time I climb it, I lose my identity – I forget my name. The only things that matter are the next instant and the next move. I have found a few such intense and enjoyable moments while writing this book. I hope you find a few while reading it.

D. S. Drumheller

Belle Fourche Buttress
Devils Tower, Wyoming
August 1996

Nomenclature

Frequently used symbols are defined, their units are shown in square brackets, and the pages on which they first appear are shown in parentheses.

a	Axial acceleration [m/s^2], (74).
$\mathbf{a}$, a_k	Acceleration vector [m/s^2], (30).
$\mathbf{b}$, b_k	Specific body force [N/kg], (60).
c	Spatial wave velocity [m/s], (139).
c_0	Velocity of sound [m/s], (101).
c_0	Hugoniot parameter [m/s], (192).
c_b	Bulk velocity [m/s], (371).
c_v	Specific heat at constant volume [J/kg-K], (263).
c_p	Specific heat at constant mean normal stress [J/kg-K], (272).
c_T	Specific heat at constant axial stress [J/kg-K], (268).
c_S	Spatial shock velocity [m/s], (175).
c_{0T}	Transverse wave velocity [m/s], (116).
C	Material wave velocity [m/s], (137).
C_I	Isothermal velocity [m/s], (313).
C_S	Material shock velocity [m/s], (175).
C^η	Isentropic longitudinal stiffness [Pa], (260).
C^ϑ	Isothermal longitudinal stiffness [Pa], (266).
CJ	Chapman–Jouguet state (461).
$\mathcal{D}$	Dissipation [W/kg], (238).
D_{km}	Deformation rate tensor [1/s], (47).
$\mathbf{D}()$, $D_k()$	Direction function (16).
e	Volume strain (45).
e_{ijk}	Alternator (12).
E	Axial component of triaxial strain (51).
$\tilde{E}$	Lateral component of triaxial strain (51).
E^0	Young's modulus [Pa], (91).
E_{kn}	Lagrangian strain tensor (41).
E'_{kn}	Deviatoric strain tensor (46).
$\mathcal{E}$	Specific internal energy [J/kg], (234).
F	Axial component of triaxial deformation gradient (50).
$\tilde{F}$	Lateral component of triaxial deformation gradient (50).
$F^E, \tilde{F}^E$	Elastic components of F and $\tilde{F}$ (400).
$F^P, \tilde{F}^P$	Plastic components of F and $\tilde{F}$ (400).
F_{km}	Deformation gradient tensor (37).
F^{-1}_{km}	Inverse deformation gradient (39).
$F_\pm$	Initial and Hugoniot values of F (175).
$\mathcal{G}$	Specific Gibbs energy [J/kg], (317).

$\mathcal{H}$ Specific enthalpy [J/kg], (316).
H_ξ Rate of volume fraction [1/s], (428).
I_H, II_H, III_H Invariants of the tensor H_{ij} (24).
$\mathbf{i}$, i_k Unit coordinate vector (9).
$J_\pm$ Riemann invariants [m/s], (119).
$J_\pm$ Riemann integrals [m/s], (138).
$\mathcal{K}$ Specific kinetic energy [m^2/s^2], (236).
$\mathcal{K}^\eta$ Isentropic bulk stiffness [Pa], (271).
$\mathcal{K}^\vartheta$ Isothermal bulk stiffness [Pa], (271).
L Axial component of triaxial velocity gradient [1/s], (57).
$\tilde{L}$ Lateral component of triaxial velocity gradient [1/s], (57).
L_{km} Velocity gradient tensor [1/s], (47).
M Mach number (333)
$\mathbf{n}$ Unit vector to surface (61).
N Mole number [moles/kg], (321).
p Pressure [Pa], (66).
q Axial component of heat flux [W/m^2], (237).
$\mathbf{q}$, q_k Heat flux vector [W/m^2], (237).
$\mathcal{Q}$ Chemical reaction energy [J/kg], (457).
$\mathcal{Q}$ Artificial viscosity [Pa], (490).
r Specific external heat supply [W/kg], (237).
R Gas constant [8.3143 J/K-mole], (321).
s Hugoniot parameter (192).
S Surface [m^2], (14).
t Time [s], (29).
$\mathbf{t}$, t_k Traction vector [Pa], (61).
T_{km} Stress tensor [Pa], (62).
T'_{km} Deviatoric stress tensor [Pa], (66).
T Axial component of triaxial stress [Pa], (70).
T^0 Reference value of axial component of triaxial stress [Pa], (92).
$T_\pm$ Initial and Hugoniot values of the axial component of triaxial stress [Pa], (189).
$\tilde{T}$ Lateral component of triaxial stress [Pa], (70).
$\mathcal{T}$ Axial component of dissipative stress [Pa], (253).
u Axial displacement [m], (74).
$\mathbf{u}$, u_k Displacement vector [m], (29).
U_S Shock velocity [m/s], (175).
$\mathbf{v}$, v_k Velocity vector [m/s], (30).
v Axial velocity [m/s], (74).
V Volume [m^3], (14).
V_0 Reference volume [m^3], (44).
W_{km} Spin tensor [1/s], (47).
x Axial position [m], (74).
$\mathbf{x}$, x_k Cartesian coordinate [m], (9).
$\mathbf{x}$, x_k Position vector [m], (28).
X Axial reference position [m], (74).
$\mathbf{X}$, X_k Reference position vector [m], (28).

Y Yield stress [Pa], (392).
z Wave impedance [kg/m^2-s], (137).
z_0 Acoustic impedance [kg/m^2-s], (119).
z_S Shock impedance [kg/m^2-s], (184).
α Longitudinal coefficient of thermal expansion [1/K], (268).
α Distention ratio (438).
$\alpha_{k(m)}(t)$ Rotation component (32).
$\alpha_{k(m)}$ Direction cosine (18).
β Volumetric coefficient of thermal expansion [1/K], (271).
γ Shear strain (43).
γ Adiabatic gas constant (98).
Γ Grüneisen coefficient (262).
δ_{km} Kronecker delta (13).
ε Longitudinal strain (42).
η Specific entropy [J/kg-K], (238).
ζ Characteristic coordinate [s], (105).
θ_k Angle of rotation (19).
ϑ Absolute temperature [K], (238).
λ Lamé constant [Pa], (91).
λ Chemical reaction variable (457).
Λ Stretch (40).
Λ_ξ Interface pressure [Pa], (429).
μ Lamé constant, shear modulus [Pa], (91).
ν Poisson's ratio (91).
ξ Characteristic coordinate [s], (105).
ρ Mass density [kg/m^3], (60).
ρ Partial density of mixture constituent [kg/m^3], (422).
ρ_0 Density in reference state [kg/m^3], (76).
$\bar{\rho}$ True density of mixture constituent [kg/m^3], (422).
σ Mean normal stress [Pa], (66).
φ Volume fraction (422).
ψ Specific Helmholtz energy [J/kg], (300).

CHAPTER 0

Introduction

The aim of this book is to describe the physics of nonlinear stress waves in fluids and solids. These waves cause changes not only in force and motion but also in heat and temperature. We introduce our topic by describing two experiments that have greatly contributed to our understanding of nonlinear waves.

The study of waves is important to virtually every branch of science and engineering. Indeed, waves are also important to everyday life. Sound waves allow us to hear, and electromagnetic waves allow us to see. In this book, we restrict our study to one important class of waves often called *stress waves*. These waves propagate through gases, liquids, and solids. Stress waves in gases and liquids are usually called *pressure waves*. This nomenclature is derived from the fact that internal forces in solid bodies are represented by a stress tensor, and internal forces in inviscid liquids and gases are represented by a special form of stress called pressure. In Chapter 1, we shall learn the distinction between stress and pressure.

Sound is the most commonly experienced pressure wave. Indeed, anyone standing in a thunderstorm knows that lightening is seen before the report of its thunder is heard. This simple observation reveals the single, most important feature of any wave – its finite velocity. We know there is always a lapse of time between the cause of a sound wave and when we hear it. This time delay is called *causality*. Because light waves travel at 3×10^8 m/s and sound waves travel at 345 m/s, causality requires that we see the lightening before we hear the thunder.

In the late seventeenth century, Isaac Newton attempted an analysis of the speed of sound. Because Newton assumed that sound was a purely mechanical phenomenon, his analysis was wrong. More than a hundred years passed before Laplace understood this error and corrected Newton's analysis. Laplace had the advantage of a century of development of the field of thermodynamics. He knew that sound not only causes the pressure in air to oscillate, it also causes the temperature to oscillate. Thus Laplace grasped an essential feature of stress waves. Stress wave phenomena are a manifestation of two fields acting together. One is the mechanical field represented by force and motion. The other is the thermal field represented by heat and temperature. Newton did not understand the difference between heat and temperature. Laplace did.

Sound is a linear wave. This statement has several implications. Perhaps most importantly, it means that sound waves do not interact with each other. Thus, for example, when two men speak simultaneously, their individual voices retain their identity. If sound were a nonlinear wave, the bizarre possibility exists that two men's voices might combine to form the tonal qualities of a woman's voice. The linear wave has another important attribute. It can be represented by a system of linear equations. Linear equations contain products of variables and constants, but they do not contain powers of variables or products of one variable with another. The analysis of linear equations draws heavily from one important advantage of these systems; that is, the sum of two solutions of a linear equation is itself a solution. This is the *superposition principle*, which is fundamental to linear wave analysis. Nonlinear systems do not offer us this advantage. We cannot add solutions of nonlinear equations to obtain new solutions, and when two nonlinear waves interact, each is changed. Often nonlinear waves combine to form a new wave and never separate. We recount such an event in the opening quote of the preface. On the first single-handed circumnavigation of the globe and shortly after rounding Cape Horn, Joshua Slocum encountered "... a tremendous wave, the culmination, it seemed, of many waves...." Captain Slocum observed the combination of nonlinear waves in one of its most frightening forms, a tidal wave.

Two types of experiments have contributed enormously to our understanding of nonlinear stress waves. These experiments are the *shock-tube* experiment and the *flyer-plate* experiment. The shock-tube experiment is used primarily to study the effects of nonlinear waves in gases. The flyer-plate experiment is usually used to study nonlinear waves in solids and liquids. In their most basic forms, both experiments are one dimensional. The primary aim of this book is to develop a fundamental understanding of the physics of stress waves that can be extended to ever more complex materials and geometries. Much of this goal can be achieved by studying these two experiments. Thus wherever possible, we also limit the analysis in this text to simple one-dimensional geometries. This allows us to concentrate upon the physics of nonlinear waves rather than the methods of wave analysis in complex geometries. However, some exceptions must be made. As a superficial inspection of Chapter 1 reveals, some fundamental concepts in this book are presented in their three-dimensional forms. This is particularly true when we introduce stress and deformation. We must do this because even the simple statement, *inviscid fluids only support pressure while solids support stress*, cannot be understood without this three-dimensional approach.

Before we start with our analysis of nonlinear waves in gases, liquids, and solids, let us take a moment to discuss both the shock-tube and flyer-plate experiments. In this discussion, we will concentrate upon the wave phenomena that are observed in these experiments, and give only a minimal description of the apparatus, because specific information about the apparatus is well documented in other sources. The physical phenomena revealed by these two experiments are varied and rich in content. Thus these descriptions may seem complex at first reading. However, if you return to them from time to time as we advance through the text, the apparent complexity of these experiments will vanish as you gain an understanding of the underlying principles that govern the formation and propagation of stress waves.

0.1 Shock-Tube Experiment

Let us turn now to the first of our two experments, the shock tube. A shock tube is a long hollow cylinder filled with gas (see Figure 0.1). The diameter of the tube is constant, and the inside surface of the tube is polished to a mirror-smooth quality. Let x denote the

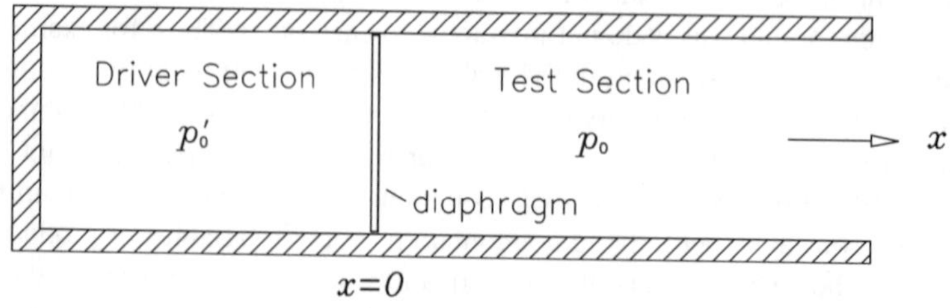

Figure 0.1 The shock-tube experiment.

position along the axis of symmetry of the tube. The gas is separated into two regions by a thin diaphragm at $x = 0$. To the left of the diaphragm is the driver section, and to the right is the test section. Often different gases are used in each section, but to simplify the discussion, we assume identical gases are used in both the driver and test sections. Initially, the gas is at rest with a uniform temperature. The pressure in the driver section is p_0', and the pressure in the test section is p_0, where $p_0' > p_0$. At time $t = 0$, the diaphragm is suddenly released. We assume that the mass and thickness of the diaphragm are negligible and do not affect the flow of the gas.

The walls of the tube are essentially rigid compared to the motion of the gas. We observe that the motion of every portion of the gas is parallel to the axis of the tube. Because of the mirror quality of the inside surface, we also find that across any cross section perpendicular to the x axis of the tube, the velocity of the gas is constant unless we are very close to the wall. Essentially, this means that at any instant in time the velocity at any point in the tube is only a function of the axial position x. Thus the flow of the gas in the tube is one dimensional. At any instant in time the pressure p at any point in the tube is also a function of x alone.

Suppose the diaphragm is released at time $t = 0$. After an increment of time Δt, we measure both the position of the diaphragm and the pressure at every position x in the tube. Then we wait another increment of time Δt and measure these quantities again. When the measurements at each time are graphed we find that the diaphragm (dotted line) moves to the right, while two pressure disturbances (solid line) propagate away from the diaphragm (see Figure 0.2). One pressure disturbance travels to the right into the test section, and the

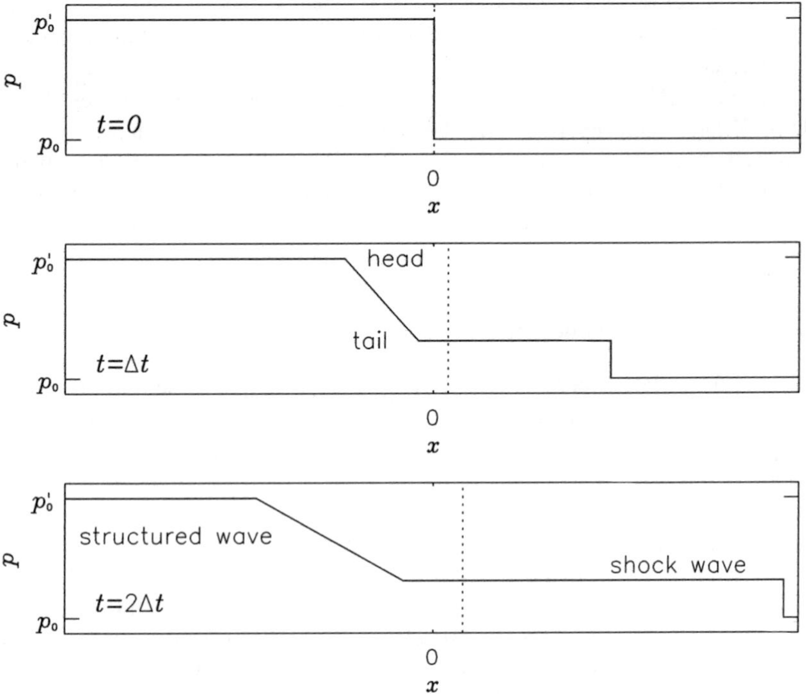

Figure 0.2 Three pressure profiles (—) and position of the diaphragm (···) in a shock tube.

other disturbance travels to the left into the driver section. The disturbance in the test section has a sharp rise or jump in pressure. It is called a *shock wave*, and it is this wave that gives the shock tube its name. The shock wave propagates at a constant velocity. The velocity of the shock wave is the ratio of its position x divided by the time t. We observe that the velocity of propagation of this shock wave is always greater than the speed of sound in the gas. In contrast, the disturbance that propagates into the driver section does not cause a sudden jump in pressure. Moreover, different portions of this wave travel at different velocities. Because it has a more complex shape that evolves with time, this disturbance is called a *structured wave*. This particular structured wave forms a ramp. The leading portion of the ramp is called the *head*, and the trailing portion is called the *tail*. Indeed, we observe that the head of the structured wave travels at the speed of sound, and in experiments where p_0' is very large, we even find that the tail travels backwards into the test section. We also observe that the pressure rise due to the shock wave is always less than the pressure drop due to the structured wave. It is important to recognize that the diaphragm moves with the gas, whereas the pressure waves do not. The motion of the gas is not the same as the motion of the waves.

Because the shock wave causes an increase in pressure in the test section, it is also called a *compression wave*. Similarly, because the ramp wave causes a drop in pressure in the driver section, it is called a *rarefaction wave*. In general, shock waves in gases are always compression waves and never rarefaction waves. We also find that structured compression waves can exist in gases, but they quickly evolve into shock waves. Another very important observation can be made. If the x scale is magnified in the neighborhood of the shock wave, we find that the shock wave is not a discontinuous jump in pressure. It too is actually a structured wave, albeit with a very rapid rise in pressure. However, unlike the rarefaction wave in the driver section, the structure of this wave does not evolve with time. Hence it is called a *steady wave*. We shall find that the terms shock wave and steady wave can be used somewhat interchangeably.

Large temperature changes are also associated with each of these waves. Temperatures of 9,000 K have been reported. The shock wave causes the temperature to increase, and the rarefaction wave causes the temperature to decrease (see Figure 0.3). If the gas on each side of the diaphragm has the same initial temperature, then the temperatures across the diaphragm will diverge upon release. Even though the shock wave causes the smaller change in pressure, it causes the larger change in temperature.

0.2 Flyer-Plate Experiment

A flyer-plate experiment is shown in Figure 0.4. The diagram of the experiment represents thin disks in a cylindrical geometry. A thin impactor is accelerated towards a target plate. Either explosives or smooth-bored guns are commonly used to accelerate the impactor. In this case an aluminum impactor is accelerated by a gun towards a crystalline-calcite target. At the instant of impact as measured by the trigger pins, two shock waves are generated at the impact plane. One shock wave propagates into the target, and the other propagates into the impactor. For diagnostic purposes, a transparent lithium-fluoride window is attached to the backside of the calcite target. The shock wave traveling through the target eventually enters this transparent window. As it does, it causes the interface between the target and the window to move. This motion is measured with an optical interferometer. Because the calcite target is thin, the shock wave arrives at the measuring point before other

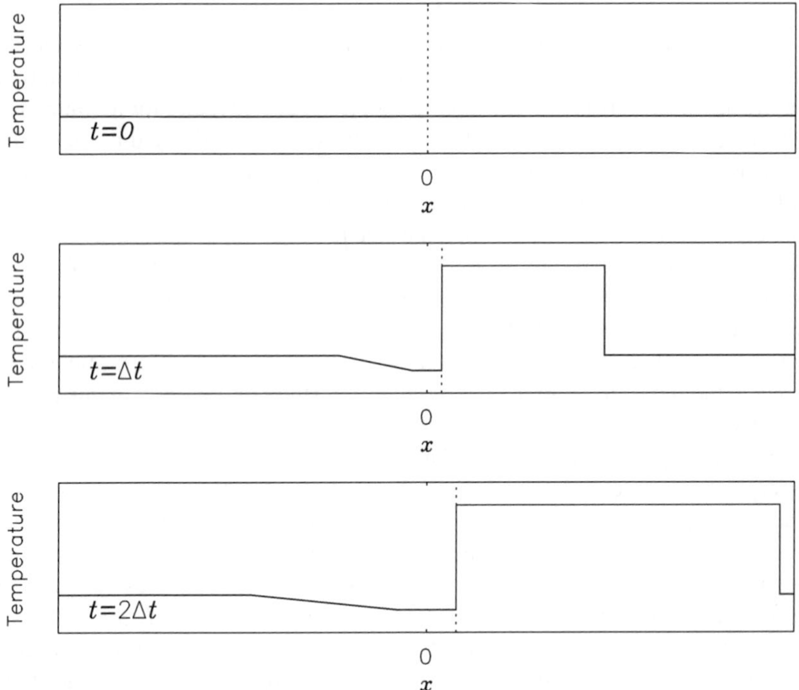

Figure 0.3 Three temperature profiles in a shock tube.

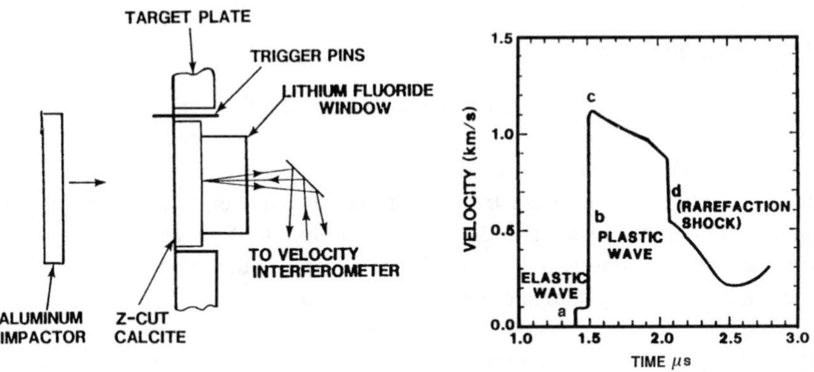

Figure 0.4 The flyer-plate experiment. Reprinted from Chhabildas, L. C., Survey of diagnostic tools used in hypervelocity impact studies. In *Hypervelocity Impact, Proceedings of the 1986 Symposium, San Antonio, Texas, 21–24 October 1986*, ed. Anderson Jr., C. E., Copyright 1986, p. 209, with kind permission from Elsevier Science Ltd, The Boulevard, Langford Lane, Kidlington 0X5 1GB, UK.

waves arrive from the circumferential edges of the calcite target. Let δt denote the period of time between the arrival of the shock wave and the arrival of the edge waves. During δt, we observe that the edges of the target do not influence the measured interface motion caused by the shock wave. This is another manifestation of causality, and for a limited amount of time the target plate behaves as if it had finite thickness but infinite diameter. We observe one-dimensional horizontal motion in the central portion of the target plate. During δt, the reflections between the impactor and target plates also produce a rarefaction wave that follows the leading shock wave through the calcite target. In Chapter 2, we shall study just how these reflections arise, but for now let us concentrate on the shapes of the shock and rarefaction waves as they exit the calcite target and enter the lithium-fluoride window.

The velocity data gathered with the interferometer are graphed against the time after impact in Figure 0.4. We shall find that these data are sufficient to determine the stress in the target without the aid of direct measurements of stress or pressure. In this case the stress is 18 GPa or about 180,000 times the pressure of the atmosphere. Notice that the leading shock wave has split into two waves. The first wave, labeled a, is called an elastic wave. The second wave, labeled b, is called a plastic wave. Both these waves are shock waves, and they split apart at the point where the stress exceeds the strength of the calcite crystal, causing realignment of the atoms and fluid-like flow of this solid. The rarefaction wave is even more interesting. Although the rarefaction wave in the shock-tube experiment is a smooth continuous ramp, here another shock wave is present at d. It appears at another stress threshold where the crystalline lattice undergoes an additional realignment called a *polymorphic phase transformation*. What is most interesting about this threshold is that although it influences the rarefaction wave, it has no effect on the plastic compression wave b. This data record reveals much about the physical structure of calcite, and we will return to it again at the end of this book. Indeed, it is important for you to recognize that stress waves are tools that are often used to investigate the behavior of materials under extreme temperatures and pressures that are unattainable by other experimental methods.

0.3 Plan of the Book

Recall that stress waves cause changes in both force and motion as well as temperature and heat. However, to simplify our presentation, we shall ignore temperature and heat initially. The first two chapters of this book contain a *mechanical* theory of waves, while temperature and heat are included in the last chapters. In Chapter 1, we define a three-dimensional Cartesian coordinate system. Motion of a material is described in this coordinate system using two methods. In the first method, we attach ourselves to a fixed Cartesian coordinate system and watch the material move past us. This is called the *spatial representation*. In the second method, we attach ourselves to a point on the material and watch the coordinate system move past us. This is called the *material representation*. Using these representations we decompose the motion into three components: translation, rotation, and deformation gradient. The deformation gradient is then used to define strain. Next, we describe both external and internal forces acting on the material. The internal forces give rise to the definition of stress. Next, the two laws of motion are derived. The first law of motion is the equation of conservation of mass. This law ensures that motion of the material does not destroy or create mass. The second law of motion is the equation of balance of momentum. This is Newton's second law of motion that interconnects force, mass, and acceleration.

To form a complete theory of mechanical waves, an additional relationship linking the deformation gradient to the stress is required. This relationship is called a *constitutive model.* In Chapter 2, we introduce the constitutive model for an elastic material. In an elastic material the stress is *only* a function of the deformation gradient. Hooke's law for a linear-elastic material and the law for the adiabatic compression of an ideal gas are both special cases of the elastic material.

Our primary purpose is to develop a comprehension of *nonlinear* waves. This goal is best served by first studying *linear* waves. Therefore we simplify the system of nonlinear equations for nonlinear waves to a single linear equation describing linear waves. This equation is called the *linear wave equation.* We use three methods to solve the linear wave equation: the d'Alembert solution, the Riemann invariant, and the method of characteristics. These are not the only methods available for solving the linear wave equation. However, unlike other solution methods, each of these methods can be applied to our main purpose, the solution of the equations for nonlinear waves.

The first nonlinear wave we study is called the *simple* wave. The simple wave is a structured wave. It can be a compression wave, a rarefaction wave, or both. We study how simple waves evolve and interact, and we study how simple waves become shock waves. The smooth solutions for certain simple waves in an elastic material evolve to solutions with discontinuous jumps in velocity and stress. This leads us to derive the mechanical jump conditions for the shock wave. This is followed with a careful examination of wave–wave interactions; that is, we study how structured waves and shock waves interact. We end Chapter 2 with a critical observation. Measurements show that shock waves do not exist as discontinuous jumps in velocity and stress. As our discussion of the shock tube revealed, they are actually steady waves with a smooth structure. We find that the theory for mechanical waves cannot yield a solution for this structured steady wave. Indeed, although the theory of mechanical waves is very useful, it is too specialized to accurately represent the structure of the shock wave.

Our solution to this dilemma is found in Chapter 3, where the thermal field is introduced into the description of the stress wave. This chapter is entitled *thermomechanics* because we are primarily interested in how the thermal field interacts with the mechanical field to produce the shock wave. We find that energy is the unifying concept that connects these two fields. Two laws are introduced that govern energy. The *first law of thermodynamics* states that energy cannot be created or destroyed. From this law we derive the equation of the balance of energy. The *second law of thermodynamics* describes the transformation of energy between the mechanical field and the thermal field. From this law we derive special constraints that must be satisfied by constitutive models of materials. To derive these constraints we define the concept of a thermodynamic process. After studying the basic thermodynamic processes found in most thermodynamics textbooks, we study a new process called the *Rayleigh-line process.* The Rayleigh-line process is the smooth steady-wave solution for the shock wave.

As revealed in Chapter 3, thermomechanics is really the study of constitutive models. However, constitutive theory is quite complex, and Chapter 4 is devoted to a more detailed examination of several important constitutive models. These include ideal gases, Mie–Grüneisen solids, elastic-plastic solids, fluid-saturated and dry porous solids, mixtures of fluids, explosives, and solids with polymorphic phase transformations.

Throughout this book we solve systems of nonlinear equations. These solutions are constructed using combinations of graphical methods, analytical integration of ordinary and partial differential equations, and numerical integration of ordinary differential equations. For the most part we avoid using numerical methods for solving partial differential equations. In the world outside of textbooks, this is not always possible, and complex numerical procedures are often used to integrate the partial differential equations of the theory of nonlinear waves. Fortunately, computer software is available that can integrate these systems of equations in three spatial dimensions. Some software even accommodates special constitutive models written by the user. However, to generate accurate solutions with these programs, the user must be familiar with the variety of ways in which these programs can be abused. For this reason, in Appendix A we review the basic aspects of numerical computation of stress waves.

CHAPTER 1

Fundamentals

To analyze waves in nonlinear materials, we must derive the equations governing the motion of such materials. In this chapter we define three types of motion: translation, rotation, and deformation. We define a quantity called the deformation gradient. This quantity is used to describe volume strain, longitudinal strain, and shear strain. Then stress and pressure are discussed. We introduce these concepts in three spatial dimensions and then specialize the results to one spatial dimension. The chapter closes with a derivation of two important laws of motion: the conservation of mass and the balance of momentum.

1.1 Index Notation

We begin by introducing a compact and convenient way to describe quantities in three spatial dimensions with a Cartesian coordinate system. We also show how to transform quantities between different Cartesian coordinate systems.

1.1.1 *Cartesian Coordinates and Vectors*

A Cartesian coordinate system has three straight coordinates that are *mutually perpendicular* to each other (see Figure 1.1). The coordinates are named x_1, x_2, and x_3. The subscripts 1, 2, and 3 are called *indices*. When we let k represent any of these indices, we can refer to the coordinates by the compact notation x_k. Each coordinate x_k also has a shaded arrow that represents a *unit coordinate vector* $\mathbf{i}_k$. We use a boldface symbol to denote a vector. The length of each $\mathbf{i}_k$ is equal to unity.

Now consider another vector $\mathbf{v}$ that is drawn with arbitrary angles to the three coordinates. We draw a box with sides parallel to $\mathbf{i}_k$ and with $\mathbf{v}$ as a diagonal. The lengths of the three sides of this rectangular box are labeled v_k. The quantities v_k are called the *scalar components* of the vector $\mathbf{v}$. The vector $\mathbf{v}$ can be expressed in terms of its scalar components v_k and the unit coordinate vectors $\mathbf{i}_k$ as

$$\mathbf{v} = v_1\mathbf{i}_1 + v_2\mathbf{i}_2 + v_3\mathbf{i}_3 = \sum_{k=1}^{k=3} v_k\mathbf{i}_k.$$

We can eliminate the need for the summation symbol by assuming that the index is to be summed when it appears twice in an expression. Thus we can write the last equation as

$$\mathbf{v} = v_k\mathbf{i}_k.$$

Now consider the vector equation

$$\mathbf{w} = \mathbf{u} + \mathbf{v}.$$

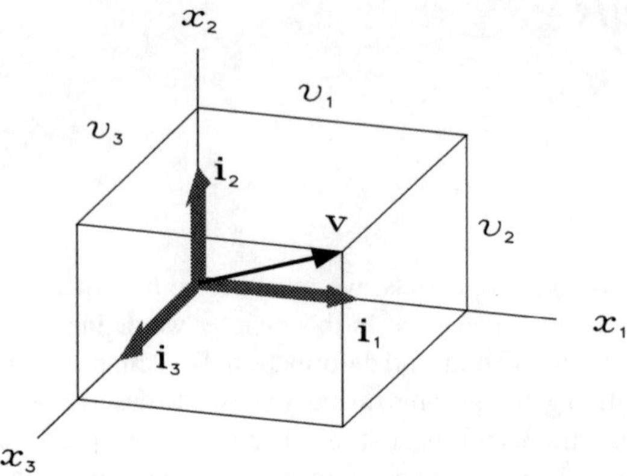

Figure 1.1 A Cartesian coordinate system and a vector **v**.

In index notation we have

$$w_k\mathbf{i}_k = u_n\mathbf{i}_n + v_m\mathbf{i}_m.$$

The three subscripts k, n, and m appear twice in each of the separate expressions, and each expression implies a summation. Therefore,

$$w_1\mathbf{i}_1 + w_2\mathbf{i}_2 + w_3\mathbf{i}_3 = u_1\mathbf{i}_1 + u_2\mathbf{i}_2 + u_3\mathbf{i}_3 + v_1\mathbf{i}_1 + v_2\mathbf{i}_2 + v_3\mathbf{i}_3.$$

We rearrange this equation to obtain

$$(w_1 - u_1 - v_1)\mathbf{i}_1 + (w_2 - u_2 - v_2)\mathbf{i}_2 + (w_3 - u_3 - v_3)\mathbf{i}_3 = \mathbf{0}.$$

In index notation

$$(w_k - u_k - v_k)\mathbf{i}_k = \mathbf{0}.$$

Each Cartesian component of this vector equation must be zero. Therefore, the following three equations must hold:

$$w_1 = u_1 + v_1,$$
$$w_2 = u_2 + v_2,$$
$$w_3 = u_3 + v_3.$$

We can write the previous three equations in the more compact index notation

$$w_k = u_k + v_k.$$

In this equation, k only appears once in each of the three expressions and summation is not implied. However, it is implied that this equation must hold for $k = 1, 2,$ and 3.

You can also write equations that contain combinations of these two cases. For example, when you write

$$a_m = b_{mk}c_k,$$

this is the compact notation for the three equations

$$a_1 = b_{11}c_1 + b_{12}c_2 + b_{13}c_3,$$
$$a_2 = b_{21}c_1 + b_{22}c_2 + b_{23}c_3,$$
$$a_3 = b_{31}c_1 + b_{32}c_2 + b_{33}c_3.$$

1.1.2 *Products of Vectors*

We have seen how two vectors can be added. Multiplication is also possible. In this section we define both the *cross product* and the *dot product* of two vectors. Then the *scalar triple product* of three vectors is defined. Expressions for the length of a vector and the angle between two vectors are required before we can define these products of vectors.

Consider the vector $\mathbf{v}$. Its length is denoted by $|\mathbf{v}|$. This is also called the *magnitude* of $\mathbf{v}$. We can use Pythagoras's Theorem to obtain the length of $\mathbf{v}$ in terms of the components v_k:

$$|\mathbf{v}| = \sqrt{v_1^2 + v_2^2 + v_3^2} = (v_k v_k)^{1/2}. \tag{1.1}$$

The lengths of the unit coordinate vectors $\mathbf{i}_k$, being one, are now expressed as

$$|\mathbf{i}_k| = 1.$$

Next, consider the two vectors $\mathbf{u}$ and $\mathbf{v}$ (see Figure 1.2). Their lengths are $|\mathbf{u}|$ and $|\mathbf{v}|$, and they form an angle γ. We can represent the angle γ by the symbol $\{\mathbf{u}, \mathbf{v}\}$. When you look at this symbol, it is helpful to remember that $\{\mathbf{u}, \mathbf{v}\}$ represents the angle measured *from* $\mathbf{u}$ *to* $\mathbf{v}$. Therefore, for example,

$$\{\mathbf{u}, \mathbf{v}\} = -\{\mathbf{v}, \mathbf{u}\}.$$

Cross Product The *cross product* or *vector product* of $\mathbf{u}$ and $\mathbf{v}$ is denoted as $\mathbf{u} \times \mathbf{v}$. Let $\mathbf{c}$ be the vector

$$\mathbf{c} = \mathbf{u} \times \mathbf{v}.$$

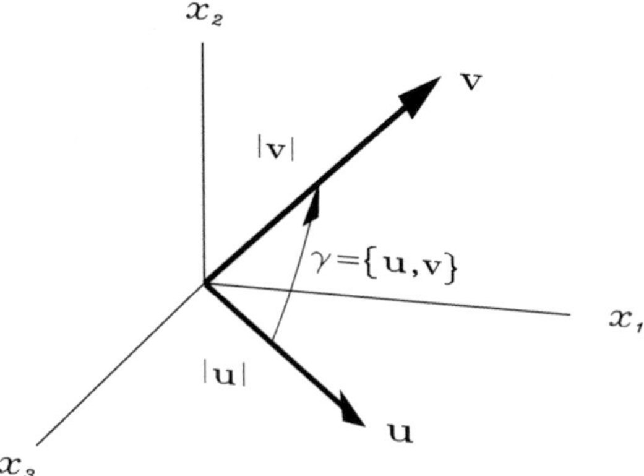

Figure 1.2 A Cartesian coordinate system with two vectors $\mathbf{u}$ and $\mathbf{v}$.

We define the vector **c** to be perpendicular to both **u** and **v** so that (**u**, **v**, **c**) forms a right-handed coordinate system. The length of **c** is defined to be

$$|\mathbf{c}| \equiv |\mathbf{u}||\mathbf{v}||\sin\{\mathbf{u}, \mathbf{v}\}|. \tag{1.2}$$

The value of the sine depends on the sign of the angle $\{\mathbf{u}, \mathbf{v}\}$:

$$\sin\{\mathbf{u}, \mathbf{v}\} = -\sin\{\mathbf{v}, \mathbf{u}\}.$$

Thus the cross product is not commutative because

$$\mathbf{u} \times \mathbf{v} = -\mathbf{v} \times \mathbf{u}.$$

When we select **u** and **v** to be unit vectors of the Cartesian coordinate system, we find that

$$\begin{array}{lll} \mathbf{i}_1 \times \mathbf{i}_1 = \mathbf{0}, & \mathbf{i}_2 \times \mathbf{i}_2 = \mathbf{0}, & \mathbf{i}_3 \times \mathbf{i}_3 = \mathbf{0}, \\ \mathbf{i}_1 \times \mathbf{i}_2 = \mathbf{i}_3, & \mathbf{i}_2 \times \mathbf{i}_3 = \mathbf{i}_1, & \mathbf{i}_3 \times \mathbf{i}_1 = \mathbf{i}_2, \\ \mathbf{i}_2 \times \mathbf{i}_1 = -\mathbf{i}_3, & \mathbf{i}_3 \times \mathbf{i}_2 = -\mathbf{i}_1, & \mathbf{i}_1 \times \mathbf{i}_3 = -\mathbf{i}_2. \end{array}$$

We can express these values of the cross products of the unit vectors with a more compact notation

$$\mathbf{i}_k \times \mathbf{i}_m = e_{kmn}\mathbf{i}_n,$$

where the definition of e_{kmn} is

$$e_{kmn} = \begin{cases} 0 & \text{when any two of the indices } k, m, n \text{ are equal,} \\ 1 & \text{when } kmn = 123 \text{ or } 312 \text{ or } 231, \\ -1 & \text{otherwise.} \end{cases}$$

The term e_{kmn} is called the *alternator* or *permutation symbol.*

The cross product is distributive:

$$\mathbf{u} \times (\mathbf{v} + \mathbf{w}) = \mathbf{u} \times \mathbf{v} + \mathbf{u} \times \mathbf{w}.$$

By employing distributivity and expanding the vectors **u** and **v** in component form, it is easily demonstrated that we can represent the cross product by the determinant

$$\mathbf{u} \times \mathbf{v} = \det \begin{bmatrix} \mathbf{i}_1 & \mathbf{i}_2 & \mathbf{i}_3 \\ u_1 & u_2 & u_3 \\ v_1 & v_2 & v_3 \end{bmatrix}.$$

Using the alternator, we can also write

$$\mathbf{u} \times \mathbf{v} = u_k v_m e_{kmn}\mathbf{i}_n.$$

We can also evaluate the angle between two vectors with the cross product. From Eq. (1.2), we obtain

$$|\sin\{\mathbf{u}, \mathbf{v}\}| = \frac{|\mathbf{u} \times \mathbf{v}|}{|\mathbf{u}||\mathbf{v}|}.$$

Dot Product Another kind of vector product is the *dot product* or *scalar product*. We denote the dot product of $\mathbf{u}$ and $\mathbf{v}$ as $\mathbf{u} \cdot \mathbf{v}$. It is defined to be the scalar quantity

$$\mathbf{u} \cdot \mathbf{v} \equiv |\mathbf{u}||\mathbf{v}| \cos\{\mathbf{u}, \mathbf{v}\}. \tag{1.3}$$

Because the value of the cosine is independent of the sign of the angle $\{\mathbf{u}, \mathbf{v}\}$, we see that

$$\cos\{\mathbf{u}, \mathbf{v}\} = \cos\{\mathbf{v}, \mathbf{u}\}.$$

The dot product is commutative:

$$\mathbf{u} \cdot \mathbf{v} = \mathbf{v} \cdot \mathbf{u}.$$

The dot product is also distributive:

$$\mathbf{u} \cdot (\mathbf{v} + \mathbf{w}) = \mathbf{u} \cdot \mathbf{v} + \mathbf{u} \cdot \mathbf{w}.$$

Suppose for a moment that $\mathbf{u}$ and $\mathbf{v}$ are two of the unit vectors of the Cartesian coordinate system. Because the unit coordinate vectors are mutually perpendicular,

$$\mathbf{i}_k \cdot \mathbf{i}_m = \begin{cases} 1, & k = m, \\ 0, & k \neq m. \end{cases}$$

We can express this result more conveniently with the *Kronecker delta* δ_{km}, which is defined to be

$$\delta_{km} \equiv \begin{cases} 1, & k = m, \\ 0, & k \neq m. \end{cases}$$

Then

$$\mathbf{i}_k \cdot \mathbf{i}_m = \delta_{km}. \tag{1.4}$$

This result shows us that the dot product of two arbitrary vectors $\mathbf{u}$ and $\mathbf{v}$ is

$$\mathbf{u} \cdot \mathbf{v} = u_k \mathbf{i}_k \cdot v_m \mathbf{i}_m = u_k v_m (\mathbf{i}_k \cdot \mathbf{i}_m) = u_k v_m \delta_{km},$$

which yields

$$\mathbf{u} \cdot \mathbf{v} = u_k v_k.$$

From Eq. (1.1), we notice that the dot product can be used to evaluate the length of a vector:

$$|\mathbf{v}| = (\mathbf{v} \cdot \mathbf{v})^{1/2} = (v_k v_k)^{1/2}. \tag{1.5}$$

And from Eq. (1.3), we find that it can also be used to evaluate the angle between two vectors:

$$\cos\{\mathbf{u}, \mathbf{v}\} = \frac{\mathbf{u} \cdot \mathbf{v}}{[(\mathbf{u} \cdot \mathbf{u})(\mathbf{v} \cdot \mathbf{v})]^{1/2}} = \frac{u_k v_k}{(u_n u_n v_m v_m)^{1/2}}. \tag{1.6}$$

Notice the particularly useful result that when two vectors are perpendicular their dot product is zero.

Scalar Triple Product The *scalar triple product* is the following combination of the dot product and the cross product:

$$\mathbf{u} \times \mathbf{v} \cdot \mathbf{w} = \mathbf{u} \cdot \mathbf{v} \times \mathbf{w} = \det \begin{bmatrix} u_1 & u_2 & u_3 \\ v_1 & v_2 & v_3 \\ w_1 & w_2 & w_3 \end{bmatrix}.$$

You can easily verify the determinant representation of the scalar triple product by writing each vector in component form and carrying out the multiplication. For this process to be meaningful, the cross product must be evaluated before the scalar product. In terms of the alternator we have

$$\mathbf{u} \times \mathbf{v} \cdot \mathbf{w} = \mathbf{u} \cdot \mathbf{v} \times \mathbf{w} = u_k v_m w_n e_{kmn}.$$

1.1.3 *Areas and Volumes*

You can use the cross product to evaluate the area of a parallelogram. Consider the parallelogram formed by the vectors **u** and **v** (see Figure 1.3). Its sides form two equal obtuse angles and two equal acute angles. Let S denote the area of this parallelogram. The area S is equal to the product of the lengths of two adjacent sides and the magnitude of the sine of the angle between them. Thus

$$S = |\mathbf{u} \times \mathbf{v}|. \tag{1.7}$$

When we add a third vector **w** to this figure, which is out of the plane formed by **u** and **v**, a parallelepiped is formed (see Figure 1.4). The volume V of this parallelepiped is the product of the area S and the altitude. The altitude of the parallelepiped is the length $|\mathbf{w}|$ times the cosine of the angle between the vector **w** and a vector normal to the area S. Because $\mathbf{u} \times \mathbf{v}$ is normal to S and has a magnitude equal to S, the volume V is equal to the magnitude of the scalar triple product:

$$V = |\mathbf{w} \cdot \mathbf{u} \times \mathbf{v}|$$

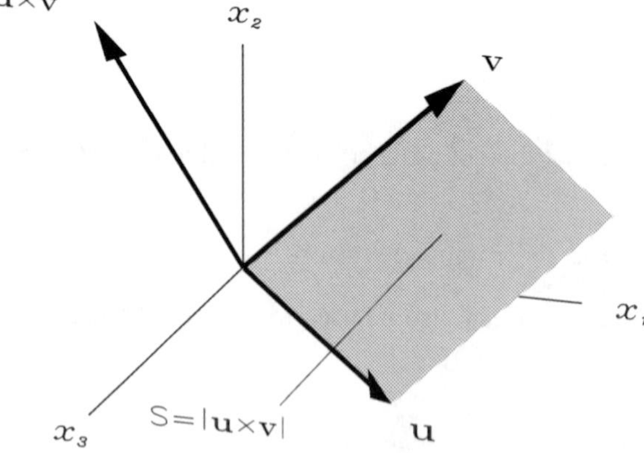

Figure 1.3 The parallelogram formed by **u** and **v**. The area of the parallelogram is $S = |\mathbf{u} \times \mathbf{v}|$, and the vector $\mathbf{u}\times\mathbf{v}$ is perpendicular to the area S.

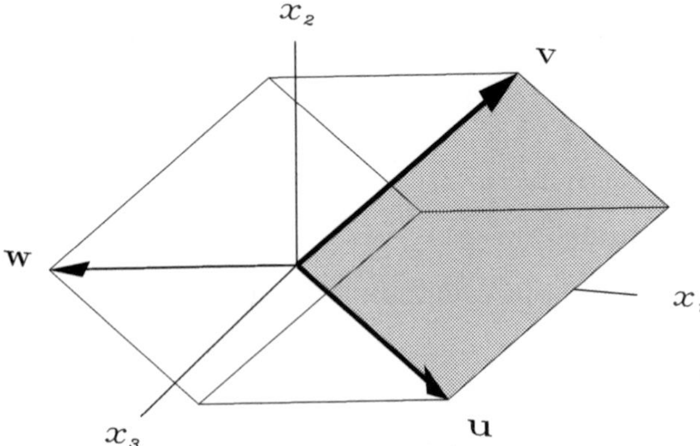

Figure 1.4 The parallelepiped formed by **u**, **v**, and **w**.

or

$$V = \left| \det \begin{bmatrix} u_1 & u_2 & u_3 \\ v_1 & v_2 & v_3 \\ w_1 & w_2 & w_3 \end{bmatrix} \right|. \tag{1.8}$$

Because of the absolute value signs, the volume V has the same value regardless of the order in which the vectors **w**, **u**, and **v** appear in the scalar triple product.

1.1.4 *Evaluation of a Determinant*

The definition of the scalar triple product leads to the following useful expression for a determinant:

$$\det \begin{bmatrix} u_1 & u_2 & u_3 \\ v_1 & v_2 & v_3 \\ w_1 & w_2 & w_3 \end{bmatrix} = u_k v_m w_n e_{kmn}.$$

You can rewrite this as

$$\det[c_{pq}] = \det \begin{bmatrix} c_{11} & c_{12} & c_{13} \\ c_{21} & c_{22} & c_{23} \\ c_{31} & c_{32} & c_{33} \end{bmatrix} = c_{1k} c_{2m} c_{3n} e_{kmn}.$$

If any two rows of this determinant are interchanged, the sign of the value of the determinant is reversed. Thus, for example,

$$\det \begin{bmatrix} c_{21} & c_{22} & c_{23} \\ c_{11} & c_{12} & c_{13} \\ c_{31} & c_{32} & c_{33} \end{bmatrix} = -c_{1k} c_{2m} c_{3n} e_{kmn}.$$

Because of this possibility, we use the alternator to obtain (see Exercise 1.11)

$$\det[c_{pq}] e_{rst} = c_{rk} c_{sm} c_{tn} e_{kmn}. \tag{1.9}$$

Notice that the elements c_{pq} form a square matrix $[c_{pq}]$. Now we recall a special property of square matrices. Suppose we have two additional square matrices $[a_{ij}]$ and $[b_{kl}]$ such that

$$c_{pq} = a_{pm}b_{mq}.$$

It is an important result of matrix algebra that

$$\det[c_{pq}] = \det[a_{ij}]\det[b_{kl}]. \tag{1.10}$$

We shall find this property particularly useful later in this chapter.

1.1.5 *Direction of a Vector*

Consider the vector $\mathbf{v}$ shown as a solid arrow in Figure 1.5. The length of this vector is $|\mathbf{v}|$. The shaded arrow represents the vector $\mathbf{D}(\mathbf{v})$ that has unit length and the same orientation as $\mathbf{v}$. Using the scalar components of $\mathbf{D}(\mathbf{v})$ and the unit coordinate vectors $\mathbf{i}_k$, we can write

$$\mathbf{D}(\mathbf{v}) = D_k(\mathbf{v})\mathbf{i}_k.$$

Because $\mathbf{D}(\mathbf{v})$ is a unit vector,

$$D_k(\mathbf{v})D_k(\mathbf{v}) = |\mathbf{D}(\mathbf{v})|^2 = 1.$$

The vector $\mathbf{D}(\mathbf{v})$ is called the *direction vector* of $\mathbf{v}$. We can represent the vector $\mathbf{v}$ in terms of its length and direction as

$$\mathbf{v} = |\mathbf{v}|\mathbf{D}(\mathbf{v}),$$

which in component form is

$$v_k = |\mathbf{v}|D_k(\mathbf{v}).$$

Because $\mathbf{D}(\mathbf{v})$ and $\mathbf{i}_k$ are unit vectors, their dot product equals the angle between them:

$$\mathbf{D}(\mathbf{v}) \cdot \mathbf{i}_k = \cos\{\mathbf{D}(\mathbf{v}), \mathbf{i}_k\}.$$

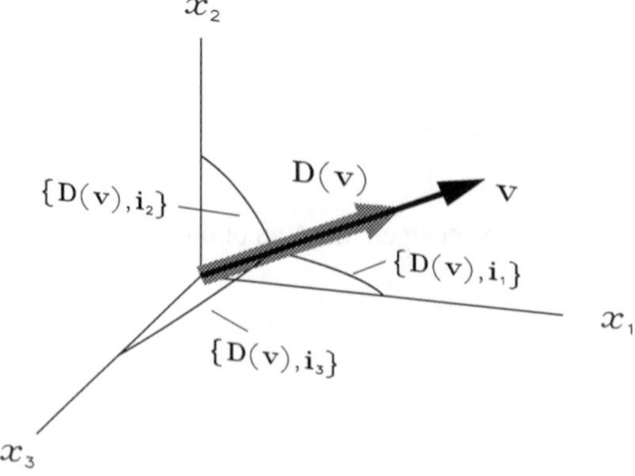

Figure 1.5 A vector $\mathbf{v}$ and its direction vector $\mathbf{D}(\mathbf{v})$.

We also see that

$$\mathbf{D}(\mathbf{v}) \cdot \mathbf{i}_k = D_m(\mathbf{v})\mathbf{i}_m \cdot \mathbf{i}_k = D_k(\mathbf{v}).$$

Therefore, the scalar components of the direction vector $\mathbf{D}(\mathbf{v})$ are equal to the cosines of the angles between the vector $\mathbf{v}$ and the unit coordinate vectors $\mathbf{i}_k$,

$$D_k(\mathbf{v}) = \cos\{\mathbf{D}(\mathbf{v}), \mathbf{i}_k\}.$$

The components $D_k(\mathbf{v})$ are called the *direction cosines* of $\mathbf{v}$. Thus

$$\mathbf{D}(\mathbf{v}) = \cos\{\mathbf{D}(\mathbf{v}), \mathbf{i}_k\}\mathbf{i}_k. \tag{1.11}$$

1.1.6 *Change of Coordinates*

So far we have considered only one coordinate system x_k. Other Cartesian coordinate systems can be constructed within this coordinate system. In this section we examine the relationships between these different coordinate systems. We examine how quantities described in one coordinate system can be described in another coordinate system. The process of changing from one coordinate system to another is called a *coordinate transformation.*

Consider three new mutually perpendicular unit vectors $\mathbf{i}_{(m)}$ placed into our Cartesian coordinate system (see Figure 1.6). We use parentheses about the subscripts to distinguish these vectors from our original vectors $\mathbf{i}_k$. Because the new unit vectors are required to be mutually perpendicular,

$$\mathbf{i}_{(k)} \cdot \mathbf{i}_{(m)} = \delta_{km}, \tag{1.12}$$

they form a second set of unit vectors for a second Cartesian coordinate system, $x_{(k)}$. The new set of unit vectors has the same origin as the original set. Thus the new coordinate system is rotated but not translated with respect to the original coordinates.

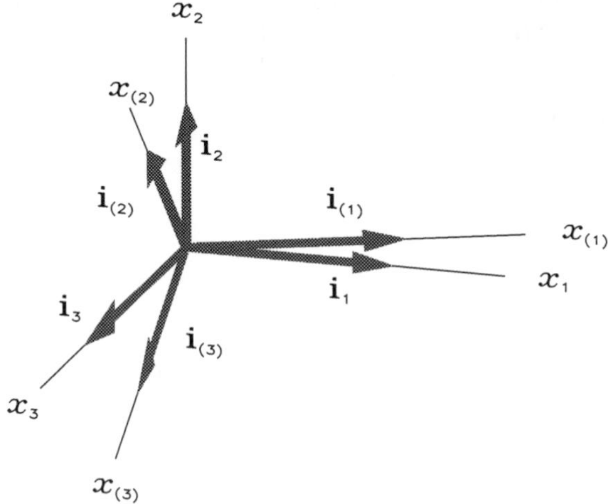

Figure 1.6 Two coordinate systems with two sets of unit vectors $\mathbf{i}_k$ and $\mathbf{i}_{(k)}$.

Using Eq. (1.11), we can represent either set of unit vectors in terms of the other set as

$$\mathbf{i}_{(k)} = \cos\{\mathbf{i}_{(k)}, \mathbf{i}_m\}\mathbf{i}_m,$$
$$\mathbf{i}_k = \cos\{\mathbf{i}_k, \mathbf{i}_{(m)}\}\mathbf{i}_{(m)}.$$

The direction cosines in these expressions connect the unit vectors of the two Cartesian coordinate systems. For convenience, we adopt the following notation for the direction cosines between $\mathbf{i}_k$ and $\mathbf{i}_{(m)}$:

$$\alpha_{k(m)} \equiv \cos\{\mathbf{i}_k, \mathbf{i}_{(m)}\}. \tag{1.13}$$

Thus the relationships between the unit vectors can be rewritten as

$$\mathbf{i}_{(k)} = \alpha_{(k)m}\mathbf{i}_m \tag{1.14}$$

and

$$\mathbf{i}_k = \alpha_{k(m)}\mathbf{i}_{(m)}. \tag{1.15}$$

The direction cosines are denoted by the function $\alpha_{k(m)}$ that has two indices. It can be represented in matrix form as

$$[\alpha_{k(m)}] = \begin{bmatrix} \cos\{\mathbf{i}_1, \mathbf{i}_{(1)}\} & \cos\{\mathbf{i}_1, \mathbf{i}_{(2)}\} & \cos\{\mathbf{i}_1, \mathbf{i}_{(3)}\} \\ \cos\{\mathbf{i}_2, \mathbf{i}_{(1)}\} & \cos\{\mathbf{i}_2, \mathbf{i}_{(2)}\} & \cos\{\mathbf{i}_2, \mathbf{i}_{(3)}\} \\ \cos\{\mathbf{i}_3, \mathbf{i}_{(1)}\} & \cos\{\mathbf{i}_3, \mathbf{i}_{(2)}\} & \cos\{\mathbf{i}_3, \mathbf{i}_{(3)}\} \end{bmatrix}. \tag{1.16}$$

The components of $\alpha_{k(m)}$ have several important and useful properties. Because the value of the cosine function is independent of the sign of its argument,

$$\alpha_{k(m)} = \alpha_{(m)k}.$$

However, it is important to notice that generally

$$\alpha_{k(m)} \neq \alpha_{m(k)}, \qquad m \neq k.$$

As an example, in Figure 1.6 let $k = 2$ and $m = 1$. The angle $\{\mathbf{i}_2, \mathbf{i}_{(1)}\}$ is the negative of the angle $\{\mathbf{i}_{(1)}, \mathbf{i}_2\}$, whereas the angle $\{\mathbf{i}_2, \mathbf{i}_{(1)}\}$ is not equal to the angle $\{\mathbf{i}_1, \mathbf{i}_{(2)}\}$.

From Eqs. (1.4) and (1.12), we obtain additional useful results

$$\mathbf{i}_{(k)} \cdot \mathbf{i}_{(n)} = \alpha_{(k)m}\mathbf{i}_m \cdot \alpha_{(n)r}\mathbf{i}_r$$
$$\delta_{kn} = \alpha_{(k)m}\alpha_{(n)m} \tag{1.17}$$

and

$$\mathbf{i}_k \cdot \mathbf{i}_n = \alpha_{k(m)}\mathbf{i}_{(m)}\alpha_{n(r)}\mathbf{i}_{(r)}$$
$$\delta_{kn} = \alpha_{k(m)}\alpha_{n(m)}. \tag{1.18}$$

We obtain another set of relationships by noticing that any point $\mathbf{x}$ can be represented in both coordinate systems as

$$\mathbf{x} = x_m\mathbf{i}_m = x_{(r)}\mathbf{i}_{(r)}.$$

Using Eq. (1.14) to replace $\mathbf{i}_{(r)}$ gives

$$(x_m - x_{(r)}\alpha_{(r)m})\mathbf{i}_m = \mathbf{0}.$$

Each vector component of this equation must be zero. Thus

$$x_m = x_{(r)}\alpha_{(r)m}. \tag{1.19}$$

A similar derivation using Eq. (1.15) gives

$$x_{(r)} = \alpha_{(r)m}x_m. \tag{1.20}$$

From the last two results we obtain the useful identities

$$\frac{\partial x_m}{\partial x_{(r)}} = \frac{\partial x_{(r)}}{\partial x_m} = \alpha_{(r)m}. \tag{1.21}$$

1.1.7 *Rotations about a Coordinate Axis*

The coordinate transformation we have considered is composed of the rotation of one coordinate system about another. Both coordinate systems have a common origin. This type of transformation from the x_k coordinate system to the $x_{(k)}$ coordinate system can be conveniently separated into a sequence of rotations about each of the x_k coordinates. In Figures 1.7, 1.8, and 1.9 we illustrate separate rotations θ_k about each of the coordinate axes x_k. In each case, the rotation is about a different axis of the original coordinate system x_k. We now evaluate the direction cosines for each of these rotations.

Rotation about $\mathbf{x_1}$ We show this rotation in Figure 1.7. From Eq. (1.16), the direction cosines for this rotation are

$$[\alpha_{n(m)}] = \begin{bmatrix} 1 & 0 & 0 \\ 0 & \cos\theta_1 & \cos(\pi/2+\theta_1) \\ 0 & \cos(\pi/2-\theta_1) & \cos\theta_1 \end{bmatrix} = \begin{bmatrix} 1 & 0 & 0 \\ 0 & \cos\theta_1 & -\sin\theta_1 \\ 0 & \sin\theta_1 & \cos\theta_1 \end{bmatrix}. \tag{1.22}$$

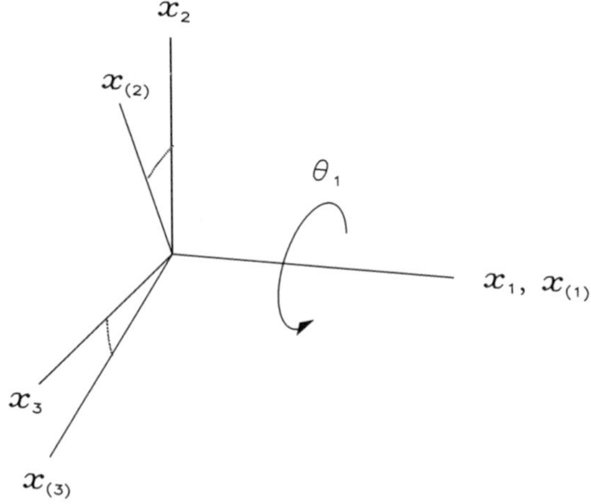

Figure 1.7 Rotation θ_1 about the coordinate axis x_1.

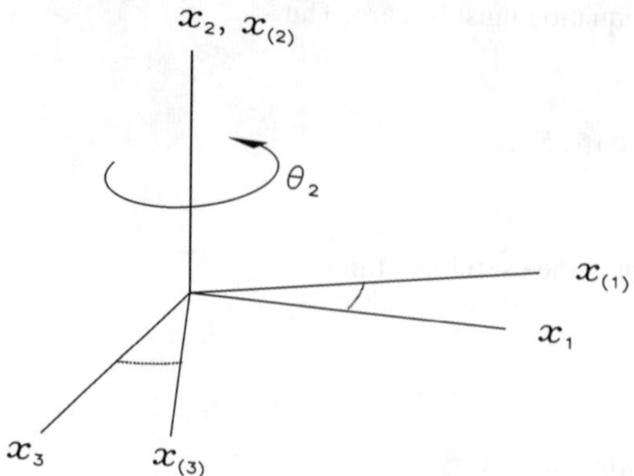

Figure 1.8 Rotation θ_2 about the coordinate axis x_2.

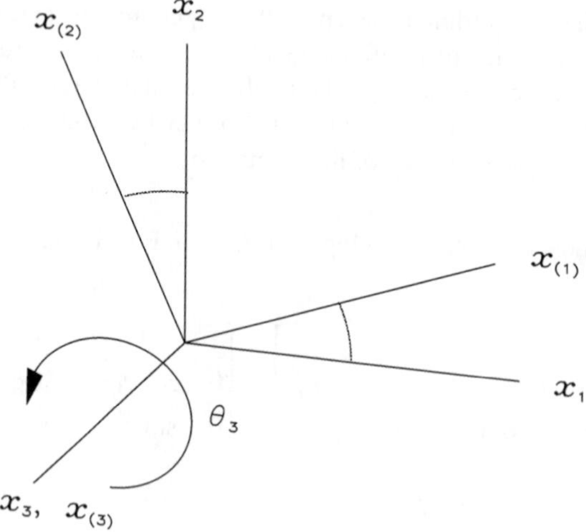

Figure 1.9 Rotation θ_3 about the coordinate axis x_3.

Rotation about x_2 We show this rotation in Figure 1.8. From Eq. (1.16), the direction cosines for this rotation are

$$[\alpha_{n(m)}] = \begin{bmatrix} \cos\theta_2 & 0 & \cos(\pi/2-\theta_2) \\ 0 & 1 & 0 \\ \cos(\pi/2+\theta_2) & 0 & \cos\theta_2 \end{bmatrix} = \begin{bmatrix} \cos\theta_2 & 0 & \sin\theta_2 \\ 0 & 1 & 0 \\ -\sin\theta_2 & 0 & \cos\theta_2 \end{bmatrix}. \tag{1.23}$$

Rotation about x_3 We show this rotation in Figure 1.9. From Eq. (1.16), the direction cosines for this rotation are

$$[\alpha_{n(m)}] = \begin{bmatrix} \cos\theta_3 & \cos(\pi/2+\theta_3) & 0 \\ \cos(\pi/2-\theta_3) & \cos\theta_3 & 0 \\ 0 & 0 & 1 \end{bmatrix} = \begin{bmatrix} \cos\theta_3 & -\sin\theta_3 & 0 \\ \sin\theta_3 & \cos\theta_3 & 0 \\ 0 & 0 & 1 \end{bmatrix}. \tag{1.24}$$

1.1.8 *Orientation Lemma*

In this section we present a result that will be useful when we discuss shear stress and shear strain resulting from one-dimensional motion. Consider a point P in a Cartesian coordinate system x_k. At the point P, we wish to draw two intersecting and mutually perpendicular direction vectors $\mathbf{D}(\mathbf{A})$ and $\mathbf{D}(\mathbf{B})$ such that

$$|D_1(\mathbf{A})D_1(\mathbf{B})| = \text{maximum}.$$

Our objective here is to determine the orientation of these direction vectors.

We illustrate these vectors in Figure 1.10. You can construct this figure by first drawing the direction vector $\mathbf{D}(\mathbf{A})$ at an arbitrary angle θ to the x_1 axis. For convenience we have placed the vector $\mathbf{D}(\mathbf{A})$ in the x_1–x_2 plane, but you could place it at any relative orientation to x_2 and x_3. Thus

$$D_1(\mathbf{A}) = \cos\theta.$$

Next, draw the direction vector $\mathbf{D}(\mathbf{B})$ perpendicular to $\mathbf{D}(\mathbf{A})$. The dashed circle illustrates the possible positions of the end point of $\mathbf{D}(\mathbf{B})$. Examination of Figure 1.10 shows that the direction cosines of all possible choices of the vector $\mathbf{D}(\mathbf{B})$ must have values that are bounded by the following limits:

$$-\sin\theta \leq D_1(\mathbf{B}) \leq \sin\theta.$$

Therefore, we find that

$$D_1^2(\mathbf{A}) + D_1^2(\mathbf{B}) \leq 1. \tag{1.25}$$

Moreover, for any choice of θ, the maximum value of the product of the cosines is

$$|D_1(\mathbf{A})D_1(\mathbf{B})| = \frac{1}{2}|\sin 2\theta|,$$

where we have used the trigonometric identity $\sin(2\theta) = 2\sin\theta\cos\theta$. This result occurs when both $\mathbf{D}(\mathbf{A})$ and $\mathbf{D}(\mathbf{B})$ are in a plane that also contains the x_1 axis. By varying our

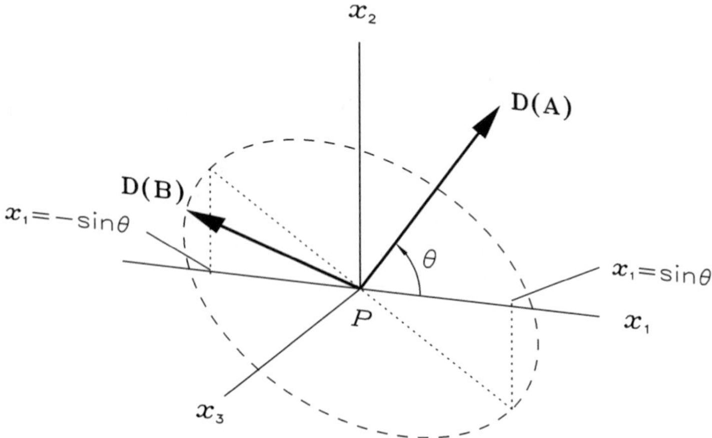

Figure 1.10 Two perpendicular vectors **A** and **B** in the coordinate system x_k. When the direction of **A** is fixed, the dashed circle represents the locus of all possible end points of the vector **B**.

original choice of the angle θ, we see that the maximum value of the product of the cosines is

$$|D_1(\mathbf{A})| = |D_1(\mathbf{B})| = \left(\frac{1}{2}\right)^{1/2}. \tag{1.26}$$

Thus the plane of the direction vectors $\mathbf{D(A)}$ *and* $\mathbf{D(B)}$ *must contain the* x_1 *axis, and both vectors must make a* 45° *angle with the* x_1 *axis.*

1.1.9 *Transformations for Coordinate Rotations*

Using the relationships we have derived in this section, we can now examine how three important classes of functions behave during a transformation involving a rotation of coordinates. These functions are called scalars, vectors, and tensors.

Vector Transformations If we know the components of a vector in the x_k coordinate system, we can compute the components in the $x_{(k)}$ coordinate system. Consider the vector $\mathbf{w}$, which can be expressed in either set of coordinates as

$$\mathbf{w} = w_{(m)}\mathbf{i}_{(m)} = w_k\mathbf{i}_k.$$

If the components w_k are known, but the components $w_{(m)}$ are unknown, we can substitute Eq. (1.15), $\mathbf{i}_k = \alpha_{k(m)}\mathbf{i}_{(m)}$, to obtain

$$w_{(m)}\mathbf{i}_{(m)} = w_k\alpha_{k(m)}\mathbf{i}_{(m)}.$$

Then equating each vector component allows us to compute $w_{(m)}$ from both w_k and the direction cosines $\alpha_{k(m)}$ as

$$w_{(m)} = \alpha_{(m)k}w_k. \tag{1.27}$$

This result is called a *vector transformation* of the components w_k of the coordinate system x_k into the components $w_{(m)}$ of the coordinate system $x_{(m)}$. A similar analysis gives

$$w_k = \alpha_{k(m)}w_{(m)}. \tag{1.28}$$

As we should suspect, the values of the components change during a coordinate rotation. Thus we say that the components of the vector $\mathbf{w}$ are *variant* under coordinate transformation.

Scalar Transformations By definition, scalar quantities do not change during a coordinate transformation. These quantities are said to be *invariant* under coordinate transformation. The dot product of two vectors $\mathbf{u}$ and $\mathbf{v}$ is a scalar quantity, which we defined in Eq. (1.3). We can show that the dot product is invariant during rotation of coordinates by using Eq. (1.28),

$$\mathbf{u}\cdot\mathbf{v} = u_kv_k = \alpha_{k(m)}u_{(m)}\alpha_{k(n)}v_{(n)}.$$

From Eq. (1.17) we have

$$\mathbf{u}\cdot\mathbf{v} = u_kv_k = u_{(m)}v_{(n)}\delta_{mn} = u_{(m)}v_{(m)}.$$

The components of **u** and **v** from both coordinate systems yield the same value for the dot product. Because the dot product is invariant under coordinate transformation, from Eq. (1.5), the length of a vector is invariant, and from Eq. (1.6) the angle between two vectors is invariant. Thus both the length of a vector and the angle formed by two vectors are scalar quantities.

Tensor Transformations Consider the relationship

$$w_m = H_{mn}u_n \tag{1.29}$$

where w_m and u_n are vectors. The function H_{mn} has two indices and connects two vectors. In the $x_{(k)}$ coordinate system, this expression is

$$w_{(m)} = H_{(mn)}u_{(n)}. \tag{1.30}$$

We can evaluate the scalar components of $H_{(mn)}$ from the scalar components of H_{mn} by substitution of Eq. (1.28) in Eq. (1.29) to obtain

$$\alpha_{m(i)}w_{(i)} = H_{mn}\alpha_{n(r)}u_{(r)}.$$

We multiply this result by $\alpha_{(k)m}$ and use Eq. (1.17) to obtain

$$\delta_{ki}w_{(i)} = w_{(k)} = [\alpha_{(k)m}H_{mn}\alpha_{n(r)}]u_{(r)}.$$

Then comparing this expression to Eq. (1.30), we find that

$$H_{(kr)} = \alpha_{(k)m}H_{mn}\alpha_{n(r)}. \tag{1.31}$$

A function that obeys this transformation is called a *second-order tensor* or simply a *tensor*. Through a similar derivation we can show that

$$H_{kr} = \alpha_{k(m)}H_{(mn)}\alpha_{(n)r}. \tag{1.32}$$

In general, tensor components are variant during coordinate transformation. However, one tensor has components that are invariant under coordinate transformation. Consider the tensor

$$H_{mn} = H\delta_{mn},$$

where H is a scalar quantity. A coordinate transformation of this tensor yields

$$H_{(kr)} = \alpha_{(k)m}H\delta_{mn}\alpha_{n(r)} = H\delta_{kr}.$$

Thus

$$H_{(mn)} = H_{mn} = H\delta_{mn}.$$

Scalar Invariants of Tensors We can combine the components of a tensor to form scalar quantities that are invariant during coordinate transformations. This is an extremely important property of the tensor. For example, we shall discover that deformation of a body is represented by a tensor. In turn, the volume of this body is determined by a scalar invariant of this tensor. This is important because the volume of a body cannot depend upon the orientation of the coordinate system.

A second-order tensor has a set of only three invariants. Let us define these invariants here. You should understand that some authors define a different set of invariants; however, it can be shown that any of these other invariants can be expressed as a scalar function of the three invariants we shall define here. The *first invariant* is called the *trace*. For the second-order tensor H_{mn} the trace I_H is defined to be

$$I_H \equiv H_{kk}. \tag{1.33}$$

We now show that the trace is invariant during coordinate transformation.

In the coordinate system $x_{(m)}$, the trace is

$$I_H = H_{(kk)} = \alpha_{(k)m} H_{mn} \alpha_{n(k)}.$$

Substitution of Eq. (1.18) yields

$$I_H = H_{(kk)} = H_{mm}.$$

Thus the trace of the transformed tensor $H_{(km)}$ equals the trace of the original tensor H_{km}, and I_H is invariant under coordinate transformation.

The *second invariant* II_H of a tensor is

$$II_H \equiv \frac{1}{2} H_{kr} H_{kr}. \tag{1.34}$$

In the coordinate system $x_{(m)}$, the second invariant is

$$II_H = \frac{1}{2} H_{(kr)} H_{(kr)} = \frac{1}{2} \alpha_{(k)m} H_{mn} \alpha_{n(r)} \alpha_{(k)s} H_{st} \alpha_{t(r)}.$$

Substitution of Eq. (1.18) yields

$$II_H = \frac{1}{2} H_{(kr)} H_{(kr)} = \frac{1}{2} H_{mn} H_{mn}.$$

Thus the second invariant II_H of the transformed tensor $H_{(kr)}$ equals the second invariant of the original tensor H_{mn}.

The *third invariant* III_H of a tensor is the *determinant* of the tensor

$$III_H \equiv \det[H_{km}], \tag{1.35}$$

where $\det[H_{km}]$ denotes the determinant of the square matrix of the tensor components,

$$\det[H_{km}] = \begin{vmatrix} H_{11} & H_{12} & H_{13} \\ H_{21} & H_{22} & H_{23} \\ H_{31} & H_{32} & H_{33} \end{vmatrix}.$$

We leave it as an exercise to show that

$$III_H = \det[H_{km}] = \det[H_{(km)}]$$

(see Exercise 1.17). Thus the determinant of the tensor $H_{(km)}$ is invariant under coordinate transformation.

1.1.10 *Exercises*

1.1 Determine which of the following vectors are perpendicular:

$$\mathbf{u} = \mathbf{i}_1 + 2\mathbf{i}_2 + 3\mathbf{i}_3, \qquad \mathbf{v} = 3\mathbf{i}_1 + 2\mathbf{i}_2 + \mathbf{i}_3, \qquad \mathbf{w} = \mathbf{i}_1 + 2\mathbf{i}_2 - 7\mathbf{i}_3.$$

1.2 Find the two unit vectors that are simultaneously perpendicular to the following vectors:

$$\mathbf{u} = -10\mathbf{i}_1 + \mathbf{i}_2 + 4\mathbf{i}_3, \qquad \mathbf{v} = \mathbf{i}_1 + 10\mathbf{i}_2 + 4\mathbf{i}_3.$$

1.3 Show that

(a) $\delta_{km} e_{kmn} = 0.$

(b) $\delta_{km} \delta_{kn} = \delta_{mn}.$

(c) $\delta_{km} a_{kn} = a_{mn}.$

1.4 If $T_{km} e_{kmn} = 0$, show that $T_{km} = T_{mk}$.

1.5 By employing distributivity and expanding the vectors **u** and **v** in component form, show that the cross product can be represented by the determinant

$$\mathbf{u} \times \mathbf{v} = \det \begin{bmatrix} \mathbf{i}_1 & \mathbf{i}_2 & \mathbf{i}_3 \\ u_1 & u_2 & u_3 \\ v_1 & v_2 & v_3 \end{bmatrix}.$$

1.6 By using both the cross product and the dot product, determine the value of the angle between the following two vectors:

$$\mathbf{u} = \mathbf{i}_1 + \mathbf{i}_2, \qquad \mathbf{v} = \mathbf{i}_2 + \mathbf{i}_3.$$

1.7 Determine the length and direction vector of

$$\mathbf{v} = \sqrt{2}\mathbf{i}_1 - \sqrt{7}\mathbf{i}_2 + 4\mathbf{i}_3.$$

1.8 Use the results of Exercise 1.5 to demonstrate that

$$\mathbf{u} \cdot \mathbf{v} \times \mathbf{w} = \det \begin{bmatrix} u_1 & u_2 & u_3 \\ v_1 & v_2 & v_3 \\ w_1 & w_2 & w_3 \end{bmatrix}.$$

1.9 Given that $\mathbf{u} \times \mathbf{v} = u_k v_m e_{kmn} \mathbf{i}_n$, show that

$$\mathbf{u} \times \mathbf{v} \cdot \mathbf{w} = u_k v_m w_n e_{kmn}.$$

1.10 A triangle has its three corners at points (1,0,0), (0,1,0), and (0,0,1) (see Figure 1.11).

(a) Write three vectors that are parallel to each edge of the triangle.

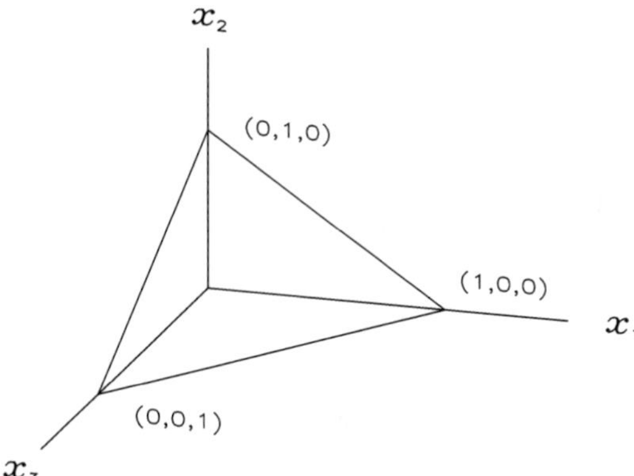

Figure 1.11 Exercise 1.10.

(b) By using the cross product, determine the area of the triangle.
(c) Using the result of part (a), determine the unit vector that is normal to this surface and points from the origin towards this surface.

1.11 Show that $\det[c_{pq}]e_{rst} = c_{rk}c_{sm}c_{tn}e_{kmn}$.

1.12 The three vectors

$$\mathbf{u} = \mathbf{i}_1, \qquad \mathbf{v} = \mathbf{i}_2 + \mathbf{i}_3, \qquad \mathbf{w} = \mathbf{i}_2$$

define a parallelepiped.
(a) Draw it.
(b) Determine its volume by using the scalar triple product.

1.13 Consider the following vector in the Cartesian coordinate system x_k:

$$\mathbf{v} = \mathbf{i}_1.$$

A second Cartesian coordinate system $x_{(k)}$ is obtained by a rotation of θ_3 about x_3. Show that

$$\mathbf{v} = \cos\theta_3 \mathbf{i}_{(1)} - \sin\theta_3 \mathbf{i}_{(2)}.$$

Now consider the vector

$$\mathbf{w} = \mathbf{i}_{(1)}.$$

Show that

$$\mathbf{w} = \cos\theta_3 \mathbf{i}_1 + \sin\theta_3 \mathbf{i}_2.$$

1.14 Consider the two relationships

$$x_{(k)} = \alpha_{(k)m} x_m, \qquad x_m = \alpha_{m(k)} x_k.$$

Because

$$\alpha_{(k)m} = \alpha_{m(k)},$$

this implies that the inverse of the direction-cosine matrix is equal to its transpose,

$$[\alpha_{(k)m}]^{-1} = [\alpha_{(k)m}]^T.$$

Verify this result for a coordinate rotation about x_1 using Eq. (1.22).

1.15 Consider this book laying on a table. The coordinate systems x_k and $x_{(k)}$ are attached to the table and the book, respectively.
(a) Rotate the book first about x_1 by $\theta_1 = \pi/2$, follow this by a rotation about $x_{(2)}$ by $\theta_{(2)} = \pi/2$, and finally apply a rotation about $x_{(3)}$ by $\theta_{(3)} = \pi/2$. Write the matrix form of the direction cosines $\alpha_{k(m)}$ for the final position of the book.
(b) After returning the book to its original configuration, rotate it about x_3 by $\theta_3 = \pi/2$, then about $x_{(2)}$ by $\theta_{(2)} = \pi/2$, and finally about $x_{(1)}$ by $\theta_{(1)} = \pi/2$. Write the matrix form of the direction cosines $\alpha_{k(m)}$ for the final position of the book.
(c) Compare these two results. Are rotations commutative?
(d) Try this experiment with this book.

Hint: Be careful to properly transpose the direction cosine matrices, and notice that we prescribe each rotation with respect to the current position of the book $x_{(k)}$ rather than the original position x_k. This allows direct application of Eqs. (1.22)–(1.24).

1.16 Using the results of Exercise 1.13, show that

$$v_k v_k = v_{(k)} v_{(k)}, \qquad w_k w_k = w_{(k)} w_{(k)}.$$

1.17 Consider two Cartesian coordinate systems with a common origin. The unit vectors of these coordinate systems are i_m and $i_{(k)}$ where

$$i_{(k)} = \alpha_{(k)m} i_m.$$

(a) Show that $\mathbf{i}_{(1)} \times \mathbf{i}_{(2)} \cdot \mathbf{i}_{(3)} = \det[\alpha_{(k)m}]$.

(b) Show that this scalar triple product implies that $\det[\alpha_{(k)m}] = 1$.

(c) Use these results and Eq. (1.10) to verify that the determinant of a tensor is invariant:

$$\det[H_{(km)}] = \det[H_{km}].$$

1.18 Consider the tensor

$$[H_{km}] = \begin{bmatrix} 1 & 0 & 0 \\ 0 & 0 & 1 \\ 0 & 1 & 0 \end{bmatrix}.$$

(a) Evaluate I_H, II_H, and III_H.

(b) Evaluate $H_{(kr)}$ for the coordinate transformation

$$\alpha_{n(m)} = \begin{bmatrix} 1 & 0 & 0 \\ 0 & \cos\theta & -\sin\theta \\ 0 & \sin\theta & \cos\theta \end{bmatrix}.$$

(c) Show that the invariants of $H_{(km)}$ are equal to I_H, II_H, and III_H.

1.19 Given the tetrahedron in Figure 1.12,

(a) use the scalar triple product to show that

$$|\mathbf{a}||\mathbf{b}| = |\mathbf{A} \times \mathbf{B} \cdot \mathbf{i}_1|.$$

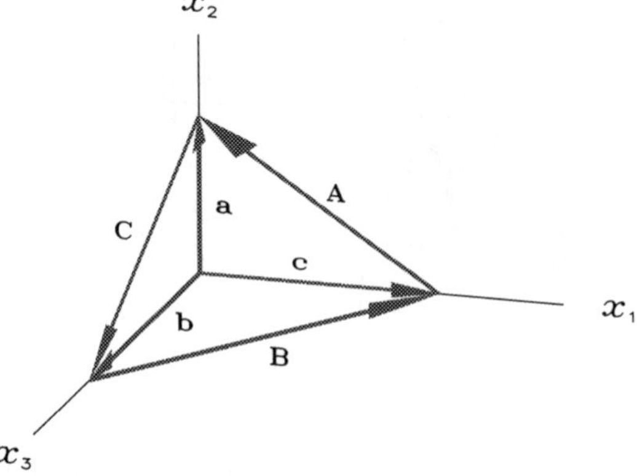

Figure 1.12 Exercise 1.19.

(b) Generalize this result to obtain

$$S_k = S|[n_k]|,$$

where S is the triangular area defined by $\frac{1}{2}|\mathbf{A} \times \mathbf{B}|$, S_k are the triangular areas defined by

$$[S_k] = \frac{1}{2}\left|\begin{bmatrix} \mathbf{a} \times \mathbf{b} \\ \mathbf{b} \times \mathbf{c} \\ \mathbf{a} \times \mathbf{c} \end{bmatrix}\right|,$$

and $\mathbf{n}$ is a vector of unit length that is normal to S.

1.2 Motion

Consider a body occupying a region of a Cartesian coordinate system. The material at any discrete point within this body moves as the body moves and changes shape. By following the motion at every point of material within the body, you can determine the displacement, velocity, and acceleration of these points.

The *current configuration* of a body is defined when we prescribe the position of every point within the body at the current time t. The motion of a body is defined when we relate its current configuration to a convenient *reference configuration*. In Figure 1.13 we show a body both at a time in the past when it was in the reference configuration and at the present time when it is in the current configuration. The vector $\mathbf{X}$ denotes the *reference position* of a point of material within the body when it is in the reference configuration. For convenience, the material at the reference point $\mathbf{X}$ is called the *material point* $\mathbf{X}$. As the body moves to its current configuration at time t, the material point $\mathbf{X}$ moves along the path in space illustrated by the dashed line. The vector $\mathbf{x}$ is the *current position* of the material point $\mathbf{X}$.

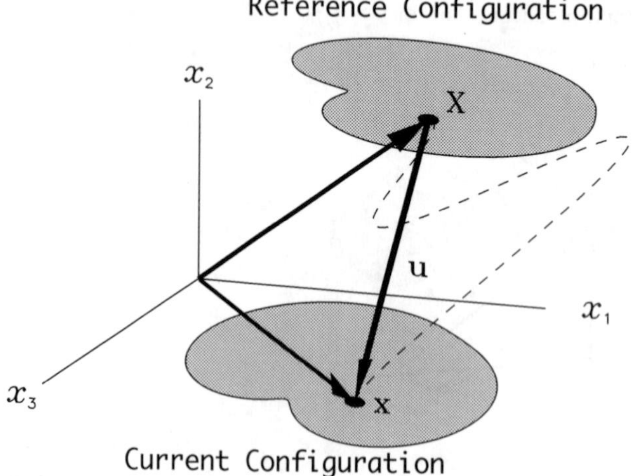

Figure 1.13 The material point $\mathbf{X}$ in the reference configuration and in the current configuration. The dashed curve is the path followed by the material point $\mathbf{X}$. The vectors $\mathbf{x}$, $\mathbf{X}$, and $\mathbf{u}$ are respectively the current position, reference position, and displacement of the material point $\mathbf{X}$.

The *displacement* $\mathbf{u}$ of the material point $\mathbf{X}$ is its current position relative to its reference position:

$$\mathbf{u} = \mathbf{x} - \mathbf{X},$$

or in index notation,

$$u_k = x_k - X_k. \qquad (1.36)$$

1.2.1 Motion and Inverse Motion

We describe the motion of the body with a Cartesian coordinate system that is fixed in space. We express the current position of every material point $\mathbf{X}$ within the body as a function of its reference position and time:

$$x_k = \hat{x}_k(X_m, t). \qquad (1.37)$$

The term X_m in the argument of this function means that the function depends on each of the scalar components X_1, X_2, and X_3 of the vector $\mathbf{X}$. This function, which gives the current position at time t of the material point $\mathbf{X}$, is called the *motion*.

You can also prescribe the reference position of the material point $\mathbf{X}$ as a function of its current position at time t:

$$X_k = X_k(x_m, t). \qquad (1.38)$$

This function is called the *inverse motion*.

1.2.2 Material and Spatial Descriptions

The motion $\hat{x}_k(X_m, t)$ and the inverse motion $X_k(x_m, t)$ are examples of two types of functions. The motion is written as a function of the reference position $\mathbf{X}$, and the inverse motion is written as a function of the current position $\mathbf{x}$. You can write other quantities such as the displacement $\mathbf{u}$ either as a function of $(\mathbf{X}, t)$ or $(\mathbf{x}, t)$, and it is often very important to make a clear distinction between these two types of functions. For this reason, we use a circumflex to indicate that $\hat{u}_k$ describes the displacement in terms of $\mathbf{X}$ and t, and we omit the circumflex to indicate that u_k describes the displacement in terms of $\mathbf{x}$ and t. Each of these descriptions has a name. In the *Lagrangian* description, which is also called the *material* description, we express the function in terms of the independent variables $\mathbf{X}$ and t. In the *Eulerian* description, which is also called the *spatial* description, we express the function in terms of the independent variables $\mathbf{x}$ and t.

Let us consider a simple example to help clarify the difference between the material and spatial descriptions as well as our use of the circumflex. Suppose a body is described by the following motion:

$$x_1 = X_1(1 + \psi t), \qquad x_2 = X_2, \qquad x_3 = X_3,$$

where ψ is a prescribed constant. The only nonzero component of displacement is

$$u_1 = x_1 - X_1.$$

The material description of this component of displacement is

$$u_1 = \hat{u}_1(X_1, t) = X_1 \psi t,$$

and the spatial description of this component of displacement is

$$u_1 = u_1(x_1, t) = x_1 \psi t/(1 + \psi t).$$

Obviously, the material and spatial representations of the displacement are two different functions that yield the same value of u_1. Thus, the use of the circumflex is superfluous in these expressions. It is also unnecessary to use it in an expression such as $E = \hat{u}_1^2$, because either function will yield the same value for E. However, suppose instead that we want to evaluate the partial derivative of the displacement with respect to time. Then we have two possibilities:

$$\frac{\partial u_1}{\partial t} = \psi X_1/(1 + \psi t) = \psi x_1/(1 + \psi t)^2$$

or

$$\frac{\partial \hat{u}_1}{\partial t} = \psi X_1 = \psi x_1/(1 + \psi t).$$

Clearly, it is important to know which form of the function we should use when evaluating the partial derivative with respect to time. This is not true for all partial derivatives. Consider the following operations:

$$\frac{\partial \hat{u}_1}{\partial X_1} = \frac{\partial}{\partial X_1}(X_1 \psi t) = \psi t,$$

$$\frac{\partial u_1}{\partial X_1} = \frac{\partial u_1}{\partial x_1}\frac{\partial \hat{x}_1}{\partial X_1} = \frac{\partial}{\partial x_1}\left(\frac{x_1 \psi t}{1 + \psi t}\right)\frac{\partial}{\partial X_1}[X_1(1 + \psi t)] = \psi t,$$

where the chain-rule expansion has been used to evaluate $\partial u_1/\partial X_1$. Here our result does not depend upon which function we use. However, we shall often use the circumflex in such situations because, even though it is superfluous, it usually adds clarity to our expressions.

1.2.3 *Displacement, Velocity, and Acceleration*

By substituting Eq. (1.37) into Eq. (1.36), we can express the displacement explicitly as a function of **X** and t:

$$u_k = \hat{x}_k(X_m, t) - X_k \equiv \hat{u}_k(X_m, t). \tag{1.39}$$

The velocity at time t of the material point **X** is defined to be the rate of change of the current position of the material point:

$$v_k = \frac{\partial}{\partial t}\hat{x}_k(X_m, t) = \frac{\partial}{\partial t}\hat{u}_k(X_m, t) \equiv \hat{v}_k(X_m, t). \tag{1.40}$$

We use the physical dimensions of m/s to represent velocity. The acceleration at time t of the material point **X** is defined to be the rate of change of its velocity:

$$a_k = \frac{\partial}{\partial t}\hat{v}_k(X_m, t) \equiv \hat{a}_k(X_m, t).$$

We use the physical dimensions of m/s^2 to represent acceleration. Displacement **u**, velocity **v**, and acceleration **a** are vectors, and, consequently, they transform as vectors during a coordinate transformation.

Spatial Descriptions In Eq. (1.40), the velocity is expressed in terms of the time derivative of the material description of the displacement. Let us consider how to express the velocity in terms of the spatial description of the displacement. Substituting the motion, Eq. (1.37), into the spatial description of the displacement, we obtain the equation

$$u_k = u_k(\hat{x}_n(X_m, t), t).$$

To evaluate the velocity, we take the time derivative of this expression with **X** held fixed:

$$v_k = \frac{\partial u_k}{\partial t} + \frac{\partial u_k}{\partial x_n}\frac{\partial \hat{x}_n}{\partial t} = \frac{\partial u_k}{\partial t} + \frac{\partial u_k}{\partial x_n} v_n. \tag{1.41}$$

This equation contains the velocity on the right side. By replacing the index k with index n, it can be written as

$$v_n = \frac{\partial u_n}{\partial t} + \frac{\partial u_n}{\partial x_m} v_m.$$

The second term on the right is often called the *convective velocity.* We use this expression to replace the term v_n on the right side of Eq. (1.41), obtaining the equation

$$v_k = \frac{\partial u_k}{\partial t} + \frac{\partial u_k}{\partial x_n}\left(\frac{\partial u_n}{\partial t} + \frac{\partial u_n}{\partial x_m} v_m\right).$$

We can now use Eq. (1.41) to replace the term v_m on the right side of this equation. Continuing in this way, we obtain the series

$$v_k = \frac{\partial u_k}{\partial t} + \frac{\partial u_k}{\partial x_n}\frac{\partial u_n}{\partial t} + \frac{\partial u_k}{\partial x_n}\frac{\partial u_n}{\partial x_m}\frac{\partial u_m}{\partial t} + \cdots. \tag{1.42}$$

This series gives the velocity in terms of derivatives of the spatial description of the displacement.

We leave it as an exercise to show that the acceleration is given in terms of derivatives of the spatial description of the velocity by (see Exercise 1.23)

$$a_k = \frac{\partial v_k}{\partial t} + \frac{\partial v_k}{\partial x_n} v_n. \tag{1.43}$$

The second term on the right is often called the *convective acceleration.*

Total Derivative For any arbitrary smooth function $\phi = \phi(x_n, t)$ written in the spatial description, the arguments in the preceding paragraph lead us to conclude that

$$\frac{\partial \hat{\phi}}{\partial t} = \frac{\partial \phi}{\partial t} + \frac{\partial \phi}{\partial x_n} v_n.$$

The derivative on the left is expressed in the material description whereas the derivatives on the right are expressed in the spatial description. We have already seen that this result is central to the description of velocity and acceleration. You should remember this result because it will be used frequently throughout the remainder of this work. This is often called the *total derivative* of ϕ.

Velocity and Acceleration in Linear Elasticity In *linear elasticity*, spatial derivatives of the displacement are assumed to be "small." This means we can neglect terms containing products of derivatives of the displacement. From Eqs. (1.42) and (1.43), we see that in linear elasticity, the velocity and acceleration are related to the spatial description of the displacement by the simple expressions

$$v_k = \frac{\partial u_k}{\partial t}, \qquad a_k = \frac{\partial^2 u_k}{\partial t^2}. \tag{1.44}$$

1.2.4 *Pure Translation and Rotation of a Body*

Pure translation and rotation are special types of motion. We examine these types of motion in this section.

Pure Translation of a Body Consider the following special type of displacement:

$$\mathbf{u} = \mathbf{u}(t).$$

Because this displacement is not a function of either the current or reference position, every material point $\mathbf{X}$ experiences the same displacement. This type of motion is called *pure translation*. The motion resulting from pure translation is

$$x_k = X_k + u_k(t).$$

The velocity and acceleration resulting from pure translation are functions of time t alone.

Pure Rotation of a Body Consider a body at time $t = 0$. It is described by two Cartesian coordinate systems x_k and $x_{(m)}$. At time $t = 0$, the two coordinate systems are coincident, and the reference position of point $\mathbf{X}$ in the body has the same value in both coordinate systems,

$$x_m = x_{(m)} = X_m = X_{(m)}.$$

We let the coordinate system x_m be fixed in space and the coordinate system $x_{(m)}$ be fixed to the body. Now we let time advance as we rotate the body. The point $\mathbf{X}$ in the body moves to the current position $\mathbf{x}$. Because the body and the coordinate system $x_{(m)}$ rotate together, the value of $x_{(m)}$ does not change. Therefore,

$$x_{(m)} = X_m \tag{1.45}$$

for all times. The current position $\mathbf{x}$ in the fixed coordinate system x_k is given by the coordinate transformation

$$x_k = \alpha_{k(m)}(t)\, x_{(m)} = \alpha_{k(m)}(t)\, X_m. \tag{1.46}$$

This is the motion of $\mathbf{X}$. Because the Cartesian system $x_{(m)}$ is rotating, the direction cosines in this relationship are expressed as functions of time. This type of motion is called *pure rotation*. To evaluate the inverse motion, we multiply it by $\alpha_{k(r)}(t)$ to obtain

$$\alpha_{k(r)}(t)\, x_k = \alpha_{k(r)}(t)\, \alpha_{k(m)}(t)\, X_m.$$

Then we apply Eq. (1.17), which yields

$$X_r = \alpha_{k(r)}(t)\, x_k. \tag{1.47}$$

Body Rotation versus Coordinate Rotation Compare the equation for the position of a rotating body in a stationary coordinate system, Eq. (1.46),

$$x_k = \alpha_{k(m)}(t)\, X_m,$$

to the transformation of the position of a stationary body during a coordinate rotation [see Eq. (1.28)],

$$x_k = \alpha_{k(m)} x_{(m)}.$$

Although these equations have similar forms, they represent quite different processes. Because the first equation represents rotation of a body in a stationary coordinate system, this process causes the body to undergo velocity and acceleration. However, the second equation represents a coordinate transformation of the position of a body, and because the body is stationary, the velocity and acceleration are zero. Because the rotation of a body is not the same as the rotation of a coordinate system, we denote the rotation of a body by $\alpha_{k(m)}(t)$ and the rotation of a coordinate system by $\alpha_{k(m)}$. This is important because quantities that are invariant during a coordinate transformation are not necessarily invariant during a body rotation.

Example of Pure Rotation of a Body Consider the body illustrated in Figure 1.14. The dotted outline is the reference configuration of the body. We rotate it about the x_3 axis to the current configuration, which is the shaded region. The current position of the point $\mathbf{x}$ in the body always appears in the same position with respect to the rotating coordinate system $x_{(m)}$. The angle of rotation about the x_3 axis is $\theta_3(t)$. The direction cosines for rotation about

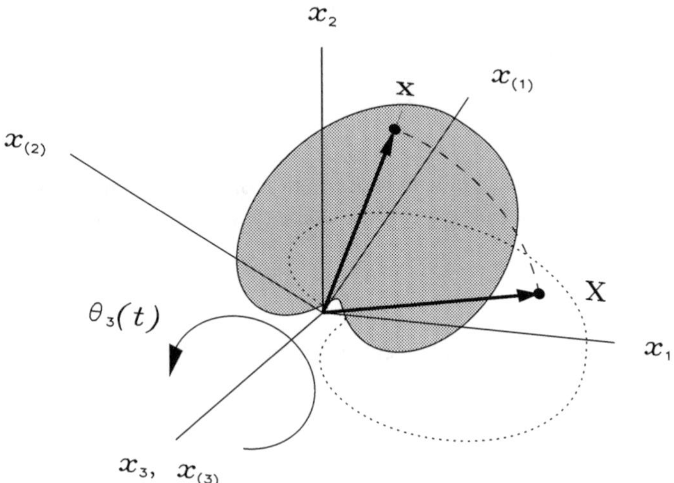

Figure 1.14 Pure rotation of a body about x_3 from its reference configuration at $t = 0$ (dotted outline) to its current configuration (shaded region). The x_k coordinate system is fixed in space, and the $x_{(m)}$ coordinate system rotates with the body. The dashed curve is the path followed by the point $\mathbf{x}$.

the x_3 axis are given by Eq. (1.24), which when substituted for $\alpha_{k(m)}(t)$ in Eq. (1.46) yields

$$\begin{bmatrix} x_1 \\ x_2 \\ x_3 \end{bmatrix} = \begin{bmatrix} \cos\theta_3 & -\sin\theta_3 & 0 \\ \sin\theta_3 & \cos\theta_3 & 0 \\ 0 & 0 & 1 \end{bmatrix} \begin{bmatrix} X_1 \\ X_2 \\ X_3 \end{bmatrix}. \tag{1.48}$$

The inverse of this relationship is given by Eq. (1.47),

$$\begin{bmatrix} X_1 \\ X_2 \\ X_3 \end{bmatrix} = \begin{bmatrix} \cos\theta_3 & \sin\theta_3 & 0 \\ -\sin\theta_3 & \cos\theta_3 & 0 \\ 0 & 0 & 1 \end{bmatrix} \begin{bmatrix} x_1 \\ x_2 \\ x_3 \end{bmatrix}. \tag{1.49}$$

Notice that the matrix of the direction cosines in Eq. (1.49) is the transpose of the one in Eq. (1.48). From these results, we also notice that the distance between the point **x** and the axis of rotation is constant, that is,

$$(x_1 x_1 + x_2 x_2)^{1/2} = (X_1 X_1 + X_2 X_2)^{1/2}.$$

We can evaluate the velocity of the point **x** from Eqs. (1.48) by taking the partial derivative of $\hat{x}_k$ with respect to t:

$$\begin{aligned} v_1 &= (-X_1 \sin\theta_3 - X_2 \cos\theta_3)\frac{d\theta_3}{dt}, \\ v_2 &= (X_1 \cos\theta_3 - X_2 \sin\theta_3)\frac{d\theta_3}{dt}, \\ v_3 &= 0. \end{aligned} \tag{1.50}$$

This is the material description of the velocity. We substitute Eqs. (1.49) to obtain

$$v_1 = -x_2 \frac{d\theta_3}{dt}, \qquad v_2 = x_1 \frac{d\theta_3}{dt}, \qquad v_3 = 0, \tag{1.51}$$

which is the spatial description of the velocity. The magnitude of the velocity is

$$|\mathbf{v}| = \left[X_1^2 + X_2^2\right]^{1/2} \frac{d\theta_3}{dt},$$

which is the distance between **x** and the axis of rotation times the angular rate of rotation.

We can also evaluate the dot product of the velocity **v** and the current position **x** and easily show that

$$v_1 x_1 + v_2 x_2 + v_3 x_3 = 0.$$

This means the direction of the velocity vector **v** is perpendicular to **x**.

1.2.5 *Exercises*

1.20 Consider a cube with sides of length l. Its reference configuration at $t = 0$ is shown in Figure 1.15. The motion of this cube is

$$x_1 = X_1/(1 + \psi t^2), \qquad x_2 = X_2(1 + \psi t^2), \qquad x_3 = X_3,$$

where ψ is a known constant.

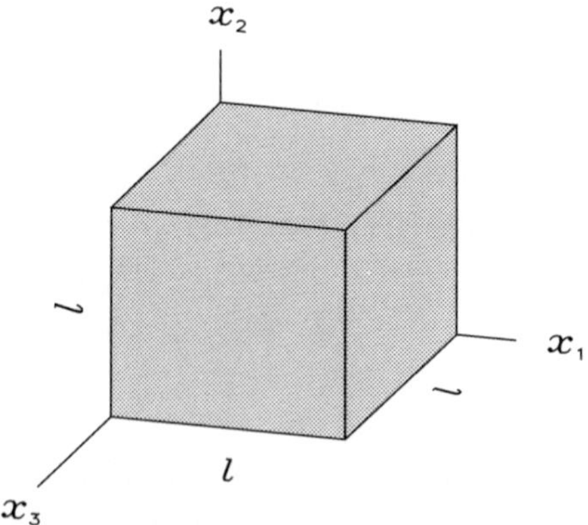

Figure 1.15 Exercise 1.20

(a) Determine the inverse motion.
(b) Determine the material description of the displacements.
(c) Determine the spatial description of the displacements.
(d) What are the current positions of each of the eight corners of this cube at $t = 2$?
(e) What are the displacements of the corners when $t = 2$ and $\psi = 1$?

1.21 In Exercise 1.20:
(a) Determine the material descriptions of velocity and acceleration.
(b) Determine the spatial descriptions of velocity and acceleration.
(c) What are the velocities and accelerations of each of the eight corners of the cube when $t = 2$ and $\psi = 1$?

1.22 Consider a cube with sides of length l. Its reference configuration at $t = 0$ is shown in Figure 1.15. The motion is

$$x_1 = X_1 + \psi X_2^2 t^2, \qquad x_2 = X_2, \qquad x_3 = X_3,$$

where ψ is a known constant. Repeat the previous two exercises using this motion.

1.23 Show that the acceleration in the spatial description is [see Eq. (1.43)]

$$a_k = \frac{\partial v_k}{\partial t} + v_m \frac{\partial v_k}{\partial x_m},$$

and demonstrate that the right side of this equation transforms as a vector.

1.24 A body is rotating at a constant angular rate of

$$\omega = \frac{d\theta_1}{dt}$$

about the x_1 axis.
(a) Determine the motion and the inverse motion.
(b) Determine the spatial and material descriptions of the velocity $\mathbf{v}$ and the acceleration $\mathbf{a}$.
(c) Let $\mathbf{r}$ be the vector

$$\mathbf{r} = x_2 \mathbf{i}_2 + x_3 \mathbf{i}_3.$$

Express $|\mathbf{v}|$ and $|\mathbf{a}|$ in terms of ω and $|\mathbf{r}|$.
(d) Show that $|\mathbf{r}|$ of any material point $\mathbf{X}$ does not change with time.

(e) Using the spatial description, evaluate $\mathbf{v} \cdot \mathbf{r}$ and $\mathbf{a} \times \mathbf{r}$. What does this say about the directions of $|\mathbf{v}|$ and $|\mathbf{a}|$?

1.25 Consider a body in pure rotation. Its motion is given by

$$x_k = \alpha_{k(m)}(t) X_m.$$

Let X_m and $\bar{X}_m$ be any two material points in this body. The current positions of these material points are x_m and $\bar{x}_m$. Show that the distance between x_m and $\bar{x}_m$ does not change with time.

1.3 One-Dimensional Motion

We have described motion in three spatial dimensions. The importance of doing this will become clear as we proceed through the remainder of this chapter to study deformation, strain, and stress. However, in this section, we pause briefly to consider a special type of motion called *one-dimensional motion*. Often one-dimensional motion refers to motion that is solely a function of x_1 and t. *Here we use a special definition of one-dimensional motion. Not only is it motion that is solely a function x_1 and t, but in addition, it is motion that is symmetric about the x_1 axis.* Thus x_1 is said to be an *axis of symmetry.* Throughout much of this book we restrict our discussion to this special type of motion. By doing this we greatly reduce the mathematical complexity of the development while retaining most of the physical complexity of nonlinear wave motion that is understood today. Indeed, with the exception of gas dynamics, experimental observation of nonlinear waves has been largely limited to one-dimensional motion.

The displacement components u_k describe the motion of a body in a Cartesian coordinate system x_k. Consider a second coordinate system $x_{(m)}$ that is obtained by a rotation through an angle of θ_1 about the x_1 axis (see Figure 1.7). The displacement components in the second coordinate system are given by the vector transformation

$$u_{(m)} = \alpha_{(m)k} u_k.$$

For motion that is symmetric about x_1, we require that

$$u_{(m)} = u_m, \tag{1.52}$$

for all possible orientations of the coordinate system $x_{(m)}$. Using this symmetry requirement, we find that

$$u_m = \alpha_{(m)k} u_k.$$

The direction cosines for rotation about the x_1 axis are given by Eq. (1.22) in Section 1.1.7. We substitute these direction cosines to obtain

$$\begin{aligned} u_1 &= u_1, \\ u_2 &= u_2 \cos\theta_1 + u_3 \sin\theta_1, \\ u_3 &= -u_2 \sin\theta_1 + u_3 \cos\theta_1. \end{aligned}$$

The first of these expressions identically satisfies the symmetry requirement. The remaining expressions must be satisfied for all possible values of θ_1. Therefore, let us select $\theta_1 = \pi/2$ so that the last two relationships become

$$u_2 = u_3, \qquad u_3 = -u_2,$$

which can only be satisfied when

$$u_2 = u_3 = 0.$$

Thus, for one-dimensional motion as we have defined it here, the only nonzero component of displacement is u_1. Under these conditions, the current positions x_k are related to the reference positions X_k by [see Eq. (1.39)]

$$x_1 = u_1 + X_1, \qquad x_2 = X_2, \qquad x_3 = X_3.$$

This motion has only one nonzero component of velocity,

$$v_1 = \frac{\partial \hat{u}_1}{\partial t},$$

and one nonzero component of acceleration,

$$a_1 = \frac{\partial \hat{v}_1}{\partial t}.$$

1.4 Deformation

Consider any two neighboring points in a body. The most general motion causes the straight line connecting these points to translate, rotate, and change length. Sometimes motion leaves the lengths of all lines connecting all possible sets of points in the body unchanged. This special type of motion is called *rigid-body motion*. Rigid bodies do not change shape or volume. The cases of pure translation and rotation we presented in the previous section are examples of rigid-body motion. In contrast, when the lengths of some or all of these lines do change, we say that such motion causes *deformation* of the body. It is important to identify the properties of motion that result in deformation. We can do this by deriving a quantity that only depends upon deformation. This is achieved in two steps. In the first step, we use the motion to derive the *deformation gradient tensor*. This quantity does not change during pure translation of the body, but it does change during pure rotation and deformation. In the second step, we use the deformation gradient to derive the *Lagrangian strain tensor*. The Lagrangian strain tensor does not change during either pure translation or pure rotation of the body. It changes only during deformation.

1.4.1 *Deformation Gradient Tensor*

We show two neighboring material points in a body with reference positions $\mathbf{X}$ and $\mathbf{X} + d\mathbf{X}$ in Figure 1.16. The current positions of these two material points are $\mathbf{x}$ and $\mathbf{x} + d\mathbf{x}$. In terms of the motion, $x_k = \hat{x}_k(X_m, t)$, the vector $d\mathbf{x}$ is related to the vector $d\mathbf{X}$ by

$$dx_k = \frac{\partial \hat{x}_k}{\partial X_m} dX_m,$$

where we evaluate $\mathbf{x}$ and $\mathbf{x} + d\mathbf{x}$ at the same time t. We define the deformation gradient F_{km} to be

$$F_{km} \equiv \frac{\partial \hat{x}_k}{\partial X_m} \tag{1.53}$$

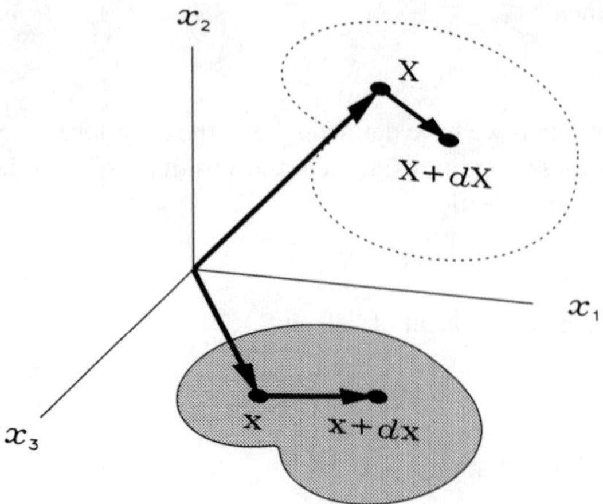

Figure 1.16 Two neighboring points of a body in the reference configuration (dotted outline) and the current configuration (shaded region).

so that

$$dx_k = F_{km} dX_m. \tag{1.54}$$

The deformation gradient F_{km} is a function that transforms the vector $d\mathbf{X}$ in the reference configuration into the vector $d\mathbf{x}$ in the current configuration. It is a tensor (see Section 1.1.9). Notice that the deformation gradient does not have physical dimensions.

During a change of coordinates, the deformation gradient transforms according to Eqs. (1.31) and (1.32) [see Exercise 1.26]:

$$\begin{aligned} F_{(kr)} &= \alpha_{(k)m} F_{mn} \alpha_{n(r)}, \\ F_{kr} &= \alpha_{k(m)} F_{(mn)} \alpha_{(n)r}. \end{aligned} \tag{1.55}$$

The three invariants of the deformation gradient tensor are [see Eqs. (1.33)–(1.35)]

$$I_F = F_{kk}, \qquad II_F = \frac{1}{2} F_{km} F_{km}, \qquad III_F = \det[F_{km}].$$

From the expression for the displacement, Eq. (1.39), we can also express the deformation gradient as

$$F_{km} = \frac{\partial}{\partial X_m}(\hat{u}_k + X_k) = \frac{\partial \hat{u}_k}{\partial X_m} + \delta_{km}. \tag{1.56}$$

Deformation Gradient for the Reference Configuration In the reference configuration, the displacement is zero, and

$$F_{km} = \delta_{km}.$$

In matrix notation, the deformation gradient tensor of the reference configuration is

$$[F_{km}] = \begin{bmatrix} 1 & 0 & 0 \\ 0 & 1 & 0 \\ 0 & 0 & 1 \end{bmatrix}. \tag{1.57}$$

Inverse of Deformation Gradient Because the inverse motion [Eq. (1.38)] exists, we can also write

$$dX_m = \frac{\partial X_m}{\partial x_k} dx_k. \tag{1.58}$$

We define the *inverse deformation gradient* to be

$$F_{mk}^{-1} \equiv \frac{\partial X_m}{\partial x_k}. \tag{1.59}$$

We obtain a relationship between the deformation gradient and the inverse deformation gradient by noticing that

$$dx_k = \frac{\partial \hat{x}_k}{\partial X_m} dX_m = \frac{\partial \hat{x}_k}{\partial X_m} \frac{\partial X_m}{\partial x_r} dx_r.$$

Then

$$dx_k = F_{km} F_{mr}^{-1} dx_r,$$

which implies that

$$F_{km} F_{mr}^{-1} = \delta_{kr}. \tag{1.60}$$

In other words, the inverse deformation gradient may be computed from the inverse of the deformation gradient.

Through the use of Cramer's rule, it is also possible to obtain Eq. (1.58) directly from Eq. (1.54). The result is

$$F_{ij}^{-1} = \frac{\mathrm{cof}[F_{ji}]}{\det[F_{mk}]}, \tag{1.61}$$

where cof $[F_{ji}]$ denotes the cofactor of component F_{ji} of the matrix. You should carefully notice the transposition of indices (i, j) between the right and left sides of this expression. Because we require the inverse motion to exist, this solution for F_{ij}^{-1} must exist, and therefore the determinant of $[F_{mk}]$ must be nonzero. We also require the deformation gradient to be a continuous function. Therefore, because $\det[F_{mk}]$ is positive in the reference configuration, it cannot become negative without passing through zero. Thus, we conclude that

$$\det[F_{km}] > 0. \tag{1.62}$$

Deformation Gradient During Pure Translation of a Body For pure translation the displacement $\mathbf{u}$ is a function of t only (see Section 1.2.4). From Eq. (1.56), it follows that for pure translation

$$F_{km} = \delta_{km}. \tag{1.63}$$

Thus pure translation does not alter the deformation gradient tensor. Using this result, we find that the vectors representing the relative positions of the two neighboring points in the reference and current configurations are identical:

$$dx_k = \delta_{km} dX_m = dX_k.$$

Deformation Gradient During Pure Rotation of a Body Consider a body in a reference configuration **X**. To examine how pure rotation affects the deformation gradient, let us first deform this body with the prescribed motion

$$x_m^0 = \hat{x}_m^0(X_n, t). \tag{1.64}$$

During this motion the material point **X** moves to the position $\mathbf{x}^0$. The deformation gradient is [see Eq. (1.53)]

$$F_{mn}^0 = \frac{\partial \hat{x}_m^0}{\partial X_n}.$$

Now we examine how a pure rotation affects this deformation gradient. The rotation moves the point $\mathbf{x}^0$ to the position $\mathbf{x}$. From Eq. (1.46) we have

$$x_k = \alpha_{(k)m}(t)\, x_m^0. \tag{1.65}$$

Using the chain rule, the deformation gradient for the combined motion is

$$F_{kn} = \frac{\partial x_k}{\partial x_m^0} \frac{\partial \hat{x}_m^0}{\partial X_n} = \alpha_{(k)m}(t) F_{mn}^0. \tag{1.66}$$

We emphasize that rotation of a body is different than a coordinate transformation. Whereas the previous expression yields the change in the deformation gradient during rotation of a body, notice that during transformation of coordinates a different change in the deformation gradient occurs [see Eq. (1.55)]. This is because the deformation gradient relates the vector $d\mathbf{x}$ in the current configuration to the vector $d\mathbf{X}$ in the reference configuration. Both the current and the reference configurations are affected by a coordinate transformation, while only the current configuration moves during rotation of a body. The description of the vector $d\mathbf{X}$ is altered by a coordinate transformation but not by rotation of the body.

1.4.2 *Strain Tensor*

Translation of a body causes the motion **x** to change while the deformation gradient F_{km} remains constant [see Eq. (1.63)]. Rotation of a body causes both the motion and the deformation gradient to change [see Eq. (1.66)]. In this section we derive a quantity that changes only during deformation. This quantity is called *strain*. It depends upon the distance between two neighboring material points in a body.

The squares of the distances between two neighboring material points in the current and reference configurations are given by the dot products:

$$|d\mathbf{x}|^2 = dx_k dx_k, \qquad |d\mathbf{X}|^2 = dX_k dX_k.$$

In terms of their lengths and directions, these vectors are

$$dx_k = |d\mathbf{x}| D_k(d\mathbf{x}), \qquad dX_k = |d\mathbf{X}| D_k(d\mathbf{X}).$$

We define the *stretch* Λ to be the ratio of these lengths,

$$\Lambda \equiv \frac{|d\mathbf{x}|}{|d\mathbf{X}|}. \tag{1.67}$$

Using the deformation gradient, we can link the square of the distance in the current configuration to the square of the distance in the reference configuration as follows:

$$|d\mathbf{x}|^2 = F_{mk}dX_k F_{mn}dX_n = |d\mathbf{X}|^2 F_{mk}F_{mn}D_k(d\mathbf{X})D_n(d\mathbf{X}).$$

The change in the square of the distance is

$$\begin{aligned}\frac{|d\mathbf{x}|^2 - |d\mathbf{X}|^2}{|d\mathbf{X}|^2} &= \Lambda^2 - 1 = (F_{mk}F_{mn} - \delta_{kn})D_k(d\mathbf{X})D_n(d\mathbf{X}) \\ &= 2E_{kn}D_k(d\mathbf{X})D_n(d\mathbf{X}),\end{aligned} \tag{1.68}$$

where

$$E_{kn} \equiv \frac{1}{2}(F_{mk}F_{mn} - \delta_{kn}) \tag{1.69}$$

is called the *Lagrangian strain tensor*. Like the deformation gradient, the Lagrangian strain tensor does not have physical dimensions. By substituting Eq. (1.56), which expresses F_{km} in terms of u_k, we can write the Lagrangian strain tensor in terms of the displacement as

$$E_{kn} = \frac{1}{2}\left(\frac{\partial \hat{u}_k}{\partial X_n} + \frac{\partial \hat{u}_n}{\partial X_k} + \frac{\partial \hat{u}_m}{\partial X_k}\frac{\partial \hat{u}_m}{\partial X_n}\right).$$

We leave it as an exercise to show that the strain E_{kn} transforms as a tensor according to the relationships (see Exercise 1.32):

$$E_{(kr)} = \alpha_{(k)m}E_{mn}\alpha_{n(r)}, \tag{1.70}$$

$$E_{kr} = \alpha_{k(m)}E_{(mn)}\alpha_{(n)r}. \tag{1.71}$$

The three invariants of the strain tensor are

$$I_E = E_{kk}, \qquad II_E = \frac{1}{2}E_{km}E_{km}, \qquad III_E = \det[E_{km}].$$

Strain in Linear Elasticity In linear elasticity the gradients of the displacement **u** are small, and we can neglect terms containing products of these derivatives. Thus the strain simplifies to

$$E_{kn} = \frac{1}{2}\left(\frac{\partial \hat{u}_k}{\partial X_n} + \frac{\partial \hat{u}_n}{\partial X_k}\right). \tag{1.72}$$

Let us expand one of the partial derivatives in this result using the motion [Eq. (1.37)] and the chain rule:

$$\frac{\partial \hat{u}_k}{\partial X_n} = \frac{\partial u_k}{\partial x_m}\frac{\partial \hat{x}_m}{\partial X_n} = \frac{\partial u_k}{\partial x_m}\frac{\partial}{\partial X_n}(\hat{u}_m + X_m) = \frac{\partial u_k}{\partial x_m}\left(\frac{\partial \hat{u}_m}{\partial X_n} + \delta_{mn}\right).$$

Neglecting the term containing products of the derivative of the displacement, we obtain

$$\frac{\partial \hat{u}_k}{\partial X_n} = \frac{\partial u_k}{\partial x_n}.$$

Substitution of this result into Eq. (1.72) shows us that, in linear elasticity, the strain is

$$E_{kn} = \frac{1}{2}\left(\frac{\partial u_k}{\partial x_n} + \frac{\partial u_n}{\partial x_k}\right). \tag{1.73}$$

Strain During Pure Translation and Rotation of a Body Pure translation and rotation cause the motion and the deformation gradient to change. We now show that strain remains unchanged during pure translation and rotation. During pure translation the deformation gradient is [see Eq. (1.63)]

$$F_{mk} = \delta_{mk}.$$

The resulting strain is

$$E_{kn} = \frac{1}{2}(\delta_{mk}\delta_{mn} - \delta_{kn}) = 0.$$

Next consider a body in a deformed configuration described by the deformation gradient F^0_{rk}, which yields the strain

$$E^0_{kn} = \frac{1}{2}\left(F^0_{rk}F^0_{rn} - \delta_{kn}\right).$$

We subject this body to a pure rotation in which the change in the deformation gradient is given by Eq. (1.66),

$$F_{rk} = \alpha_{(r)m}(t)F^0_{mk}.$$

After this rotation, the strain is

$$E_{kn} = \frac{1}{2}\left[\alpha_{(r)m}(t)\,\alpha_{(r)s}(t)\,F^0_{mk}F^0_{sn} - \delta_{kn}\right].$$

Substitution of Eq. (1.17) yields

$$E_{kn} = \frac{1}{2}\left(\delta_{ms}F^0_{mk}F^0_{sn} - \delta_{kn}\right).$$

Thus

$$E_{kn} = E^0_{kn},$$

and the Lagrangian strain does not change during rotation of the body. Only motion with deformation causes the strain E_{kn} to change. In contrast, a coordinate transformation changes the strain components according to the tensor transformation in Eq. (1.70). The coordinate transformation affects the strain components because it alters the descriptions of both the current and the reference configurations. However, pure rotation of the body affects only the current configuration.

1.4.3 *Longitudinal Strain*

The *longitudinal strain* ε is a measure of the change in length of a line element in a body. Here we show how the longitudinal strain can be expressed in terms of the strain tensor.

The longitudinal strain ε of a line element $d\mathbf{X}$ relative to a reference configuration is defined to be its change in length divided by its length in the reference configuration,

$$\varepsilon \equiv \frac{|d\mathbf{x}| - |d\mathbf{X}|}{|d\mathbf{X}|}.$$

In terms of the stretch Λ, Eq. (1.67), the longitudinal strain ε is

$$\Lambda = 1 + \varepsilon. \tag{1.74}$$

Solving for $|d\mathbf{x}|$ in terms of ε and substituting the result into Eq. (1.68), we obtain the relation

$$2\varepsilon + \varepsilon^2 = 2E_{mn}D_m(d\mathbf{X})D_n(d\mathbf{X}). \tag{1.75}$$

When the Lagrangian strain tensor E_{mn} is known at a material point, we can solve this equation for the longitudinal strain ε of the line element $d\mathbf{X}$. From the definition of the longitudinal strain [Eq. (1.74)], we see that it depends only on scalar quantities. Therefore, longitudinal strain is invariant during a coordinate transformation. This means that the value we obtain for the longitudinal strain of a line element in the body is independent of the coordinate system that we use to describe the body.

Longitudinal Strain in Linear Elasticity Equation (1.75) becomes very simple in linear elasticity. If we assume that ε is small, we can neglect the quadratic term to obtain

$$\varepsilon = E_{mn}D_m(d\mathbf{X})D_n(d\mathbf{X}). \tag{1.76}$$

The change in length of the line element $d\mathbf{X}$ is given by the product of the direction cosines of the line element in the reference configuration and the strain tensor.

1.4.4 *Shear Strain*

The *shear strain* γ is a measure of the change in angle between two intersecting line elements in a body during deformation. Here we show how the shear strain between two perpendicular line elements that are arbitrarily oriented in the body can be expressed in terms of the strain tensor. In Figure 1.17 we show two perpendicular vectors $d\mathbf{A}$ and $d\mathbf{B}$ that represent line elements in the reference configuration of a body. In the current configuration

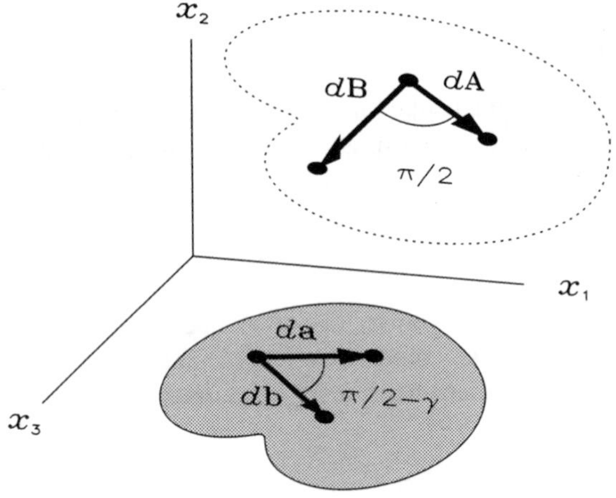

Figure 1.17 Two perpendicular line elements in a body in the reference configuration and the current configuration.

these line elements are $d\mathbf{a}$ and $d\mathbf{b}$. Using the direction cosines and deformation gradient, we have

$$dA_k = D_k(d\mathbf{A})|d\mathbf{A}|, \qquad dB_k = D_k(d\mathbf{B})|d\mathbf{B}|$$

and

$$da_m = F_{mk} D_k(\mathbf{A})|d\mathbf{A}|, \qquad db_m = F_{mk} D_k(\mathbf{B})|d\mathbf{B}|.$$

The angle between the vectors $d\mathbf{a}$ and $d\mathbf{b}$ is

$$\cos(\pi/2 - \gamma) = \frac{d\mathbf{a} \cdot d\mathbf{b}}{|d\mathbf{a}||d\mathbf{b}|} = \frac{da_m db_m}{(da_r da_r)^{1/2}(db_n db_n)^{1/2}},$$

which becomes

$$\sin\gamma = \frac{F_{mp} F_{mq} D_p(d\mathbf{A}) D_q(d\mathbf{B})}{[F_{rs} F_{rt} D_s(d\mathbf{A}) D_t(d\mathbf{A})]^{1/2} [F_{nj} F_{ni} D_j(d\mathbf{B}) D_i(d\mathbf{B})]^{1/2}}.$$

We substitute the definition of strain [Eq. (1.69)] to obtain

$$\sin\gamma = \frac{(2E_{pq} + \delta_{pq}) D_p(d\mathbf{A}) D_q(d\mathbf{B})}{[(2E_{st} + \delta_{st}) D_s(d\mathbf{A}) D_t(d\mathbf{A})]^{1/2} [(2E_{ji} + \delta_{ji}) D_j(d\mathbf{B}) D_i(d\mathbf{B})]^{1/2}}.$$

Because $d\mathbf{A}$ and $d\mathbf{B}$ are perpendicular unit vectors, we see that $D_k(d\mathbf{A}) D_k(d\mathbf{B}) = 0$, $D_k(d\mathbf{A}) D_k(d\mathbf{A}) = D_k(d\mathbf{B}) D_k(d\mathbf{B}) = 1$, and

$$\sin\gamma = \frac{2E_{pq} D_p(d\mathbf{A}) D_q(d\mathbf{B})}{[2E_{st} D_s(d\mathbf{A}) D_t(d\mathbf{A}) + 1]^{1/2} [2E_{ji} D_j(d\mathbf{B}) D_i(d\mathbf{B}) + 1]^{1/2}}. \tag{1.77}$$

Because shear strain γ measures the change of angle between two vectors, it is a scalar quantity and invariant during coordinate transformation. Thus the value of the shear strain between two intersecting line elements of a body is independent of the coordinate system we use to describe the body.

Shear Strain in Linear Elasticity In linear elasticity, both γ and E_{km} are small so that the last equation becomes

$$\gamma = 2E_{pq} D_p(d\mathbf{A}) D_q(d\mathbf{B}).$$

The change in angle between two perpendicular line elements in the reference configuration is given by twice the product of the strain tensor and the direction cosines of the two line elements.

1.4.5 *Volume Strain*

The *volume strain* e is a measure of the change of volume of an element of the body during deformation. To examine how the volume of a body changes during deformation, consider the three mutually perpendicular line elements $d\mathbf{A}$, $d\mathbf{B}$, and $d\mathbf{C}$ in the reference configuration (see Figure 1.18):

$$d\mathbf{A} = dA\mathbf{i}_1, \qquad d\mathbf{B} = dB\mathbf{i}_2, \quad d\mathbf{C} = dC\mathbf{i}_3. \tag{1.78}$$

These three line elements define the sides of an infinitesimal rectangular box of volume

$$dV_0 = dA\, dB\, dC.$$

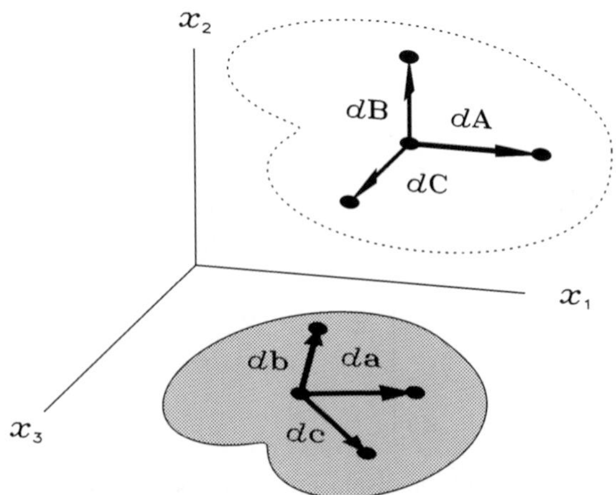

Figure 1.18 Three mutually perpendicular line elements $d\mathbf{A}$, $d\mathbf{B}$, and $d\mathbf{C}$ in the reference configuration and the deformed elements $d\mathbf{a}$, $d\mathbf{b}$, and $d\mathbf{c}$ in the current configuration.

We require dA, dB, and dC to be positive; thus dV_0 is positive. In the current configuration, this rectangular box deforms to a parallelepiped with sides $d\mathbf{a}$, $d\mathbf{b}$, and $d\mathbf{c}$. The volume dV of this parallelepiped is $dV = |d\mathbf{a} \cdot d\mathbf{b} \times d\mathbf{c}|$ [see Eq. (1.8)],

$$dV = \left| \det \begin{bmatrix} da_1 & db_1 & dc_1 \\ da_2 & db_2 & dc_2 \\ da_3 & db_3 & dc_3 \end{bmatrix} \right|,$$

where we have transposed the elements of this matrix without affecting the value of its determinant. Expressing the line elements $d\mathbf{a}$, $d\mathbf{b}$, and $d\mathbf{c}$ in terms of the deformation gradient F_{km} [see Eq. (1.53)] and the line elements $d\mathbf{A}$, $d\mathbf{B}$, and $d\mathbf{C}$, we obtain

$$dV = \left| \det \begin{bmatrix} F_{11}dA & F_{12}dB & F_{13}dC \\ F_{21}dA & F_{22}dB & F_{23}dC \\ F_{31}dA & F_{32}dB & F_{33}dC \end{bmatrix} \right|.$$

We have postulated that the inverse motion $\mathbf{X}(\mathbf{x}, t)$ must exist. As a result, Eq. (1.62) states that the determinant of $[F_{km}]$ must be positive. Because dV_0 is also positive, we obtain

$$dV = \det[F_{km}]\, dV_0 = III_F\, dV_0. \tag{1.79}$$

Suppose for a moment that we drop the requirement for the existence of the inverse motion. This result suggests that the body could turn "inside out" to produce negative values of dV. This is an undesirable result.

The *volume strain* e is defined to be

$$e \equiv \frac{dV - dV_0}{dV_0} = III_F - 1. \tag{1.80}$$

This is also called the *dilatation*. Because the value of the volume strain e depends only on the third invariant of F_{km}, it is independent of the coordinate system used to describe the body. We can also show that the volume strain e does not change during either pure translation or pure rotation of the body [see Exercise 1.35]. By similar arguments we can show that

$$\frac{dV_0}{dV} = \det\left[F_{km}^{-1}\right]. \tag{1.81}$$

Volume Strain in Linear Elasticity Using the relationship between the deformation gradient and the displacement Eq. (1.56), we can express the volume strain as

$$e = \det\begin{bmatrix} \frac{\partial \hat{u}_1}{\partial X_1} + 1 & \frac{\partial \hat{u}_1}{\partial X_2} & \frac{\partial \hat{u}_1}{\partial X_3} \\ \frac{\partial \hat{u}_2}{\partial X_1} & \frac{\partial \hat{u}_2}{\partial X_2} + 1 & \frac{\partial \hat{u}_2}{\partial X_3} \\ \frac{\partial \hat{u}_3}{\partial X_1} & \frac{\partial \hat{u}_3}{\partial X_2} & \frac{\partial \hat{u}_3}{\partial X_3} + 1 \end{bmatrix} - 1.$$

In linear elasticity the partial derivatives of the displacements are small. When we expand the determinant, we can ignore terms containing products of the partial derivatives to obtain

$$e = \frac{\partial \hat{u}_1}{\partial X_1} + \frac{\partial \hat{u}_2}{\partial X_2} + \frac{\partial \hat{u}_3}{\partial X_3}. \tag{1.82}$$

In terms of the linear elastic strain tensor, Eq. (1.72), we have

$$e = E_{kk} = I_E. \tag{1.83}$$

1.4.6 *Deviatoric Strain*

Consider the special strain tensor E_{mn}^d,

$$E_{mn}^d = \delta_{mn} E,$$

where E is an arbitrary scalar quantity. From Eq. (1.77), the corresponding shear strain is

$$\sin\gamma = \frac{2ED_p(d\mathbf{A})D_p(d\mathbf{B})}{[2ED_s(d\mathbf{A})D_s(d\mathbf{A}) + 1]^{1/2}[2ED_j(d\mathbf{B})D_j(d\mathbf{B}) + 1]^{1/2}}.$$

Because the vectors $d\mathbf{A}$ and $d\mathbf{B}$ are perpendicular we have

$$\sin\gamma = 0, \tag{1.84}$$

and deformation from E_{mn}^d does not result in shear strain.

Now consider a general strain tensor E_{mn}. We now know that the following portion of E_{mn} does not produce shear strain:

$$E_{mn}^d \equiv \frac{1}{3} E_{kk}\delta_{mn}.$$

The difference between E_{mn} and E_{mn}^d is called the *deviatoric strain tensor* E'_{mn},

$$E'_{mn} \equiv E_{mn} - \frac{1}{3} E_{kk}\delta_{mn}. \tag{1.85}$$

The invariants of the deviatoric strain tensor are

$$I_{E'} = E'_{kk} = 0, \qquad II_{E'} = \frac{1}{2} E'_{km} E'_{km}, \qquad III_{E'} = \det[E'_{km}].$$

Notice that because of our definition of E^d_{mn}, the first invariant of the deviatoric strain tensor is zero.

Deviatoric Strain in Linear Elasticity In linear elasticity the volume strain is equal to the trace of the strain tensor, $e = E_{kk}$ [Eq. (1.83)]. Thus the deviatoric strain is

$$E'_{mn} = E_{mn} - \frac{1}{3}e\delta_{mn}.$$

In linear elasticity, we see that deviatoric strain does not contribute to volume strain.

1.4.7 *Velocity Gradient, Deformation Rate, and Spin*

In this section we define three new quantities that are related to the spatial gradient of the velocity and describe the rate of deformation and rotation of the body. We begin with the definition of the *velocity gradient tensor* L_{km},

$$L_{km} \equiv \frac{\partial v_k}{\partial x_m}.$$

This tensor has physical dimensions of 1/time. Using the chain rule expansion of the partial derivative in this expression, we write

$$L_{km} = \frac{\partial \hat{v}_k}{\partial X_r}\frac{\partial X_r}{\partial x_m}.$$

Noticing the definition of the deformation gradient [Eq. (1.53)] and the inverse deformation gradient [Eq. (1.59)], we find that

$$L_{km} = \frac{\partial \hat{F}_{kr}}{\partial t}F^{-1}_{rm}. \tag{1.86}$$

Next we separate L_{km} into two additive parts. The first part is called the *deformation rate tensor*,

$$D_{km} \equiv \frac{1}{2}(L_{km} + L_{mk}), \tag{1.87}$$

and the second part is called the *spin tensor*,

$$W_{km} \equiv \frac{1}{2}(L_{km} - L_{mk}). \tag{1.88}$$

Thus

$$L_{km} = D_{km} + W_{km}.$$

Deformation Rate in Linear Elasticity In linear elasticity the strain is given by Eq. (1.73):

$$E_{km} = \frac{1}{2}\left(\frac{\partial u_k}{\partial x_m} + \frac{\partial u_m}{\partial x_k}\right).$$

Using Eq. (1.44), we evaluate the partial derivative of this expression to obtain

$$\frac{\partial}{\partial t}\hat{E}_{km} = \frac{1}{2}\left(\frac{\partial v_k}{\partial x_m} + \frac{\partial v_m}{\partial x_k}\right),$$

$$\frac{\partial}{\partial t}\hat{E}_{km} = D_{km}.$$

In linear elasticity the rate of strain is equal to the deformation rate tensor.

1.4.8 *Exercises*

1.26 Given that $dx_k = F_{km} dX_m$, where dx_k and dX_m transform as vectors, show that F_{mn} transforms as

$$F_{(kr)} = \alpha_{(k)m} F_{mn} \alpha_{n(r)}.$$

1.27 Use Eq. (1.61) to show that $F_{km} F_{mr}^{-1} = \delta_{kr}$. Hint: When a matrix has identical rows or columns, its determinant is zero.

1.28 Show that

$$F_{km}^{-1} = \delta_{km} - \frac{\partial u_k}{\partial x_m}.$$

1.29 Use Eq. (1.61) to find the inverse of

$$M = \begin{bmatrix} 1 & 0 & 0 \\ A & 1 & 0 \\ 0 & 0 & 1 \end{bmatrix}.$$

1.30 Consider a cube with sides of unit length. The motion of this cube is

$$x_1 = X_1, \qquad x_2 = X_2(1 - X_1\psi t), \qquad x_3 = X_3,$$

where ψ is a prescribed constant.

(a) Determine the deformation gradient.
(b) Determine the inverse deformation gradient in the spatial description.
(c) Determine the inverse deformation gradient in the material description.
(d) Does F_{ij}^{-1} exist everywhere for all values of ψt; that is, does F_{ij}^{-1} have finite values for all values of ψt?
(e) Draw the deformed cube at $\psi t = 1$.

1.31 The determinant of the deformation gradient $\det[F_{rs}]$ is composed of distinct components F_{ij}. Show that the partial derivative of this determinant with respect to the component F_{ij} is

$$\frac{\partial}{\partial F_{ij}}\{\det[F_{rs}]\} = \mathrm{cof}[F_{ij}],$$

where $\mathrm{cof}[F_{ij}]$ is the cofactor of the matrix component F_{ij}.

1.32 Show that the tensor transformations, Eqs. (1.70)–(1.71), follow directly from Eqs. (1.69) and (1.55).

1.33 Consider the motion

$$x_1 = X_1 + X_2\psi t, \qquad x_2 = X_2, \qquad x_3 = X_3,$$

where ψ is a prescribed constant.

(a) Determine F_{ij} and E_{ij}.
(b) Determine the volume strain.
(c) Determine the longitudinal strain in the x_k directions.
(d) Determine the shear strain between the two perpendicular vectors

$$d\mathbf{A} = \mathbf{i}_1, \quad d\mathbf{B} = \mathbf{i}_2$$

for every point in the body.

1.34 Consider the motion

$$x_1 = X_1(1 + \psi t), \qquad x_2 = X_2, \qquad x_3 = X_3,$$

where ψ is a prescribed constant.

(a) Determine F_{ij} and E_{ij}.
(b) Determine the volume strain.
(c) Determine the longitudinal strain in the x_k directions.
(d) Determine the shear strain between the two perpendicular vectors

$$d\mathbf{A} = \mathbf{i}_1/\sqrt{2} - \mathbf{i}_2/\sqrt{2}, \qquad d\mathbf{B} = \mathbf{i}_1/\sqrt{2} + \mathbf{i}_2/\sqrt{2}$$

for every point in the body.

1.35 Show that the volume strain e is unaffected by either pure translation or rotation.

1.36 Consider a body that has undergone the following motion:

$$x_1 = X_1(1 + \psi_1), \qquad x_2 = X_2(1 - \psi_2), \qquad x_3 = X_3,$$

where the ψ_ks are prescribed constants.

(a) Determine F_{ij}.
(b) Determine the volume strain by Eq. (1.80).
(c) Determine the volume strain by the linear-elastic approximation Eq. (1.82).
(d) Show that the difference between these volume strain computations is $-\psi_1\psi_2$.

1.37 Determine the deviatoric strain for Exercises 1.33 and 1.34.

1.38 Determine the spin for Exercises 1.33 and 1.34.

1.39 *Irrotational motion* is defined to be motion where $W_{km} = 0$. Show that this statement is equivalent to

$$\nabla \times \mathbf{v} = \mathbf{0},$$

where ∇ is defined to be the vector operator

$$\nabla(\cdot) = \frac{\partial(\cdot)}{\partial x_k}\mathbf{i}_k.$$

1.5 One-Dimensional Deformation

In Section 1.3 we defined a special type of motion called one-dimensional motion. Recall that, in one-dimensional motion, u_1 is the only nonzero component of displacement,

$$u_1 = u_1(x_1, t).$$

This motion results in the following deformation gradient tensor:

$$[F_{km}] = \begin{bmatrix} F & 0 & 0 \\ 0 & 1 & 0 \\ 0 & 0 & 1 \end{bmatrix}. \tag{1.89}$$

This tensor is called *uniaxial deformation.* It is one of three types of one-dimensional deformation that we shall consider in this section. The other two are called *spherical* and *triaxial* deformation. Like one-dimensional motion, these three types of deformation are invariant with respect to a rotation of coordinates about the x_1 axis. Here we present a catalog of useful results that we shall draw upon later during our analysis of nonlinear waves.

1.5.1 *Triaxial Deformation*

Consider a motion that results in the following deformation gradient tensor:

$$[F_{km}] = \begin{bmatrix} F & 0 & 0 \\ 0 & \tilde{F} & 0 \\ 0 & 0 & \tilde{F} \end{bmatrix}. \tag{1.90}$$

The scalar components F and $\tilde{F}$ have arbitrary values. The resulting deformation is called *triaxial deformation. Spherical deformation* is a special case of Eq. (1.90) that is obtained by setting $\tilde{F} = F$, and *uniaxial deformation* is a special case of Eq. (1.90) that is obtained by setting $\tilde{F} = 1$. We shall show that the scalar components F_{km} of triaxial deformation are invariant during coordinate rotation about x_1. For this reason, $F_{11} = F$ is called the *axial component* of the deformation and $F_{22} = F_{33} = \tilde{F}$ are called the *lateral components* of the deformation. The three invariants of the deformation gradient for triaxial deformation are

$$I_F = F + 2\tilde{F}, \qquad II_F = \frac{1}{2}F^2 + \tilde{F}^2, \qquad III_F = F\tilde{F}^2.$$

The inverse deformation gradient for triaxial deformation is

$$\left[F_{km}^{-1}\right] = \begin{bmatrix} 1/F & 0 & 0 \\ 0 & 1/\tilde{F} & 0 \\ 0 & 0 & 1/\tilde{F} \end{bmatrix}. \tag{1.91}$$

You can verify this by substituting this expression and Eq. (1.90) into Eq. (1.60).

Rotation of Coordinates about x_1 We now show that the x_1 axis is an axis of symmetry for triaxial deformation. Consider a second coordinate system $x_{(k)}$ that we obtain by rotation through an angle of θ_1 about the x_1 axis (see Figure 1.7). The deformation gradient in the coordinate system $x_{(k)}$ is $F_{(km)}$. We evaluate the components of the deformation gradient in $x_{(k)}$ using the tensor transformation [see Eq. (1.55)]

$$F_{(km)} = \alpha_{(k)r} F_{rs} \alpha_{s(m)}.$$

We substitute the deformation gradient [Eq. (1.90)] and the direction cosines [Eq. (1.22)] to obtain

$$[F_{(km)}] = \begin{bmatrix} 1 & 0 & 0 \\ 0 & \cos\theta_1 & \sin\theta_1 \\ 0 & -\sin\theta_1 & \cos\theta_1 \end{bmatrix} \begin{bmatrix} F & 0 & 0 \\ 0 & \tilde{F} & 0 \\ 0 & 0 & \tilde{F} \end{bmatrix} \begin{bmatrix} 1 & 0 & 0 \\ 0 & \cos\theta_1 & -\sin\theta_1 \\ 0 & \sin\theta_1 & \cos\theta_1 \end{bmatrix}. \quad (1.92)$$

Notice that we have transposed the first matrix in this expression to obtain $\alpha_{(k)r}$ from $\alpha_{r(k)}$. The result simplifies to

$$[F_{(km)}] = \begin{bmatrix} F & 0 & 0 \\ 0 & \tilde{F} & 0 \\ 0 & 0 & \tilde{F} \end{bmatrix}.$$

Thus the scalar components of this deformation gradient tensor are invariant during a rotation of coordinates about the x_1 axis. The x_1 axis is an axis of symmetry, and an identical condition of deformation exists in all coordinate systems obtained by rotation about x_1.

Volume Strain The volume strain for triaxial deformation is [see Eq. (1.80)]

$$e = III_F - 1 = F\tilde{F}^2 - 1. \quad (1.93)$$

Thus

$$1 + e = \frac{dV}{dV_0} = F\tilde{F}^2. \quad (1.94)$$

Both the current volume dV and reference volume dV_0 must be positive and nonzero. Therefore,

$$F > 0. \quad (1.95)$$

Furthermore, $\tilde{F}$ cannot be zero. We require that $\tilde{F}$ is a continuous function that cannot suddenly jump from a positive to a negative value without going through zero. Therefore, because $\tilde{F} = 1$ in the reference configuration [see Eq. (1.57)],

$$\tilde{F} > 0. \quad (1.96)$$

Strain Tensor In the coordinate system x_k, the strain tensor is [see Eq. (1.69)]

$$[E_{km}] = \begin{bmatrix} E & 0 & 0 \\ 0 & \tilde{E} & 0 \\ 0 & 0 & \tilde{E} \end{bmatrix}, \quad (1.97)$$

where

$$E = \frac{1}{2}(F^2 - 1), \qquad \tilde{E} = \frac{1}{2}(\tilde{F}^2 - 1). \quad (1.98)$$

From Eqs. (1.95) and (1.96), we get

$$E > -\frac{1}{2}, \qquad \tilde{E} > -\frac{1}{2}.$$

This strain tensor is also invariant during a rotation of coordinates about x_1.

The three invariants of the strain tensor for triaxial deformation are

$$I_E = E + 2\tilde{E} = \frac{1}{2}(F^2 - 1) + (\tilde{F}^2 - 1),$$
$$II_E = \frac{1}{2}E^2 + \tilde{E}^2 = \frac{1}{8}(F^2 - 1)^2 + \frac{1}{4}(\tilde{F}^2 - 1)^2,$$
$$III_E = E\tilde{E}^2 = \frac{1}{8}(F^2 - 1)(\tilde{F}^2 - 1)^2.$$

Deviatoric Strain From the definition of deviatoric strain, Eq. (1.85), the matrix form of the deviatoric strain tensor for triaxial deformation is

$$\left[E'_{km}\right] = \begin{bmatrix} \frac{2}{3}(E - \tilde{E}) & 0 & 0 \\ 0 & -\frac{1}{3}(E - \tilde{E}) & 0 \\ 0 & 0 & -\frac{1}{3}(E - \tilde{E}) \end{bmatrix}.$$

For spherical deformation where $E = \tilde{E}$ notice that the deviatoric strain is zero.

The three invariants of the deviatoric strain tensor for triaxial deformation are

$$I_{E'} = 0,$$
$$II_{E'} = \frac{1}{3}(E - \tilde{E})^2 = \frac{1}{12}(F^2 - \tilde{F}^2)^2,$$
$$III_{E'} = \frac{2}{27}(E - \tilde{E})^3 = \frac{1}{108}(F^2 - \tilde{F}^2)^3.$$

Longitudinal Strain Consider a line element $d\mathbf{X}$ with direction $\mathbf{D}(d\mathbf{X})$ (see Figure 1.19):

$$d\mathbf{X} = \mathbf{D}(d\mathbf{X})|d\mathbf{X}|.$$

The current position of $d\mathbf{X}$ is

$$d\mathbf{x} = \mathbf{D}(d\mathbf{x})|d\mathbf{x}|.$$

The vectors $d\mathbf{X}$ and $d\mathbf{x}$ make angles θ and β, respectively, with the x_1 axis:

$$D_1(d\mathbf{X}) = \cos\theta, \quad D_1(d\mathbf{x}) = \cos\beta.$$

The longitudinal strain of line element $d\mathbf{X}$ is [see Eq. (1.75)]

$$2\varepsilon + \varepsilon^2 = 2ED_1^2(d\mathbf{X}) + 2\tilde{E}\left[D_2^2(d\mathbf{X}) + D_3^2(d\mathbf{X})\right].$$

Because $D_m(d\mathbf{X})D_m(d\mathbf{X}) = 1$, we have

$$2\varepsilon + \varepsilon^2 = 2(E - \tilde{E})D_1^2(d\mathbf{X}) + 2\tilde{E}.$$

Using Eq. (1.98), we obtain

$$\frac{(1+\varepsilon)^2}{\tilde{F}^2} = \bar{F}^2\cos^2\theta + \sin^2\theta,$$

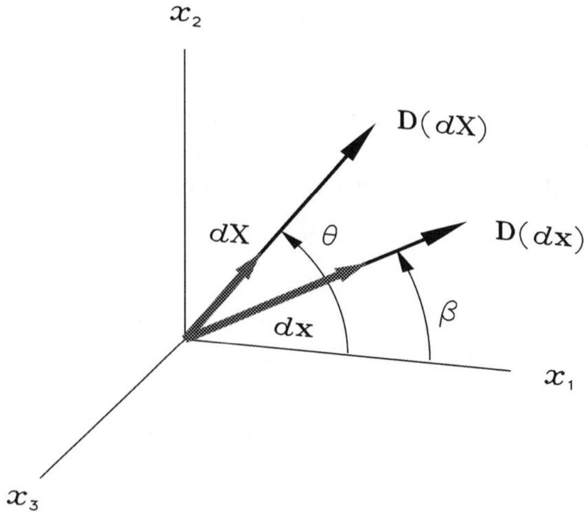

Figure 1.19 A line element $d\mathbf{X}$ and direction vector $\mathbf{D}(d\mathbf{X})$ make an angle θ with the x_1 axis. Triaxial deformation causes this line element to rotate and elongate to $d\mathbf{x}$ with direction $\mathbf{D}(d\mathbf{x})$ that makes an angle β with the x_1 axis.

where

$$\bar{F} \equiv \frac{F}{\tilde{F}}. \tag{1.99}$$

Similarly to Eq. (1.74) we define the stretch parameter $\bar{\Lambda}$ to be

$$\bar{\Lambda} \equiv \frac{1+\varepsilon}{\tilde{F}} = \frac{\Lambda}{\tilde{F}}$$

so that

$$\bar{\Lambda} = (\bar{F}^{\,2}\cos^2\theta + \sin^2\theta)^{1/2}. \tag{1.100}$$

Now we evaluate the angle β in terms of $\bar{\Lambda}$, $\bar{F}$, and θ. From Eq. (1.54) we have

$$dx_1 = F\,dX_1, \qquad dx_2 = \tilde{F}\,dX_2, \qquad dx_3 = \tilde{F}\,dX_3.$$

From the first of these expressions we obtain

$$D_1(d\mathbf{x})|d\mathbf{x}| = F D_1(d\mathbf{X})|d\mathbf{X}|.$$

Using Eq. (1.67) we obtain

$$\cos\beta = \frac{F\cos\theta}{\Lambda}. \tag{1.101}$$

After dividing both the numerator and denominator by $\tilde{F}$, we find that

$$\cos\beta = \frac{\bar{F}\cos\theta}{\bar{\Lambda}}. \tag{1.102}$$

We can graphically represent these results in two ways. The stretch $\bar{\Lambda}$ can be plotted either as a function of θ or as a function of β. We show a polar plot of $\bar{\Lambda}$ against θ in Figure 1.20 for three values of $\bar{F}$. As a special case consider uniaxial strain where $\bar{F} = 1$. When $F = 1$, then $\bar{\Lambda} = 1$, and the plot is a circle of unit radius (solid line). When $F < 1$, the volume decreases and the material is in a configuration of *compression* (dotted line). When $F > 1$, the volume increases and the material is in a configuration of *expansion* (dashed line).

We show a polar plot of $\bar{\Lambda}$ against β in Figure 1.21. Notice the difference between this plot and the previous one. When we plot $\bar{\Lambda}$ as a function of θ, the stretch of the line element $d\mathbf{X}$ is represented as a function of its angle with the x_1 axis in the reference configuration. Therefore, Figure 1.20 is a material description of the stretch. Similarly, when we plot $\bar{\Lambda}$ as a function of β, the stretch is represented as a function of the angle between $d\mathbf{x}$ and the x_1 axis in the current configuration, and Figure 1.21 is a spatial description of the stretch.

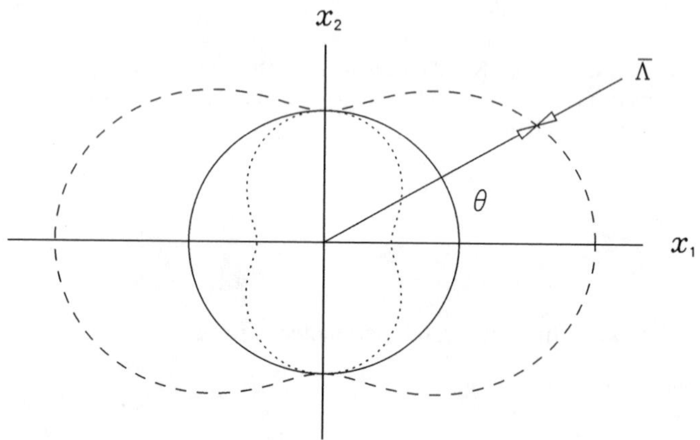

Figure 1.20 The stretch $\bar{\Lambda}$ as a function of θ during uniaxial deformation for $\bar{F} = 0.5$ (···), $\bar{F} = 1.0$ (—), and $\bar{F} = 2.0$ (– –).

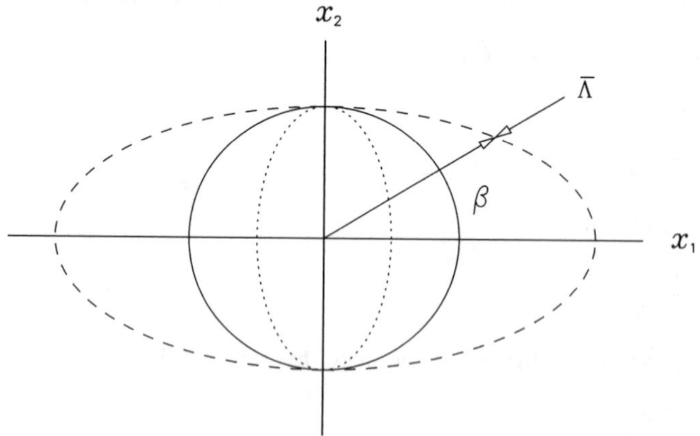

Figure 1.21 The stretch $\bar{\Lambda}$ as a function of β during uniaxial deformation for $\bar{F} = 0.5$ (···), $\bar{F} = 1.0$ (—), and $\bar{F} = 2.0$ (– –).

Shear Strain The shear strain between two perpendicular line elements $d\mathbf{A}$ and $d\mathbf{B}$ is [see Eq. (1.77)]

$$\begin{aligned}\sin\gamma = &\{2ED_1(d\mathbf{A})D_1(d\mathbf{B}) + 2\tilde{E}[D_2(d\mathbf{A})D_2(d\mathbf{B}) + D_3(d\mathbf{A})D_3(d\mathbf{B})]\}\\ &\div\left\{2ED_1^2(d\mathbf{A}) + 2\tilde{E}\left[D_2^2(d\mathbf{A}) + D_3^2(d\mathbf{A})\right] + 1\right\}^{1/2}\\ &\div\left\{2ED_1^2(d\mathbf{B}) + 2\tilde{E}\left[D_2^2(d\mathbf{B}) + D_3^2(d\mathbf{B})\right] + 1\right\}^{1/2}.\end{aligned}$$

Because $d\mathbf{A}$ and $d\mathbf{B}$ are mutually perpendicular unit vectors where $D_m(d\mathbf{A})D_m(d\mathbf{A}) = D_m(d\mathbf{B})D_m(d\mathbf{B}) = 1$ and $D_m(d\mathbf{A})D_m(d\mathbf{B}) = 0$, we find that

$$\sin\gamma = \frac{2(E - \tilde{E})D_1(d\mathbf{A})D_1(d\mathbf{B})}{\left[2(E - \tilde{E})D_1^2(d\mathbf{A}) + 2\tilde{E} + 1\right]^{1/2}\left[2(E - \tilde{E})D_1^2(d\mathbf{B}) + 2\tilde{E} + 1\right]^{1/2}}.$$

Letting

$$\bar{E} \equiv \frac{E - \tilde{E}}{\tilde{F}^2} = \frac{1}{2}\left(\frac{F^2}{\tilde{F}^2} - 1\right) = \frac{1}{2}(\bar{F}^2 - 1), \tag{1.103}$$

we obtain

$$\sin\gamma = \frac{2\bar{E}\,D_1(d\mathbf{A})D_1(d\mathbf{B})}{\left[2\bar{E}\,D_1^2(d\mathbf{A}) + 1\right]^{1/2}\left[2\bar{E}\,D_1^2(d\mathbf{B}) + 1\right]^{1/2}}.$$

The square of the sine of the shear strain is

$$\sin^2\gamma = \frac{\phi}{\eta + \phi}, \tag{1.104}$$

where we have defined

$$\phi \equiv 4\bar{E}^2 D_1^2(d\mathbf{A})D_1^2(d\mathbf{B})$$

and

$$\eta \equiv 1 + 2\bar{E}\left[D_1^2(d\mathbf{A}) + D_1^2(d\mathbf{B})\right].$$

For any nonzero value of $\bar{E}$, the shear strain changes as we change the directions of $d\mathbf{A}$ and $d\mathbf{B}$. We now examine Eq. (1.104) to determine the orientation of the two line elements $d\mathbf{A}$ and $d\mathbf{B}$ that yield the greatest value of shear strain $|\gamma| = \gamma_{\max}$. Because $d\mathbf{A}$ and $d\mathbf{B}$ are perpendicular, Eq. (1.25) shows us that $D_1^2(d\mathbf{A}) + D_1^2(d\mathbf{B}) \leq 1$. Furthermore, because $\bar{F} > 0$ and $\bar{E} > -1/2$ [see Eq. (1.103)], then $\eta > 0$. Consequently, $0 \leq \sin^2\gamma \leq 1$ and, for a prescribed nonzero value of $\bar{E}$, the maximum value of $\sin^2\gamma$ occurs when ϕ is maximum, or equivalently, when

$$|D_1(d\mathbf{A})D_1(d\mathbf{B})| = \text{maximum}.$$

From the orientation lemma in Section 1.1.8 we find that the maximum shear strain occurs when both $d\mathbf{A}$ and $d\mathbf{B}$ are contained in a plane that also contains the x_1 axis (see Figure 1.22). Because the components of the deformation gradient for triaxial strain are independent of a rotation about x_1, we have conveniently placed the vectors $d\mathbf{A}$ and $d\mathbf{B}$ in the x_1–x_2 plane. The angle θ represents the rotation about the x_3 axis, and the direction cosines are

$$D_1(d\mathbf{A}) = \cos\theta, \qquad D_1(d\mathbf{B}) = -\sin\theta.$$

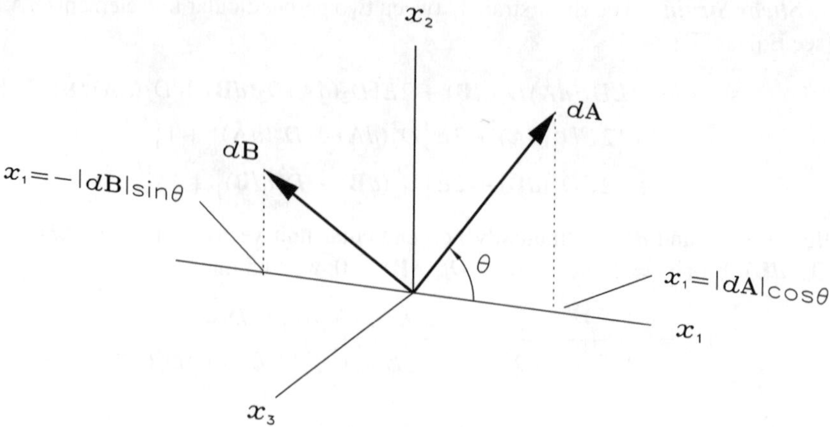

Figure 1.22 Two perpendicular line elements in uniaxial strain. The two mutually perpendicular vectors $d\mathbf{A}$ and $d\mathbf{B}$ reside in the x_1–x_2 plane.

The value of shear strain between $d\mathbf{A}$ and $d\mathbf{B}$ is

$$\sin\gamma = \frac{-\bar{E}\,\sin(2\theta)}{(2\bar{E}\,\cos^2\theta + 1)^{1/2}(2\bar{E}\,\sin^2\theta + 1)^{1/2}}, \tag{1.105}$$

where we have used the trigonometric identity $\sin(2\theta) = 2\sin\theta\cos\theta$. The maximum magnitudes of γ are

$$\gamma = \begin{cases} -\gamma_{\max}, & \text{at } \theta = -\frac{3}{4}\pi, \frac{1}{4}\pi, \\ \gamma_{\max}, & \text{at } \theta = -\frac{1}{4}\pi, \frac{3}{4}\pi, \end{cases}$$

where

$$\gamma_{\max} = \sin^{-1}\left(\frac{\bar{E}}{\bar{E}+1}\right) = \sin^{-1}\left(\frac{\bar{F}^2 - 1}{\bar{F}^2 + 1}\right). \tag{1.106}$$

In Figure 1.23 we show the orientations of the line elements that yield the maximum values of the magnitude of the shear strain. These line elements are shown as functions of the angle θ. For example, $d\mathbf{A}(-\frac{1}{4}\pi)$ and $d\mathbf{B}(-\frac{1}{4}\pi)$ are the two perpendicular line elements in the reference configuration when $\theta = -\frac{1}{4}\pi$. Deformation causes these line elements to shift to $d\mathbf{a}(-\frac{1}{4}\pi)$ and $d\mathbf{b}(-\frac{1}{4}\pi)$. It is left as an exercise for you to show that when $\theta = -\frac{1}{4}\pi$, then $\beta = \frac{1}{4}\pi - \frac{1}{2}\gamma_{\max}$, where β is given by Eq. (1.102) (see Exercise 1.46). Notice how the elements in the reference configuration form legs of an "X" pattern. During deformation these legs rotate in opposite directions, and the "X" pattern is extended in the x_1 direction. In this illustration, we have selected $\bar{F} = 2$; however, any value $\bar{F} > 1$ produces similar results. When $\bar{F} < 1$, the "X" pattern is shortened in the x_1 direction (see Exercise 1.47).

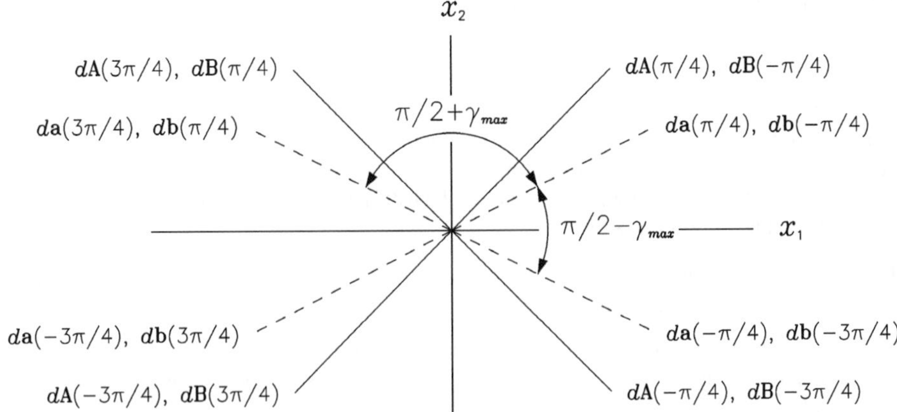

Figure 1.23 The line elements $d\mathbf{A}(\theta)$ and $d\mathbf{B}(\theta)$ in the reference configuration (—) and the line elements $d\mathbf{a}(\theta)$ and $d\mathbf{b}(\theta)$ in the current configuration (– –) that yield the maximum magnitude of the shear strain. The value of the deformation gradient is $\bar{F} = 2$.

Velocity Gradient, Deformation Rate, and Spin We evaluate the velocity gradient for triaxial deformation from Eq. (1.86):

$$[L_{km}] = \begin{bmatrix} L & 0 & 0 \\ 0 & \tilde{L} & 0 \\ 0 & 0 & \tilde{L} \end{bmatrix}, \tag{1.107}$$

where

$$L = \frac{1}{F}\frac{\partial \hat{F}}{\partial t}, \qquad \tilde{L} = \frac{1}{\tilde{F}}\frac{\partial \hat{\tilde{F}}}{\partial t}. \tag{1.108}$$

We determine the deformation gradient and spin from Eqs. (1.87) and (1.88):

$$[D_{km}] = L_{km}, \qquad [W_{km}] = 0.$$

Because the spin is zero, triaxial deformation results in irrotational motion.

1.5.2 *Exercises*

1.40 Consider the motion

$$x_k = X_k(1 + \psi_k t).$$

Determine the values of the constants ψ_k that produce:

(a) spherical deformation,
(b) uniaxial deformation,
(c) triaxial deformation.

1.41 Consider the motion

$$x_1 = X_1(1 + \psi_1 t) + X_2\psi_2 t, \qquad x_2 = X_2(1 + \psi_3 t), \qquad x_3 = X_3(1 + \psi_3 t).$$

Repeat the previous exercise.

1.42 Consider a spherical ball with the motion

$$x_k = X_k(1 + \psi t),$$

where ψ is an arbitrary constant. The center of the ball is at $X_k = 0$.

(a) Show that the direction of motion is radial with respect to $X_k = 0$.
(b) Show that the deformation is spherical.
(c) Show that any axis through the center of the ball is an axis of symmetry.
(d) Determine the volume strain.

1.43 Execute the matrix multiplications in Eq. (1.92) to show that the x_1 axis is an axis of symmetry for triaxial deformation.

1.44 Consider a cube with circles of unit radius drawn on each of its faces (see Figure 1.24). Graph the shape of each of the circles after the cube is subjected to uniaxial deformation with $F = 1.5$.

1.45 Consider a cube with squares drawn on each of its faces. Each square is oriented at 45° to the edges of the cube (see Figure 1.25). The sides of some of the squares rotate as the cube

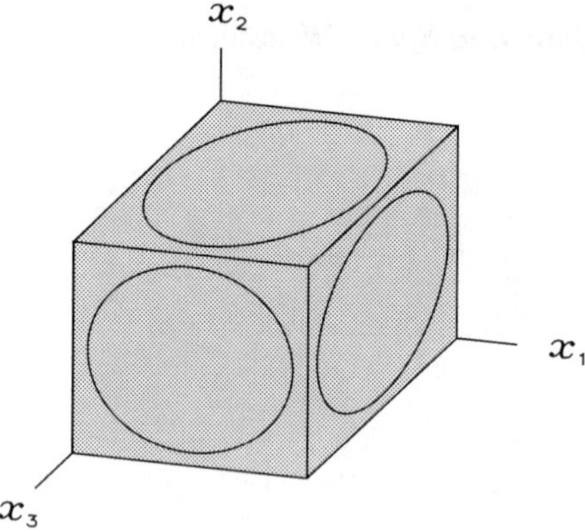

Figure 1.24 Exercise 1.44.

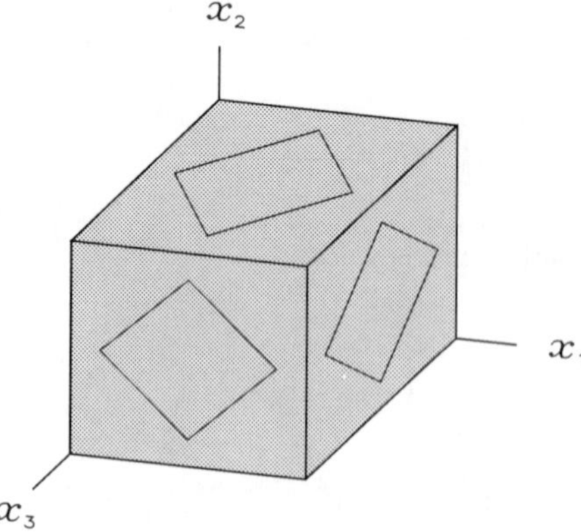

Figure 1.25 Exercise 1.45

is uniaxially deformed to $F = 1.5$. Which squares deform and what are each of the angles formed by the sides of the deformed squares?

1.46 Using Eqs. (1.106), (1.100), and (1.102), show that $\beta = \frac{1}{2}\gamma_{\max} - \frac{1}{4}\pi$ when $\theta = -\frac{1}{4}\pi$. Hint: First show that $\cos(2\beta) = \sin\gamma_{\max}$.

1.47 Redraw Figure 1.23 for $\bar{F} = 0.5$.

1.48 Use the inversion formula Eq. (1.61) to derive the inverse deformation gradient for triaxial strain. Specialize the results to obtain the inverse deformation gradients for uniaxial strain and for spherical strain.

1.49 Repeat Exercise 1.45 for a cube subjected to triaxial strain. Assume the cube is composed of a material whose volume cannot change. The axial component of the deformation gradient is $F = 1.5$.

1.6 Stress

Two kinds of forces are important to waves in nonlinear materials: *external forces* and *internal forces*. The action of gravity on a body is an example of an external force, and deforming a body results in internal forces called tractions. In this section, we discuss both types of forces.

1.6.1 *Body Force*

A body of volume V is shown in its current configuration in a Cartesian coordinate system x_k (see Figure 1.26). The *mass* of the body is m. We use the physical dimension of kg for mass. The volume V is called a *material volume* because it always contains the same mass m. As the boundaries of the body move and deform, the material volume V moves and deforms with the body. At the material point X_k inside this material volume, we select a material element ΔV of the material volume. The mass of this material element is Δm. This element is also described in the material coordinates. Thus, as the body moves and deforms, the material element moves and deforms with the body and always contains the

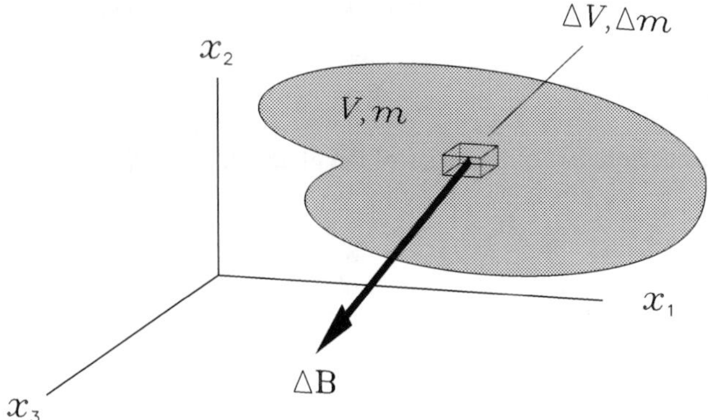

Figure 1.26 A body of volume V and mass m that contains a material element ΔV of mass Δm. The external body force ΔB acts on the material element of the body.

same element of mass Δm. The average mass per unit volume of this material element is

$$\bar{\rho} \equiv \frac{\Delta m}{\Delta V}.$$

As we let the volume of this element approach zero, the average mass per unit volume $\bar{\rho}$ becomes ρ, where

$$\rho = \lim_{\Delta V \to 0} \left(\frac{\Delta m}{\Delta V} \right).$$

The scalar quantity ρ is called the *mass density* or *density* of the body. We assume that ρ is always finite and positive. We assign to it the physical dimensions of kg/m^3.

The external body force acting on the material element is represented by the vector $\Delta \mathbf{B}$. The average external body force per unit mass is given by the ratio

$$\bar{\mathbf{b}} = \frac{\Delta \mathbf{B}}{\Delta m} = \frac{\Delta \mathbf{B}}{\bar{\rho}\, \Delta V}.$$

As we let the volume of the element approach zero, the average body force per unit mass $\bar{\mathbf{b}}$ becomes $\mathbf{b}$, where

$$\mathbf{b} = \lim_{\Delta V \to 0} \left(\frac{\Delta \mathbf{B}}{\bar{\rho}\, \Delta V} \right).$$

The vector $\mathbf{b}$ is called the *specific external body force* or sometimes simply the *body force*. Here we employ the adjective "specific" to denote a quantity that is normalized to the mass of the body. We use the physical dimensions of N/kg for $\mathbf{b}$.

Inertial Force Newton's second law of motion states that the sum of the forces acting on the volume element is equal to the rate of change of linear momentum of the volume element,

$$\sum \mathbf{f} = \frac{\partial}{\partial t} \widehat{(\Delta m \bar{\mathbf{v}})},$$

where $\sum \mathbf{f}$ represents the sum of all of the external and internal forces acting on the element, and the vector $\bar{\mathbf{v}}$ represents the average velocity of the mass of the element. The mass Δm is constant. Therefore,

$$\sum \mathbf{f} = \Delta m \bar{\mathbf{a}},$$

where $\bar{\mathbf{a}}$ is the average acceleration of the mass of the element. We rewrite this expression as

$$\sum \mathbf{f} + \Delta \mathbf{B} = \mathbf{0},$$

where we have cast the inertial term as a particular type of body force

$$\Delta \mathbf{B} = -\Delta m \bar{\mathbf{a}}.$$

Thus

$$\bar{\mathbf{b}} = \frac{\Delta \mathbf{B}}{\Delta m} = -\bar{\mathbf{a}}$$

and

$$\mathbf{b} = -\lim_{\Delta V \to 0} \bar{\mathbf{a}} = -\mathbf{a}.$$

We see that acceleration of the material element can be represented as a specific external body force **b** that is equal and opposite to the acceleration vector **a**. We call this the *inertial force*. When we include the inertial force in a free-body diagram, Newton's second law of motion requires the sum of the forces in the diagram to be zero.

1.6.2 *Traction*

Suppose that we use an imaginary plane to divide the body in Figure 1.26 into two parts (see Figure 1.27). We draw the free-body diagrams for the portions of the body to the left and right of the imaginary plane. The vectors drawn on the free-body diagrams represent the internal forces acting across the imaginary plane. These forces result from deformation of the body and represent the portion of the body on one side of the imaginary plane acting against the portion of the body on the opposite side of the imaginary plane.

Consider an area element ΔS of the imaginary plane (see Figure 1.28). We draw this area element ΔS on both the left and right free-body diagrams. The element ΔS on the left free-body diagram has a unit vector **n** that is normal to ΔS and points away from the interior of the left free-body diagram. The unit vector $-\mathbf{n}$ on the right free-body diagram is also normal to ΔS and points away from the interior of the right diagram. The force acting on the area element ΔS of the left diagram is **f**. From Newton's third law of motion, the force acting on the area element ΔS of the right diagram is $-\mathbf{f}$. For each free-body diagram, the average force per unit area acting on ΔS is

$$\bar{\mathbf{t}}(\mathbf{n}) \equiv \frac{\mathbf{f}}{\Delta S}, \quad \text{left free-body,}$$

$$\bar{\mathbf{t}}(-\mathbf{n}) \equiv \frac{-\mathbf{f}}{\Delta S}, \quad \text{right free-body.}$$

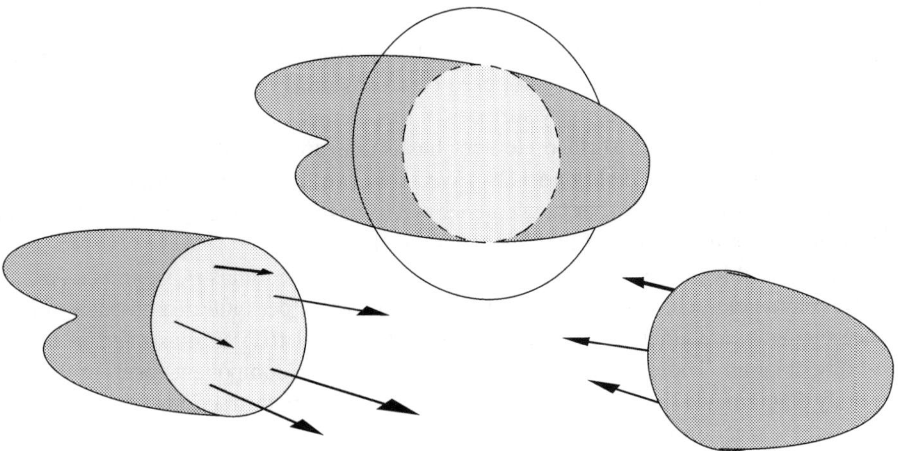

Figure 1.27 An imaginary plane passing through a body and the left and right free-body diagrams.

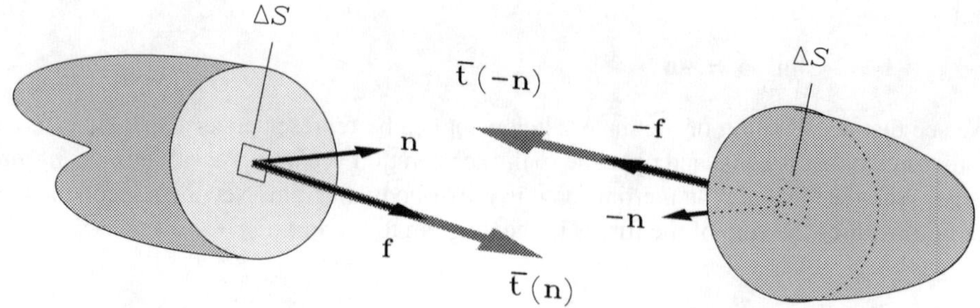

Figure 1.28 The free-body diagrams with the same area element ΔS illustrated on both the left and right diagrams.

We use the notation $\bar{\mathbf{t}}(\mathbf{n})$ to imply a functional relationship between the vectors $\bar{\mathbf{t}}$ and $\mathbf{n}$. This means the average force acting across the area element of the imaginary plane is a function of the direction of the normal to the imaginary plane. In the limit as ΔS approaches zero, the average force per unit area becomes $\bar{\mathbf{t}} \to \mathbf{t}$ where

$$\mathbf{t}(\mathbf{n}) = \lim_{\Delta S \to 0} \left(\frac{\mathbf{f}}{\Delta S} \right), \quad \text{left free-body,}$$

$$\mathbf{t}(-\mathbf{n}) = \lim_{\Delta S \to 0} \left(\frac{-\mathbf{f}}{\Delta S} \right), \quad \text{right free-body,}$$

and

$$\mathbf{t}(\mathbf{n}) = -\mathbf{t}(-\mathbf{n}). \tag{1.109}$$

The function $\mathbf{t}(\mathbf{n})$, which represents the distribution of force on both the left and right free-body diagrams, is called the *traction*. We use physical dimensions of N/m^2 for the traction. This is a unit of pressure, which is also called the Pascal (Pa).

1.6.3 *Cauchy Stress*

In the previous section we describe the tractions acting on an imaginary plane. The material element in Figure 1.26 is composed of six imaginary surfaces. We now describe the tractions acting on the imaginary surfaces of this element. The element is a rectangular block of material and each surface element has a unit normal that is parallel to one of the coordinate vectors $\mathbf{i}_k$ (see Figure 1.29). We show two free-body diagrams of the same volume element of the body. For convenience we have moved the origin of the Cartesian coordinate system to a corner of the volume element.

The three visible planes of the volume element have unit normals $\mathbf{i}_1$, $\mathbf{i}_2$, and $\mathbf{i}_3$. Assume that the volume element is small and that the average force per unit area on each of these surface elements are equal to the tractions $\mathbf{t}(\mathbf{i}_1)$, $\mathbf{t}(\mathbf{i}_2)$, and $\mathbf{t}(\mathbf{i}_3)$ as illustrated in the left free-body diagram. These traction vectors can be written in component form. In the right free-body diagram, we show the scalar components of the surface tractions. The component forms of these traction vectors are

$$\begin{aligned}
\mathbf{t}(\mathbf{i}_1) &= T_{11}\mathbf{i}_1 + T_{12}\mathbf{i}_2 + T_{13}\mathbf{i}_3, \\
\mathbf{t}(\mathbf{i}_2) &= T_{21}\mathbf{i}_1 + T_{22}\mathbf{i}_2 + T_{23}\mathbf{i}_3, \\
\mathbf{t}(\mathbf{i}_3) &= T_{31}\mathbf{i}_1 + T_{32}\mathbf{i}_2 + T_{33}\mathbf{i}_3.
\end{aligned}$$

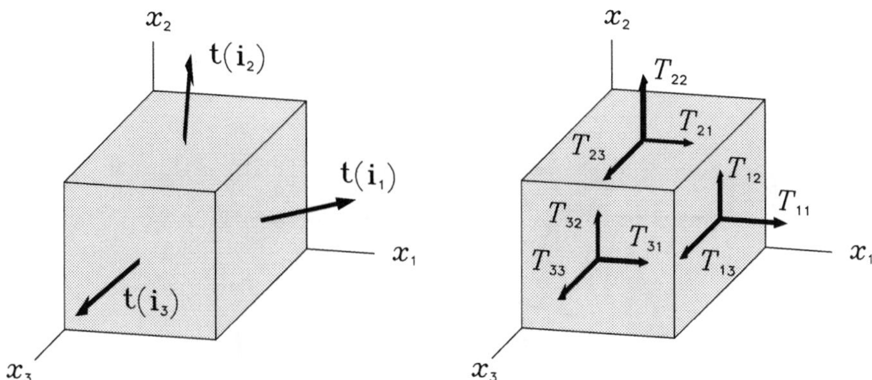

Figure 1.29 Two free-body diagrams of the rectangular element of the body. In the left diagram we show the surface tractions on the three visible surfaces. In the right diagram we show the scalar components of the surface tractions.

In index notation these three equations are

$$\mathbf{t}(\mathbf{i}_k) = T_{km}\mathbf{i}_m. \tag{1.110}$$

The quantity T_{km} is the traction component acting in the direction $\mathbf{i}_m$ and acting on a surface element with normal $\mathbf{i}_k$. The scalar quantities T_{km} are called *stress components*. Collectively, they form a matrix called *stress*. Because this definition of stress was first proposed by the mathematician Cauchy, T_{km} is also called the *Cauchy stress*. We use the physical dimensions of Pa for the Cauchy stress.

The stress components T_{11}, T_{22}, and T_{33} act normal to their respective surface elements. Consequently, these components are called *normal stress components*. When these components are positive the volume element is stretched, and when they are negative the volume element is compressed. The remaining components of the stress act tangentially to their respective surface elements. They are called *shear stress components*.

Tetrahedron Argument

Stress is defined in Eq. (1.110). This equation suggests that stress is a tensor, because it transforms one set of vectors $\mathbf{i}_m$ into another set of vectors $\mathbf{t}(\mathbf{i}_k)$. However, the form of Eq. (1.110) is not equivalent to the general form of a tensor transformation [see Eq. (1.29)]. Using the tetrahedron argument first presented by Cauchy, we now show that stress is a tensor.

Consider another material element of the body (see Figure 1.30). This volume element is enclosed by four surface elements with unit normals $-\mathbf{i}_1$, $-\mathbf{i}_2$, $-\mathbf{i}_3$, and $\mathbf{n}$. They form a tetrahedron. Three of the surface elements are hidden from view and intersect at the origin of the coordinate system. We denote the area of each surface element by ΔS_1, ΔS_2, ΔS_3, and ΔS, respectively. From elementary geometry, we have shown (see Exercise 1.19) that

$$\Delta S_k = n_k \Delta S.$$

The volume ΔV of the tetrahedron is

$$\Delta V = \frac{1}{3} h \Delta S,$$

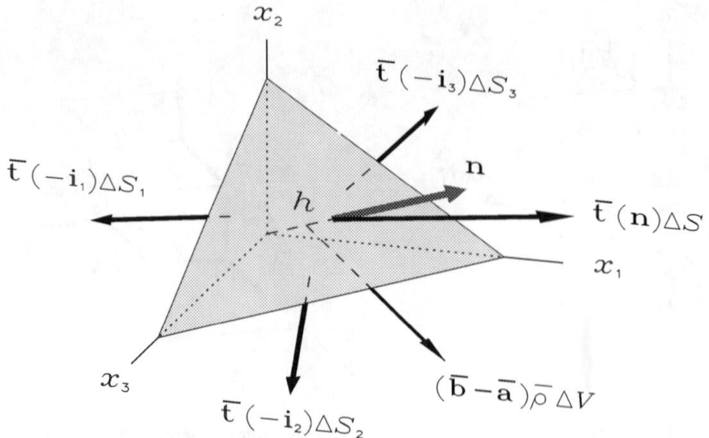

Figure 1.30 The tetrahedral element of the body with surface tractions.

where h is the height of the tetrahedron measured in the direction $\mathbf{n}$. Each surface element has a traction acting on it, namely $\bar{\mathbf{t}}(-\mathbf{i}_1)$, $\bar{\mathbf{t}}(-\mathbf{i}_2)$, $\bar{\mathbf{t}}(-\mathbf{i}_3)$, and $\bar{\mathbf{t}}(\mathbf{n})$. A body force $\bar{\rho}\,\Delta V\bar{\mathbf{b}}$ and an inertial force $-\bar{\rho}\,\Delta V\bar{\mathbf{a}}$ also act on the tetrahedron (see Section 1.6.1). Because we have included an inertial force that is equal and opposite to the average acceleration of the volume element, Newton's second law of motion requires the sum of the forces in the free-body diagram to be zero:

$$\bar{\mathbf{t}}(\mathbf{n})\Delta S+\bar{\mathbf{t}}(-\mathbf{i}_1)n_1\Delta S+\bar{\mathbf{t}}(-\mathbf{i}_2)n_2\Delta S+\bar{\mathbf{t}}(-\mathbf{i}_3)n_3\Delta S+\frac{1}{3}\bar{\rho}\,h\Delta S(\bar{\mathbf{b}}-\bar{\mathbf{a}})=\mathbf{0}.$$

Dividing by ΔS, taking the limit as $h\to 0$, and applying Eq. (1.109), we obtain

$$\mathbf{t}(\mathbf{n})=\mathbf{t}(\mathbf{i}_1)n_1+\mathbf{t}(\mathbf{i}_2)n_2+\mathbf{t}(\mathbf{i}_3)n_3,$$

which in index notation is

$$\mathbf{t}(\mathbf{n})=\mathbf{t}(\mathbf{i}_k)n_k.$$

After substitution of Eq. (1.110), we see that

$$\mathbf{t}(\mathbf{n})=T_{km}n_k\mathbf{i}_m,$$

which yields

$$t_m(\mathbf{n})=T_{km}n_k. \tag{1.111}$$

This is an important result. When the components of the stress tensor at a point are known, we can determine the traction acting on any imaginary plane passing through this point. Also, when we compare this result to Eq. (1.29) in Section 1.1.9, we discover that *stress is a tensor*. Thus, during a change of coordinates, stress transforms according to Eqs. (1.31) and (1.32):

$$\begin{aligned}T_{(kr)}&=\alpha_{(k)m}T_{mn}\alpha_{n(r)},\\ T_{kr}&=\alpha_{k(m)}T_{(mn)}\alpha_{(n)r}.\end{aligned} \tag{1.112}$$

Because the stress is a tensor, it has three scalar invariants [see Eqs. (1.33)–(1.35)]. They are

$$I_T = T_{kk}, \qquad II_T = \frac{1}{2}T_{km}T_{km}, \qquad III_T = \det[T_{km}]. \tag{1.113}$$

Symmetry of Cauchy Stress

The stress tensor T_{km} has nine scalar components. Later in this chapter [see Eq. (1.157)], we show that the stress tensor is symmetric,

$$T_{km} = T_{mk},$$

and has only six independent components.

Cauchy Stress During Pure Rotation

We began the discussion of stress by considering a material element in the current configuration of the body (see Figure 1.26). We passed an imaginary plane through the body to obtain the free-body diagrams of the current configuration, and we used the tetrahedron argument to show that the Cauchy stress T_{km} is a tensor that describes the traction **t** acting on a surface element with unit normal **n** in the current configuration [Eq. (1.111)],

$$t_m(\mathbf{n}) = T_{km}n_k. \tag{1.114}$$

Suppose that a body reaches the current configuration in two steps. Starting in the reference configuration, we first subject it to a load that results in a surface traction $\mathbf{t}^0$ acting on the surface element with unit normal $\mathbf{n}^0$. The Cauchy stress for this configuration is T^0_{rs}, where

$$t^0_s(\mathbf{n}^0) = T^0_{rs}n^0_r.$$

Next, we subject the body to a pure rotation, which changes the unit normal from $\mathbf{n}^0$ to **n** and the tractions from $\mathbf{t}^0$ to **t**.

To examine how the stress changes during this pure rotation of the body, we must express the changes of the unit normal and the traction in terms of the rotation $\alpha_{k(r)}(t)$. The effect of pure rotation on $\mathbf{n}^0$ is easily determined. It simply transforms as a vector

$$n^0_r = \alpha_{r(k)}(t)n_k.$$

The traction **t** is also a vector, and therefore

$$t_m = \alpha_{(m)s}(t)t^0_s.$$

That is, during pure rotation, the internal forces of a body rotate with the body.

We combine the previous three expressions to obtain

$$t_m = \alpha_{s(m)}(t)T^0_{rs}\alpha_{r(k)}(t)n_k.$$

By comparison to Eq. (1.114), we find that

$$T_{km} = \alpha_{(k)r}(t)T^0_{rs}\alpha_{s(m)}(t). \tag{1.115}$$

This is equivalent to the expression for the change in stress components during a coordinate transformation, Eq. (1.112). *Thus, for the Cauchy stress, a coordinate transformation and a body rotation are equivalent processes.* Recall that the components of the deformation gradient and the Lagrangian strain transform as tensors only during a coordinate transformation. Because of their connection to the reference configuration, they obey different transformations during a body rotation. In contrast, the Cauchy stress has no connection to the reference configuration.

1.6.4 *Pressure and Deviatoric Stress*

The components of stress, T_{11}, T_{22}, and T_{33}, are perpendicular to the surface elements on which they act (see Figure 1.29). They are called the *normal* components of stress. *By definition, fluids at rest only support normal components of stress where* $T_{11} = T_{22} = T_{33}$. Thus, in fluids at rest, the shear stress components are zero. This is true regardless of the orientation of the coordinate system. We now show that when

$$T_{mn} = -p\delta_{mn},$$

the shear stresses are zero in all coordinate systems.

Consider a second Cartesian coordinate system $x_{(k)}$. The stress tensor in this coordinate system is $T_{(kr)}$. The tensor transformation for these stress components is [see Eq. (1.17)]

$$T_{(kr)} = -p\alpha_{(k)m}\delta_{mn}\alpha_{n(r)} = -p\delta_{(kr)}.$$

We see that the shear stresses are zero in all coordinate systems and that the normal components of stress are always equal to $-p$. The scalar component p is called the *pressure*. Because of its association with fluid behavior, p is also called the *hydrostatic pressure*. Our use of the minus sign in this definition means that positive pressure causes compression and a reduction in volume.

For stress tensors that contain nonzero values of both normal and shear stresses, we define a term that is similar to pressure. It is called the *mean normal stress* σ:

$$\sigma \equiv \frac{1}{3}T_{kk}. \tag{1.116}$$

As a special case, we notice that

$$\sigma = -p \quad \text{when } T_{km} = -p\delta_{km}.$$

We define the *deviatoric stress* T'_{km} to be the difference between T_{km} and the mean normal stress σ:

$$T'_{km} \equiv T_{km} - \sigma\delta_{km}. \tag{1.117}$$

Notice the similarity between this definition and deviatoric strain [see Eq. (1.85)]. The invariants of the deviatoric stress tensor are

$$I_{T'} = T'_{kk} = 0, \qquad II_{T'} = \frac{1}{2}T'_{km}T'_{km}, \qquad III_{T'} = \det\left[T'_{km}\right].$$

1.6.5 *Exercises*

1.50 Consider a rectangular parallelepiped at rest with sides of length 3 mm, 4 mm, and 5 mm (see Figure 1.31). The forces in Newtons on each of the three visible faces of the cube are

$$\mathbf{f}_1 = 12\mathbf{i}_1 - 4\mathbf{i}_2, \qquad \mathbf{f}_2 = x\mathbf{i}_1 + 30\mathbf{i}_2 + y\mathbf{i}_3, \qquad \mathbf{f}_3 = -20\mathbf{i}_2 - 40\mathbf{i}_3,$$

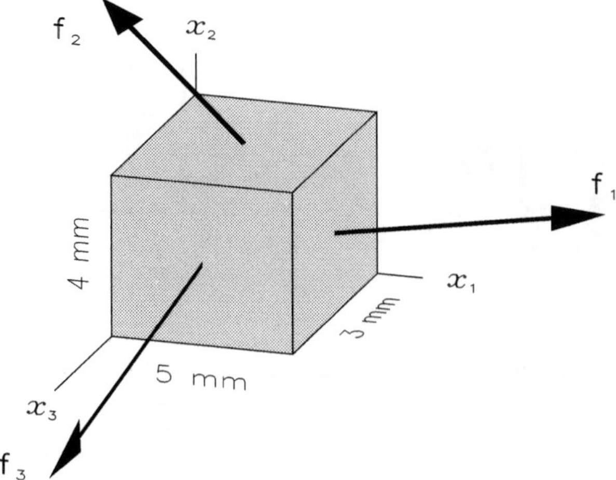

Figure 1.31 Exercise 1.50.

where x and y are unknown. Assume these forces arise from stresses that are uniformly distributed over each of the faces of the cube.

(a) Determine the tractions on each face in terms of the unknowns x and y.
(b) Determine the components of the Cauchy stress in terms of the unknowns x and y.
(c) Given that the stress tensor must be symmetric, determine x and y.
(d) Determine the forces on the three faces that are not visible.

1.51 Suppose in the previous exercise that a plane is passed through the cube. The normal to this plane is

$$\mathbf{n} = \frac{1}{\sqrt{3}}\mathbf{i}_1 + \frac{1}{\sqrt{3}}\mathbf{i}_2 + \frac{1}{\sqrt{3}}\mathbf{i}_3.$$

(a) Determine the traction acting on this plane.
(b) Determine the component of the traction that acts normal to this plane.
(c) Determine the component of the traction that acts along a tangent to this plane.

1.52 Suppose that the components of stress at a point are

$$[T_{ij}] = \begin{bmatrix} T_{11} & 2 & 1 \\ 2 & 0 & 2 \\ 1 & 2 & 0 \end{bmatrix}.$$

(a) Determine the value of T_{11} that results in a traction-free plane through this point.
(b) Determine the normal of this plane.

1.7 One-Dimensional Stress

We have already defined one-dimensional motion and studied the deformation that results from this special type of motion. Special forms of stress are also present during one-dimensional motion. In this section we consider two special stress tensors: spherical stress and triaxial stress. Like one-dimensional motion and deformation, these stress tensors are invariant with respect to a rotation of coordinates about the x_1 axis. Before studying

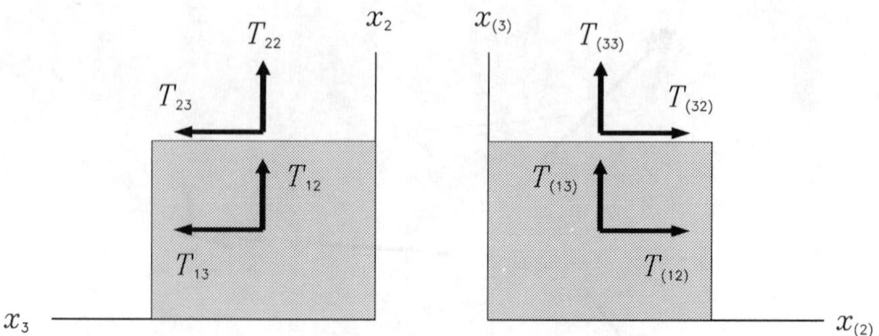

Figure 1.32 Two illustrations of the same volume element. In the left illustration we show several of the stress components for the coordinate system x_k. We do not show the x_1 axis, which projects out of the page. In the right illustration we show these stress components in the same volume element using another coordinate system $x_{(k)}$, which we have rotated $-\pi/2$ about the x_1 axis.

these configurations of stress in detail, let us see how one-dimensional motion affects the configuration of the stress.

Newton's laws of motion link stress and motion together. When motion is one dimensional, the stress must also be one dimensional. Thus the stress tensor must only be a function of x_1 and t, and it must be symmetric with respect to a coordinate rotation about x_1. We illustrate the implications of this symmetry requirement in Figure 1.32, where the same volume element is illustrated twice using two different coordinate systems. The two coordinate systems differ only by a rotation about the x_1 axis. From the figure we see that

$$\begin{aligned} T_{12} &= T_{(13)}, & T_{13} &= -T_{(12)}, \\ T_{23} &= -T_{(32)}, & T_{22} &= T_{(33)}. \end{aligned} \tag{1.118}$$

For the stress tensor to be symmetric about x_1,

$$T_{km} = T_{(km)}.$$

Thus, by applying this result to the right-hand sides of Eq. (1.118), we obtain

$$\begin{aligned} T_{12} &= T_{13}, & T_{13} &= -T_{12}, \\ T_{23} &= -T_{32}, & T_{22} &= T_{33}. \end{aligned}$$

The first two relations show that

$$T_{12} = T_{13} = 0.$$

Because the stress tensor is symmetric, $T_{23} = T_{32}$; when we combine this with the previous result that $T_{23} = -T_{32}$, we find that

$$T_{23} = 0.$$

Hence the shear components of the stress tensor are zero. The stress tensor has the following form:

$$[T_{km}] = \begin{bmatrix} T & 0 & 0 \\ 0 & \tilde{T} & 0 \\ 0 & 0 & \tilde{T} \end{bmatrix}, \tag{1.119}$$

where

$$T = T_{11}$$

and

$$\tilde{T} = T_{22} = T_{33}.$$

This is the most general form of one-dimensional stress. In the next section we consider a special form of Eq. (1.119) where $T = \tilde{T}$. This special case of one-dimensional stress is called *spherical stress*. Then we study the case where $T \neq \tilde{T}$. This form of one-dimensional stress is called *triaxial stress*.

1.7.1 *Spherical Stress*

Consider the material point $\mathbf{X}$ in a body described by the coordinate system x_k. This body is subject to the following stress tensor:

$$T_{km} = -p\delta_{km}. \tag{1.120}$$

This is called *spherical stress*. The scalar p is called the *pressure*. The mean normal stress σ for spherical stress is equal to the negative of the pressure [see Eq. (1.116)]:

$$\sigma = \frac{1}{3}T_{kk} = -p.$$

The distinction between pressure and mean normal stress is sometimes confusing. We use the term pressure for fluids where shear stress is zero, and we use the term mean normal stress for solids that have nonzero shear stresses.

The three invariants for spherical stress are

$$I_T = -3p, \qquad II_T = \frac{3}{2}p^2, \qquad III_T = -p^3.$$

The tractions produced by spherical stress are [see Eq. (1.111)]

$$t_m(\mathbf{n}) = T_{km}n_k = -p\delta_{km}n_k = -pn_m.$$

For example, if $\mathbf{n} = \mathbf{i}_1$, then

$$t_1(\mathbf{i}_1) = -p, \qquad t_2(\mathbf{i}_1) = 0, \qquad t_3(\mathbf{i}_1) = 0.$$

Thus the tractions act in directions that are perpendicular to the surfaces of the volume element (see Figure 1.33). When p is positive, the tractions point toward the element.

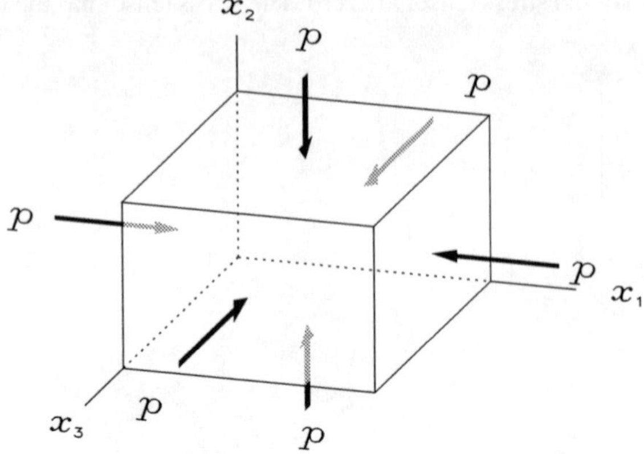

Figure 1.33 Surface tractions $t_m(\mathbf{n}) = -pn_m$ resulting from spherical stress.

Transformation to Other Coordinate Systems Consider another coordinate system $x_{(k)}$ and transform the components of the stress tensor to this new coordinate system. We obtain [see Eq. (1.112)]

$$T_{(km)} = -\alpha_{(k)i}\, p\delta_{ij}\alpha_{j(m)} = -p\delta_{km}.$$

Thus the scalar components of the stress tensor for spherical stress are invariant, and a condition of spherical stress exists in all Cartesian coordinate systems.

Deviatoric Stress The deviatoric stress for spherical stress is identically zero,

$$T'_{km} = -p\delta_{km} - \frac{1}{3}I_T\delta_{km} = 0.$$

1.7.2 *Triaxial Stress*

Consider the material point $\mathbf{X}$ in a body described by the coordinate system x_k. We have shown that one-dimensional motion results in the following stress tensor [see Eq. (1.119)]:

$$[T_{km}] = \begin{bmatrix} T & 0 & 0 \\ 0 & \tilde{T} & 0 \\ 0 & 0 & \tilde{T} \end{bmatrix}. \tag{1.121}$$

The scalar components T and $\tilde{T}$ have arbitrary values. This is called *triaxial stress*. Spherical stress is a special case of Eq. (1.121) obtained by setting $\tilde{T} = T$. Another special case is obtained when $\tilde{T} = 0$. This is called *uniaxial stress*.

During a change of coordinates, triaxial stress transforms according to [see Eq. (1.112)]

$$T_{(km)} = \alpha_{(k)1}T\alpha_{1(m)} + \alpha_{(k)2}\tilde{T}\alpha_{2(m)} + \alpha_{(k)3}\tilde{T}\alpha_{3(m)}.$$

Noticing that $\alpha_{(k)n}\alpha_{(m)n} = \delta_{km}$ [see Eq. (1.17) in Section 1.1.6], we obtain

$$T_{(km)} = (T - \tilde{T})\alpha_{(k)1}\alpha_{1(m)} + \tilde{T}\delta_{km}. \tag{1.122}$$

We will show that the scalar components of triaxial stress are invariant during coordinate rotation about x_1. For this reason, $T_{11} = T$ is called the *axial component* and $T_{22} = T_{33} = \tilde{T}$ are called the *lateral components*. The three invariants of the triaxial stress tensor are

$$I_T = T + 2\tilde{T}, \quad II_T = \frac{1}{2}T^2 + \tilde{T}^2, \qquad III_T = T\tilde{T}^2.$$

The tractions produced by triaxial stress are [see Eq. (1.111)]

$$\begin{aligned} t_1(\mathbf{i}_1) &= -t_1(-\mathbf{i}_1) = T_{11} = T, \\ t_2(\mathbf{i}_2) &= -t_2(-\mathbf{i}_2) = T_{22} = \tilde{T}, \\ t_3(\mathbf{i}_3) &= -t_3(-\mathbf{i}_3) = T_{33} = \tilde{T}. \end{aligned}$$

When considering the volume element of the body, these tractions act in directions that are perpendicular to the surface elements, pointing away from these surfaces when the respective stress components are positive (Figure 1.34).

Deviatoric Stress The mean normal stress σ is

$$\sigma = \frac{1}{3}(T + 2\tilde{T}). \tag{1.123}$$

From the definition of deviatoric stress, Eq. (1.117), the matrix form of the deviatoric stress tensor for triaxial stress is

$$\left[T'_{km}\right] = \begin{bmatrix} \frac{2}{3}(T - \tilde{T}) & 0 & 0 \\ 0 & -\frac{1}{3}(T - \tilde{T}) & 0 \\ 0 & 0 & -\frac{1}{3}(T - \tilde{T}) \end{bmatrix}. \tag{1.124}$$

The three invariants of the deviatoric stress tensor are

$$I_{T'} = 0, \qquad II_{T'} = \frac{1}{3}(T - \tilde{T})^2, \qquad III_{T'} = \frac{2}{27}(T - \tilde{T})^3. \tag{1.125}$$

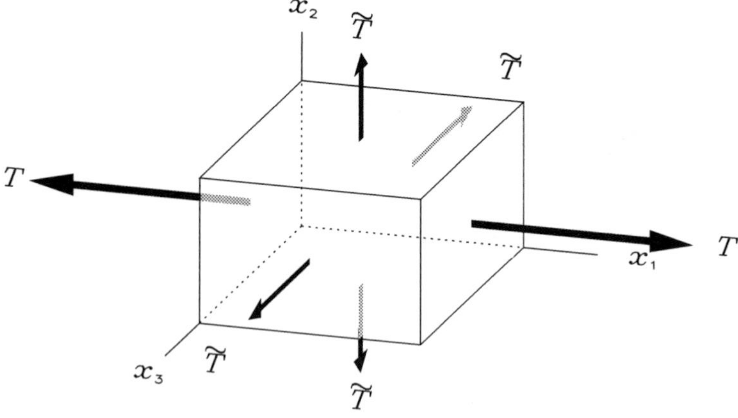

Figure 1.34 Surface tractions resulting from triaxial stress.

Maximum Shear Stress In the coordinate system x_k, the shear stresses are zero. Consider a second coordinate system $x_{(n)}$ that has the same origin but has an arbitrary orientation with respect to x_k. The shear stresses in $x_{(n)}$ are [see Eq. (1.122)]

$$T_{(km)} = (T - \tilde{T})\alpha_{(k)1}\alpha_{1(m)}, \qquad k \neq m. \tag{1.126}$$

Let $\tau_{\max}$ denote the maximum magnitude of the shear stress. We next determine the value of $\tau_{\max}$ and the orientation of the coordinate system $x_{(n)}$ that yields $|T_{(km)}| = \tau_{\max}$.

The maximum shear stress occurs when

$$|\alpha_{(k)1}\alpha_{1(m)}| = |D_1(\mathbf{i}_{(k)})D_1(\mathbf{i}_{(m)})| = \text{maximum},$$

and we use the orientation lemma in Section 1.1.8 to determine $\tau_{\max}$. This lemma requires that $\mathbf{i}_{(k)}$ and $\mathbf{i}_{(m)}$ lie in a plane that also contains the x_1 axis. Because the scalar components of the triaxial stress tensor are unaffected by rotation about the x_1 axis, any plane that contains the x_1 axis will give identical results. For convenience, we use the x_1–x_2 plane. Without loss of generality, we assume that $k = 1$ and $m = 2$ to determine the coordinates that give $|T_{(12)}| = \tau_{\max}$. The angle θ determines the orientation of $\mathbf{i}_{(1)}$ and $\mathbf{i}_{(2)}$ with respect to the x_1 axis (see Figure 1.35). The direction cosines are

$$\alpha_{(1)1} = \cos\theta, \qquad \alpha_{1(2)} = -\sin\theta. \tag{1.127}$$

The resulting expression for the shear stress is

$$T_{(12)}(\theta) = -\frac{1}{2}(T - \tilde{T})\sin(2\theta), \tag{1.128}$$

where we have used the trigonometric identity $\sin(2\theta) = 2\sin\theta\cos\theta$. The maximum value of $|T_{(12)}(\theta)|$ is

$$\tau_{\max} = \frac{1}{2}|T - \tilde{T}|. \tag{1.129}$$

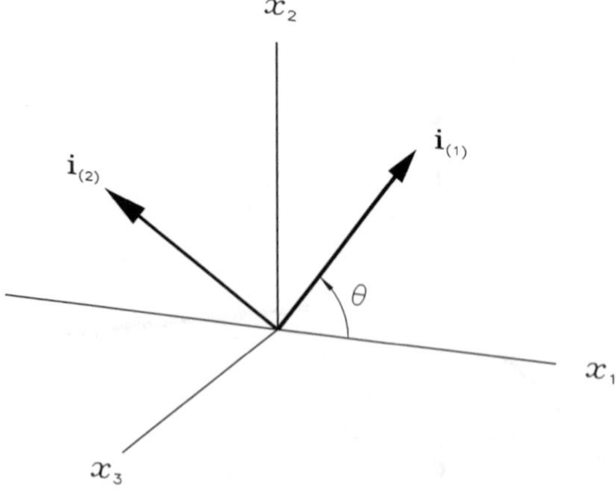

Figure 1.35 The unit vectors $\mathbf{i}_{(1)}$ and $\mathbf{i}_{(2)}$ of the rotated coordinate system.

There are four values of θ where $|T_{(12)}| = \tau_{\max}$,

$$|T_{(12)}(\theta)| = \tau_{\max}, \qquad \text{at } \theta = -\frac{3}{4}\pi,\ -\frac{1}{4}\pi,\ \frac{1}{4}\pi,\ \frac{3}{4}\pi.$$

1.7.3 *Exercises*

1.53 We subject a body to triaxial stress. The axis of symmetry of the stress is parallel to the x_1 axis. The axial and lateral components of stress are $T = 3$ MPa and $\tilde{T} = -3$ MPa.

(a) Determine the maximum shear stress.

(b) Determine the mean normal stress.

(c) Determine the normal stresses in the coordinate systems that yield maximum magnitudes of shear stress.

1.54 Repeat the previous exercise for $T = 3$ MPa and $\tilde{T} = 0$ MPa.

1.55 For each of the previous two exercises, determine the matrix of stress components $[T_{(km)}]$ for a coordinate system that is rotated 45° about x_3.

1.56 Consider the following stress configuration:

$$[T_{km}] = \begin{bmatrix} -10 & 0 & 0 \\ 0 & -2 & 0 \\ 0 & 0 & -2 \end{bmatrix}.$$

(a) Determine the matrix of stress components in the coordinate systems that are obtained by rotation about x_3 where $\theta_3 = \pi/4, 3\pi/4, 5\pi/4$, and $7\pi/4$.

(b) Determine the stress invariants I_T, II_T, and III_T for $\theta_3 = 0, \pi/4, 3\pi/4, 5\pi/4$, and $7\pi/4$.

1.57 A configuration of triaxial stress exists in the x_k coordinate system. The axis of symmetry is x_1. Suppose that a second coordinate system $x_{(k)}$ is obtained by rotation about x_2. Determine the matrix of the stress components for this new coordinate system.

1.8 Laws of Motion

Recall that stress waves are a manifestation of the mechanical field and the thermal field acting together. The mechanical field is represented by force and motion, and the thermal field is represented by heat and temperature. Remember that, throughout this chapter and the next, we ignore heat and temperature and instead focus upon force and motion. Motion is composed of translation, rotation, and deformation. Deformation is represented by the deformation gradient and the strain. Forces are represented by the stress and the external body force. Although we have described force and motion, we have not described how force causes motion or how motion results in force. Part of the link between force and motion depends on properties of the body. For example, apples and acorns resist deformation in quite different ways. However, part of this link is independent of the properties of the body. Apples and acorns fall from trees in much the same way.

In this section we derive two important laws that link motion and force in a body. These laws of motion are universal and independent of the properties of the body. The first law is called the *conservation of mass*. It connects the mass density of a body to its motion. The second law is called the *balance of momentum*. This is Newton's second law of motion that links force to mass and motion. We shall derive each of these laws in a variety of ways. First, we derive them for one-dimensional motion. This allows us to minimize the mathematics and appeal to our intuition. The one-dimensional derivations will be presented in both the material and spatial descriptions. After completing the one-dimensional derivations, we will derive the three-dimensional forms of the laws of motion. Here we shall employ a more elegant mathematical approach.

1.8.1 *Simplified Notation: One-Dimensional Motion*

Let us restrict our discussion to one-dimensional motion, which is only a function of x_1. Because u_1, v_1, and a_1 are the only nonzero components of displacement, velocity, and acceleration, it is convenient to simplify the notation by elimination of the Cartesian indices where possible. Thus the current position x_1 is denoted by x and the reference position X_1 is denoted by X. Then the motion is

$$x = \hat{x}(X, t).$$

The displacement u_1 is denoted by u,

$$u = x - X.$$

The velocity v_1 is denoted by v,

$$v = \frac{\partial \hat{x}}{\partial t}.$$

The acceleration a_1 is denoted by a,

$$a = \frac{\partial \hat{v}}{\partial t}.$$

The deformation gradient tensor for this motion is [Eq. (1.53)]

$$[F_{km}] = \begin{bmatrix} F & 0 & 0 \\ 0 & 1 & 0 \\ 0 & 0 & 1 \end{bmatrix},$$

where the F_{11} component of this deformation gradient tensor is

$$F = \frac{\partial \hat{x}}{\partial X} = 1 + \frac{\partial \hat{u}}{\partial X}. \tag{1.130}$$

The chain rule leads to the relationship

$$\frac{\partial \hat{u}}{\partial X} = F \frac{\partial u}{\partial x},$$

and substitution of Eq. (1.130) yields

$$\frac{\partial \hat{u}}{\partial X} = \left(1 + \frac{\partial \hat{u}}{\partial X}\right) \frac{\partial u}{\partial x} = \left[1 + \left(1 + \frac{\partial \hat{u}}{\partial X}\right) \frac{\partial u}{\partial x}\right] \frac{\partial u}{\partial x}.$$

With repeated substitution of Eq. (1.130), we find that

$$\frac{\partial \hat{u}}{\partial X} = \frac{\partial u}{\partial x} + \left(\frac{\partial u}{\partial x}\right)^2 + \cdots. \tag{1.131}$$

The derivative of Eq. (1.130) with respect to time is

$$\frac{\partial \hat{F}}{\partial t} = \frac{\partial \hat{v}}{\partial X}. \tag{1.132}$$

The strain tensor for uniaxial strain is [see Eq. (1.97)]

$$[E_{km}] = \begin{bmatrix} E & 0 & 0 \\ 0 & 0 & 0 \\ 0 & 0 & 0 \end{bmatrix},$$

where the E_{11} component of the strain tensor is

$$E = \frac{1}{2}(F^2 - 1). \tag{1.133}$$

The volume strain for uniaxial deformation is [see Eq. (1.94)]

$$\frac{dV}{dV_0} = F. \tag{1.134}$$

The velocity gradient, Eq. (1.107), is

$$[L_{km}] = \begin{bmatrix} L & 0 & 0 \\ 0 & 0 & 0 \\ 0 & 0 & 0 \end{bmatrix},$$

where the $L = L_{11}$ component is

$$L = \frac{\partial v}{\partial x} = \frac{1}{F}\frac{\partial \hat{F}}{\partial t}. \tag{1.135}$$

The deformation rate tensor has one nonzero component $D = D_{11}$,

$$D = \frac{\partial v}{\partial x} = \frac{1}{F}\frac{\partial \hat{F}}{\partial t},$$

while the spin tensor W_{km} is identically zero.

In this notation, the motion u, deformation gradient F, and velocity gradient L *appear* to be scalar quantities. We emphasize that they are, however, components of vectors and tensors. Thus the uniaxial component of deformation F results in both longitudinal and shear strains. Moreover, the normal components of triaxial stress T and $\tilde{T}$ result in shear stresses.

1.8.2 *Conservation of Mass: One-Dimensional Motion*

In this section we derive the first of the laws of motion. This law links the mass density of a body to its motion. The material volume of the current configuration of a body is denoted by V. In this configuration, the total mass of a body is denoted by m, and the distribution of mass throughout the body is described by the density ρ. We defined density in Section 1.6 as the mass per unit volume. The density of a body can change with both position and time, and it can be expressed in either the material description as a function of the reference position $\mathbf{X}$ and the time t or in the spatial description as a function of the current position $\mathbf{x}$ and time t. Regardless of how we express the density, the total mass m is equal to the integral of the density ρ over the material volume V,

$$m = \int_V \rho \, dV.$$

We postulate that *the total mass of a body is conserved.* When we integrate the density over the material volume at any time in the past, present, or future, we must obtain the same value of m.

Let us evaluate this volume integral for the reference configuration. If we denote the density and volume of the reference configuration as ρ_0 and V_0, then

$$m = \int_{V_0} \rho_0 \, dV_0.$$

It is obvious that any motion that changes the material volume of a body will also change the density. It is also important for you to recognize that motion can change the density without changing the total volume of the body. We next examine how motion and density are coupled together.

Material Description

The relationship between the density of a body and its motion is called the *equation of conservation of mass.* We can derive it in one of two ways. We can either describe the density and the motion as functions of the material coordinate X or we can describe the density and the motion as functions of the spatial coordinate x. The first method is called the *material description*; the second method is called the *spatial description.* In this section we use the material description to obtain the equation of conservation of mass. In the next section we use the spatial description.

Consider a body undergoing one-dimensional motion. We show the reference and current configurations of an infinitesimal material element of this body in Figure 1.36. The two cross-hatched regions represent imaginary walls that constrain the motion of the body to the x direction much like a piston in a cylinder. However the body actually extends to infinity in the lateral directions, and our illustration merely serves to emphasize the one-dimensional motion of the body. The reference position of the center of this volume element is X. The thickness is dX. The dashed lines represent faces of the volume element in the reference

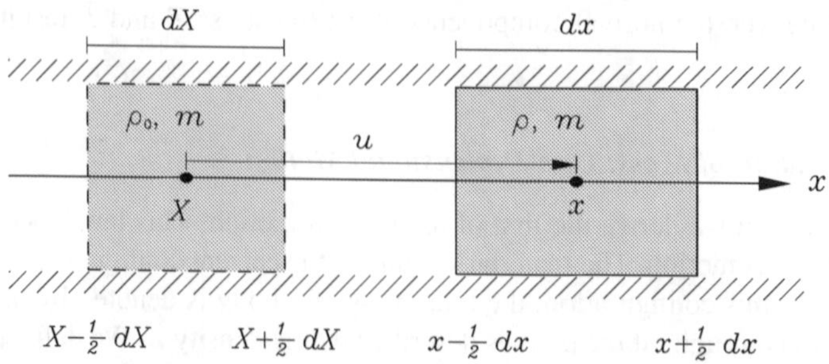

Figure 1.36 The reference position (dashed outline) and current position (solid outline) of a material element of a body undergoing one-dimensional motion.

configuration. The two vertical faces perpendicular to the x axis have an area of dS. The density of the material in the reference configuration is $\hat{\rho}_0(X, t)$, and the mass of the element is dm. The mass is equal to the density times the volume of the element:

$$dm = \rho_0 \, dX \, dS.$$

Because this is a material element, the motion of the body displaces the center of this element by an amount u from the reference position X to the current position x. Deformation causes the thickness to change from dX to dx. Solid lines represent the faces of the material element in the current configuration. Because the motion is one dimensional, the areas of the vertical faces in the current configuration are also equal to dS. The density of the material in the current configuration is $\hat{\rho}(X, t)$. During displacement from the reference configuration to the current configuration, the mass of the material element does not change. Thus

$$\rho_0 \, dX \, dS = \rho \, dx \, dS,$$

which can also be written as

$$\rho_0 \, dV_0 = \rho \, dV. \tag{1.136}$$

Dividing this result by $\rho \, dV_0$ and substituting Eq. (1.134), we obtain

$$\frac{\rho_0}{\rho} = F. \tag{1.137}$$

This is the material form of the equation of conservation of mass. We obtain another useful form of the conservation of mass by taking the partial derivative of this result with respect to time and substituting Eq. (1.132):

$$\frac{\rho_0}{\rho^2} \frac{\partial \hat{\rho}}{\partial t} = -\frac{\partial \hat{v}}{\partial X}. \tag{1.138}$$

Spatial Description

In the material description of the equation of conservation of mass, we follow the motion of a material element of the body. This element always contains the same mass of material. In the spatial description we use a volume element that is fixed in space and observe the material as it moves through the fixed element.

In Figure 1.37 we show an element that is fixed in space. Its thickness is dx. The faces of the volume element are denoted by dashed lines, and the two faces that are perpendicular to the x axis have areas dS. The light shaded regions on each side of the fixed element represent portions of the body that are entering and leaving the fixed element. The velocity of the material at the center of the volume element is $v(x, t)$. Using a first-order Taylor's series approximation, we can estimate the velocity at the left and right boundaries of the volume element as

$$\begin{aligned} \text{left velocity} &= v - \frac{\partial v}{\partial x} \frac{dx}{2}, \\ \text{right velocity} &= v + \frac{\partial v}{\partial x} \frac{dx}{2}. \end{aligned} \tag{1.139}$$

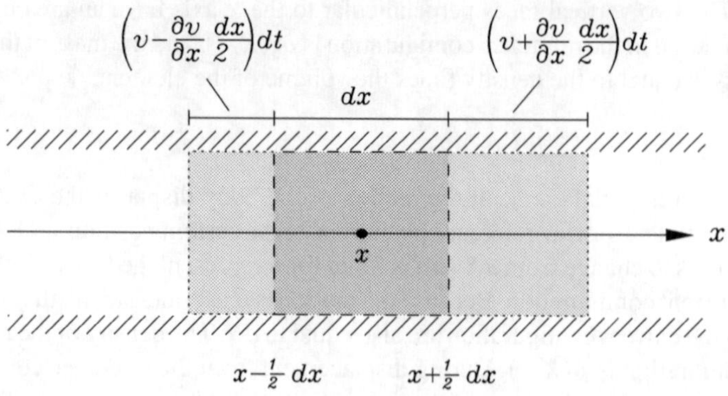

Figure 1.37 The dark shaded region defines a volume element that is fixed in space. A body undergoing one-dimensional motion moves through this fixed element. The light shaded regions move with the body. When the velocity is positive during a small increment of time dt, the light shaded portion of the body to the left enters the fixed element, and the one to the right leaves the fixed element.

Thus, during an increment of time dt, the volumes of the body that are entering the fixed element from the left and leaving the fixed element on the right are

$$\begin{aligned} \text{volume entering} &= \left(v - \frac{\partial v}{\partial x}\frac{dx}{2}\right) dt\, dS, \\ \text{volume leaving} &= \left(v + \frac{\partial v}{\partial x}\frac{dx}{2}\right) dt\, dS. \end{aligned} \tag{1.140}$$

The density of the material within the fixed element is $\rho(x, t)$. Using another first-order Taylor's approximation, the values of the density at the left and right faces of the fixed element are

$$\begin{aligned} \text{density entering} &= \left(\rho - \frac{\partial \rho}{\partial x}\frac{dx}{2}\right) dt\, dS, \\ \text{density leaving} &= \left(\rho + \frac{\partial \rho}{\partial x}\frac{dx}{2}\right) dt\, dS. \end{aligned} \tag{1.141}$$

We determine the masses of material that are entering and leaving the fixed volume by taking the products of these densities and volumes. The difference between the mass entering and the mass leaving must be equal to the change in mass of the fixed element during the time increment dt:

$$\begin{aligned} \frac{\partial(\rho dx dS)}{\partial t} dt = &\left(\rho - \frac{\partial \rho}{\partial x}\frac{dx}{2}\right)\left(v - \frac{\partial v}{\partial x}\frac{dx}{2}\right) dt\, dS \\ &- \left(\rho + \frac{\partial \rho}{\partial x}\frac{dx}{2}\right)\left(v + \frac{\partial v}{\partial x}\frac{dx}{2}\right) dt\, dS. \end{aligned}$$

The quantities dx and dS do not change with respect to time. When we expand the products in this expression and divide by $dt\,dx\,dS$, we obtain

$$\frac{\partial \rho}{\partial t} + \frac{\partial(\rho v)}{\partial x} = 0. \tag{1.142}$$

This is the spatial form of the equation of conservation of mass.

The material form [Eq. (1.138)] and the spatial form [Eq. (1.142)] of the equation of conservation of mass are equivalent statements. We demonstrate this by deriving the spatial form directly from the material form. Noticing that

$$\frac{\partial v}{\partial X} = \frac{\partial \hat{x}}{\partial X}\frac{\partial v}{\partial x} = F\frac{\partial v}{\partial x},$$

the material form of the conservation of mass, Eq. (1.138), becomes

$$\frac{\partial \hat{\rho}}{\partial t} + \rho\frac{\partial v}{\partial x} = 0.$$

When we recall that

$$\frac{\partial \hat{\rho}}{\partial t} = \frac{\partial \rho}{\partial t} + v\frac{\partial \rho}{\partial x},$$

we obtain the spatial form of conservation of mass.

1.8.3 *Balance of Linear Momentum: One-Dimensional Motion*

The linear momentum of a volume element of a body is the product of the mass and velocity of that element. For a volume element of material with density ρ, velocity $\mathbf{v}$, and volume dV, the linear momentum is

$$\text{momentum} = \rho\mathbf{v}\,dV.$$

Newton's second law of motion states that the rate of change of linear momentum of a volume element is equal to the sum of the forces acting on the element. The equation representing Newton's second law of motion is called the *equation of balance of linear momentum.* We will derive both the material and spatial forms of this equation.

Material Description

Consider the current configuration of the material element in Figure 1.36. Remember that this element moves with the material in the body. Triaxial stresses act on this element. We illustrate the resulting tractions in Figure 1.38. The three-dimensional representation of this element illustrates both the axial and lateral tractions. These tractions act on the centers of each face of the element. The lateral tractions, which result from the lateral stress $\tilde{T}$, act in equal and opposite pairs. The sum of the lateral tractions is zero. The axial tractions, which result from the axial stress T, act at the centers of the left and right faces of the element. We recall that the areas of these faces are both equal to dS. The axial stress at the center of the element is $T(x,t)$. We use a first-order Taylor's series approximation to evaluate the axial tractions at the positions of the left and right faces. The magnitudes of the axial tractions

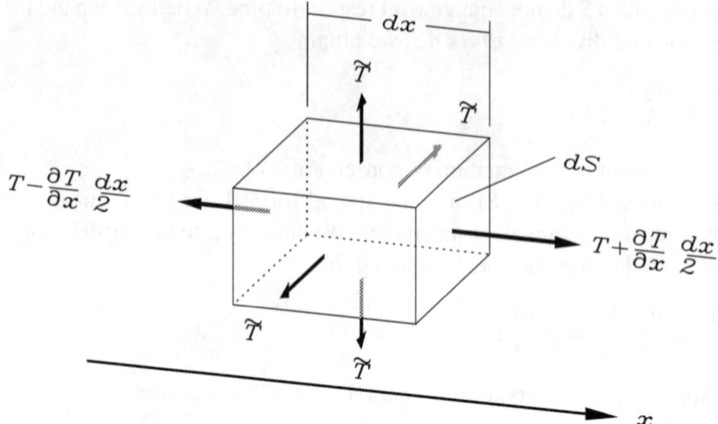

Figure 1.38 Triaxial stress acting on the material element in the current configuration.

are

$$\text{left traction} = T - \frac{\partial T}{\partial x}\frac{dx}{2},$$

$$\text{right traction} = T + \frac{\partial T}{\partial x}\frac{dx}{2}.$$

The net force on the element acts in the direction of the x axis and equals the sum of the axial tractions times the area dS:

$$\text{net force} = \frac{\partial T}{\partial x}\, dx\, dS.$$

The net force is equal to the spatial gradient of the axial stress times the volume of the element.

Newton's second law of motion states that the rate of change of linear momentum of the volume element is equal to the net force acting on the element. When we include a body force b acting in the x direction, we obtain

$$\frac{\partial}{\partial t}(\hat{\rho}\hat{v}\, d\hat{x})\, dS = \frac{\partial T}{\partial x}\, dx\, dS + \rho b\, dx\, dS,$$

where we have noticed that dS is a constant. From the material description of the conservation of mass, we know that the product $\rho\, dx = \rho_0\, dX$ is not a function of time. Therefore, we can divide this expression by $dx\, dS$ to obtain

$$\rho a = \frac{\partial T}{\partial x} + \rho b. \tag{1.143}$$

Notice that

$$\frac{\partial T}{\partial x} = \frac{\partial X}{\partial x}\frac{\partial \hat{T}}{\partial X} = \frac{1}{F}\frac{\partial \hat{T}}{\partial X} = \frac{\rho}{\rho_0}\frac{\partial \hat{T}}{\partial X},$$

and therefore

$$\rho_0 a = \frac{\partial \hat{T}}{\partial X} + \rho_0 b. \tag{1.144}$$

This is the material form of the equation of balance of linear momentum.

Spatial Description

We can derive the spatial form of the equation of balance of linear momentum by using the fixed element in Figure 1.37. We determine the net force acting on the element from a free-body diagram that is similar to Figure 1.38. The only differences in the diagram are the fixed positions of the left and right faces. Once again, the net force is

$$\text{net force} = \frac{\partial T}{\partial x}\, dx\, dS.$$

As mass enters and leaves the fixed volume, it transports momentum to and from the volume. This is called *momentum flux*. The momentum flux across the left and right faces during a small interval of time dt is equal to the product of the density, velocity, and volume. Using Eqs. (1.139)–(1.141), the momentum flux is

$$\text{left momentum flux} = \left(\rho - \frac{\partial \rho}{\partial x}\frac{dx}{2}\right)\left(v - \frac{\partial v}{\partial x}\frac{dx}{2}\right)^2 dS,$$

$$\text{right momentum flux} = -\left(\rho + \frac{\partial \rho}{\partial x}\frac{dx}{2}\right)\left(v + \frac{\partial v}{\partial x}\frac{dx}{2}\right)^2 dS.$$

We include a negative sign in the second expression because a positive value of velocity at the right face indicates a loss of momentum from the fixed element.

The rate of change of momentum of the fixed element is equal to the momentum flux through the left face minus the momentum flux through the right face plus the net force and body force:

$$\begin{aligned}\frac{\partial(\rho v\, dx\, dS)}{\partial t} &= \left(\rho - \frac{\partial \rho}{\partial x}\frac{dx}{2}\right)\left(v - \frac{\partial v}{\partial x}\frac{dx}{2}\right)^2 dS \\ &\quad - \left(\rho + \frac{\partial \rho}{\partial x}\frac{dx}{2}\right)\left(v + \frac{\partial v}{\partial x}\frac{dx}{2}\right)^2 dS + \frac{\partial T}{\partial x}\, dx\, dS + \rho b\, dx\, dS.\end{aligned}$$

Because this is a fixed element, both dx and dS are independent of time. After expansion of the products and division by $dx\, dS$, we obtain

$$\frac{\partial(\rho v)}{\partial t} = \frac{\partial}{\partial x}(T - \rho v^2) + \rho b - \frac{\partial \rho}{\partial x}\left(\frac{\partial v}{\partial x}\right)^2 \frac{dx^2}{4}.$$

We take the limit of this equation as $dx \to 0$ to obtain

$$\frac{\partial(\rho v)}{\partial t} = \frac{\partial}{\partial x}(T - \rho v^2) + \rho b. \tag{1.145}$$

This is the spatial form of the equation of balance of linear momentum. The left side is the rate of change of linear momentum in the fixed element. The right side contains the three forces, T, ρv^2, and b. The axial stress T and the body force b are familiar terms, but the term ρv^2 is new. It is called the *radiation pressure*. Notice that ρv^2 is always positive.

We obtain another form of the equation of balance of linear momentum by substituting the equation of conservation of mass, Eq. (1.142), into Eq. (1.145):

$$\rho\left(\frac{\partial v}{\partial t} + v\frac{\partial v}{\partial x}\right) = \frac{\partial T}{\partial x} + \rho b. \tag{1.146}$$

The radiation pressure does not appear in this form. The left side of this relationship is the density ρ times the acceleration a. Thus the material form [Eq. (1.143)] and the spatial form [Eq. (1.146)] of the equation of balance of linear momentum are identical statements.

1.8.4 *Transport Theorem: Three-Dimensional Motion*

In the preceding sections we derived the equation of conservation of mass and the equation of balance of linear momentum for a body undergoing one-dimensional motion. In the following sections we derive these equations again for a body undergoing a general three-dimensional motion. In this derivation, we employ a more elegant mathematical approach. This approach is based upon the *Gauss theorem*, the *Leibniz rule for differentiation of an integral*, and the *transport theorem*.

Gauss Theorem

Here we simply state the Gauss theorem in index notation. Consider a smooth, closed surface $\mathcal{S}$ that encloses the volume $\mathcal{V}$ (see Figure 1.39). Let $\mathbf{n}$ denote the unit vector that is perpendicular to $\mathcal{S}$ and directed outward. The Gauss theorem for a vector field $\mathbf{u}$ states that

$$\int_{\mathcal{V}} \frac{\partial u_k}{\partial x_k}\, dV = \int_{\mathcal{S}} u_k n_k \, dS \tag{1.147}$$

for continuous integrands. You should understand that the volume $\mathcal{V}$ enclosed by the surface $\mathcal{S}$ can contain all or part of the physical volume of a material. This volume can move, and the motion of the volume can be independent of the motion of the body.

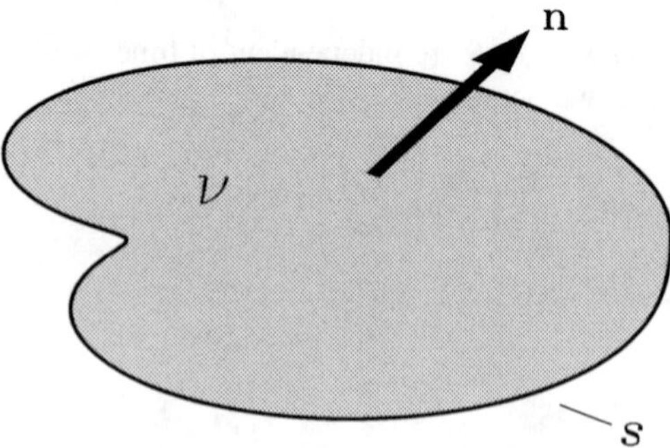

Figure 1.39 A closed surface $\mathcal{S}$ containing volume $\mathcal{V}$. The unit vector $\mathbf{n}$ is perpendicular to $\mathcal{S}$.

Leibniz Rule for Differentiation of an Integral

Suppose we integrate a continuous and differentiable scalar field $\phi = \phi(\mathbf{x}, t)$ over the volume $\mathcal{V}$ in Figure 1.39. The integral

$$f(t) = \int_{\mathcal{V}} \phi \, dV$$

is a scalar-valued function of time. Let the velocity of every point on the surface $\mathcal{S}$ be given by the vector $\mathbf{w}(x, t)$. The Leibniz rule for differentiation of $f(t)$ is

$$\frac{df(t)}{dt} = \int_{\mathcal{V}} \frac{\partial \phi}{\partial t} dV + \int_{\mathcal{S}} \phi w_k n_k \, dS. \tag{1.148}$$

Transport Theorem

We shall now derive the transport theorem from the Leibniz rule. The Leibniz rule applies to a volume $\mathcal{V}$ enclosed by a surface $\mathcal{S}$ with an arbitrary velocity $\mathbf{w}$. Let us apply this rule to a material volume V that is enclosed by a surface S. Because S moves with the material, we find that

$$\mathbf{w} = \mathbf{v},$$

where $\mathbf{v}$ is the velocity of the material. Thus

$$\frac{d}{dt} \int_V \phi \, dV = \int_V \frac{\partial \phi}{\partial t} dV + \int_S \phi v_k n_k \, dS.$$

By applying the Gauss theorem to the surface integral in this expression, we obtain

$$\frac{d}{dt} \int_V \phi \, dV = \int_V \left[\frac{\partial \phi}{\partial t} + \frac{\partial (\phi v_k)}{\partial x_k} \right] dV. \tag{1.149}$$

This is the transport theorem for the material volume V. We shall use it to derive the laws of motion in three dimensions.

1.8.5 *Conservation of Mass: Three-Dimensional Motion*

Because a material volume always contains the same material, the total mass of the material in a material volume is constant:

$$\frac{d}{dt} \int_V \rho \, dV = 0.$$

By using the transport theorem, we obtain the result

$$\int_V \left[\frac{\partial \rho}{\partial t} + \frac{\partial (\rho v_k)}{\partial x_k} \right] dV = 0.$$

This equation must hold for every material volume. We assume that the integrand is continuous. The equation can be satisfied only if the integrand vanishes at each point:

$$\frac{\partial \rho}{\partial t} + \frac{\partial (\rho v_k)}{\partial x_k} = 0. \tag{1.150}$$

This is the spatial description of the equation of conservation of mass in three dimensions.

1.8.6 *Balance of Linear Momentum: Three-Dimensional Motion*

Newton's second law states that the force acting on a particle is equal to the rate of change of its linear momentum:

$$\mathbf{F} = \frac{d}{dt}(m\mathbf{v}). \tag{1.151}$$

We can determine the equation of balance of linear momentum for a material by postulating that the rate of change of the total linear momentum of the material contained in a material volume is equal to the total force acting on the material volume:

$$\frac{d}{dt}\int_V \rho\mathbf{v}\,dV = \int_S \mathbf{t}\,dS + \int_V \rho\mathbf{b}\,dV. \tag{1.152}$$

The first term on the right side is the force exerted on the surface S by the traction vector $\mathbf{t}$. The second term on the right side is the force exerted on the material by the specific external body force $\mathbf{b}$. Writing this equation in index notation and using $t_m(\mathbf{n}) = T_{km}n_k$ [see Eq. (1.111)] we obtain

$$\frac{d}{dt}\int_V \rho v_m\,dV = \int_S T_{km}n_k\,dS + \int_V \rho b_m\,dV. \tag{1.153}$$

We leave it as an exercise for you to show that by using the Gauss theorem, the transport theorem, and the equation of conservation of mass, this equation can be expressed in the form

$$\int_V \left(\rho a_m - \frac{\partial T_{km}}{\partial x_k} - \rho b_m\right) dV = 0, \tag{1.154}$$

where the acceleration is

$$a_m = \frac{\partial v_m}{\partial t} + \frac{\partial v_m}{\partial x_k}v_k.$$

Equation (1.154) must hold for every material volume of the material. If we assume that the integrand is continuous, we obtain the equation

$$\rho a_m = \frac{\partial T_{km}}{\partial x_k} + \rho b_m. \tag{1.155}$$

This is the spatial description of the equation of balance of linear momentum in three dimensions.

1.8.7 *Balance of Angular Momentum: Three-Dimensional Motion*

In one-dimensional motion, rotation is not possible. Consequently, our derivation of the balance of momentum in one dimension was restricted to *linear* momentum. In three dimensions, rotation or *angular motion* is possible. Angular motion gives rise to *angular momentum.* To examine the conditions that ensure the balance of angular momentum, we take the cross product of Newton's second law for a particle and the position vector $\mathbf{x}$ of the particle. The result is

$$\begin{aligned}\mathbf{x} \times \mathbf{F} &= \mathbf{x} \times \frac{d}{dt}(m\mathbf{v}) \\ &= \frac{d}{dt}(\mathbf{x} \times m\mathbf{v}).\end{aligned}$$

The term on the left is the moment exerted by the force $\mathbf{F}$. The term $\mathbf{x} \times m\mathbf{v}$ is called the angular momentum of the particle. This equation states that the moment is equal to the rate of change of the angular momentum.

The equation of balance of angular momentum for a volume of material is obtained by postulating that the rate of change of the total angular momentum of a material volume is equal to the total moment exerted on the material volume:

$$\frac{d}{dt}\int_V (\mathbf{x} \times \rho\mathbf{v})\, dV = \int_S (\mathbf{x} \times \mathbf{t})\, dS + \int_V (\mathbf{x} \times \rho\mathbf{b})\, dV. \tag{1.156}$$

We leave it as an exercise for you to show that this postulate implies that the stress tensor is symmetric:

$$T_{km} = T_{mk}. \tag{1.157}$$

1.8.8 *Exercises*

1.58 Consider the vector

$$\mathbf{u} = \frac{x_1\psi t}{1+\psi t}\mathbf{i}_1 - x_2\psi t\mathbf{i}_2.$$

This vector is defined over a cubic region bounded by the six sides $x_1 = 0$, $x_2 = 0$, $x_3 = 0$, $x_1 = (\psi t)^2$, $x_2 = 1$, and $x_3 = 1$.

(a) Substitute $\mathbf{u}$ into the Gauss theorem. Using the cubic boundary, execute both the volume integration and the surface integration to verify the Gauss theorem for $\mathbf{u}$.

(b) Using the process of part (a), verify the Leibniz rule for differentiation of an integral for each component of $\mathbf{u}$.

(c) Suppose $\mathbf{u}$ is the displacement of a material. Assume that at $t = 0$ the boundaries of this material are defined by a cube with sides at $x_1 = 0$, $x_2 = 0$, $x_3 = 0$, $x_1 = 1$, $x_2 = 1$, and $x_3 = 1$. Determine the motion and velocity of each boundary segment of this material volume.

(d) Substitute each component of $\mathbf{u}$ into the transport theorem. Execute the two volume integrations of the transport theorem over the material volume defined in part (c). Verify the transport theorem for this motion.

1.59 Newton's second law of motion yields the following equation [see Eq. (1.153)]:

$$\frac{d}{dt}\int_V \rho v_m\, dV = \int_S T_{km}n_k\, dS + \int_V \rho b_m\, dV.$$

Use the Gauss theorem, the transport theorem, and the equation of conservation of mass to obtain the equation of balance of linear momentum,

$$\rho a_m = \frac{\partial T_{km}}{\partial x_k} + \rho b_m.$$

1.60 Use the definition of the cross product of two vectors to show that, for a particle of mass m,

$$\mathbf{x} \times \frac{d}{dt}(m\mathbf{v}) = \frac{d}{dt}(\mathbf{x} \times m\mathbf{v}).$$

1.61 The postulate of angular momentum is [see Eq. (1.156)]

$$\frac{d}{dt}\int_V (\mathbf{x} \times \rho\mathbf{v})\, dV = \int_S (\mathbf{x} \times \mathbf{t})\, dS + \int_V (\mathbf{x} \times \rho\mathbf{b})\, dV.$$

Show that this implies

$$T_{km} = T_{mk}.$$

1.62 Consider a body undergoing one-dimensional motion. The left and right boundaries of this body are located at $x = 0$ and $x = L$. The stresses on the left and right boundaries are denoted by T_l and T_r. The body force is zero. Show that the change of momentum of the body is given by

$$\frac{d}{dt}\int_0^L \rho v_m\, dX = T_r - T_l,$$

where the integration is performed over the entire length of the body.

CHAPTER 2

Mechanical Waves

The sea provides the most intuitive image we have of wave motion. As a wave passes, we notice that seaweed floating in the water moves slightly forward and then returns, and the motion of the wave seems only weakly connected to the motion of the water. Sometimes several waves merge into a single wave of much greater energy. The stronger wave travels at a slower speed, which allows other weaker waves behind it to catch up and add to its strength. Sometimes, these waves destroy ships at sea (see the opening quote of the preface). We know that wave motion causes changes in both the mechanical field and the thermal field, but much of what we know about wave motion in solids and fluids can be explained with a purely mechanical theory. In this chapter we study mechanical waves. To do this, we first define the elastic material. This material is described by a constitutive equation that only contains the mechanical variables associated with force and motion. We show that a linear-elastic material yields one type of solution called a linear wave. A nonlinear-elastic material yields a variety of solutions that include compression waves, rarefaction waves, and shock waves. To study these waves, we use the d'Alembert solution, the Riemann integral, the method of characteristics, the theory of singular surfaces, and the steady-wave solution.

2.1 Elastic Material

It is a common observation that individual materials react differently to applied loads. For example, we know that it is easy to reduce the volume of a container of air by fifty percent, whereas it is very difficult to reduce the volume of a block of steel by one percent. Thus, the same configuration of stress in two different bodies can result in entirely different deformations. Every material has its own unique behavior, which we can characterize through a *constitutive model.* A constitutive model is a function or set of functions that interconnect stress, deformation gradient, and strain in an individual material. Constitutive models can also contain thermal variables such as heat and temperature, but in this chapter we shall restrict our discussion to mechanical variables. As an example, consider a material that is incapable of supporting shear stress. Then the Cauchy stress is always spherical:

$$T_{ij} = -p\delta_{ij}.$$

Suppose we assume that the pressure p and the volume strain $e = (dV - dV_0)/dV_0$ are related by

$$p = -Ke,$$

where K is a positive constant called the *bulk stiffness.*

This particular constitutive model has several properties that are important requirements for *all* constitutive models. Notice that because K is positive, the pressure increases as the volume decreases. We generally observe that real materials resist deformation ($K > 0$) rather than assist deformation ($K < 0$). This is an arguable condition when applied to materials undergoing a phase change such as the transformation of ice to water. If K is large, this model describes a material that is difficult to compress; if K is small, the same model describes a material that is easy to compress. Another important feature of this model is that pressure and volume are scalars and remain unchanged during a slow translation and rotation of the body where we can ignore the effects of acceleration. In the absence of external forces, the internal forces and deformation in all real materials are unaffected by slow rotation. More general constitutive models that contain vector and tensor quantities must conform to this requirement. Also, we require that coordinate transformations cannot change the constitutive model, because the description of the material must not be affected by our choice of Cartesian coordinate systems. Although all constitutive models should have these properties, this particular constitutive model has several other properties that are quite specialized and not generally applicable. This model is linear, which means p is proportional to e; it is a purely mechanical description, which means the pressure depends only on e and not on some additional quantity such as temperature; and it contains only values of pressure and volume evaluated at one point in time and position.

In this section we examine a constitutive model that is considerably more general than the simple example above, but it is still very specialized. This is the constitutive model for an *elastic material*. The stress at the material point $\mathbf{X}$ in an elastic material is a single valued and continuous function of the deformation gradient tensor F_{km} at the same material point $\mathbf{X}$. It is not a function of either temperature or past values of F_{km}, and it is not a function of F_{km} at neighboring points. The elastic material also has *perfect memory*. This means that when we change the stress in an elastic material and then reverse the stress to the initial value, the material will change volume and shape and then return to its initial volume and shape. The elastic material is a very specialized constitutive model, but it is also very useful and, with the exception of our analysis of steady waves, constitutes an integral part of the discussion throughout the remainder of this chapter. In this section we discuss several types of elastic materials. One type is called a linear-elastic material. The others are called nonlinear-elastic materials. By definition, the term elastic applies to both the linear- and nonlinear-elastic descriptions. However, in common usage, the term elastic is often assumed to mean linear elastic. To avoid confusion, we will continue to use the terms linear elastic and nonlinear elastic.

2.1.1 *Linear-Elastic Material*

For a linear-elastic material, the displacement gradients are assumed to be small. The constitutive model for a linear-elastic material is usually written in terms of T_{km} and E_{km}. We assume that the stress at a point in the material at a time t depends only on the strain at that point at time t, and we express the stress T_{km} as a function of E_{ij}:

$$T_{km} = T_{km}(E_{ij}).$$

We expand the component T_{km} as a power series in terms of the components of strain:

$$T_{km} = T^0_{km} + c_{kmij}E_{ij} + d_{kmijrs}E_{ij}E_{rs} + \cdots,$$

where the coefficients T^0_{km}, c_{kmij}, d_{kmijrs}, ... are constants. Because we assume the strains are small, only the terms up to the first order are retained in this expansion. For a linear-elastic material,

$$T_{km} - T_{km}^0 = c_{kmij} E_{ij}. \tag{2.1}$$

The constants c_{kmij} are called the *linear-elastic stiffness coefficients* of the material. They have physical dimensions of stress. Because each index k, m, i, and j can assume three values, there are $3 \times 3 \times 3 \times 3 = 81$ constants. The values of some of these constants are coupled together. This interdependence arises in part because the stress tensor is symmetric: $T_{km} = T_{mk}$. Because the gradients of the displacement are small, the strain tensor is given by Eq. (1.73):

$$E_{kn} = \frac{1}{2}\left(\frac{\partial u_k}{\partial x_n} + \frac{\partial u_n}{\partial x_k}\right),$$

which is also symmetric. To illustrate the interdependence of the elastic constants, consider the following two components of stress evaluated from Eq. (2.1):

$$\begin{aligned} T_{12} - T_{12}^0 &= c_{1211}E_{11} + c_{1222}E_{22} + c_{1233}E_{33} + (c_{1212} + c_{1221})E_{12} \\ &\quad + (c_{1223} + c_{1232})E_{23} + (c_{1213} + c_{1231})E_{13}, \\ T_{21} - T_{21}^0 &= c_{2111}E_{11} + c_{2122}E_{22} + c_{2133}E_{33} + (c_{2112} + c_{2121})E_{12} \\ &\quad + (c_{2123} + c_{2132})E_{23} + (c_{2113} + c_{2131})E_{13}. \end{aligned}$$

Because the strain is symmetric, the stress T_{12} does not depend upon c_{1212} and c_{1221} separately, but only on the combination $c_{1212} + c_{1221}$. Moreover, these relationships must also hold for any combination of strain. Suppose that all strain components are zero except for E_{11}. Because the stress is symmetric, $T_{12} - T_{12}^0 = T_{21} - T_{21}^0$ and $c_{1211} = c_{2111}$. Similar arguments show that $c_{1222} = c_{2122}$, $c_{1233} = c_{2133}$, $c_{1212} + c_{1221} = c_{2112} + c_{2121}$, $c_{1223} + c_{1232} = c_{2123} + c_{2132}$, and $c_{1213} + c_{1231} = c_{2113} + c_{2131}$. Thus the equations for T_{12} and T_{21} contain only 6 independent combinations of 18 constants. Similar arguments show there are at most 36 independent combinations of constants in Eq. (2.1). We shall find that even fewer independent combinations exist for special types of materials. This is our next topic of discussion.

2.1.2 *Isotropic, Linear-Elastic Material*

If we apply uniform tractions to two opposite faces of a block of wood, the deformation we produce will depend on the grain direction of the wood. Clearly, the deformation we obtain if the grain direction is perpendicular to the two faces is different from what we obtain if the grain direction is parallel to the two faces. The deformation depends on the grain direction. A material for which the relationship between stress and deformation is independent of the orientation of the material is called *isotropic*. Wood is an example of an *anisotropic* material.

We can show there are only two independent elastic constants for a linear-elastic material that is isotropic. A complete derivation of the relation for an isotropic material is beyond the scope of this chapter, but we can demonstrate the kinds of arguments that can be used. If we solve Eq. (2.1) for the strain components in terms of the stresses, we can express the results as

$$E_{ij} = h_{ijkm}\left(T_{km} - T_{km}^0\right), \tag{2.2}$$

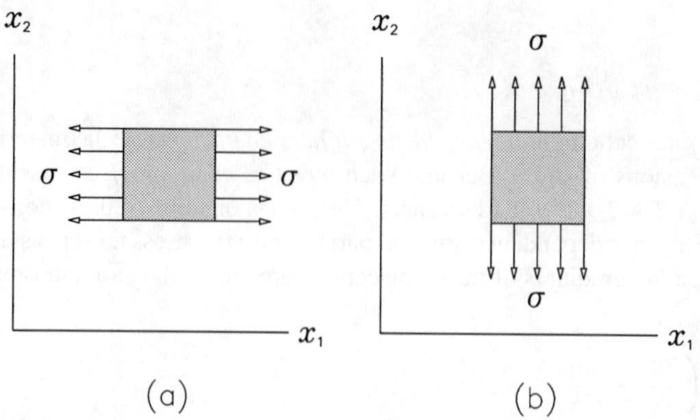

Figure 2.1 (a) The stress $T_{11} = \sigma$. (b) The stress $T_{22} = \sigma$.

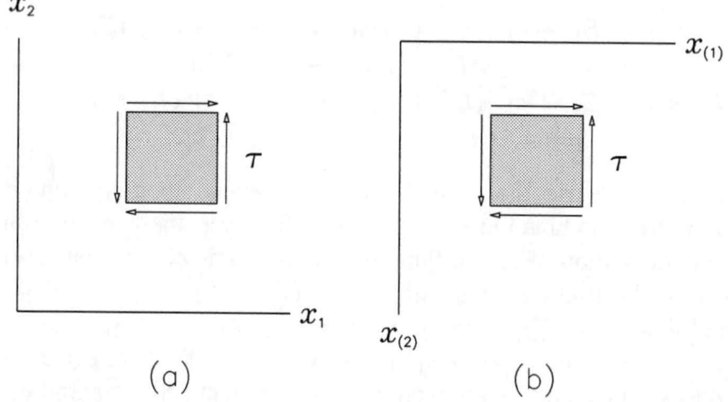

Figure 2.2 (a) A uniform shear stress. (b) Reorienting the coordinate system.

where the h_{ijkm} are called the *linear-elastic compliance coefficients*. This inversion is possible only when each configuration of stress is uniquely related to a particular configuration of strain. We assume this is always true. Now consider a block of material that is free of stress, $T_{km}^0 = 0$. Suppose that we subject this block of material to the uniform normal stress σ shown in Figure 2.1a, so that $T_{11} = \sigma$. The other stress components are zero. From Eq. (2.2), the equation for the strain component E_{11} is

$$E_{11} = h_{1111} T_{11} = h_{1111} \sigma. \tag{2.3}$$

If we apply the stress σ shown in Fig. 2.1b instead, the equation for the strain component E_{22} is

$$E_{22} = h_{2222} T_{22} = h_{2222} \sigma. \tag{2.4}$$

If the material is isotropic, the strain E_{11} in Eq. (2.3) must be equal to the strain E_{22} in Eq. (2.4). Therefore, $h_{1111} = h_{2222}$.

As a second example, suppose we subject the block of material to the uniform shear stress τ shown in Figure 2.2a, so that $T_{12} = T_{21} = \tau$. All the other stress components are zero. From Eq. (2.2), the strain component E_{11} is

$$E_{11} = h_{1112}T_{12} + h_{1121}T_{21} = (h_{1112} + h_{1121})\tau. \tag{2.5}$$

If the material is isotropic, the constants in Eq. (2.2) must have the same values for any orientation of the coordinate system relative to the material. Let us reorient the coordinate system as shown in Fig. 2.2b. With this orientation, the stress $T_{(12)} = T_{(21)} = -\tau$, and the strain component $E_{(11)}$ is

$$E_{(11)} = h_{1112}T_{(12)} + h_{1121}T_{(21)} = -(h_{1112} + h_{1121})\tau. \tag{2.6}$$

Because the x_1 and $x_{(1)}$ directions are the same, the strain E_{11} in Eq. (2.5) must equal the strain $E_{(11)}$ in Eq. (2.6). Therefore, $h_{1112} + h_{1121} = 0$.

By continuing with arguments of this kind, it can be shown that the relations between the stress and strain components for an isotropic, linear-elastic material must be of the form

$$T_{km} - T^0_{km} = \lambda\delta_{km}E_{jj} + 2\mu E_{km}, \tag{2.7}$$

where δ_{km} is the Kronecker delta and λ and μ are called the *Lamé* constants, which have physical dimensions of stress. Notice that summation of indices is implied and $E_{jj} = E_{11} + E_{22} + E_{33}$. We require the Lamé constants to be positive:

$$\lambda > 0, \qquad \mu \geq 0. \tag{2.8}$$

Instead of the Lamé constants, λ and μ, two constants called *Young's modulus* E^0 and *Poisson's ratio* ν are often used to describe the isotropic, linear-elastic material. In terms of these constants, the relations between the strain components and the stress components are

$$E_{km} = -\frac{\nu}{E^0}\delta_{km}\left(T_{jj} - T^0_{jj}\right) + \left(\frac{1+\nu}{E^0}\right)\left(T_{km} - T^0_{km}\right). \tag{2.9}$$

Appendix B contains a table listing various relationships between λ, μ, E^0, and ν. We also list the values of these constants for various materials in Appendix B.

Fluids and Solids

The constant μ is also called the *shear modulus.* When the shear modulus of a material is nonzero, all the components of stress can be nonzero. *Materials for which $\mu \neq 0$ are called solids.* In contrast, when the shear modulus of a material is zero, all of the shear stresses are identically zero, and the stress tensor has the form

$$T_{km} = -p\delta_{km}.$$

Thus all components of the stress tensor are uniquely described by the pressure p. *Materials for which $\mu = 0$ are called fluids*

Uniaxial Deformation of a Linear-Elastic Material

During uniaxial deformation, the deformation gradient is [see Eq. (1.89)]

$$[F_{km}] = \begin{bmatrix} F & 0 & 0 \\ 0 & 1 & 0 \\ 0 & 0 & 1 \end{bmatrix}.$$

The only nonzero component of strain resulting from F_{km} is $E_{11} = E = \frac{1}{2}(F^2 - 1)$. By substituting this strain component into the constitutive relation Eq. (2.7), we obtain

$$\begin{aligned} T_{11} - T_{11}^0 &= T - T^0 = (\lambda + 2\mu)E, \\ T_{22} - T_{22}^0 &= T_{33} - T_{33}^0 = \tilde{T} - \tilde{T}^0 = \lambda E, \end{aligned} \tag{2.10}$$

where $T^0 = T_{11}^0$ and $\tilde{T}^0 = T_{22}^0 = T_{33}^0$ denote the values of stress in the reference configuration. All other components of stress are zero, and *uniaxial deformation of an isotropic, linear-elastic material results in a triaxial-stress configuration.* For fluids, the Lamé constant $\mu = 0$. The stress T and the reference stress T^0 are spherical, and $T = \tilde{T} = -p$ (see Section 1.7.1). There are no shear stresses in a fluid, regardless of the orientation of the coordinate system. For solids, the Lamè constant $\mu \neq 0$. In Section 1.7 we show how triaxial stress results in shear stress.

We can obtain another form of these constitutive relations by writing the axial component of uniaxial deformation as

$$F = 1 + \Delta F.$$

The axial component of strain is

$$E = \frac{1}{2}(F^2 - 1) = \frac{1}{2}(2\Delta F + \Delta F^2).$$

Because the displacement gradient is assumed to be small, we ignore the term ΔF^2. Thus

$$\begin{aligned} T - T^0 &= (\lambda + 2\mu)(F - 1), \\ \tilde{T} - \tilde{T}^0 &= \lambda(F - 1). \end{aligned} \tag{2.11}$$

You can also write these expressions as

$$\begin{aligned} T - T^0 &= (\lambda + 2\mu)\frac{\partial \hat{u}}{\partial X}, \\ \tilde{T} - \tilde{T}^0 &= \lambda \frac{\partial \hat{u}}{\partial X}, \end{aligned} \tag{2.12}$$

and

$$\begin{aligned} T - T^0 &= (\lambda + 2\mu)\frac{\partial u}{\partial x}, \\ \tilde{T} - \tilde{T}^0 &= \lambda \frac{\partial u}{\partial x}, \end{aligned} \tag{2.13}$$

where Eq. (1.131) has been used and $(\partial u/\partial x)^2 \approx 0$.

2.1.3 *Nonlinear-Elastic Material*

Constructing constitutive models is complicated by the fact that the Cauchy stress T_{km} is described with respect to a free-body diagram of the current configuration of the body, whereas the deformation gradient F_{km} and the strain E_{km} are described by relating the motion to the reference configuration of the body. To illustrate how complications can arise, consider a body with a known configuration of stress. Suppose we subject this body to a pure rotation. Recall that any pure rotation of a body leaves the Lagrangian strain E_{km} unchanged, but the Cauchy stress T_{km} changes [see Eq. (1.115)]. As a result, unless, as in the linear-elastic theory, we restrict the displacement gradients to be small, we cannot write a constitutive model such as

$$T_{km} = \mathcal{G}(E_{ij}),$$

because this relationship does not allow T_{km} to change unless E_{km} changes. Hence this type of relationship contradicts our previous results for the pure rotation of a body. Indeed, we find that this constitutive model is too special. However, if we replace it by a more general constitutive model such as

$$T_{km} = \mathcal{G}(F_{ij}),$$

we shall find that it is possible to construct relationships that do not contradict our previous results. Indeed, construction of constitutive models would be more convenient if stress, deformation gradient, and strain were all defined with respect to either the current configuration or the reference configuration. One solution to this problem is to define a special form of stress that describes the surface tractions with respect to a free-body diagram of the reference configuration. It is called the Piola–Kirchhoff stress†. Another solution is to write the Cauchy stress as

$$T_{km} = \rho F_{mj} \frac{\partial \psi_F}{\partial F_{kj}}, \tag{2.14}$$

where the function ψ_F is of a special form that satisfies

$$\psi_F(F_{kn}) = \psi_E\left(\frac{1}{2}[F_{mi}F_{mj} - \delta_{ij}]\right).$$

Notice that ψ_E can be written as a function of E_{ij} alone. The functions $\psi_F(F_{kn})$ and $\psi_E(E_{ij})$ are called *strain-energy* potential functions. We leave it as an exercise for you to show that the left and right sides of Eq. (2.14) undergo equivalent transformations during pure rotation of a body.

Nonlinear-Elastic Fluid

Recall that ψ_F is a function of the deformation gradient. Suppose we assume that the strain-energy function ψ_F is a function of density alone. We can make this assumption because, as we shall now show, the density can be written as a function of F_{rs}. By combining Eq. (1.79),

$$\frac{dV}{dV_0} = \det[F_{rs}],$$

and Eq. (1.136),

$$\rho_0 dV_0 = \rho dV,$$

we obtain

$$\det[F_{rs}] = \frac{\rho_0}{\rho}. \tag{2.15}$$

† See L. E. Malvern, *Introduction to the Mechanics of a Continuous Medium* (Prentice Hall, 1969).

By expanding the derivative of ψ_F in Eq. (2.14), we find that

$$T_{km} = -\frac{\rho^3}{\rho_0} F_{mj} \frac{\partial\{\det[F_{rs}]\}}{\partial F_{kj}} \frac{d\psi_F}{d\rho}.$$

In Exercise 1.31 we evaluated the partial derivative of the determinant $\det[F_{rs}]$. When we substitute that result along with Eqs. (1.61) and (1.60) into this expression for the stress, we obtain

$$T_{km} = -\rho^2 \frac{d\psi_F}{d\rho} \delta_{km}. \tag{2.16}$$

This is a spherical stress configuration. We conclude that if the strain energy has a restricted dependence on the density alone, shear stresses are absent and the material is a fluid. Whereas, if the strain energy has a general dependence on all of the components of F_{rs}, shear stresses exist, and the material is a solid.

Small-Strain Response of a Nonlinear-Elastic Material

When the displacement gradients are small, we can approximate the strain as

$$E_{rs} = \frac{1}{2}(F_{rs} + F_{sr}) - \delta_{rs}. \tag{2.17}$$

Because $F_{mj} \to \delta_{mj}$, Eq. (2.14) becomes

$$T_{km} = \rho_0 \frac{\partial \psi_F}{\partial F_{km}}.$$

We can rewrite this result as

$$T_{km} = \rho_0 \frac{\partial \psi_E}{\partial E_{rs}} \frac{\partial E_{rs}}{\partial F_{km}}.$$

By substituting Eq. (2.17) and recalling that the stress tensor is symmetric, we obtain

$$T_{km} = \rho_0 \frac{\partial \psi_E}{\partial E_{km}}.$$

When we expand this result as a power series in strain, we obtain a relationship for the small-strain response of the nonlinear-elastic material that is similar to the one obtained for the linear-elastic material,

$$T_{km} = T_{km}^0 + c_{kmij} E_{ij} + d_{kmijrs} E_{ij} E_{rs} + \cdots.$$

The elastic constants are related to the potential function. For example,

$$c_{kmij} = \frac{\partial T_{km}}{\partial E_{ij}} = \rho_0 \frac{\partial^2 \psi_E}{\partial E_{ij} \partial E_{km}},$$

where the derivatives are evaluated at $E_{ij} = 0$. There are 81 separate elements of c_{kmij}. However, the order of differentiation of the potential function can be reversed without changing the value of the result. Thus $c_{kmij} = c_{ijkm}$. Now if we assume the strain is small, then the terms containing products of the strain are negligible. It follows that

$$T_{km} = T_{km}^0 + c_{kmij} E_{ij}. \tag{2.18}$$

Because the stress T_{km} and the strain E_{ij} are symmetric, we can use the same arguments employed for the linear-elastic material to show that only 21 independent combinations of c_{ijkm} appear in Eq. (2.18). Furthermore, if the material is isotropic, we again find that only two independent combinations of c_{kmij} appear in Eq. (2.18). These results are identical to the isotropic, linear-elastic material, but only in the small-strain approximation.

Uniaxial Deformation of a Nonlinear-Elastic Material

By Eq. (2.14), the nonzero components of the Cauchy stress are

$$\begin{aligned} T = T_{11} &= \rho F_{11}\frac{\partial \psi_F}{\partial F_{11}}, \\ \tilde{T} = T_{22} = T_{33} &= \rho\frac{\partial \psi_F}{\partial F_{22}} = \rho\frac{\partial \psi_F}{\partial F_{33}}, \end{aligned} \tag{2.19}$$

where the partial derivatives are evaluated for

$$[F_{km}] = \begin{bmatrix} F & 0 & 0 \\ 0 & 1 & 0 \\ 0 & 0 & 1 \end{bmatrix}.$$

Therefore, during uniaxial deformation, we can write the axial and lateral components of the Cauchy stress as functions of the axial deformation F:

$$T = T(F), \qquad \tilde{T} = \tilde{T}(F). \tag{2.20}$$

Suppose we expand the Cauchy stress in a Taylor's series about some arbitrary value of the deformation gradient $F = F_0$ to obtain

$$\begin{aligned} T - T^0 &= \frac{dT}{dF}(F - F_0) + \frac{1}{2}\frac{d^2T}{dF^2}(F - F_0)^2 + \cdots, \\ \tilde{T} - \tilde{T}^0 &= \frac{d\tilde{T}}{dF}(F - F_0) + \frac{1}{2}\frac{d^2\tilde{T}}{dF^2}(F - F_0)^2 + \cdots, \end{aligned}$$

where the derivatives are evaluated at $F = F_0$. The derivatives dT/dF and $d\tilde{T}/dF$ are called the *elastic stiffnesses* at $F = F_0$. We assume the nonlinear-elastic material becomes a linear-elastic material when $F \approx 1$. This implies that

$$\left(\frac{dT}{dF}\right)_{F=1} = \lambda + 2\mu, \qquad \left(\frac{d\tilde{T}}{dF}\right)_{F=1} = \lambda. \tag{2.21}$$

Thus, small changes of uniaxial strain about the reference configuration produce stresses that are determined by the Lamé constants. At other values of deformation gradient, the elastic stiffnesses of the nonlinear-elastic model differ from the Lamé constants. However, at all values of F, we require that the elastic stiffnesses must always be positive:

$$\frac{dT}{dF} > 0, \qquad \frac{d\tilde{T}}{dF} > 0. \tag{2.22}$$

This means that an increase in stress will always stretch the material and a decrease in stress will always compress the material. Coupled with this requirement, we also impose

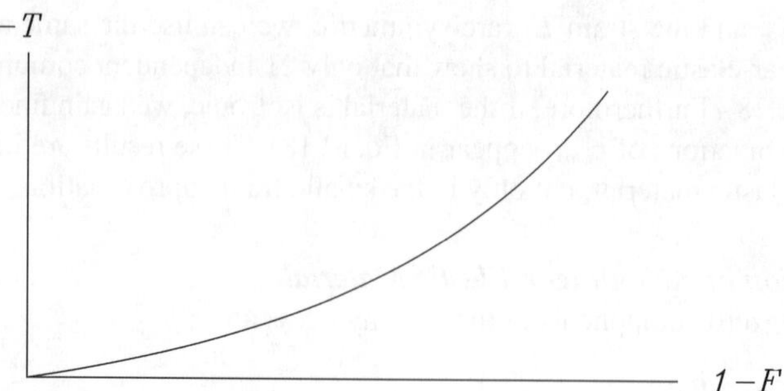

Figure 2.3 The response curve of a nonlinear-elastic material.

the condition

$$\frac{d^2T}{dF^2} \le 0. \tag{2.23}$$

This means that the nonlinear-elastic material becomes stiffer as it is compressed and softer as it is stretched. This is true for most but not all materials, and sometimes we will drop this requirement. These inequalities have important implications for the propagation of nonlinear waves.

The conditions on the first and second derivatives of the function $T = T(F)$ produce a characteristic shape to the curve representing this function (see Figure 2.3). Typically, we are interested in large compressive stresses that result when the deformation gradient F drops below $F = 1$. In such cases, $T \le 0$. We graph the negative of the stress $-T$ as a function of the change of the deformation gradient with respect to the reference configuration [see Eq. (1.134)]:

$$1 - F = \frac{\rho - \rho_0}{\rho} = \frac{V_0 - V}{V_0}.$$

This quantity represents the relative change in density and volume of the material. The constraints on the derivatives of $T = T(F)$ result in a curve that has a positive slope and a concave side that faces upward. Compressive loading of the nonlinear-elastic material causes the stress to move up this curve. When the load is removed, the stress returns *down the same path*. Thus T is uniquely determined by F.

Bulk Response of a Nonlinear-Elastic Material

From our discussion of triaxial stress (see Section 1.7.2) we recall that the normal stress in the x_1 direction is

$$T_{11} = T = \sigma + T',$$

where the mean normal stress is [see Eq. (1.123)]

$$\sigma = \frac{1}{3}(T + 2\tilde{T}) \tag{2.24}$$

and the deviatoric stress is [see Eq. (1.124)]

$$T' = \frac{2}{3}(T - \tilde{T}).$$

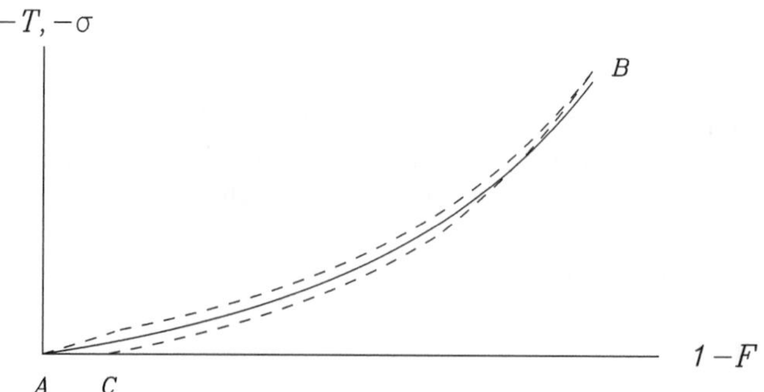

Figure 2.4 The mean normal stress σ of a typical nonlinear-elastic solid (—) compared to the uniaxial stress T (- - -).

In gases and liquids under static conditions, the deviatoric stress is zero. In solids, it is not zero. From experiments on solids, we know that as large compressive loads are applied, the magnitude of the mean normal stress $|\sigma|$ in a solid increases without bound while the deviatoric stress T' remains bounded. We illustrate this for a typical solid such as steel (see Figure 2.4) The solid curve represents the mean normal stress σ. It is called the *bulk response* of the material. The dashed curve represents the axial component T of the Cauchy stress. Consider a material that is initially at point A. We apply a compressive load and cause the material to follow the upper dashed curve to point B. Notice that, as we load the material, the mean normal stress shown by the solid curve continually increases. The deviatoric stress, represented by the vertical distance between the solid and the upper dashed curves, increases initially, but then remains constant. We now remove the load and cause the stress to follow the lower dashed curve from point B to point C. Notice that the material does not return to point A. Therefore, what we have illustrated *is not* an elastic material, because the stress T is not a unique function of F. We shall describe these kinds of material in detail in Chapter 4. For the present, during large compression where

$$|T'| \ll |\sigma|,$$

it is reasonable to approximate the dashed curves as being coincident with the solid curve. Thus we shall approximate the uniaxial response of this material by the bulk response, and because the bulk response is a unique function of F, the approximation

$$T \approx \sigma(F) \tag{2.25}$$

represents a nonlinear-elastic material.

This approximation, as determined by Eqs. (2.21) and (2.24), yields

$$\left(\frac{d\sigma}{dF}\right)_{F=1} = \frac{1}{3}[(\lambda + 2\mu) + 2\lambda].$$

Thus

$$\left(\frac{dT}{dF}\right)_{F=1} \approx \left(\frac{d\sigma}{dF}\right)_{F=1} = \lambda + \frac{2}{3}\mu. \tag{2.26}$$

The quantity $\lambda + \frac{2}{3}\mu$ is called the *bulk stiffness*.

2.1.4 *Adiabatic Gas*

We shall discuss the constitutive model for an ideal gas in detail in Chapter 4. Here we examine a special type of material behavior for the ideal gas. It is observed during rapid compression and expansion, where there is insufficient time for the gas to gain or lose heat. The term adiabatic is often used to describe a process in which heat cannot flow. Thus we call this constitutive model the *adiabatic gas*.

The constitutive model for the adiabatic gas is defined to be

$$\frac{p}{p_0} \equiv \left(\frac{\rho}{\rho_0}\right)^{\gamma}. \tag{2.27}$$

The density and pressure of the reference configuration are ρ_0 and p_0. The constant γ is called the *adiabatic gas constant.*

The adiabatic gas is a nonlinear-elastic fluid because it is a special case of Eq. (2.16). The Cauchy stress of the adiabatic gas is spherical,

$$T = \tilde{T} = -p.$$

By employing the deformation gradient $F = \rho_0/\rho = V/V_0$, we obtain

$$T = -p_0 F^{-\gamma}. \tag{2.28}$$

Let us see when this model satisfies the conditions imposed by Eqs. (2.22) and (2.23). The first derivative of the stress is

$$\frac{dT}{dF} = \gamma p_0 F^{-\gamma-1},$$

and the second derivative is

$$\frac{d^2T}{dF^2} = -\gamma(\gamma+1)p_0 F^{-\gamma-2}.$$

We find that Eqs. (2.22) and (2.23) are satisfied when both the pressure p_0 and the gas constant γ are positive. Using Eq. (2.21), and recalling that $T = \tilde{T}$, we find that the Lamé constants for the adiabatic gas are

$$\lambda = \gamma p_0, \qquad \mu = 0. \tag{2.29}$$

2.1.5 *Exercises*

2.1 Show that for an isotropic, linear-elastic material $h_{1111} = h_{3333}$ and $h_{3312} + h_{3321} = 0$.

2.2 Discuss why we should write the potential function ψ_E solely as a function of the scalar invariants I_E, II_E, and III_E.

2.3 The Lamé constants of an isotropic, linear-elastic material are $\lambda = 61.3$ GPa and $\mu = 24.9$ GPa. The components of the strain tensor at a material point $\mathbf{X}$ are

$$[E_{km}] = \begin{bmatrix} 0.002 & -0.001 & 0 \\ -0.001 & 0.001 & 0 \\ 0 & 0 & 0.004 \end{bmatrix}.$$

Determine the components of the stress tensor T_{km} at $\mathbf{X}$.

2.4 The potential function ψ_E is a scalar, but its derivative $\partial\psi_E/\partial E_{ij}$ is not. Show that this derivative transforms as a tensor during a rotation of coordinate systems.

2.5 We can write the constitutive model for an isotropic, linear-elastic material either as

$$T_{km} - T^0_{km} = \lambda\delta_{km}E_{jj} + 2\mu E_{km}$$

or as

$$E_{km} = -\frac{\nu}{E^0}\delta_{km}\left(T_{jj} - T^0_{jj}\right) + \left(\frac{1+\nu}{E^0}\right)\left(T_{km} - T^0_{km}\right).$$

(a) Solve the first equation for E_{km}, and obtain expressions for λ and μ in terms of E^0 and ν.
(b) Solve the second equation for T_{km}, and obtain expressions for E^0 and ν in terms of λ and μ.

2.6 Consider the expression for the Cauchy stress,

$$T_{km} = \rho F_{mj}\frac{\partial\psi_F}{\partial F_{kj}}.$$

Show that the left and right sides of this expression undergo equivalent transformations during pure rotation of a body.

2.7 Verify that the strain-energy potential ψ_F for an adiabatic gas is

$$\frac{d\psi_F}{d\rho} = p_0\rho^{\gamma-2}/\rho_0^{\gamma}.$$

2.2 One-Dimensional Nonlinear-Elastic Equations

The conservation of mass and the balance of linear momentum are the laws of motion that apply to all materials. However, the nonlinear-elastic constitutive model only applies to one special type of material discussed in this book. Together these relationships form a system of equations called the *nonlinear-elastic equations.* We can write the nonlinear-elastic equations in either the material or spatial forms. Here we summarize the one-dimensional forms of these equations.

2.2.1 *Material Description*

When we omit the term containing the body force, the material description of the conservation of mass, Eq. (1.137), and balance of linear momentum, Eq. (1.144), become

$$\frac{\rho_0}{\rho} = F, \tag{2.30}$$

$$\rho_0\frac{\partial\hat{v}}{\partial t} = \frac{\partial\hat{T}}{\partial X}. \tag{2.31}$$

For a nonlinear-elastic material, the relationship between the stress T and the deformation gradient F is given by Eq. (2.20),

$$T = T(F), \tag{2.32}$$

where from Eq. (1.132)

$$\frac{\partial\hat{F}}{\partial t} = \frac{\partial\hat{v}}{\partial X}. \tag{2.33}$$

These four equations contain four unknown dependent variables, $\hat{\rho}(X,t)$, $\hat{v}(X,t)$, $\hat{F}(X,t)$, and $\hat{T}(X,t)$. Because products of the dependent variables appear, they form a set of nonlinear, partial differential equations. Because of Eq. (2.32), only nonlinear-elastic materials are described by this system. Nonlinear-elastic materials include the linear-elastic material and the adiabatic gas as special cases.

2.2.2 *Spatial Description*

When we omit the term containing the body force, the spatial forms of the conservation of mass, Eq. (1.142), and balance of linear momentum, Eq. (1.145), become

$$\frac{\partial \rho}{\partial t} = -\frac{\partial(\rho v)}{\partial x}, \tag{2.34}$$

$$\frac{\partial(\rho v)}{\partial t} = \frac{\partial}{\partial x}(T - \rho v^2). \tag{2.35}$$

For a nonlinear-elastic material, the relationship between the stress T and the deformation gradient F is given by Eq. (2.20),

$$T = T(F), \tag{2.36}$$

where from Eq. (1.137)

$$\frac{\rho_0}{\rho} = F. \tag{2.37}$$

These four equations contain four unknown dependent variables, $\rho(x,t)$, $v(x,t)$, $F(x,t)$, and $T(x,t)$. Because products of the dependent variables appear, they form a set of nonlinear, partial differential equations. Because of Eq. (2.36), only nonlinear-elastic materials are described by this system.

We have shown that Eqs. (2.30)–(2.33) and Eqs. (2.34)–(2.37) are equivalent statements. We can derive the material equations from the spatial equations, and conversely we can derive the spatial equations from the material equations. Notice that products of the dependent variables appear in both formulations of the nonlinear-elastic equations. In future sections, we shall examine solutions to these equations that are called *nonlinear waves.* However, in the next section, we shall start with a simplified system of equations in which products of the dependent variables do not appear. These are *linear equations*, and the solutions to these equations are called *linear waves*.

We notice that the lateral component of stress $\tilde{T}$ does not appear in the nonlinear-elastic equations. Thus, for nonlinear-elastic materials, the lateral component of stress does not influence the wave solutions. We shall find that this conclusion does not hold for other constitutive models. In subsequent chapters we study nonlinear waves that are strongly influenced by the lateral stress.

2.3 Wave Equations

In the previous section we summarized the nonlinear-elastic equations. Solutions to these equations are called nonlinear waves. As noted in Chapter 0, nonlinear waves exist in a rich variety of forms that exhibit complex interactions with each other and with the physical boundaries of the body. It is our intention to study these phenomena in detail. However, our introduction to the theory of waves is best served by first studying a simplified system of equations for the linear-elastic material. Solutions to these equations are called linear waves. Although linear waves are far less complex than nonlinear waves, they still exhibit many important features of wave propagation.

You will recall that the nonlinear-elastic equations are composed of the equation of conservation of mass, the equation of balance of linear momentum, and the constitutive model for a nonlinear-elastic material. We now show that for a linear-elastic material, these equations can be combined to yield a linear, second-order, partial differential equation called a wave equation. We derive the wave equation from the material form of these equations. The derivation of the wave equation from the spatial form of the nonlinear-elastic equations is left as an exercise. Following the derivation of the wave equation, we present a general solution for this equation.

2.3.1 *Material Description*

Consider a homogeneous linear-elastic body in which the stress is given by Eq. (2.12):

$$T - T^0 = (\lambda + 2\mu)\frac{\partial \hat{u}}{\partial X}.$$

When we substitute this expression into the material description of the equation of balance of linear momentum, we obtain

$$\rho_0 \frac{\partial^2 \hat{u}}{\partial t^2} = (\lambda + 2\mu)\frac{\partial^2 \hat{u}}{\partial X^2},$$

where λ and μ are constants. Let us define the constant c_0,

$$c_0 = \left(\frac{\lambda + 2\mu}{\rho_0}\right)^{1/2} > 0, \tag{2.38}$$

so that the constitutive relation for the stress becomes

$$T - T^0 = \rho_0 c_0^2 \frac{\partial \hat{u}}{\partial X}$$

and the equation of balance of linear momentum becomes

$$\frac{\partial^2 \hat{u}}{\partial t^2} = c_0^2 \frac{\partial^2 \hat{u}}{\partial X^2}.$$

This relationship is called a *wave equation*, and c_0 is called the *velocity of sound*. These results are expressed in material coordinates. Because the displacement gradients are small, we find that

$$T - T^0 = \rho_0 c_0^2 \frac{\partial u}{\partial x} \tag{2.39}$$

and

$$\frac{\partial^2 u}{\partial t^2} = c_0^2 \frac{\partial^2 u}{\partial x^2}. \tag{2.40}$$

The dependent variable in Eq. (2.40) is the displacement u. By differentiating this equation with respect to time and position, we can easily demonstrate that wave equations can be written for other dependent variables including the velocity v,

$$\frac{\partial^2 v}{\partial t^2} = c_0^2 \frac{\partial^2 v}{\partial x^2},$$

and the stress T,

$$\frac{\partial^2 T}{\partial t^2} = c_0^2 \frac{\partial^2 T}{\partial x^2}.$$

2.3.2 *The x–t Diagram*

The wave equation for the displacement,

$$\frac{\partial^2 u}{\partial t^2} = c_0^2 \frac{\partial^2 u}{\partial x^2}, \tag{2.41}$$

is a linear partial differential equation that describes the propagation of linear-elastic waves. These waves propagate at the velocity of sound c_0. Before we discuss a method for solving the wave equation, let us study a graphical representation of a propagating wave. Notice that the displacement u in the wave equation is the dependent variable that is expressed as a single valued function of the two independent variables time t and position x,

$$u = u(x, t).$$

Because we have assumed that u is small, we ignore the distinction between the material coordinate X and the spatial coordinate x. The function $u(x, t)$ is a surface in the three-dimensional space defined by the coordinates u, x, and t. An example of a surface that represents a wave propagating in the positive-x direction is shown in Figure 2.5. The surface

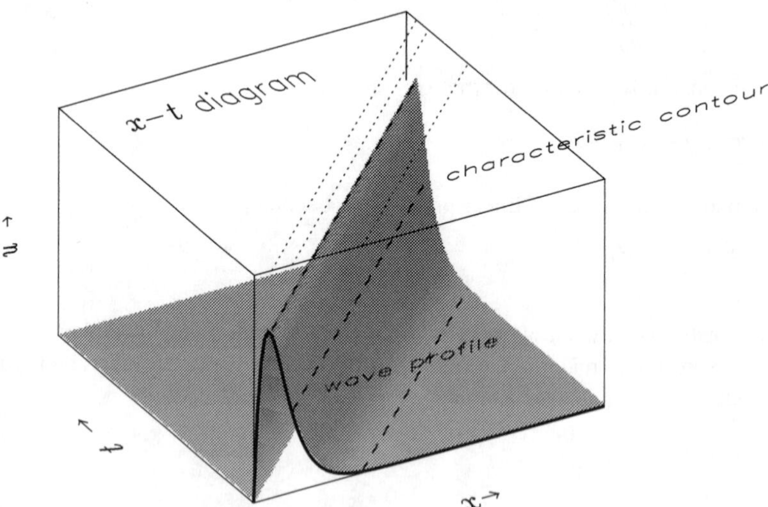

Figure 2.5 A propagating wave represented in the three-dimensional space defined by u, x, and t. The dotted lines in the x–t diagram of this surface are projections of the dashed contours for constant values of u.

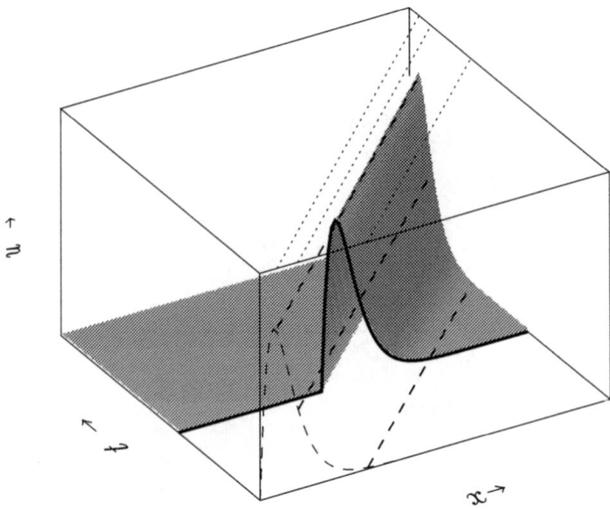

Figure 2.6 A slice of the three-dimensional space showing a wave propagating in the positive-x direction.

is enclosed in a transparent box. The front vertical face of this box is located at $t = 0$. We use a bold curve to delineate the intersection of the surface $u(x, t)$ with the vertical plane $t = 0$. The equation of this bold curve is $u = u(x, 0)$. This is called the *profile* of the wave at the time $t = 0$. Suppose we slice the surface along another vertical plane that is parallel to $t = 0$ (see Figure 2.6). This exposes another profile of the wave at a larger value of t. Notice that this profile looks like the first one except that it has shifted towards larger values of x. By continued slicing of the surface along increasing values of t, we expose new profiles that continue to shift towards larger values of x. We see that this surface represents a wave of constant shape that propagates in the positive-x direction at a constant velocity.

We have drawn dashed lines on the surface $u(x, t)$. Along each of these lines u is constant. We project the dashed lines to the top horizontal surface of the transparent box where they are shown as the dotted lines. The dotted lines are called *characteristic contours*. They are also called *characteristics*. We show three distinct contours. The characteristic contours form a two-dimensional image of the surface $u(x, t)$, which is called an *x–t diagram*. Because this wave propagates with constant shape and velocity, the characteristic contours in this x–t diagram are parallel and straight.

Wave Profiles and Wave Histories

You will recall that wave equations can be written for the displacement u, velocity v, and stress T. We can represent any of these variables with an x–t diagram. Consider another wave. In Figure 2.7 we show the x–t diagram for the velocity v of the material resulting from this wave. Each contour represents a distinct value of velocity. Remember that this is the velocity of the material and not the velocity of the wave. As we shall see, the velocity of the wave is actually the inverse of the slope of the contours. We could include the velocity v of the material in diagram labels alongside each contour, but in order to keep the diagrams simple, we will not do this. However, we can display this information in other ways. Let us see how.

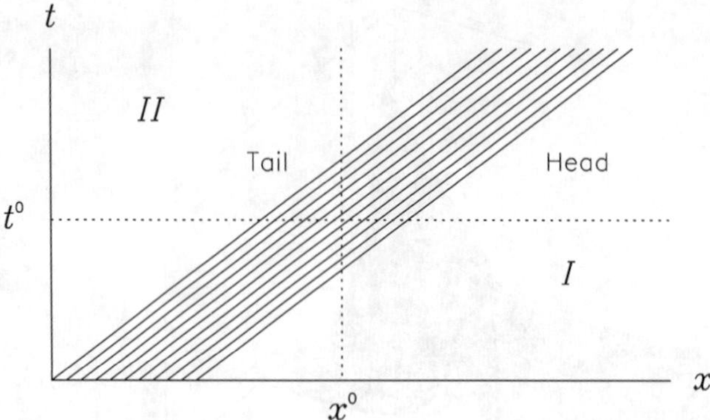

Figure 2.7 The x–t diagram of the velocity.

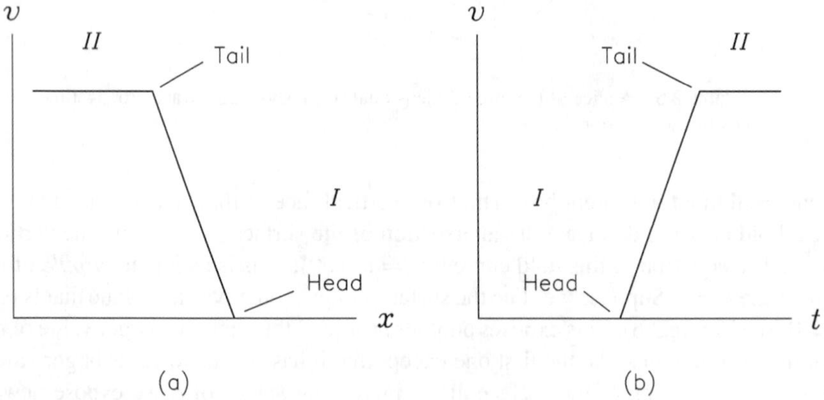

Figure 2.8 (a) A wave profile at $t = t^0$. (b) A wave history at $x = x^0$.

First we notice that the contours are concentrated in a band. The band separates two areas in the diagram where there are no contours. These are labeled regions *I* and *II*. In these regions, the velocity of the material is constant. Thus the wave is contained within the band of contours. Because this wave propagates in the positive-x direction, we say that region *I* is in front of the wave and region *II* is behind the wave. The contour adjacent to region *I* is called the *head contour* of the wave, and the contour adjacent to region *II* is called the *tail contour* of the wave. We have drawn a horizontal dotted line on the x–t diagram at the time $t = t^0$. The profile of this wave at the time $t = t^0$ is shown in Figure 2.8a. This profile shows that the velocity in region *II* is greater than that in region *I*. Between the head and tail contours of the wave, the velocity profile forms a smooth ramp. This means that the increment of velocity between each of the contours is uniform. We might also have drawn a wave profile where the velocity in region *II* is less than that in region *I*. Both profiles would match the same x–t diagram.

Once the profile at $t = t^0$ is known, the value of velocity v on each contour is known, and we can use the x–t diagram to construct a profile at any other time. Alternatively, we can also determine the velocity v as the wave propagates past the position $x = x^0$ (see Figure 2.8b).

This graph is called the *history* of the wave at the position $x = x^0$. For this example, the history of the wave appears as the mirror image of the profile of the wave. As our study of waves progresses, we shall see that this symmetry between the profile and the history of a wave is not always true for linear waves, and it is almost never true for nonlinear waves. This fact is important because most experimental data of wave phenomena are collected and presented as a history. We cannot assume that the shape of the wave history is simply the mirror image of the wave profile. Often they are dramatically different.

2.3.3 *D'Alembert Solution*

Waves can propagate in both the positive and negative directions. Thus the characteristic contours of these waves can have either positive or negative slopes. In general, complex solutions are composed of waves that propagate in both directions with characteristic contours that cross. In this section we use these concepts to derive a general solution to the wave equation.

Consider the x–t diagram in Figure 2.9. We show two new coordinates ζ and ξ that we can use to describe the displacement:

$$u = u(x, t) = \breve{u}(\zeta, \xi).$$

We use a breve ($\breve{}$) to indicate when u is written as a function of ξ and ζ. The relationships between the two coordinate systems are

$$\xi = t - \frac{x}{c_0}, \qquad \zeta = t + \frac{x}{c_0}. \tag{2.42}$$

We shall show that these new coordinates are actually the characteristic contours shown in Figures 2.5–2.7. We shall also show that these coordinates provide a convenient method for obtaining a general solution to the wave equation.

Along a ξ contour, $d\xi = 0$. So we can differentiate Eqs. (2.42) to obtain

$$\frac{dx}{dt} = c_0 \quad \text{when } d\xi = 0.$$

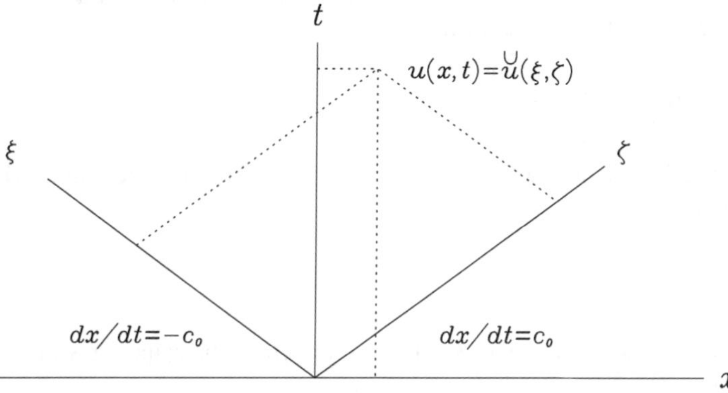

Figure 2.9 The x–t diagram with the ζ–ξ coordinate system.

Because the ζ coordinate corresponds to the contour $\xi = 0$, we find that the slope of this coordinate is c_0. Similarly, along a ζ contour, $d\zeta = 0$, and

$$\frac{dx}{dt} = -c_0 \quad \text{when } d\zeta = 0.$$

Therefore, the slope of the ξ coordinate is the negative of the slope of the ζ coordinate. Different materials have different values of the velocity of sound c_0. For materials where the velocity of sound c_0 is large, both ζ and ξ approach horizontal lines, and for materials where c_0 is small, both ζ and ξ approach vertical lines.

We shall next demonstrate the advantages of the new coordinates ζ and ξ by applying them to the wave equation. By using the chain rule, we write the partial derivative of u with respect to x as

$$\frac{\partial u}{\partial x} = \frac{\partial \xi}{\partial x}\frac{\partial \breve{u}}{\partial \xi} + \frac{\partial \zeta}{\partial x}\frac{\partial \breve{u}}{\partial \zeta} = -\frac{1}{c_0}\frac{\partial \breve{u}}{\partial \xi} + \frac{1}{c_0}\frac{\partial \breve{u}}{\partial \zeta}. \tag{2.43}$$

Then by applying the chain rule again, we determine the second partial derivative of u with respect to x in terms of derivatives with respect to ξ and ζ:

$$\begin{aligned}
\frac{\partial^2 u}{\partial x^2} &= \frac{\partial}{\partial x}\left(-\frac{1}{c_0}\frac{\partial \breve{u}}{\partial \xi} + \frac{1}{c_0}\frac{\partial \breve{u}}{\partial \zeta}\right) \\
&= \frac{\partial \xi}{\partial x}\frac{\partial}{\partial \xi}\left(-\frac{1}{c_0}\frac{\partial \breve{u}}{\partial \xi} + \frac{1}{c_0}\frac{\partial \breve{u}}{\partial \zeta}\right) + \frac{\partial \zeta}{\partial x}\frac{\partial}{\partial \zeta}\left(-\frac{1}{c_0}\frac{\partial \breve{u}}{\partial \xi} + \frac{1}{c_0}\frac{\partial \breve{u}}{\partial \zeta}\right) \\
&= \frac{1}{c_0^2}\left(\frac{\partial^2 \breve{u}}{\partial \xi^2} - 2\frac{\partial^2 \breve{u}}{\partial \xi \partial \zeta} + \frac{\partial^2 \breve{u}}{\partial \zeta^2}\right).
\end{aligned}$$

The second partial derivative of u with respect to t in terms of derivatives with respect to ξ and ζ is obtained in the same way. The result is

$$\frac{\partial^2 u}{\partial t^2} = \frac{\partial^2 \breve{u}}{\partial \xi^2} + 2\frac{\partial^2 \breve{u}}{\partial \xi \partial \zeta} + \frac{\partial^2 \breve{u}}{\partial \zeta^2}.$$

After substituting these expressions for the second partial derivatives of u into Eq. (2.41), we obtain the wave equation in the form

$$\frac{\partial^2 \breve{u}}{\partial \xi \partial \zeta} = 0. \tag{2.44}$$

We can integrate this equation twice to obtain its general solution. Integrating with respect to ζ, we obtain

$$\frac{\partial \breve{u}}{\partial \xi} = h(\xi),$$

where $h(\xi)$ is an arbitrary function of ξ. Then we integrate this equation with respect to ξ, obtaining the solution

$$u = \int h(\xi)\, d\xi + g(\zeta),$$

where $g(\zeta)$ is an arbitrary function of ζ. By defining

$$f(\xi) = \int h(\xi)\, d\xi,$$

we write the solution in the form

$$u = f(\xi) + g(\zeta), \tag{2.45}$$

where $f(\xi)$ and $g(\zeta)$ are arbitrary twice-differentiable functions. This is called the *d'Alembert solution* of the wave equation.

The d'Alembert solution represents the displacement u as the sum of two functions $f(\xi)$ and $g(\zeta)$. We shall demonstrate that $f(\xi)$ represents a wave that propagates to the right in the positive-x direction and $g(\zeta)$ represents a wave that propagates to the left in the negative-x direction. To illustrate this, consider a solution where $g(\zeta) = 0$, so that $u = f(\xi)$. If ξ is constant, then u is constant. This means the characteristic contours, which represent constant values of u in the x–t diagram, are parallel to the ζ coordinate. *Therefore, the line ξ = constant is a characteristic contour.* Because these characteristic contours are straight and parallel, $f(\xi)$ represents a wave of constant shape propagating in the positive-x direction.

A different argument can also be used to illustrate that $f(\xi)$ represents a wave propagating in the positive-x direction. We substitute Eq. (2.42) into the solution for the displacement,

$$u = f(\xi) = f(t - x/c_0).$$

At the time $t = t_0$, the profile of this wave is

$$u = f(t_0 - x/c_0).$$

Suppose that the profile at $t = t_0$ is represented by the dashed curve in Figure 2.10. The point x_0 on this profile has a value of displacement of

$$u = f(t_0 - x_0/c_0).$$

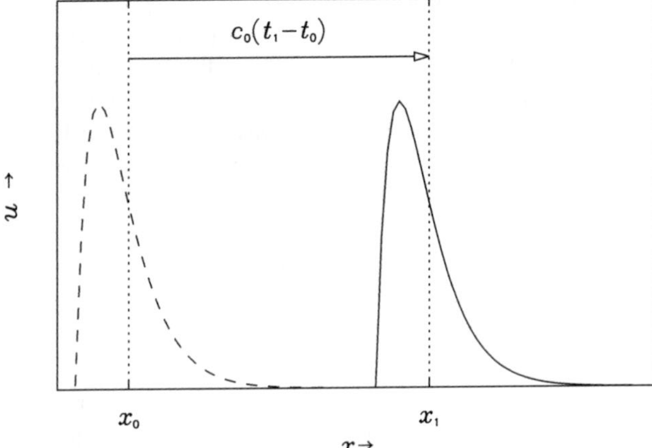

Figure 2.10 Two profiles of a wave traveling in the positive-x direction. The dashed curve is the profile at $t = t_0$, and the solid curve is the profile at $t = t_1$.

At time $t = t_1$, this value of displacement moves to x_1 where

$$f(t_1 - x_1/c_0) = f(t_0 - x_0/c_0).$$

Thus

$$t_1 - x_1/c_0 = t_0 - x_0/c_0$$

and

$$x_1 = x_0 + c_0(t_1 - t_0).$$

During the interval of time $t_1 - t_0$, the profile of the wave translates a distance $c_0(t_1 - t_0)$ to the right. By similar arguments we can show that $g(\zeta)$ represents a wave propagating in the negative-x direction.

2.3.4 *Superposition of Linear Waves*

The d'Alembert solution of the wave equation is [see Eq. (2.45)]

$$u = f(\xi) + g(\zeta).$$

The function $f(\xi)$ represents a wave propagating in the positive-x direction, and the function $g(\zeta)$ represents a wave propagating in the negative-x direction. We have assumed the first and second derivatives of f and g are defined for all values of ξ and ζ. For reasons that are beyond the scope of this text, and in the context of linear waves only, we drop this requirement and allow jump discontinuities in the values of these functions.

The wave equation for u is linear. By definition, this means that products of u do not appear in the wave equation. It is a property of a linear differential equation that the sum of two solutions of the equation is also a solution. Thus, because $f(\xi)$ and $g(\zeta)$ are solutions of the wave equation, then $u = f(\xi) + g(\zeta)$ is also a solution of the wave equation. These observations have far-reaching implications for the analysis of the propagation of linear waves. We can illustrate this point with a simple example. Consider an unbounded body. At $t = 0$, two waves are present (see the top graph in Figure 2.11). The wave to the left has a cosine profile; the wave to the right has a triangular profile. These waves represent graphs of the functions $f(t - x/c_0)$ and $g(t + x/c_0)$ of a particular d'Alembert solution. Suppose for a moment we let $g = 0$, then f represents both the triangular wave and the cosine wave. This means that both waves propagate in the positive-x direction. Similarly, if we let $f = 0$, then g represents both waves, which now propagate in the negative-x direction. However, instead of letting either f or g be zero, let us require that f describes the cosine wave on the left and that g describes the triangular wave on the right. This means the two waves approach each other. Now we can see what happens as time advances and the two waves make contact. The wave profiles at four different times show how the waves pass through each other. At each time, the profile is the sum of the individual profiles of the cosine and triangular waves. The solution at each time is said to be a *linear superposition* of the individual wave solutions.

Notice that the final shapes of the two waves at $t = 1.5$ are unaltered by their contact with each other. This illustrates a common everyday occurrence. We all are familiar with the linear superposition of human speech. When you have a conversion with a friend, the sound of that person's voice is not altered by the sound of your voice. We all take it for granted that our voice will be unaltered in a crowd, and we all expect the sound of a crowd of people

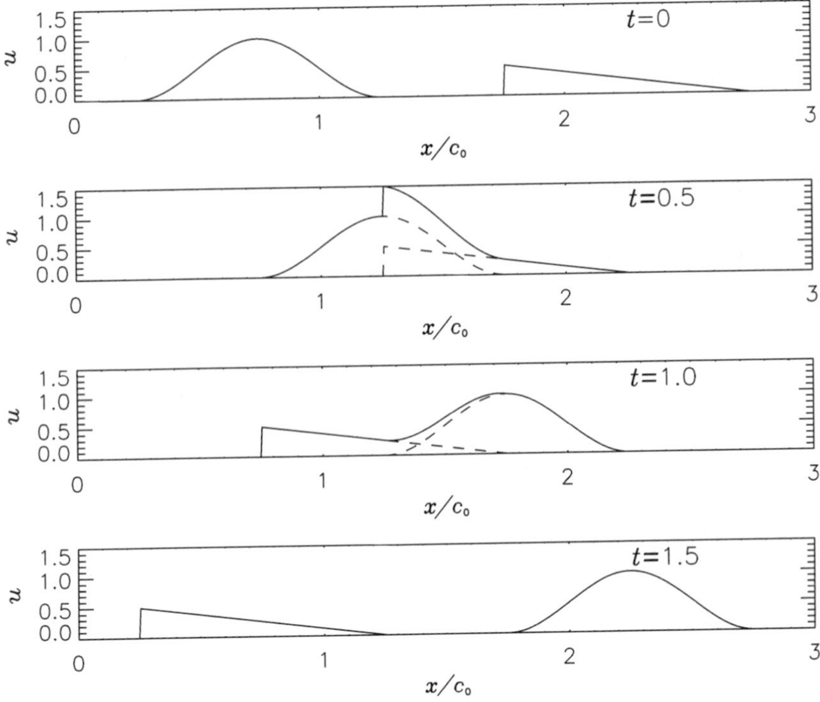

Figure 2.11 The superposition of two linear waves.

to be the sum of their individual voices. Many methods of linear analysis use the property of linear superposition to great advantage. The classical approaches include the Fourier series as well as the Fourier integral transform and the Laplace integral transform†. These topics of linear analysis are outside the scope of this book. Here we are primarily concerned with nonlinear waves, and when two nonlinear waves make contact, their individual shapes change and linear superposition cannot be used.

2.3.5 *Initial- and Boundary-Value Problems*

In this section we apply the d'Alembert solution to several problems for wave propagation in a linear-elastic material. These problems are important because not only do they provide a fundamental understanding of linear waves, they also serve as a prelude to the study of nonlinear waves.

Initial-Value Problems

Suppose that at $t = 0$ we prescribe the components of the displacement and velocity of an unbounded linear-elastic material to be

$$u(x, 0) = p(x), \qquad \frac{\partial u}{\partial t}(x, 0) = q(x), \tag{2.46}$$

† For example, see A. Bedford and D. S. Drumheller, *Introduction to Elastic Wave Propagation* (Wiley, 1994) and P. M. Morse and K. U. Ingard, *Theoretical Acoustics* (McGraw-Hill, 1968).

where $p(x)$ and $q(x)$ are known functions. These *initial conditions* cause motion of the material, which is described by the wave equation for u. To obtain the solution, we write the displacement $u(x, t)$ in terms of the d'Alembert solution

$$u(x,t) = f(\xi) + g(\zeta) = f(t - x/c_0) + g(t + x/c_0). \tag{2.47}$$

Our objective is to determine functions $f(\xi)$ and $g(\zeta)$ that satisfy the initial conditions given by $p(x)$ and $q(x)$.

We obtain an expression for the velocity from Eq. (2.47) by using the chain rule,

$$\frac{\partial u}{\partial t}(x,t) = \frac{df(\xi)}{d\xi}\frac{\partial \xi}{\partial t} + \frac{dg(\zeta)}{d\zeta}\frac{\partial \zeta}{\partial t} = \frac{df(\xi)}{d\xi} + \frac{dg(\zeta)}{d\zeta}.$$

By setting $t = 0$ in this equation and in Eq. (2.47), we can write the initial conditions, Eq. (2.46), as

$$f(\alpha) + g(\beta) = p(x), \tag{2.48}$$

$$\frac{df(\alpha)}{d\alpha} + \frac{dg(\beta)}{d\beta} = q(x), \tag{2.49}$$

where $\alpha \equiv -x/c_0$ and $\beta \equiv x/c_0$. Multiplying the second equation by dx/c_0 and integrating it from 0 to x, we obtain

$$\int_0^\alpha \frac{df}{d\alpha}(-d\alpha) + \int_0^\beta \frac{dg}{d\beta}\,d\beta = \frac{1}{c_0}\int_0^x q(\bar{x})\,d\bar{x},$$

or

$$-f(\alpha) + g(\beta) = \frac{1}{c_0}\int_0^x q(\bar{x})\,d\bar{x} - f(0) + g(0), \tag{2.50}$$

where $\bar{x}$ is an integration variable. Now we solve Eqs. (2.48) and (2.50) for the functions $f(\alpha)$ and $g(\beta)$:

$$\begin{aligned} f(\alpha) &= \frac{1}{2}p(x) - \frac{1}{2c_0}\int_0^x q(\bar{x})\,d\bar{x} + \frac{1}{2}f(0) - \frac{1}{2}g(0) \\ &= \frac{1}{2}p(-\alpha c_0) - \frac{1}{2c_0}\int_0^{-\alpha c_0} q(\bar{x})\,d\bar{x} + \frac{1}{2}f(0) - \frac{1}{2}g(0), \end{aligned}$$

$$\begin{aligned} g(\beta) &= \frac{1}{2}p(x) + \frac{1}{2c_0}\int_0^x q(\bar{x})\,d\bar{x} - \frac{1}{2}f(0) + \frac{1}{2}g(0) \\ &= \frac{1}{2}p(\beta c_0) + \frac{1}{2c_0}\int_0^{\beta c_0} q(\bar{x})\,d\bar{x} - \frac{1}{2}f(0) + \frac{1}{2}g(0). \end{aligned}$$

To obtain the solution for $u(x, t)$ from Eq. (2.47), we must now let $\alpha = t - x/c_0$ and $\beta = t + x/c_0$. The resulting solution of the initial-value problem in terms of the initial conditions is

$$u(x,t) = \frac{1}{2}p(x - c_0 t) + \frac{1}{2}p(x + c_0 t) + \frac{1}{2c_0}\int_{x-c_0 t}^{x+c_0 t} q(\bar{x})\,d\bar{x}. \tag{2.51}$$

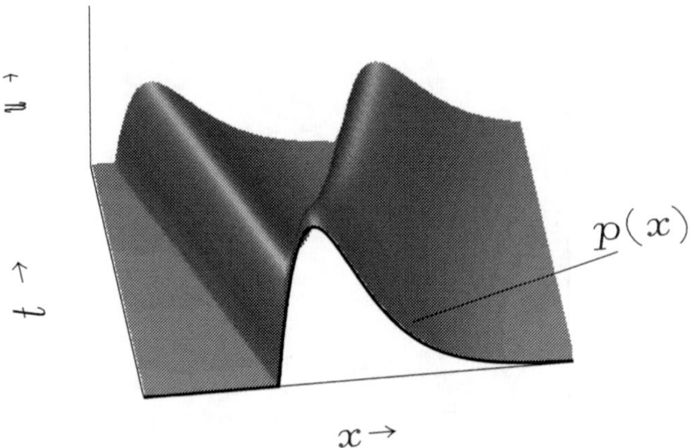

Figure 2.12 The $u(x, t)$ surface for an initial displacement condition $p(x)$, which is represented by the solid curve. Two waves propagate in opposite directions away from the initial disturbance.

Initial-Displacement Condition This solution is easy to interpret when the initial velocity is zero,

$$u(x, t) = \frac{1}{2}p(x - c_0 t) + \frac{1}{2}p(x + c_0 t). \tag{2.52}$$

This solution contains two waves that propagate in opposite directions. Each wave has an amplitude that is half the amplitude of the initial condition $u(x, 0) = p(x)$. We illustrate an example in Figure 2.12.

Initial-Velocity Condition When the initial displacement is zero, the solution is

$$u(x, t) = \frac{1}{2c_0}\int_{x-c_0 t}^{x+c_0 t} q(\bar{x})\, d\bar{x}. \tag{2.53}$$

The interpretation of this expression for the displacement is not obvious, but we can easily interpret the solution for the velocity. From the Leibniz theorem for differentiation of an integral we obtain

$$\frac{\partial}{\partial t}\int_{a(t)}^{b(t)} q(\bar{x})\, d\bar{x} = q(b)\frac{db}{dt} - q(a)\frac{da}{dt},$$

where we have noticed that q is not a function of t. Applying this relationship to the displacement solution we obtain

$$v(x, t) = \frac{\partial u}{\partial t} = \frac{1}{2}q(x + c_0 t) + \frac{1}{2}q(x - c_0 t).$$

For a prescribed initial velocity, this solution has the same interpretation as the solution for a prescribed initial displacement, Eq. (2.52).

Boundary-Value Problems

Displacement of a Boundary Suppose that a half space of linear-elastic material is initially undisturbed; that is, $u(x, t) = 0$ for $t \leq 0$. We subject the boundary at $x = 0$ to the

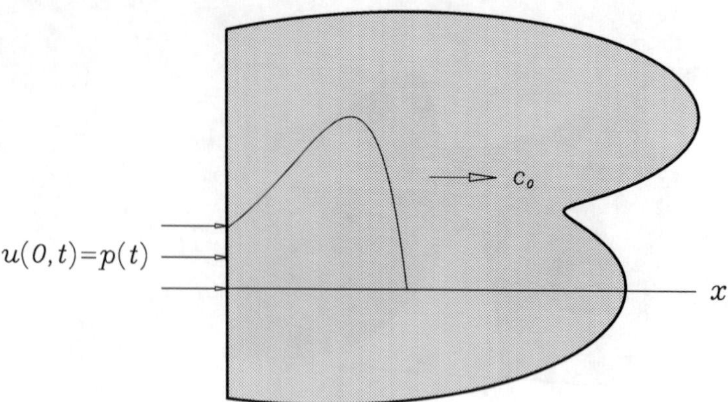

Figure 2.13 A half space with a linear wave.

displacement boundary condition

$$u(0, t) = p(t),$$

where $p(t)$ is a prescribed function of time that vanishes for $t \leq 0$. This is called a *boundary-value problem*.

We can determine the solution by expressing the displacement in terms of the d'Alembert solution. Because the half space occupies $x \geq 0$, we expect the imposed motion of the boundary to give rise to a wave propagating in the positive-x direction but not a wave propagating in the negative-x direction (see Figure 2.13). Therefore, we express the displacement in terms of the part of the d'Alembert solution that represents a wave propagating in the positive-x direction,

$$u(x, t) = f(\xi) = f(t - x/c_0).$$

By setting $x = 0$ in this expression, we can write the displacement boundary condition as

$$f(t) = p(t).$$

We see that the function $f(t)$ is equal to the prescribed function $p(t)$; so the solution for the displacement field is

$$u(x, t) = f(\xi) = p(t - x/c_0). \tag{2.54}$$

In Figure 2.14 we show a particular function $p(t)$ as a solid curve. The surface $u(x, t)$ shows the resulting displacement field and the wave propagating in the positive-x direction. Because the function $p(t)$ vanishes for $t \leq 0$, notice that $u = 0$ when $t \leq x/c_0$. We outline this region with a dashed triangle.

Stress on a Boundary Now suppose that we subject the boundary of our half space to a normal-stress boundary condition

$$T(0, t) - T^0 = \rho_0 c_0^2 p(t), \tag{2.55}$$

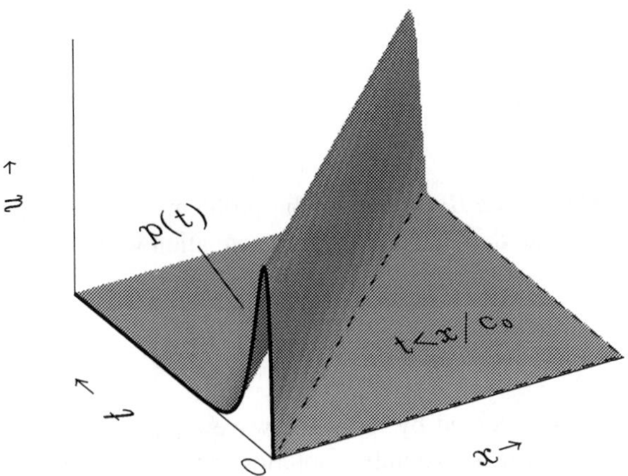

Figure 2.14 The $u(x, t)$ surface for an initial boundary condition $p(t)$. In the region ahead of the wave, where $t \leq x/c_0$, the displacement is zero.

where T^0 is a given constant and $p(t)$ is a prescribed function of time that vanishes for $t \leq 0$. From Eq. (2.39), the stress component T is related to the displacement field by

$$T - T^0 = \rho_0 c_0^2 \frac{\partial u}{\partial x};$$

hence we can write the boundary condition as

$$\frac{\partial u}{\partial x}(0, t) = p(t). \tag{2.56}$$

Let us assume that the displacement field consists of a wave propagating in the positive-x direction,

$$u(x, t) = f(\xi) = f(t - x/c_0).$$

The partial derivative of this expression with respect to x is

$$\frac{\partial u}{\partial x}(x, t) = \frac{df(\xi)}{d\xi}\frac{\partial \xi}{\partial x} = -\frac{1}{c_0}\frac{df(\xi)}{d\xi}.$$

By setting $x = 0$ in this expression, we can write the boundary condition as

$$\frac{df(t)}{dt} = -c_0 p(t). \tag{2.57}$$

We integrate this equation from 0 to t, obtaining the function $f(t)$ in terms of $p(t)$,

$$f(t) = f(0) - c_0 \int_0^t p(\bar{t})\, d\bar{t},$$

where $\bar{t}$ is an integration variable. Thus the displacement field is

$$u(x, t) = f(\xi) = f(0) - c_0 \int_0^{t - x/c_0} p(\bar{t})\, d\bar{t}.$$

The initial condition is violated unless the constant $f(0) = 0$. Therefore the solution is

$$u(x, t) = -c_0 \int_0^{t-x/c_0} p(\bar{t})\, d\bar{t}. \tag{2.58}$$

Alternative Solution for Stress on a Boundary This problem can also be approached in a simpler way. We take the partial derivative of the wave equation with respect to x and write the result as

$$\frac{\partial^2}{\partial t^2}\left(\frac{\partial u}{\partial x}\right) = c_0^2 \frac{\partial^2}{\partial x^2}\left(\frac{\partial u}{\partial x}\right). \tag{2.59}$$

We find that the strain $\partial u/\partial x$ is governed by the same wave equation that governs the displacement. Therefore, we can write a d'Alembert solution for the strain field:

$$\frac{\partial u}{\partial x}(x, t) = h(\xi) = h(t - x/c_0). \tag{2.60}$$

By setting $x = 0$ in this expression, we can write the boundary condition as

$$h(t) = p(t).$$

We see that the function $h(t)$ is equal to the prescribed function $p(t)$; so the solution for the strain field is

$$\frac{\partial u}{\partial x}(x, t) = h(\xi) = p(t - x/c_0). \tag{2.61}$$

Although this approach is simpler, it leads to the solution for the strain field instead of the displacement field.

2.3.6 *Transverse Waves*

We are primarily interested in waves that result in only one nonzero component of displacement, $u_1 = u_1(x_1, t)$. For convenience, we refer to this type of wave as a one-dimensional wave. It is also called a *longitudinal wave*. However, it is important to recognize that another type of one-dimensional wave exists. It is called a *transverse wave* because the direction of the displacement of the material is perpendicular or transverse to the direction of propagation of the wave. In this section we derive a wave equation for the propagation of a linear-transverse wave. This is accomplished in several steps. First we examine deformation of the body caused by transverse motion. We then determine the linear-elastic constitutive equation for this motion. Finally, we substitute these results into the balance of linear momentum to obtain the appropriate wave equation.

We postulate the motion of a transverse wave to be

$$x_1 = X_1, \qquad x_2 = \hat{x}_2(X_1, t), \qquad x_3 = X_3.$$

We next determine the deformation gradient, Lagrangian strain, and volume strain for this motion. Notice that the displacement associated with this motion is

$$u_1 = 0, \qquad u_2 = \hat{u}_2(X_1, t), \qquad u_3 = 0.$$

Substituting this displacement into the deformation gradient

$$F_{km} = \frac{\partial \hat{u}_k}{\partial X_m} + \delta_{km},$$

we obtain

$$[F_{km}] = \begin{bmatrix} 1 & 0 & 0 \\ F_{21} & 1 & 0 \\ 0 & 0 & 1 \end{bmatrix},$$

where

$$F_{21} = \frac{\partial \hat{u}_2}{\partial X_1}.$$

Recall that the Lagrangian strain is [see Eq. (1.69)]

$$E_{kn} \equiv \frac{1}{2}(F_{mk}F_{mn} - \delta_{kn}).$$

For transverse one-dimensional motion, we obtain the following strain components:

$$[E_{km}] = \begin{bmatrix} \frac{1}{2}F_{21}^2 & \frac{1}{2}F_{21} & 0 \\ \frac{1}{2}F_{21} & 0 & 0 \\ 0 & 0 & 0 \end{bmatrix}.$$

When we limit our discussion to linear-elastic materials, the term F_{21}^2 is negligible with respect to F_{21}. Thus the linear-elastic strain components are

$$[E_{km}] = \begin{bmatrix} 0 & \frac{1}{2}F_{21} & 0 \\ \frac{1}{2}F_{21} & 0 & 0 \\ 0 & 0 & 0 \end{bmatrix}.$$

Notice there are only two nonzero components of linear-elastic strain. Both are shear components. The volume strain is [see Eq. (1.79)]

$$dV = \det[F_{km}]dV_0 = III_F dV_0.$$

When the deformation gradient for transverse motion is substituted into this expression, we find that

$$dV = dV_0.$$

Thus, transverse motion does not cause the volume of the body to change. It only causes the body to shear.

Having described the deformation of the body due to transverse motion, let us next examine the linear-elastic constitutive equations for this motion. We substitute the results for the Lagrangian strain into the constitutive model for a linear-elastic material to obtain [see Eq. (2.7)]

$$\left[T_{km} - T_{km}^0\right] = \mu \begin{bmatrix} 0 & F_{21} & 0 \\ F_{21} & 0 & 0 \\ 0 & 0 & 0 \end{bmatrix}.$$

Hence only two components of stress are affected by this deformation:

$$T_{12} - T_{12}^0 = T_{21} - T_{21}^0 = \mu \frac{\partial u_2}{\partial x_1}, \tag{2.62}$$

where, because the gradient of the displacement is assumed to be small,

$$F_{21} = \frac{\partial \hat{u}_2}{\partial X_1} = \frac{\partial u_2}{\partial x_1}.$$

Now we use the equation of the balance of linear momentum to obtain a wave equation. If we ignore the body force, the balance of linear momentum becomes [see Eq. (1.155)]

$$\rho_0 a_2 = \frac{\partial T_{12}}{\partial x_1},$$

where, because the volume strain is zero, we have replaced ρ by ρ_0. We obtain a wave equation by substituting the constitutive equation, Eq. (2.62), into this expression. Thus

$$\frac{\partial^2 u_2}{\partial t^2} = c_{0T}^2 \frac{\partial^2 u_2}{\partial x_1^2},$$

where

$$c_{0T} = \sqrt{\frac{\mu}{\rho_0}}$$

is called the *transverse wave velocity*. Because this wave only causes shear stress in the material, c_{0T} is also called the *shear wave velocity*. Because both Lamé constants, λ and μ, are positive, it is easily demonstrated that

$$c_0 > c_{0T}.$$

The d'Alembert solution for the longitudinal wave is also the solution to the wave equation for the transverse wave; albeit, a different wave velocity must be used. In Figure 2.15 we illustrate the different motions resulting from the longitudinal and transverse waves. Two illustrations are shown. In each illustration the dots depict an array of material points. These points are uniformly spaced in the reference configuration of each material. In each illustration a wave propagates from the left to the right and displaces the material points.

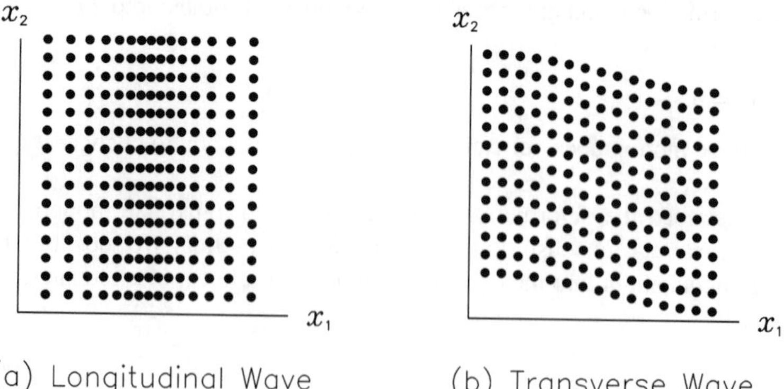

Figure 2.15 (a) Longitudinal and (b) transverse waves.

The left illustration represents a longitudinal wave that displaces the material points in the x_1 direction. Notice that the material is first compressed by the wave and then expanded. The right illustration represents a transverse wave that displaces the material points in the x_2 direction. This particular wave causes the shear strain to change from zero to a negative value and then return to zero.

Other names are often used to describe longitudinal and shear waves. Sometimes they are called P and S waves or *pressure* and *shear* waves. Sometimes they are called *dilatational* and *distortional* waves. In general, we do not discuss transverse waves in this book. Consequently, when we refer to a wave we actually mean a longitudinal or pressure wave. In the rare cases when we discuss transverse waves, we will make a careful distinction between these wave types. Moreover, in a homogeneous isotropic infinite medium, these linear waves do not interact, and you can independently determine their solutions. This is also true in a bounded medium provided the direction of wave propagation is always normal to the boundary. In contrast, when a longitudinal wave strikes a boundary at some angle other than normal incidence, both longitudinal and transverse waves will be reflected. This is also true if a transverse wave strikes a boundary at an angle. The interaction of these waves at boundaries gives rise to many interesting phenomena; however, this topic is outside the scope of this book.

2.3.7 *Exercises*

2.8 Derive a wave equation for displacement from the spatial description of the nonlinear-elastic equations.

2.9 Reconstruct Figure 2.10 to illustrate that $g(\zeta)$ represents a wave propagating in the negative-x direction.

2.10 Consider a homogeneous, infinite, linear-elastic material. A wave propagates through this material. At $x = 0$, we measure the following displacement history in this material:

$$u(0, t) = \begin{cases} te^{-t}, & t \geq 0. \\ 0, & t < 0. \end{cases}$$

(a) Write the d'Alembert solution assuming the wave propagates in the positive-x direction.
(b) Write the d'Alembert solution assuming the wave propagates in the negative-x direction.
(c) For $t = 3$ and $-10 \leq x/c_0 \leq 10$, plot the profiles for both waves on the same graph.

2.11 Consider the expression

$$u = A\cos(kx - \omega t).$$

What conditions must the constants A, k, and ω satisfy for this expression to be a solution of the one-dimensional wave equation?

2.12 Verify that

$$\frac{\partial^2 u}{\partial t^2} = \frac{\partial^2 \breve{u}}{\partial \xi^2} + 2\frac{\partial^2 \breve{u}}{\partial \xi \partial \zeta} + \frac{\partial^2 \breve{u}}{\partial \zeta^2}.$$

2.13 Consider the following d'Alembert solution for a material for which $c_0 = 1$:

$$u(x, t) = \sin\left[\frac{\pi}{4}(t - x)\right] + \sin\left[\frac{\pi}{4}(t + x)\right].$$

This solution is the linear superposition of two sinusoidal waves propagating in opposite directions. This is called a *standing wave*.

(a) Demonstrate why this name is attached to this type of solution by graphing it for values of $-8 < x < 8$ and $t = -2, -1, -0.5, 0, 0.5, 1$, and 2.

(b) Use the identity $2 \sin x \cos y = \sin(x + y) + \sin(x - y)$ to explain these results.

2.14 Consider a homogeneous, infinite, linear-elastic material subjected to the following initial conditions:

$$u(x, 0) = \begin{cases} 1 + \cos(\pi x), & -1 \le x \le 1, \\ 0, & \text{otherwise}, \end{cases}$$

$$v(x, 0) = 0.$$

Graph the wave profile of the solution u to the wave equation at $c_0 t = 0, 0.25, 0.5, 0.75, 1, 1.25$, and 1.5.

2.15 Consider a homogeneous, infinite, linear-elastic material subjected to the following initial conditions:

$$u(x, 0) = 0,$$

$$v(x, 0) = \begin{cases} c_0/50, & -10c_0 \le x \le 10c_0, \\ 0, & \text{otherwise}. \end{cases}$$

(a) Evaluate the d'Alembert solution for the displacement $u(0, t)$.

(b) Graph the solution.

(c) What is the final displacement of the material at $x = 0$?

2.16 Consider a half space of linear-elastic material that occupies the region $x \ge 0$. It is subjected to the following boundary condition at $x = 0$:

$$\frac{\partial u}{\partial x}(0, t) = \begin{cases} 1 + \cos[\pi(t - 1)], & 0 \le t \le 2, \\ 0, & \text{otherwise}. \end{cases}$$

(a) Using the d'Alembert solution, Eq. (2.60), determine the strain $\partial u / \partial x$ everywhere in the half space.

(b) Graph the wave profiles for the strain at $t = 0, 1$, and 2.

(c) What is the final displacement of the boundary?

2.17 Use the Leibniz rule for differentiation of an integral to derive Eq. (2.61) from the partial derivative of Eq. (2.58).

2.18 For the linear, transverse, one-dimensional motion discussed in this section:

(a) Show that the inverse deformation gradient is

$$\left[F_{km}^{-1}\right] = \begin{bmatrix} 1 & 0 & 0 \\ -F_{21} & 1 & 0 \\ 0 & 0 & 1 \end{bmatrix}.$$

(b) Show that the shear strain between two perpendicular line elements oriented in the x_1 and x_2 directions is

$$\sin \gamma = \frac{2E_{12}}{(2E_{11} + 1)^{1/2}} = \frac{F_{21}}{\left(F_{21}^2 + 1\right)^{1/2}}.$$

2.4 Method of Characteristics

We have used the characteristic coordinates to obtain the d'Alembert solution to the wave equation. It is also possible to use the x–t diagram of the characteristic contours to analyze complicated solutions to the wave equation. This solution technique is called the *method of characteristics*. Effective application of the method of characteristics requires several tools in addition to the x–t diagram. These are the Riemann invariants, the simple wave, and the T–v diagram. In this section, we introduce these tools and apply them to solve several important problems.

2.4.1 *Riemann Invariants*

To construct solutions to the wave equation with the x–t diagram, we need a convenient method for evaluating the dependent variables of the solution on each of the characteristic contours. For reasons that shall become apparent as we proceed, the stress T and velocity v are the most convenient variables to use in the construction of the solution. Here we derive some useful expressions for the stress and velocity.

The general solution to the wave equation is the d'Alembert solution:

$$u = f(\xi) + g(\zeta),$$

where the displacement is the sum of two arbitrary functions $f(\xi)$ and $g(\zeta)$. The d'Alembert solution for the stress $T - T^0 = \rho_0 c_0^2 \partial u/\partial x$ is

$$T - T^0 = \rho_0 c_0^2 \left[\frac{df(\xi)}{d\xi}\frac{\partial \xi}{\partial x} + \frac{dg(\zeta)}{d\zeta}\frac{\partial \zeta}{\partial x} \right] = z_0 \left[-\frac{df(\xi)}{d\xi} + \frac{dg(\zeta)}{d\zeta} \right], \tag{2.63}$$

where the constant z_0,

$$z_0 = \rho_0 c_0, \tag{2.64}$$

is called the *acoustic impedance*. Similarly, the d'Alembert solution for the velocity v is

$$v = \frac{df(\xi)}{d\xi} + \frac{dg(\zeta)}{d\zeta}.$$

We can rewrite the d'Alembert solutions for stress and velocity as

$$\begin{aligned} T - T^0 &= \frac{z_0}{2}[-J_+(\xi) + J_-(\zeta)], \\ v - v^0 &= \frac{1}{2}[J_+(\xi) + J_-(\zeta)], \end{aligned} \tag{2.65}$$

where

$$J_-(\zeta) = 2\frac{dg(\zeta)}{d\zeta} - v^0, \tag{2.66}$$

$$J_+(\xi) = 2\frac{df(\xi)}{d\xi} - v^0. \tag{2.67}$$

When Eqs. (2.65) are inverted, we obtain

$$J_-(\zeta) = v - v^0 + \frac{T - T^0}{z_0}, \tag{2.68}$$

$$J_+(\xi) = v - v^0 - \frac{T - T^0}{z_0}. \tag{2.69}$$

The two functions $J_\pm$ are called the *Riemann invariants*. They represent combinations of stress and velocity that do not change along particular characteristic contours. Because $J_-(\zeta)$ is not a function of ξ, the quantity $v + T/z_0$ is constant along the characteristic contours where $d\zeta = 0$. Similarly, the quantity $v - T/z_0$ is constant along the characteristic contours where $d\xi = 0$.

As we shall see, the Riemann invariants are extremely useful tools for analyzing linear solutions of the wave equation. These analysis methods are not restricted to the wave equation alone. We shall show that the invariants Eqs. (2.68) and (2.69) are a special solution of the Riemann integrals for large deformation of nonlinear-elastic materials. The analysis methods for linear-elastic materials that we are about to discuss will eventually be extended to nonlinear-elastic materials.

2.4.2 *Simple Wave*

A *simple wave* is a smooth solution to the wave equation that propagates in one direction. We have already discussed this type of wave. Here we give a more precise definition of a simple wave. We do this because linear superposition allows us to use simple waves as building blocks to construct more complex solutions to the wave equation. To take maximum advantage of this method, we must fully understand the properties of the simple wave.

Recall the d'Alembert solution for the velocity:

$$v = \frac{df(\xi)}{d\xi} + \frac{dg(\zeta)}{d\zeta}.$$

This solution is the sum of two arbitrary functions $df(\xi)/d\xi$ and $dg(\zeta)/d\zeta$ that represent two waves propagating in opposite directions. Let us examine a solution where

$$\frac{dg(\zeta)}{d\zeta} = \frac{v^0}{2}$$

and v^0 is a constant. This solution for v is a function of ξ only and represents a wave that propagates in the positive-x direction. From Eqs.(2.65) and (2.66), we obtain

$$\left.\begin{aligned} &\tfrac{1}{2}J_+(\xi) = v - v^0 = -\tfrac{T-T^0}{z_0} \\ &J_-(\zeta) = 0 \end{aligned}\right\} \text{positive-}x\text{ direction.} \tag{2.70}$$

Any solution where one of the Riemann invariants is a constant is called a *simple wave*. This simple wave propagates in the positive-x direction and resides in a portion of the x–t diagram where J_- is constant.

Similarly, when

$$\frac{df(\xi)}{d\xi} = \frac{v^0}{2},$$

we obtain the solution for a simple wave that propagates in the negative-x direction. For this simple wave

$$\left.\begin{aligned} &J_+(\xi) = 0 \\ &\tfrac{1}{2}J_-(\zeta) = v - v^0 = \tfrac{T-T^0}{z_0} \end{aligned}\right\} \text{negative-}x\text{ direction.} \tag{2.71}$$

From these results, we obtain several useful relationships between the stress, velocity, and strain. From Eqs. (2.70) and (2.71), the stress and velocity in simple waves are related by

$$\begin{aligned} T - T^0 &= -z_0(v - v^0), \quad \text{positive-}x\text{ direction,} \\ T - T^0 &= z_0(v - v^0), \quad \text{negative-}x\text{ direction.} \end{aligned} \tag{2.72}$$

From the constitutive model for a linear-elastic material, Eq. (2.39), the strain and velocity in simple waves are related by

$$\begin{aligned} \frac{\partial u}{\partial x} &= -\frac{v - v^0}{c_0}, \quad \text{positive-}x\text{ direction,} \\ \frac{\partial u}{\partial x} &= \frac{v - v^0}{c_0}, \quad \text{negative-}x\text{ direction.} \end{aligned} \tag{2.73}$$

We illustrate the x–t diagram of a simple wave in Figure 2.16. This simple wave propagates in the positive-x direction and is represented by a band of characteristic contours. On each contour, $d\xi = 0$. The value of $J_+(\xi)$ is constant on each contour but changes between contours. The value of $J_-(\zeta)$ is constant everywhere in the diagram. From Eq. (2.42), the slope of each of the contours is $dx/dt = c_0$. From Eq. (2.70), both the stress and velocity are constant along each of the contours. We illustrate the simple wave with a band of contours to emphasize that it can have an arbitrary wave profile consisting of many characteristic contours. The values of the Riemann invariants on each of these contours determine the

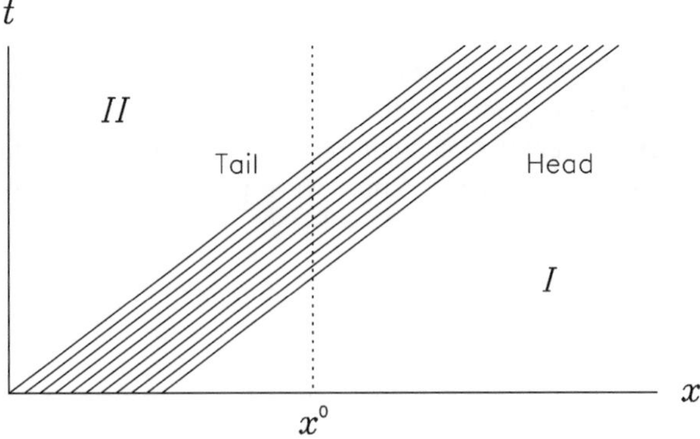

Figure 2.16 The x–t diagram of the simple wave propagating in the positive-x direction.

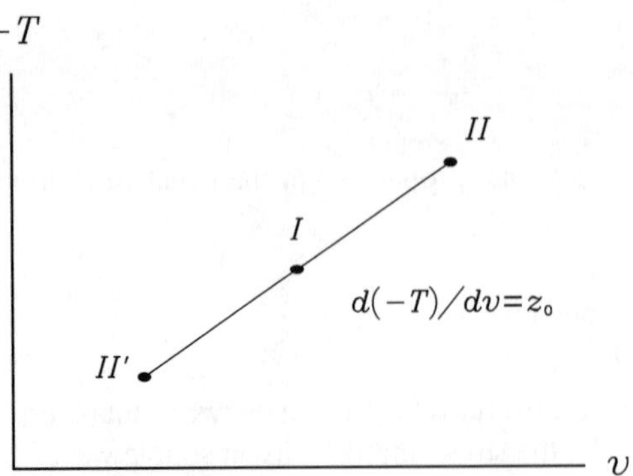

Figure 2.17 The T–v diagram for the material point x^0, which encounters a simple wave propagating in the positive-x direction.

stress T and velocity v everywhere in the x–t diagram. Within the band of contours, the velocity and stress change. Outside the band, the velocity and stress are constant. Two regions of constant stress and velocity are shown. The contours on the edges of the band are the head and tail contours. The head contour of the wave separates the interior of the wave from region *I*; the tail contour of the wave separates the interior of the wave from region *II*.

2.4.3 *The T–v Diagram*

The T–v *diagram* is another important tool used in the method of characteristics. We shall use it to determine changes in stress and velocity that result as a simple wave passes a material point.

Consider a simple wave propagating in a linear-elastic material. We can determine the stress and velocity in the body by Eq. (2.72), which can be written as:

$$-(T - T^0) = \pm z_0(v - v^0). \tag{2.74}$$

The positive sign is used when the simple wave propagates in the positive-x direction; the negative sign is used when the simple wave propagates in the negative-x direction. Consider the material point x^0 in Figure 2.16. Initially, this point is in region *I* of the x–t diagram. Let us assume that the velocity and stress in this region are given by v^0 and T^0. We construct a T–v diagram for the material at x^0 by plotting the quantity $-T$ against v (see Figure 2.17).† The initial values of T and v are represented by the point *I* in the T–v diagram. As time advances, the material at x^0 encounters the head contour of the simple wave. This causes the stress and velocity to change. Because the simple wave propagates in the positive-x direction, the changes in stress and velocity are related by

$$-(T - T^0) = z_0(v - v^0).$$

† Notice that the T–v diagram is not a plot of stress against velocity, but rather it is a plot of the *negative* stress against velocity. At first this appears to be a little awkward. However, our only alternative to this procedure is to define a stress that is positive in compression. This alternative definition of stress is adopted by some authors, but we have chosen not to do this. Such definitions tend to isolate the theory of waves from related fields of study, such as continuum mechanics.

This is the positive-sloping curve in the T–v diagram. This curve is called the *response path* of the simple wave. Its slope is equal to the acoustic impedance z_0 of the material. Notice that a simple wave, which propagates in the positive-x direction, is represented by a response path with a positive slope z_0. If the simple wave causes compression, the values of stress and velocity move up the response path; if the simple wave causes expansion, they move down the response path. After the material at x^0 encounters the tail contour of the simple wave, it moves into region *II* of the x–t diagram. If the simple wave compresses the material, the stress and velocity in region *II* of the x–t diagram are represented by point *II* in the T–v diagram. If the simple wave expands the material, the stress and velocity in region *II* of the x–t diagram are represented by point *II′* in the T–v diagram. Simple waves, which cause both compression and expansion, can even cause the material to move up and down the response path and eventually return to point *I*. In such cases, regions *I* and *II* reside at the same point in the T–v diagram. Notice that the symmetry between the direction of propagation and the slope of the response path offers us an easily remembered method for tracking the changes in stress and velocity caused by simple waves.

2.4.4 *Initial-Value Problems*

We have used the d'Alembert solution to solve the initial-value problem (see Section 2.3.5). Now that we have discussed the various tools of the method of characteristics, let us repeat this solution here to demonstrate the utility of the Riemann invariants $J_{\pm}$.

Suppose that at $t = 0$ we prescribe the components of the stress and velocity of an unbounded linear-elastic material to be

$$\frac{T(x,0) - T^0}{\rho_0 c_0^2} = p(x),$$

$$v(x,0) - v^0 = q(x),$$

where $p(x)$ and $q(x)$ are known functions. Consider the x–t diagram in Figure 2.18. We have drawn two characteristic contours that intersect at the point (x, t). When we extend these contours back in time, they intersect the x axis at $x + c_0 t$ and $x - c_0 t$. Along the contour for the wave propagating in the positive-x direction, $\xi = t - x/c_0$ is constant, and along the contour for the wave propagating in the negative-x direction, $\zeta = t + x/c_0$ is constant [see Eq. (2.42)].

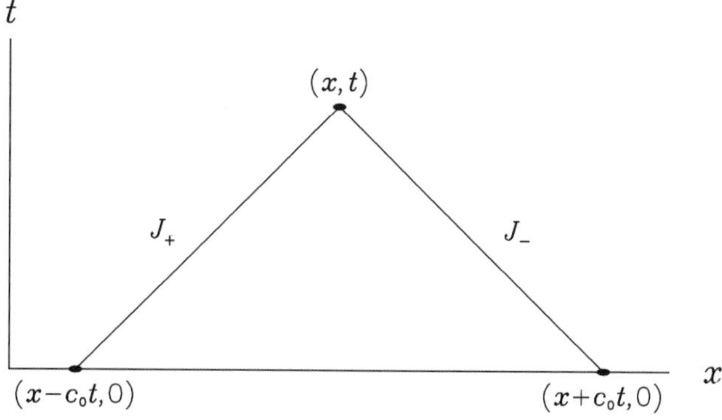

Figure 2.18 The x–t diagram for an initial-value problem.

Thus, along the positive-sloping characteristic contour, $J_+(x, t) = J_+(t - x/c_0, 0)$, and from Eq. (2.69) we obtain

$$v(x,t) - v^0 - \frac{T(x,t) - T^0}{z_0} = q(x - c_0 t) - c_0 p(x - c_0 t).$$

Similarly, along the negative-sloping characteristic contour, we obtain

$$v(x,t) - v^0 + \frac{T(x,t) - T^0}{z_0} = q(x + c_0 t) + c_0 p(x + c_0 t).$$

Solving for v and T, we find that

$$v(x,t) - v^0 = \frac{1}{2}[q(x + c_0 t) + q(x - c_0 t)] + \frac{c_0}{2}[p(x + c_0 t) - p(x - c_0 t)],$$

$$T(x,t) - T^0 = \frac{z_0}{2}[q(x + c_0 t) - q(x - c_0 t)] + \frac{\rho_0 c_0^2}{2}[p(x + c_0 t) + p(x - c_0 t)]. \tag{2.75}$$

We leave it as an exercise to show that these results are equivalent to the d'Alembert solution for the displacement, Eq. (2.51) (see Exercise 2.20).

2.4.5 *Boundary-Value Problems*

Suppose that a half space of linear-elastic material initially has the velocity $v = 0$ and the stress $T = 0$ for $t \leq 0$. We subject the boundary at $x = 0$ to the velocity boundary condition

$$v(0,t) - v^0 = r(t),$$

where $r(t)$ is a prescribed function of time that vanishes for $t \leq 0$. Consider the x–t diagram in Figure 2.19. We have drawn two characteristic contours that intersect the point (x, t). Notice that the point (x, t) is above the dashed line $x = c_0 t$. When the point (x, t) is below the dashed line so that $t < x/c_0$, both of these contours intersect the x axis, and the solution is identical to the initial-value problem. Because the initial velocity and stress are v^0 and T^0 on the x axis, we find that the solution to the initial-value problem [Eq. (2.75)] gives

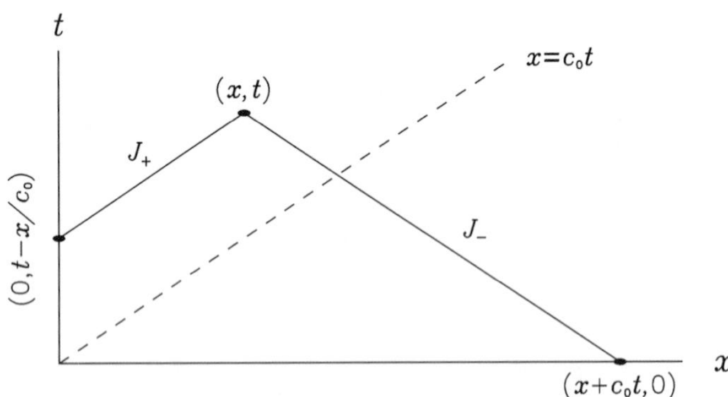

Figure 2.19 The x–t diagram for a boundary-value problem.

$v(x,t) = v^0 = 0$ and $T(x,t) = T^0 = 0$ everywhere in this region. However, when the point (x,t) is above the dashed line so that $t > x/c_0$, we see that the positive-sloping contour intersects the boundary of the half space at time $t - x/c_0$. Along this contour $\xi = t - x/c_0$ is constant, and

$$\begin{aligned} J_+(t - x/c_0) &= v(x,t) - v^0 - \frac{T(x,t) - T^0}{z_0} \\ &= v(0, t - x/c_0) - v^0 - \frac{T(0, t - x/c_0) - T^0}{z_0}. \end{aligned} \tag{2.76}$$

Along the negative-sloping contour, $J_- = 0$. Thus the solution is a simple wave propagating in the positive-x direction. From Eq. (2.72), we have

$$v - v^0 = -\frac{T - T^0}{z_0}.$$

Upon substitution of this expression into Eq (2.76) we find that

$$J_+(t - x/c_0) = 2[v(x,t) - v^0] = 2[v(0, t - x/c_0) - v^0].$$

Using the boundary condition we obtain the desired solution

$$v(x,t) - v^0 = r(t - x/c_0). \tag{2.77}$$

2.4.6 *Reflection and Transmission at an Interface*

Our previous solutions obtained by the method of characteristics involve simple application of the Riemann invariants. In this section we apply the method of characteristics to a more complex problem. To examine this problem, we use both the x–t diagram and the T–v diagram.

Suppose that two half spaces of linear-elastic material are bonded together at $x = 0$ (see Figure 2.20). Let the left half space be material L and the right half space be material R. The velocities of sound of the left and right materials are c_0^L and c_0^R, and the acoustic impedances are z_0^L and z_0^R. Initially $T = v = 0$ in the neighborhood of the interface, but, as time advances, a wave with a prescribed profile approaches the interface from the left.

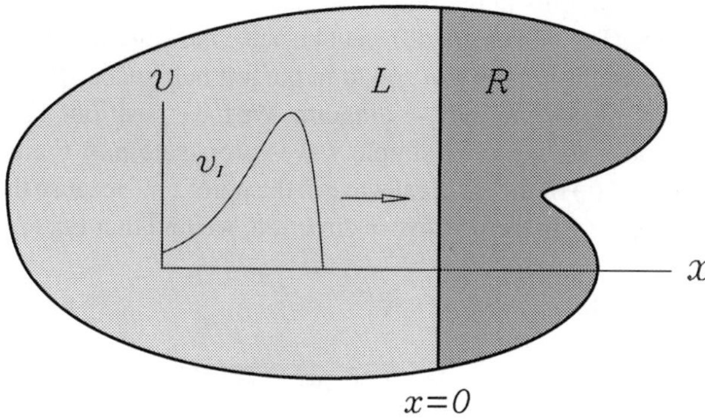

Figure 2.20 A linear wave incident on an interface at $x = 0$.

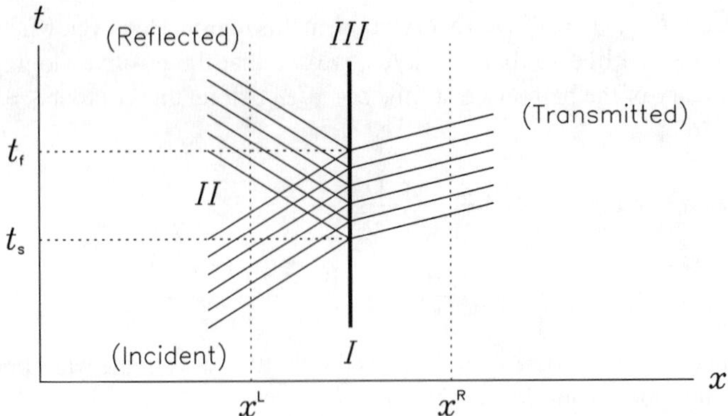

Figure 2.21 The x–t diagram for the reflection of a simple wave at an interface.

This is the *incident wave*, and its head characteristic contour contacts the interface at the time $t = t_s$.

Consider the x–t diagram for the two half spaces shown in Figure 2.21. The incident wave is the simple wave with a prescribed velocity history $v_I(t)$. It approaches the interface from the left. The interaction of the incident wave with the interface gives rise to two additional waves. One is the *reflected* wave, which propagates in the negative-x direction into material L; the other is the *transmitted* wave, which propagates in the positive-x direction into material R. The transmitted wave is a simple wave, but during the interval of time $t_s < t < t_f$ that the incident wave is in contact with the interface, the incident and reflected waves are not simple waves. We represent each of the three waves by a band of characteristic contours to emphasize that each can have an arbitrary wave profile consisting of many characteristic contours. If we know the values of the Riemann invariants on each of these contours, we can evaluate the stress T and velocity v everywhere in the x–t diagram.

Within each of the bands of contours, the velocity and stress change. Outside these bands, the velocity and stress are constant. We have illustrated three regions of constant stress and velocity. Because the stress and velocity of the material must be continuous across an interface, we have extended regions *I* and *III* across the interface between the two half spaces. Although the regions where the solution is constant extend across the interface, we have drawn the characteristic contours in the right half space with a different slope than the contours in the left half space. The slopes are different because the velocity of sound in the right half space c_0^R differs from the velocity of sound in the left half space c_0^L.

Consider the two points x^L and x^R in the x–t diagram. We have positioned these points in the left and right half spaces so that only simple waves propagate past them. Now we prescribe the velocity history $v_I(t)$ of the incident wave at the point x^L. Because the incident wave is a simple wave propagating in the positive-x direction, we find that Eq. (2.70) yields

$$J_+^L\left(t - x^L/c_0^L\right) = 2v_I(t),$$

where $J_\pm^L$ are the Riemann invariants of the left half space. This leads to the result

$$J_+^L(\xi) = 2v_I\left(t - (x - x^L)/c_0^L\right).$$

The solutions for the stress and velocity in the left half space are

$$\left.\begin{aligned} T &= \frac{z_0^L}{2}\{-2v_I\big(t-(x-x^L)/c_0^L\big)+J_-^L(\zeta)\},\\ v &= \tfrac{1}{2}\{2v_I\big(t-(x-x^L)/c_0^L\big)+J_-^L(\zeta)\}, \end{aligned}\right\} x \le 0, \tag{2.78}$$

where we see that the Riemann invariant J_-^L is an unknown function that represents the reflected wave. Let us now determine the solution for the right half space.

Because the transmitted wave is a simple wave and the right half space is initially at rest, we know that $J_-^R = 0$. Thus, the only unknown Riemann invariant for the right half space is $J_+^R(\xi)$. Indeed, we find that the solutions for the stress and velocity in the right space are

$$\left.\begin{aligned} T &= -\frac{z_0^R}{2}J_+^R(\xi),\\ v &= \tfrac{1}{2}J_+^R(\xi), \end{aligned}\right\} x \ge 0. \tag{2.79}$$

We now have the expressions for T and v in both half spaces. Therefore, our objective is to determine the two unknown Riemann invariants J_-^L and J_+^R and then evaluate the velocity histories at x^L and x^R. We achieve this by requiring the stress and velocity to be continuous across the interface. Setting $x = 0$ in the previous expressions and equating the velocities and stresses of the left and right half spaces, we obtain

$$-2v_I\big(t+x^L/c_0^L\big)+J_-^L(t) = -KJ_+^R(t),$$

$$2v_I\big(t+x^L/c_0^L\big)+J_-^L(t) = J_+^R(t),$$

where

$$K \equiv \frac{z_0^R}{z_0^L}.$$

We solve these relations for the two unknown invariants,

$$\begin{aligned} \frac{1}{2}J_-^L(t) &= \frac{1-K}{1+K}v_I\big(t+x^L/c_0^L\big),\\ \frac{1}{2}J_+^R(t) &= \frac{2}{1+K}v_I\big(t+x^L/c_0^L\big). \end{aligned} \tag{2.80}$$

Then we obtain the solution for the velocity in the left and right half spaces:

$$v(x,t) = \begin{cases} v_I\big(t-(x-x^L)/c_0^L\big)+\frac{1-K}{1+K}v_I\big(t+(x+x^L)/c_0^L\big), & x \le 0,\\ \frac{2}{1+K}v_I\big(t-x/c_0^R+x^L/c_0^L\big), & x \ge 0. \end{cases} \tag{2.81}$$

The velocity history at x^L is

$$v(x^L,t) = v_I(t)+\frac{1-K}{1+K}v_I\big(t+2x^L/c_0^L\big), \tag{2.82}$$

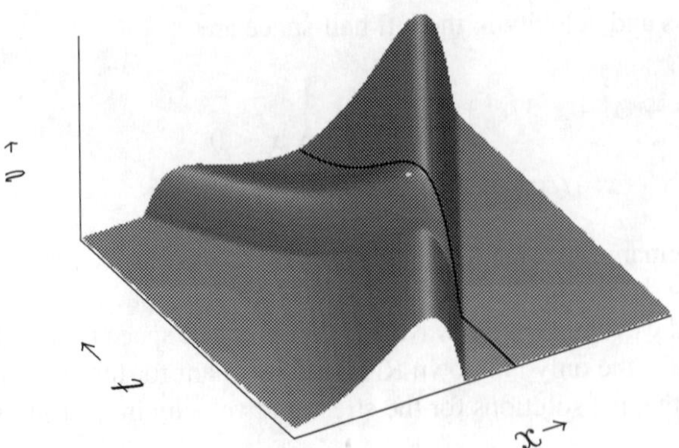

Figure 2.22 The solution surface for the velocity v for an incident wave reflected by an interface where $K = 1/3$.

and the velocity history at x^R is

$$v(x^R, t) = \frac{2}{1+K} v_I\left(t - x^R/c_0^R + x^L/c_0^L\right).$$

The expression for the velocity history at x^L is the sum of two terms. The first term $v_I(t)$ is the incident wave. The second term is the reflected wave. The history of the reflected wave is proportional to the history of the incident wave, but it is delayed by the period of time $-2x^L/c_0^L > 0$. (Remember, the interface is at $x = 0$, and $x^L < 0$.) The history of the transmitted wave is also proportional to the history of the incident wave, and it is delayed by the period of time $x^R/c_0^R - x^L/c_0^L$. We see that the reflection and transmission of a simple wave at an interface between two linear-elastic materials depends solely upon the ratio of acoustic impedances $K = z_0^R/z_0^L$. If the two materials have the same acoustic impedance, there is no reflected wave, and except for a shift in time, the histories of the transmitted and incident waves are identical.

In Figure 2.22 we illustrate the solution Eq. (2.81) for two half spaces where $c_0^L = c_0^R$ and $K = 1/3$. Notice that the profile of the reflected wave is reversed with respect to the profiles of the incident and transmitted waves. We leave it as a exercise to show that these results also apply to the case when $T^0 \neq 0$ (see Exercise 2.22).

The T–v Diagram We can also use the T–v diagram to analyze the reflection and transmission of a linear wave at an interface. In Figure 2.23 we illustrate the response paths for an interface where the acoustic impedance of the left half space is greater than the right half space, $K < 1$. Consider the material at x^L in the x–t diagram (see Figure 2.21). Initially, the material is in region *I* where $T = v = 0$. Suppose we subject this material to an incident wave that only compresses the left half space as it passes the point x^L. Because this wave propagates in the positive-x direction, we know that the stress and velocity of the left half space at x^L must move up the positive-sloping response path from point *I* to point *II* (see Figure 2.23). We determine the location of point *II* by prescribing either the stress or the velocity of the incident wave in region *II*. The slope of the response path connecting points *I* and *II* is z_0^L. Next, the material at x^L encounters the reflected wave and moves from point *II* to point *III*. Because the reflected wave propagates in the negative-x direction, we know

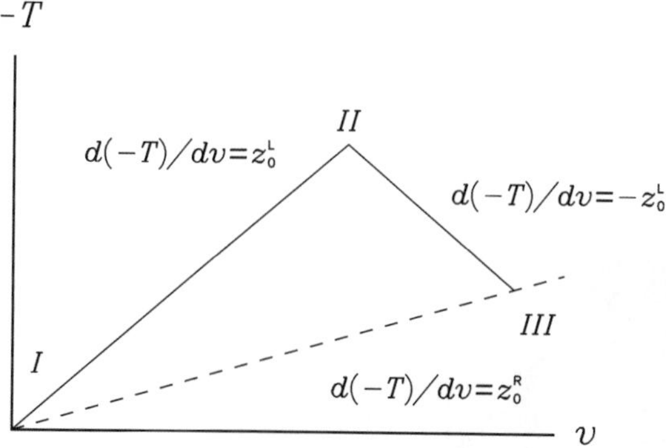

Figure 2.23 The T–v diagram for reflection and transmission at an interface when $K < 1$. We show the response paths at x^L (—) and the response path at x^R (- - -).

that the response path for the reflected wave must be negative sloping. Its slope is equal to $-z_0^L$. Meanwhile, in the right half space, the transmitted wave causes the stress and velocity to move from point *I* to point *III* along the response path for the right half space. The slope of this response path is z_0^R, where $z_0^R < z_0^L$. This diagram shows us that the left half space is first compressed by the incident wave and then expanded by the reflected wave. The right half space is compressed by the transmitted wave. In region *III* the stress and velocity at x^L and x^R must be equal. Thus the point of intersection of the positive-sloping dashed line and the negative-sloping solid line tells us the final values of stress and velocity in both half spaces.

Reflection at a Rigid Boundary

Suppose that the right half space in Figure 2.20 is rigid. That is, the left half space is initially stationary and bonded to a fixed support at $x = 0$. Let us assume that an incident wave propagates in the positive-x direction in the left half space. We obtain the solution for the velocity v in the left half space from Eq. (2.81) by letting the acoustic impedance $z_0^R \to \infty$, so that $K \to \infty$:

$$v = v_I\big(t - (x - x^L)/c_0^L\big) - v_I\big(t + (x + x^L)/c_0^L\big). \tag{2.83}$$

The velocity history at x^L is

$$v(x^L, t) = v_I(t) - v_I\big(t + 2x^L/c_0^L\big). \tag{2.84}$$

We show the solution for the velocity in the left half space in Figure 2.24. This surface is similar to that of Figure 2.22 except that we have rotated it about the vertical axis to better illustrate the reflection of the wave by the rigid boundary.

The solution Eq. (2.84) is the linear superposition of the incident wave and the reflected wave. Strictly speaking, both the incident and reflected waves are only defined for $x \leq 0$. Wave motion is impossible for $x \geq 0$ because of the rigid boundary condition. However, suppose for a moment we evaluate these solutions for positive values of x. We illustrate the resulting profiles of the solution at four different times in Figure 2.25. In each of these profiles the solid curve is the linear-superposition solution for the velocity v, and the dotted curves are the incident wave $v_I(t - (x - x^L)/c_0^L)$ and the reflected wave $-v_I(t + (x + x^L)/c_0^L)$. We only show the solid curve for $x \leq 0$ where the velocity v is defined. However, we do show

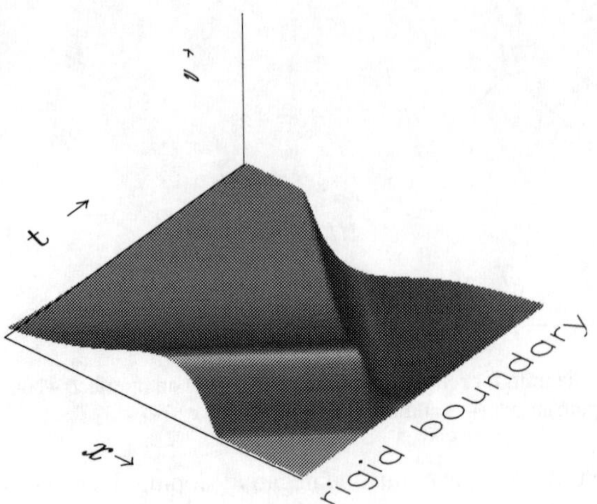

Figure 2.24 The solution surface for the velocity v for an incident wave reflected by a rigid boundary.

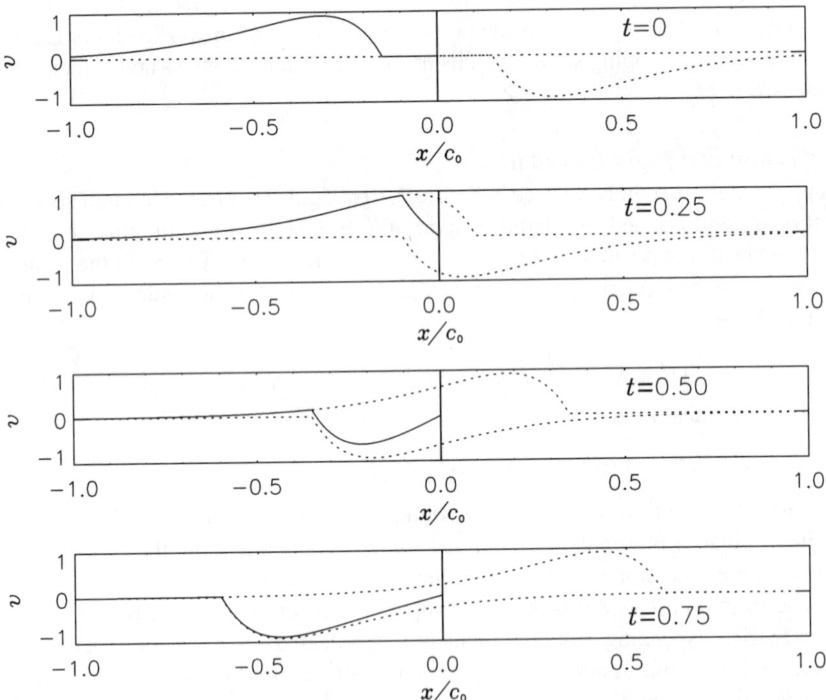

Figure 2.25 Four wave profiles illustrating the linear superposition of the incident wave and the reflected wave at a rigid boundary.

the dotted curves on both sides of the rigid boundary. By plotting the incident and reflected waves over the entire range of x, we gain a clearer picture of how linear superposition works. We can see how the incident and reflected waves merge at the rigid boundary to produce $v(0, t) = 0$. Indeed, when we compare the reflected wave to the incident wave, we see that its profile is reversed; it propagates in the opposite direction; and it has the opposite sign. This graphical technique for separating the incident and reflected waves is called *the method of images*.

We can also use an alternative approach to analyze the reflection of a simple wave at a rigid boundary. We can use the x–t diagram and the T–v diagram to determine the changes in stress and velocity resulting from the incident and reflected waves (see Figures 2.26 and 2.27). To better illustrate this procedure, we shall assume that the incident wave only results in compression of the material at x^L. The incident wave and the reflected wave separate the x–t diagram into three regions. In region *I*, $T = v = 0$. In region *II*, we prescribe the stress behind the incident wave to be $T = T_{II}$ where $T_{II} < 0$. In region *III*, the rigid boundary requires that $v = 0$. The material at x^L initially resides in region *I*. When it encounters the incident wave, it is compressed and moves from point *I* to point *II* in the T–v diagram. The material at x^L remains at this point until it encounters the reflected wave. It then moves into

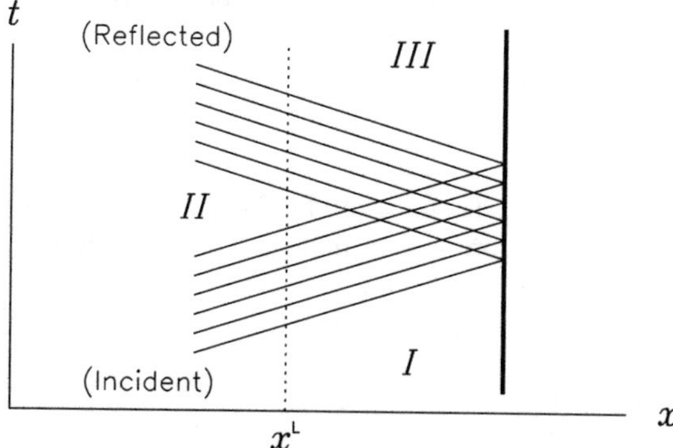

Figure 2.26 The x–t diagram for reflection at a rigid boundary.

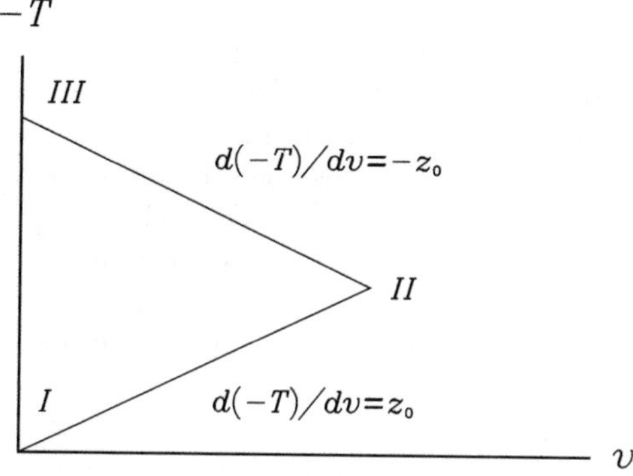

Figure 2.27 The T–v diagram for the material at x^L during the reflection of a simple wave at a rigid boundary.

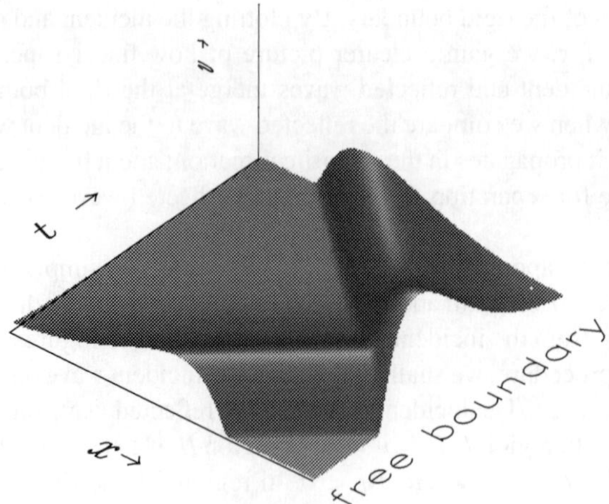

Figure 2.28 The solution surface for the velocity v for an incident wave reflected by a free boundary.

region *III*. Because region *III* contains the rigid boundary, the velocity is zero in this region. The velocity is also zero at point *III* of the T–v diagram. We notice that the reflected wave causes the stress at x^L to double as the material moves from region *II* into region *III*.

Reflection at a Free Boundary

Suppose that $z_0^R = 0$ in the right half space. Then the left half space has a free boundary at $x = 0$. Let us assume that an incident wave propagates in the positive-x direction in the left half space. We obtain the solution for the velocity v in the left half space from Eq. (2.81) by letting the acoustic impedance $z_0^R = 0$, so that $K = 0$:

$$v = v_I\big(t - (x - x^L)/c_0^L\big) + v_I\big(t + (x + x^L)/c_0^L\big). \tag{2.85}$$

The velocity history at x^L is

$$v(x^L, t) = v_I(t) + v_I\big(t + 2x^L/c_0^L\big).$$

We show this solution for v in Figure 2.28. The velocity v is the linear superposition of the incident wave and the reflected wave. We illustrate the profiles of this solution at four different times in Figure 2.29. Following the method of images as used for the rigid boundary, we plot the incident and reflected waves on both sides of the free boundary. In each of these profiles the solid curve is the linear-superposition solution for the velocity v, and the dotted curves are the incident wave $v_I(t - (x - x^L)/c_0^L)$ and the reflected wave $v_I(t + (x + x^L)/c_0^L)$. The solid curve is shown only for $x \leq 0$ where the velocity v is defined. It is the sum of the two dotted curves. When we compare the reflected wave to the incident wave, we see that its profile is reversed; it propagates in the opposite direction; and it has the same sign.

We can use an alternative approach to analyze the reflection of a wave at a free boundary. We use the x–t diagram and the T–v diagram to determine the changes in stress and velocity resulting from the incident and reflected waves. The x–t diagram for the reflection at a free boundary is identical to the diagram for the reflection at a rigid boundary (see Figure 2.26).

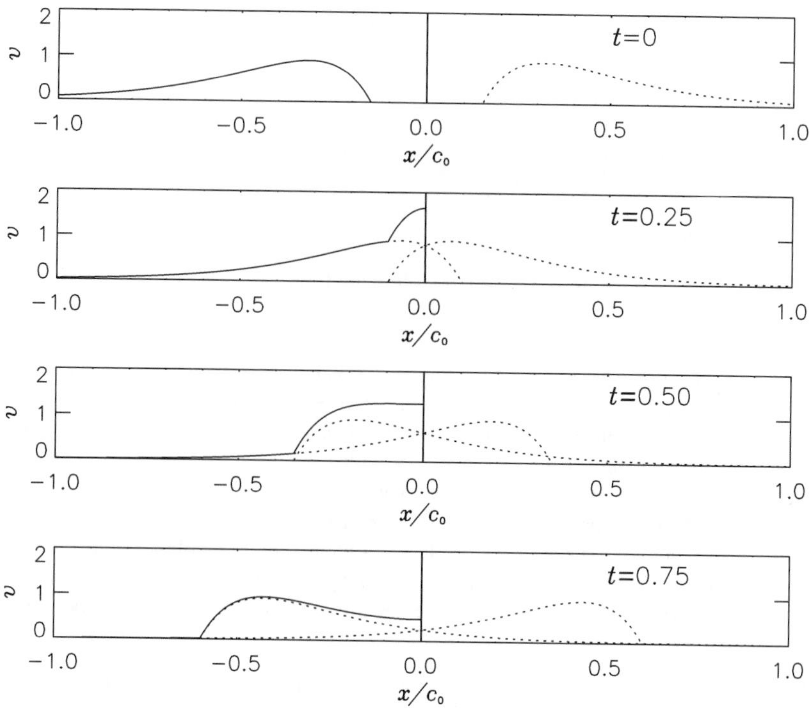

Figure 2.29 Four wave profiles illustrating the linear superposition of the incident wave and the reflected wave at a free boundary.

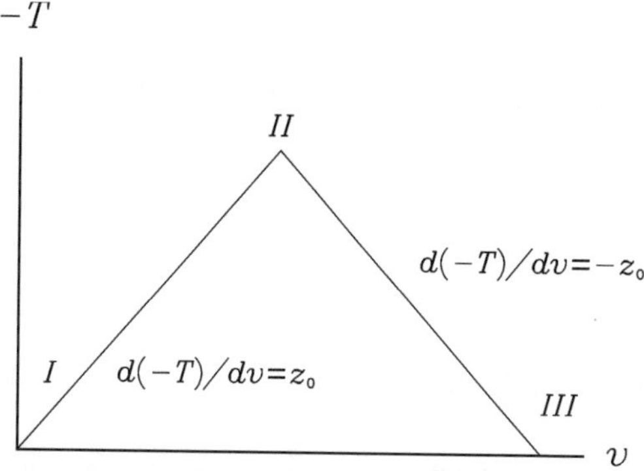

Figure 2.30 The T–v diagram for the material at x^L during the reflection of a simple wave at a free boundary.

To better illustrate this procedure, we shall assume that the incident wave only compresses the material at x^L. The incident wave and the reflected wave separate the x–t diagram into three regions. In region *I*, $T = v = 0$. In region *II*, we prescribe the stress behind the incident wave to be $T = T_{II}$ where $T_{II} < 0$. In region *III*, the free boundary condition requires that $T = 0$. We show the T–v diagram for the reflection at a free boundary in Figure 2.30. The material at x^L initially resides in region *I*. When it encounters the incident wave, it is compressed and moves from point *I* to point *II* in the T–v diagram. The material at x^L remains at this point until it encounters the reflected wave. It then moves into region *III*. Because region *III* contains the free boundary, the stress is zero in this region. The stress is also zero at point *III* of the T–v diagram. We notice that the reflected wave causes the velocity at x^L to double as the material moves from region *II* into region *III*.

2.4.7 Exercises

2.19 Consider a pressure wave with the following form:

$$p = p_0 \sin[2\pi f(t - x/c_0)],$$

which propagates in air. The parameter f is called the frequency of the wave. Suppose $f = 1{,}000$ Hz and $p_0 = 20\ \mu$Pa. For air, $\rho_0 = 1.22$ kg/m^3 and $c_0 = 345$ m/s. Typically, the ears of a young adult with good hearing can detect such a wave; albeit, this wave is at the *threshold of hearing*.

(a) Determine the expression for the velocity v of this wave.

(b) Integrate this expression to determine the displacement u.

(c) A molecule of oxygen in air travels about 65 nm before it collides with another molecule. This distance is called the mean-free-path length. Express the amplitude of the displacement in part (b) in physical dimensions of mean-free-path lengths. What do you think about this answer?

2.20 We have used both the d'Alembert solution and the method of characteristics to analyze the initial-value problem. Results for v are contained in Eqs. (2.51) and (2.75), respectively. Demonstrate that these are equivalent results.

2.21 For a linear-elastic material show that

$$E = \frac{\partial u}{\partial x}, \qquad \frac{\partial E}{\partial t} = \frac{\partial v}{\partial x}, \qquad \frac{\partial v}{\partial t} = c_0^2 \frac{\partial E}{\partial x}.$$

Consider the Riemann integrals $J_\pm(x,t) = v(x,t) \mp c_0 E(x,t)$ and the differential

$$dJ_\pm = \frac{\partial J_\pm}{\partial t} dt + \frac{\partial J_\pm}{\partial x} dx.$$

Show that

$$dJ_\pm = 0, \quad \text{when} \quad \frac{dx}{dt} = \mp c_0.$$

2.22 Derive the results for the reflection and transmission of a linear wave at an interface for the case when $T^0 \neq 0$.

2.23 We recall that Eq. (2.81) is the solution for the velocity v resulting from a linear wave that strikes the interface between two half spaces.

(a) Determine the solution for the stress T.

(b) Use this solution to determine T for a reflection of a linear wave from a rigid boundary.

(c) Compare the solution in part (b) to Eq. (2.85). Show how Figures 2.28 and 2.29, which illustrate v during a reflection from a *free* boundary, can be interpreted as illustrations of T during a reflection from a *rigid* boundary.

(d) Determine the solution for T for a reflection from a free boundary.

(e) Compare the solution in part (d) to Eq. (2.83). Show how Figures 2.24 and 2.25, which illustrate v during a reflection from a *rigid* boundary, can be interpreted as illustrations of T during a reflection from a *free* boundary.

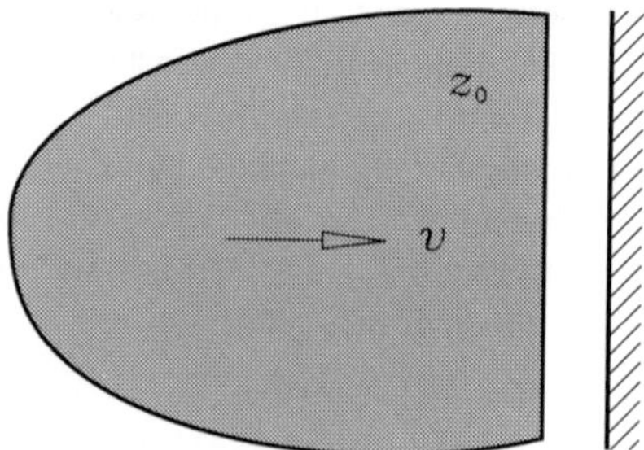

Figure 2.31 Exercise 2.25.

2.24 Consider the sinusoidal wave

$$v = A \sin[\omega(t - x/c_0)],$$

where A is a constant. Suppose this wave is incident on a free boundary of a half space that exists in the region $x \leq 0$.

(a) Determine the solution for v that results from the incident and reflected waves.

(b) Show that this solution is a standing wave, because it can be expressed as the product of a function of x and a function of t (see Exercise 2.13).

(c) Repeat parts (a) and (b) for a half space with a rigid boundary.

2.25 A stress-free half space undergoing pure translation at velocity v strikes a rigid wall (see Figure 2.31).

(a) Use the Riemann invariants $J_\pm$ to determine the stress resulting from the impact.

(b) Draw the T–v diagram for this impact. Assume the half space is nylon ($\rho_0 = 1.12$ Mg/m^3, $c_0 = 2.6$ km/s), and the velocity is $v = 10$ m/s.

2.26 Suppose the nylon half space in the preceding exercise strikes a stationary half space of aluminum ($\rho_0 = 2.7$ Mg/m^3, $c_0 = 6.42$ km/s).

(a) Draw the T–v diagram for this impact and use this diagram to determine the stress and velocity at the impact plane.

(b) Derive an analytical expression for T and v at the impact plane. Compare this solution to the results from part (a).

2.5 Riemann Integrals

In the previous section, we restricted our discussion to linear elasticity where the gradients of the displacement are assumed to be small. This simplification allowed us to combine the equation of conservation of mass, the equation of balance of linear momentum, and the constitutive model for a linear-elastic material into a single, linear, second-order partial differential equation called the wave equation. For a material with uniform properties, the velocity of sound c_0 is constant, and the linear wave equation can be integrated to yield a general solution called the d'Alembert solution. In this section we shall drop these restrictions and consider large deformation of nonlinear-elastic materials. These materials are described by the far more complex nonlinear-elastic equations, and because these equations are nonlinear, the d'Alembert solution no longer applies. However, the nonlinear-elastic equations still have solutions that can be evaluated by integration along the characteristic contours.

In the linear formulation, the characteristic contours are straight lines and the Riemann invariants are algebraic combinations of velocity and stress that are constant along these contours. In the nonlinear formulation, we shall see that the characteristic contours are curved. The curvatures of the contours are functions of the dependent variables of the solution, and the Riemann invariants are integral combinations of velocity and stress that are constant along the characteristic contours. In this section we consider large deformation, and we present both the material and spatial derivations of these solutions.

2.5.1 *Material Description*

The material forms of the nonlinear-elastic equations are summarized in Eqs. (2.30)–(2.33). We omit the first of these equations and eliminate ρ from the solution. This results in a balanced system of three equations,

$$T = T(F), \tag{2.86}$$

$$\frac{\partial \hat{F}}{\partial t} = \frac{\partial \hat{v}}{\partial X}, \tag{2.87}$$

$$\rho_0 \frac{\partial \hat{v}}{\partial t} = \frac{\partial \hat{T}}{\partial X}, \tag{2.88}$$

in three unknown dependent variables, T, F, and v. As in our analysis of the wave equation, we search for the characteristic contours of this system of equations. If these contours exist, we can integrate these equations and obtain a general solution.

Let us assume the coordinates that describe the characteristic contours are ζ and ξ:

$$\zeta = \hat{\zeta}(X, t), \qquad \xi = \hat{\xi}(X, t). \tag{2.89}$$

Differentiation yields

$$d\zeta = \frac{\partial \hat{\zeta}}{\partial X} dX + \frac{\partial \hat{\zeta}}{\partial t} dt, \qquad d\xi = \frac{\partial \hat{\xi}}{\partial X} dX + \frac{\partial \hat{\xi}}{\partial t} dt.$$

We constrain these functions with the following conditions:

$$\frac{\partial \hat{\zeta}}{\partial t} = C \frac{\partial \hat{\zeta}}{\partial X}, \qquad \frac{\partial \hat{\xi}}{\partial t} = -C \frac{\partial \hat{\xi}}{\partial X}, \tag{2.90}$$

where C is an arbitrary positive function. Combination of these two sets of equations yields

$$d\zeta = \frac{\partial \hat{\zeta}}{\partial X}(dX + C dt), \qquad d\xi = \frac{\partial \hat{\xi}}{\partial X}(dX - C dt).$$

We illustrate the coordinates ζ and ξ in the X–t diagram in Figure 2.32. The slopes of the ζ and ξ contours are obtained by alternately setting $d\zeta$ and $d\xi$ equal to zero:

$$\begin{aligned} \frac{dX}{dt} &= C, && \text{for } d\xi = 0, \\ \frac{dX}{dt} &= -C, && \text{for } d\zeta = 0. \end{aligned} \tag{2.91}$$

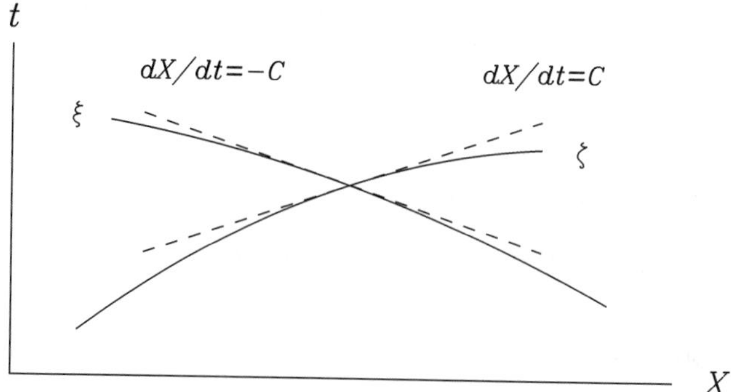

Figure 2.32 The X–t diagram with the ζ–ξ coordinate system. The dashed lines denote the slopes at the intersection of the coordinates.

To solve Eqs. (2.86)–(2.88), we define

$$C \equiv \left(\frac{1}{\rho_0}\frac{dT}{dF}\right)^{1/2} > 0. \tag{2.92}$$

Thus C is a function of $T = T(F)$, and when $F = 1$ we obtain the velocity of sound $C = c_0$ [see Eqs. (2.21) and (2.38)]. We call C the *material wave velocity*, because it defines the slope of the characteristic contours in the X–t diagram of the material description.

The nonlinear-elastic equations, Eqs. (2.86)–(2.88), contain derivatives with respect to X and t. Our objective is to express them in terms of derivatives with respect to ξ and ζ. By taking the partial derivative of the nonlinear-elastic constitutive model with respect to time, we obtain

$$\frac{\partial \hat{T}}{\partial t} = \frac{dT}{dF}\frac{\partial \hat{F}}{\partial t} = \frac{dT}{dF}\frac{\partial \hat{v}}{\partial X} = \rho_0 C^2 \frac{\partial \hat{v}}{\partial X}, \tag{2.93}$$

where we have substituted Eq. (2.87). If we define the *wave impedance* to be

$$z \equiv \rho_0 C,$$

then Eq. (2.93) and the equation of balance of linear momentum, Eq. (2.88), become

$$C\frac{\partial \hat{v}}{\partial X} = \frac{1}{z}\frac{\partial \hat{T}}{\partial t}, \qquad \frac{\partial \hat{v}}{\partial t} = \frac{C}{z}\frac{\partial \hat{T}}{\partial X}. \tag{2.94}$$

Notice that F has been eliminated from these equations, and we are left with two equations in the two unknown variables v and T. Adding these two equations we obtain

$$\frac{\partial \hat{v}}{\partial t} + C\frac{\partial \hat{v}}{\partial X} = \frac{1}{z}\left(\frac{\partial \hat{T}}{\partial t} + C\frac{\partial \hat{T}}{\partial X}\right). \tag{2.95}$$

The velocity and stress can be represented as

$$v = \check{v}(\zeta, \xi), \qquad T = \check{T}(\zeta, \xi), \tag{2.96}$$

where we have used a breve (˘) to denote that v and T are functions of ζ and ξ. Using the chain rule, we obtain

$$\frac{\partial \hat{v}}{\partial t} = \frac{\partial \breve{v}}{\partial \zeta}\frac{\partial \hat{\zeta}}{\partial t} + \frac{\partial \breve{v}}{\partial \xi}\frac{\partial \hat{\xi}}{\partial t}.$$

By similar expansion of the remaining derivatives in Eq. (2.95), we obtain

$$\left(\frac{\partial \breve{v}}{\partial \zeta} - \frac{1}{z}\frac{\partial \breve{T}}{\partial \zeta}\right)\left(\frac{\partial \hat{\zeta}}{\partial t} + C\frac{\partial \hat{\zeta}}{\partial X}\right) + \left(\frac{\partial \breve{v}}{\partial \xi} - \frac{1}{z}\frac{\partial \breve{T}}{\partial \xi}\right)\left(\frac{\partial \hat{\xi}}{\partial t} + C\frac{\partial \hat{\xi}}{\partial X}\right) = 0.$$

Substitution of Eqs. (2.90) yields

$$\frac{\partial \breve{v}}{\partial \zeta} = \frac{1}{z}\frac{\partial \breve{T}}{\partial \zeta}. \tag{2.97}$$

Similarly, when we subtract Eqs. (2.94), we obtain

$$\frac{\partial \breve{v}}{\partial \xi} = -\frac{1}{z}\frac{\partial \breve{T}}{\partial \xi}. \tag{2.98}$$

These two expressions are called the *characteristic equations* of the nonlinear-elastic equations. Whereas the nonlinear-elastic equations contain partial derivatives in both the X and t coordinates, the individual characteristic equations contain partial derivatives in only one of the characteristic coordinates. When we integrate these equations from the reference configuration where $T = T^0$ and $v = v^0$ to the current configuration, we obtain

$$\begin{aligned} J_+(\xi) &= v - v^0 - \int_{T^0}^{T} \frac{d\bar{T}}{z(\bar{T})}, \\ J_-(\zeta) &= v - v^0 + \int_{T^0}^{T} \frac{d\bar{T}}{z(\bar{T})}, \end{aligned} \tag{2.99}$$

where $\bar{T}$ is an integration variable and $J_\pm$ are arbitrary functions called *Riemann integrals*. You can easily verify these solutions by taking the partial derivatives of the first expression with respect to ζ and the second expression with respect to ξ. We obtain an alternate form of the Riemann integrals by recalling that C is a function of $T = T(F)$, and so $dT = \rho_0 C^2 dF$. Then

$$\begin{aligned} J_+(\xi) &= v - v^0 - \int_{1}^{F} C(\bar{F})\, d\bar{F}, \\ J_-(\zeta) &= v - v^0 + \int_{1}^{F} C(\bar{F})\, d\bar{F}, \end{aligned} \tag{2.100}$$

where $\bar{F}$ is an integration variable. In this case, the lower limit of integration is the value of the deformation gradient in the reference configuration, $F = 1$. Inverting these expressions we obtain

$$\begin{aligned} v - v^0 &= \frac{1}{2}[J_+(\xi) + J_-(\zeta)], \\ \int_{1}^{F} C(\bar{F})\, d\bar{F} &= \frac{1}{2}[-J_+(\xi) + J_-(\zeta)]. \end{aligned} \tag{2.101}$$

These results are similar to the d'Alembert solutions of the wave equation. However, they are *not* d'Alembert solutions. The primary distinctions between these results and the d'Alembert solutions are that X, t, ζ, and ξ are related through the differential expressions Eq. (2.90) and not through the algebraic expressions Eq. (2.42) in Section 2.3.3. Unlike the characteristic solutions we obtained from the linear wave equation, you will find that it is often very difficult to transform these nonlinear solutions between their respective representations in the ζ–ξ and X–t coordinate systems.

2.5.2 *Spatial Description*

In the previous section we derived the Riemann integral solutions for the material forms of the nonlinear-elastic equations. We now repeat this derivation using the spatial formulation. We summarized the spatial forms of the nonlinear-elastic equations in Eqs. (2.34)–(2.37). They are a balanced system of four equations,

$$
\begin{aligned}
&T = T(F), \qquad F = \frac{\rho_0}{\rho}, \\
&\frac{\partial \rho}{\partial t} + v\frac{\partial \rho}{\partial x} + \rho\frac{\partial v}{\partial x} = 0, \\
&\rho\left(\frac{\partial v}{\partial t} + v\frac{\partial v}{\partial x}\right) = \frac{\partial T}{\partial x},
\end{aligned}
\tag{2.102}
$$

in the four unknown dependent variables, ρ, T, F, and v.

The characteristic coordinates ζ and ξ can be expressed as functions of position x and time t:

$$
\zeta = \zeta(x,t), \qquad \xi = \xi(x,t).
$$

Once again, we constrain these functions with Eqs. (2.90). Using the motion $x = \hat{x}(X,t)$, we expand the partial derivatives appearing in Eqs. (2.90) to obtain

$$
\begin{aligned}
\frac{\partial \hat{\zeta}}{\partial X} &= \frac{\partial \zeta}{\partial x}\frac{\partial \hat{x}}{\partial X}, \qquad & \frac{\partial \hat{\zeta}}{\partial t} &= \frac{\partial \zeta}{\partial x}\frac{\partial \hat{x}}{\partial t} + \frac{\partial \zeta}{\partial t}, \\
\frac{\partial \hat{\xi}}{\partial X} &= \frac{\partial \xi}{\partial x}\frac{\partial \hat{x}}{\partial X}, \qquad & \frac{\partial \hat{\xi}}{\partial t} &= \frac{\partial \xi}{\partial x}\frac{\partial \hat{x}}{\partial t} + \frac{\partial \xi}{\partial t}.
\end{aligned}
$$

Noticing that $F = \partial\hat{x}/\partial X$ and $v = \partial\hat{x}/\partial t$, Eqs. (2.90) yield the following constraints on the functions $\zeta(x,t)$ and $\xi(x,t)$:

$$
\frac{\partial \zeta}{\partial t} = (c - v)\frac{\partial \zeta}{\partial x}, \qquad \frac{\partial \xi}{\partial t} = -(c + v)\frac{\partial \xi}{\partial x}, \tag{2.103}
$$

where we have defined

$$
c \equiv FC > 0. \tag{2.104}
$$

Using these results, the differentials of ζ and ξ become

$$
d\zeta = \frac{\partial \zeta}{\partial x}dx + \frac{\partial \zeta}{\partial t}dt = \frac{\partial \zeta}{\partial x}[dx - (v - c)dt]
$$

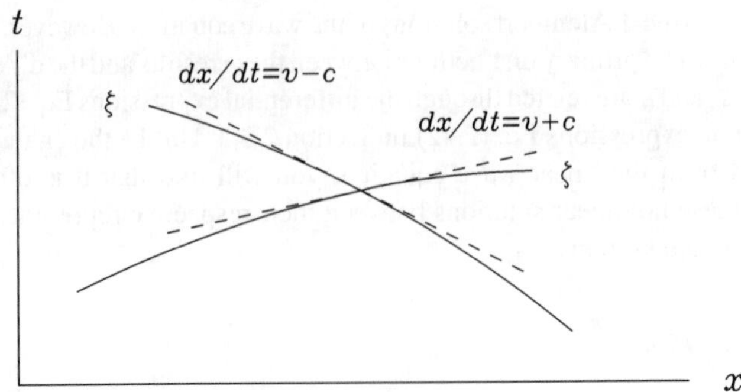

Figure 2.33 The x–t diagram with the ζ–ξ coordinate system. The dashed curves denote the slopes at the intersection of the ζ and ξ coordinates.

and

$$d\xi = \frac{\partial \xi}{\partial x}dx + \frac{\partial \xi}{\partial t}dt = \frac{\partial \xi}{\partial x}[dx - (v + c)dt].$$

We obtain the slopes of the ζ and ξ contours by alternately setting $d\zeta$ and $d\xi$ equal to zero:

$$\begin{aligned} \frac{dx}{dt} &= v + c, \quad \text{for } d\xi = 0, \\ \frac{dx}{dt} &= v - c, \quad \text{for } d\zeta = 0. \end{aligned} \tag{2.105}$$

We have illustrated the coordinates ζ and ξ in the x–t diagram in Figure 2.33. Notice how the velocity v alters the slopes of the contours. We call $c = FC$ the *spatial wave velocity* because, in conjunction with the material velocity v, it defines the slope of the characteristic contours in the x–t diagram of the spatial description.

The nonlinear-elastic equations, Eqs. (2.102), contain derivatives with respect to x and t. Our objective is to express them in derivatives with respect to ξ and ζ. Taking the partial derivative of the nonlinear-elastic constitutive model with respect to time, we obtain

$$\frac{\partial T}{\partial t} = \frac{dT}{dF}\frac{\partial F}{\partial t} = \frac{dT}{dF}\frac{\partial}{\partial t}\left(\frac{\rho_0}{\rho}\right) = -\left(\frac{\rho_0}{\rho}\right)^2\left(\frac{1}{\rho_0}\frac{dT}{dF}\right)\frac{\partial \rho}{\partial t},$$

or

$$\frac{\partial T}{\partial t} = -c^2\frac{\partial \rho}{\partial t}.$$

Similarly,

$$\frac{\partial T}{\partial x} = -c^2\frac{\partial \rho}{\partial x}.$$

After the equation of conservation of mass, Eq. (2.102), is multiplied by $-c^2$, we substitute these results to find that

$$\frac{\partial T}{\partial t} + v\frac{\partial T}{\partial x} = \rho c^2\frac{\partial v}{\partial x}.$$

Noticing that the wave impedance is

$$z = \rho_0 C = \rho F C = \rho c,$$

we rewrite the equation of conservation of mass and the equation of balance of linear momentum to obtain

$$\begin{aligned} c\frac{\partial v}{\partial x} &= \frac{1}{z}\left(\frac{\partial T}{\partial t} + v\frac{\partial T}{\partial x}\right), \\ \frac{\partial v}{\partial t} + v\frac{\partial v}{\partial x} &= \frac{c}{z}\frac{\partial T}{\partial x}. \end{aligned} \tag{2.106}$$

We add these two expressions to obtain

$$\frac{\partial v}{\partial t} + (v+c)\frac{\partial v}{\partial x} = \frac{1}{z}\left[\frac{\partial T}{\partial t} + (v+c)\frac{\partial T}{\partial x}\right].$$

Using the chain rule expansion of the partial derivatives of this equation, we find that

$$\left[\frac{\partial \breve{v}}{\partial \zeta} - \frac{1}{z}\frac{\partial \breve{T}}{\partial \zeta}\right]\left[\frac{\partial \zeta}{\partial t} + (v+c)\frac{\partial \zeta}{\partial x}\right] + \left[\frac{\partial \breve{v}}{\partial \xi} - \frac{1}{z}\frac{\partial \breve{T}}{\partial \xi}\right]\left[\frac{\partial \xi}{\partial t} + (v+c)\frac{\partial \xi}{\partial x}\right] = 0,$$

where $\breve{v}$ and $\breve{T}$ are functions of ξ and ζ as defined in Eq. (2.96). When Eqs. (2.103) are substituted into this result, we obtain

$$\frac{\partial \breve{v}}{\partial \zeta} = \frac{1}{z}\frac{\partial \breve{T}}{\partial \zeta}.$$

Similarly, by subtraction of Eqs. (2.106), we see that

$$\frac{\partial \breve{v}}{\partial \xi} = -\frac{1}{z}\frac{\partial \breve{T}}{\partial \xi}.$$

These are the same characteristic equations that we derived from the material forms of the nonlinear-elastic equations. They yield the same Riemann integrals $J_\pm$ [see Eqs. (2.99)].

2.5.3 *Initial and Boundary Conditions*

The Riemann integrals are solutions to the nonlinear-elastic equations. The solutions for the velocity v and deformation gradient F are [see Eqs. (2.101)]

$$\begin{aligned} v - v^0 &= \frac{1}{2}[J_+(\xi) + J_-(\zeta)], \\ \int_1^F C(\bar{F})\,d\bar{F} &= \frac{1}{2}[-J_+(\xi) + J_-(\zeta)]. \end{aligned}$$

These nonlinear solutions are expressed in the characteristic coordinates ξ and ζ. To obtain the solutions in terms of the material or spatial coordinates, we must solve one of two sets of equations. The choices are:

- Solve Eqs. (2.90) for $\hat{\xi}(X, t)$ and $\hat{\zeta}(X, t)$.
- Solve Eqs. (2.103) for $\xi(x, t)$ and $\zeta(x, t)$.

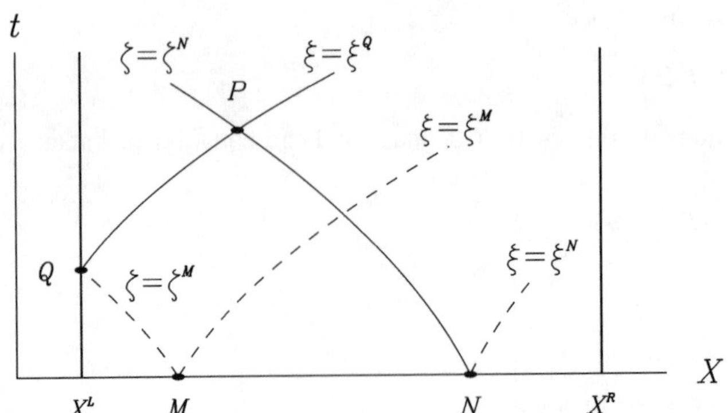

Figure 2.34 The X–t diagram showing the characteristic contours that determine the solution at the point P.

Either choice requires the solution of a pair of nonlinear, partial differential equations. At first it may appear that we have avoided the solution of the nonlinear-elastic equations by replacing them with a second set of equations that are just as difficult to solve. Clearly, in special cases when the transformations from characteristic coordinates to the material and spatial coordinates are reasonably routine, this solution method offers significant advantages over direct solution of the nonlinear-elastic equations. However, this is not the only reason we study these solutions. In fact, they offer several other important advantages. Indeed, they lead to a powerful method of numerical analysis. Later in this chapter we also use them to provide insight and intuition into the nature of nonlinear wave propagation. And, as we shall now show, they can be used to determine the initial conditions and boundary conditions that result in unique solutions of the nonlinear-elastic equations.

Consider the X–t diagram in Figure 2.34. This diagram has an interior region that is enclosed by the left boundary of the material at X^L, the right boundary at X^R, and the initial line at $t = 0$. The solution in the interior region of the diagram is determined by the conditions that we prescribe on these three sides. Our objective is to determine what initial data we must prescribe on $t = 0$ and what boundary data we must prescribe on $X = X^R$ and $X = X^L$ in order to define a unique solution in the interior region.

We can determine exactly what data are required by examining the Riemann integral solution at an arbitrary point P. The values of the characteristic coordinates at the point P are $\xi = \xi^Q$ and $\zeta = \zeta^N$. The velocity and deformation gradient at P are

$$v^P - v^0 = \frac{1}{2}[J_+(\xi^Q) + J_-(\zeta^N)],$$

$$\int_1^{F^P} C(\bar{F})\,d\bar{F} = \frac{1}{2}[-J_+(\xi^Q) + J_-(\zeta^N)].$$

Thus, the values of v and F at P are given by $J_+(\xi^Q)$ and $J_-(\zeta^N)$. Let us evaluate $J_-(\zeta^N)$ first. Suppose that we prescribe initial data composed of values of F and v everywhere along $t = 0$. From these initial data, we determine J_+ and J_- everywhere on the line $t = 0$. We notice that the points P and N are on the same contour $\zeta = \zeta^N$. Thus these initial data determine $J_-(\zeta^N)$.

To evaluate $J_+(\xi^Q)$, we notice that the points P and Q are on contour $\xi = \xi^Q$. Suppose that we prescribe the boundary data so that $v = v^Q$ at the point Q. Then

$$J_+(\xi^Q) = 2(v^Q - v^0) - J_-(\zeta^M).$$

Because $J_-(\zeta^M)$ is also determined by the initial data, this expression determines $J_+(\xi^Q)$. Alternatively, we could have prescribed $F = F^Q$ at the point Q instead of $v = v^Q$. Then

$$J_+(\xi^Q) = -2\int_1^{F^Q} C(\bar{F})\,d\bar{F} + J_-(\zeta^M).$$

Again, we obtain an expression for $J_+(\xi^Q)$. We conclude that the boundary data should contain values for either F or v, but not both. It is easily demonstrated that the solutions for F and v at any point in the interior are either determined by initial data alone or by a combination of initial data and boundary data as discussed above. *Thus the solution at any point in the interior is determined by specifying F and v as initial data and either F or v as boundary data.* Although we have obtained these results by considering the X–t diagram, it the clear that the same arguments and conclusions apply to the x–t diagram.

Domain of Dependence The concept of *domain of dependence* proves helpful to understanding initial-value and boundary-value problems. This concept is most easily introduced by returning to the initial-value problem for the linear wave equation (see Figure 2.35). This x–t diagram is similar to Figure 2.18. We have changed the labels to facilitate the discussion. Consider the solution to the wave equation at point P in this diagram. This point resides on the characteristic contours $\xi = \xi^{P_+}$ and $\zeta = \zeta^{P_-}$.

The solution at P is given by the values of the Riemann invariants $J_+(\xi^{P_+})$ and $J_-(\zeta^{P_-})$. These invariants are determined by the initial data at the two points P_+ and P_-. The value of the solution at P depends solely on the initial data at P_+ and P_-. As long as we do not change the initial data at P_+ and P_-, we can change the initial data at other points on $t = 0$ without affecting the solution at P. The domain of dependence of the solution at P only contains the two points P_+ and P_-.

Now consider the x–t diagram for an initial-value problem of the nonlinear-elastic equations [see Figure 2.36 (a)]. Like the diagram for the linear wave equation, the solution at point P depends solely upon the values of the Riemann integrals $J_+(\xi^{P_+})$ and $J_-(\zeta^{P_-})$.

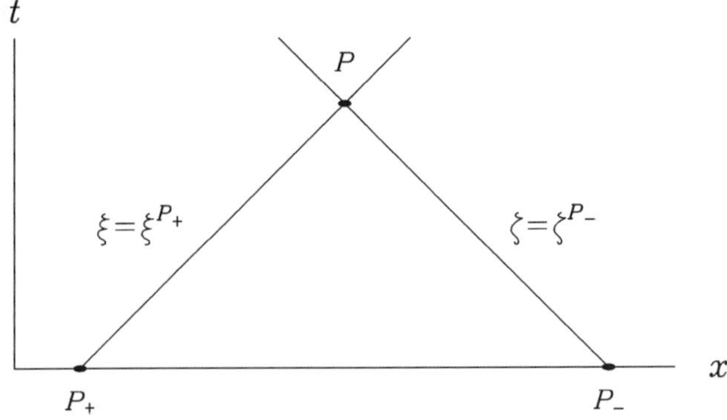

Figure 2.35 The x–t diagram for an initial-value problem of the wave equation.

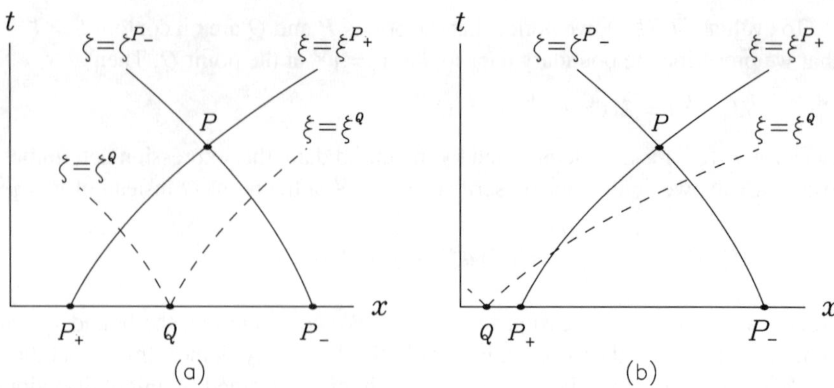

Figure 2.36 The x–t diagram for an initial-value problem of the nonlinear-elastic equations when the initial data are changed (a) between the points P_+ and P_- and (b) outside the points P_+ and P_-.

These integrals are also determined by the initial data for deformation gradient F and the velocity v on the line $t = 0$. We might be tempted to conclude that the value of the solution at P depends solely on the initial data at P_+ and P_-, but this is not true. Consider the point Q between P_+ and P_-. If we change the initial data in some small neighborhood of Q, the values of the Riemann integrals change in the neighborhood of the dashed contours $\xi = \xi^Q$ and $\zeta = \zeta^Q$. This in turn changes the shapes of the solid contours $\xi = \xi^{P_+}$ and $\zeta = \zeta^{P_-}$ so that they no longer intersect at P. Thus if we change the initial data anywhere between P_+ and P_-, we change the solution at P. Therefore, the domain of dependence of the solution at P contains the points P_+ and P_- plus the line between P_+ and P_-.

What if we change the initial data in the neighborhood of a point Q that is outside of the region between P_+ and P_- [see Figure 2.36 (b)]? This change can only influence the solution at P when the contour $\xi = \xi^Q$ intersects the contour $\xi = \xi^{P_+}$. At such an intersection the solution is multivalued. So far, we have only considered single valued solutions. We put this possibility aside for the moment but will return to it in a future section.

Range of Influence Consider the initial-value problem shown in Figure 2.37. Suppose we change the initial data in the cross-hatched neighborhood of Q. The range of influence of this neighborhood of Q is defined to be the region of the x–t diagram that changes when the initial data change. In the illustrated example, the solution at P will change, but the solutions at R and S will not change. Thus the range of influence of Q includes P but not R or S. We can also say that Q cannot influence S because there is not enough time for a wave to travel from Q to S. In Chapter 0 we referred to this as causality.

2.5.4 *Riemann Integrals for the Adiabatic Gas*

The adiabatic gas is described by the simple constitutive model [see Eq. (2.28)],

$$T = -p_0 F^{-\gamma}.$$

We shall derive the expressions for the wave velocities and Riemann integrals of an adiabatic gas in this section.

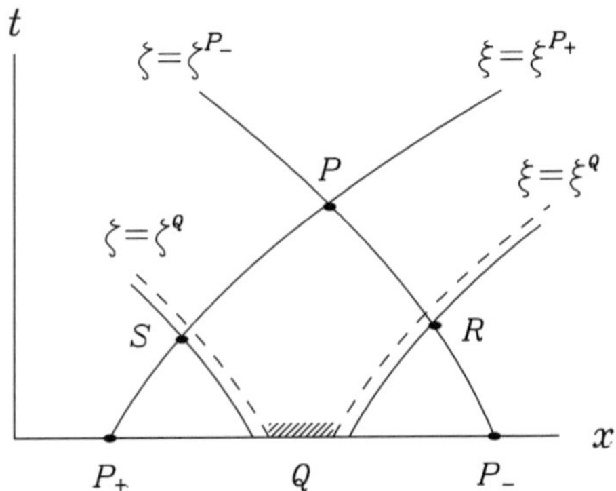

Figure 2.37 The x–t diagram for an initial-value problem. The initial data are changed in the cross-hatched neighborhood of Q.

From our analysis of the linear wave equation, we know that the velocity of sound is [see Eq. (2.38)]

$$c_0 = \left(\frac{\lambda + 2\mu}{\rho_0}\right)^{1/2} = \left(\frac{\gamma p_0}{\rho_0}\right)^{1/2}, \tag{2.107}$$

where for an adiabatic gas $\lambda = \gamma p_0$ and $\mu = 0$ [see Eq. (2.29)]. Let us see how the velocity of sound c_0 is related to the material wave velocity C and the spatial wave velocity c. First we evaluate the derivative of the stress:

$$\frac{dT}{dF} = \gamma p_0 F^{-\gamma-1}.$$

We substitute this result into the expression for the material wave velocity [see Eq. (2.92)]

$$C = \left(\frac{1}{\rho_0}\frac{dT}{dF}\right)^{1/2}$$

to obtain

$$C = c_0 F^{-\frac{\gamma+1}{2}}. \tag{2.108}$$

Moreover, the spatial wave velocity is

$$c = FC = c_0 F^{-\frac{\gamma-1}{2}}. \tag{2.109}$$

Notice that when $F = 1$, then $C = c = c_0$. We also notice that F can be eliminated from these results to obtain the following useful relationship between these three wave velocities (see Exercise 2.30):

$$\frac{C}{c_0} = \left(\frac{c}{c_0}\right)^{\frac{\gamma+1}{\gamma-1}}. \tag{2.110}$$

Next, we use these expressions to evaluate the Riemann integrals for an adiabatic gas. The integral of the material velocity used in the Riemann solution is

$$\int_1^F C(\bar{F})\, d\bar{F} = \frac{2c_0}{1-\gamma} F^{(1-\gamma)/2}\bigg|_1^F = -\frac{2}{\gamma-1}(c - c_0). \tag{2.111}$$

Thus, the Riemann integrals are

$$J_+ = v - v^0 + \frac{2}{\gamma - 1}(c - c_0), \qquad J_- = v - v^0 - \frac{2}{\gamma - 1}(c - c_0). \tag{2.112}$$

2.5.5 *Exercises*

2.27 Derive the characteristic equation

$$\frac{\partial \breve{v}}{\partial \xi} = -\frac{1}{z}\frac{\partial \breve{T}}{\partial \xi}$$

from Eq. (2.94).

2.28 Verify that the Riemann integrals

$$J_+(\xi) = v - v^0 - \int_{T^0}^{T} \frac{d\bar{T}}{z(\bar{T})},$$

$$J_-(\zeta) = v - v^0 + \int_{T^0}^{T} \frac{d\bar{T}}{z(\bar{T})}$$

are solutions to the characteristic equations by substituting them into Eqs. (2.97)–(2.98).

2.29 Derive the characteristic equation

$$\frac{\partial \breve{v}}{\partial \xi} = -\frac{1}{z}\frac{\partial \breve{T}}{\partial \xi}$$

from Eqs. (2.106).

2.30 The wave velocities for an adiabatic gas are

$$C = c_0 F^{-\frac{\gamma+1}{2}}, \qquad c = c_0 F^{-\frac{\gamma-1}{2}}.$$

Combine these relationships to obtain

$$\frac{C}{c_0} = \left(\frac{c}{c_0}\right)^{\frac{\gamma+1}{\gamma-1}}.$$

2.31 Consider the X–t diagram in Figure 2.34. Suppose this diagram represents an adiabatic gas where $p = p_0 F^{-\gamma}$. Recall that $p_0 > 0$, and $\gamma > 1$. The initial pressure and velocity are $p(X, 0) = p_0$ and $v(X, t) = 0$. Suppose, at $t > 0$, we subject the left boundary to zero pressure, $p(X^L, t) = 0$. Show that the velocity at point P is

$$v = -\frac{2c_0}{\gamma - 1},$$

where c_0 is the initial velocity of sound in the gas.

2.6 Structured Waves

A *structured wave* is a solution to the nonlinear-elastic equations in which the velocity v and deformation gradient F are continuous functions. When these functions are discontinuous, the solution is called a *shock wave.* We shall discuss shock waves in the next section. In our analysis of linear waves we make no distinction between continuous and discontinuous solutions. This seems reasonable because in nature these waves exhibit similar behavior. In contrast, nonlinear structured waves and shock waves exhibit remarkable differences. In this section, we use the Riemann integrals to analyze structured waves. Our approach parallels the methods we employed in the analysis of the linear wave equation.

2.6.1 *Simple Waves*

In our discussion of the linear wave equation we define a simple wave to be a solution in which one of the Riemann invariants of the wave equation is constant. Let us call this simple wave solution of the wave equation a *linear simple wave*. We know that a linear simple wave propagates with constant shape in one direction. The Riemann invariant of the wave equation $J_+(\xi)$ represents a linear simple wave propagating in the positive-x direction, while $J_-(\zeta)$ represents a linear simple wave propagating in the negative-x direction. We conclude that the general solution to the wave equation is the sum of two linear simple waves propagating in opposite directions.

We define the *nonlinear simple wave* in a similar manner. It is a structured wave solution to the nonlinear-elastic equations in which one of the Riemann integrals is constant. We shall show that a nonlinear simple wave has straight characteristic contours. The resulting Riemann integrals of the nonlinear-elastic equations are similar in form to the Riemann invariants of the linear wave equation. However, unlike the linear simple wave, these nonlinear solutions exist and the contours remain straight only so long as one of the Riemann integrals is a constant. This means that *the general solution of the nonlinear-elastic equations is not a sum of nonlinear simple waves.* We shall also show that the contours of a nonlinear simple wave are not parallel. They often converge and intersect. When this occurs, the simple wave solution ceases to be valid.

Consider a nonlinear simple wave where

$$J_-(\zeta) = 0.$$

The Riemann solution, Eq. (2.101), for this wave is

$$v - v^0 = \frac{1}{2} J_+(\xi),$$

$$\int_1^F C(\bar{F})\, d\bar{F} = -\frac{1}{2} J_+(\xi).$$

Thus

$$v - v^0 = -\int_1^F C(\bar{F})\, d\bar{F}\,. \tag{2.113}$$

We differentiate this result to obtain

$$\frac{dv}{dF} = -C. \tag{2.114}$$

When these results are combined with $c = FC$ [see Eq. (2.104)], we find that

$$v = \breve{v}(\xi), \quad F = \breve{F}(\xi),$$

$$C = \breve{C}(\xi), \quad c = \breve{c}(\xi).$$

Because $v + c$ and C are functions of ξ alone, the shape of the ξ contours are easily determined. To illustrate, let us select a set of scalar constants and denote them by ξ^i, where $i = 0, 1, 2, \ldots$. Then $\xi = \xi^i$ defines a set of characteristic contours. We determine the shape of these contours with respect to the material coordinates X and t by solving Eq. (2.91),

$$\frac{dX}{dt} = C(\xi^i).$$

Because we are integrating this equation subject to the condition $\xi = \xi^i$, the wave velocity $C(\xi^i)$ can be treated as a constant. The result, of course, is the equation of the straight line

$$\frac{X}{C(\xi^i)} = t - t^i,$$

where t^i is a constant of integration. Hence each characteristic contour is a straight line; albeit, each contour has a different slope. For the constant of integration, let us select $t^i = \xi^i$. Then we replace ξ^i by ξ in the previous expression to obtain

$$\xi = t - \frac{X}{C(\xi)}.$$

We combine this result with the Riemann integral solutions to obtain

$$v - v^0 = \frac{1}{2} J_+(t - X/C),$$

$$\int_1^F C(\bar{F})\, d\bar{F} = -\frac{1}{2} J_+(t - X/C).$$

These results have the appearance of the d'Alembert solution for a linear simple wave propagating in the positive-X direction. In fact, for the special case of a linear-elastic material where $C = c_0$, this is the d'Alembert solution. However, in a nonlinear-elastic material $C = \check{C}(\xi, \zeta)$. When considering the nonlinear simple wave, we obtain $C = \check{C}(\xi)$ only because of the requirement that $J_-(\zeta)$ is constant. As a result, C is only constant along the characteristic contour $\xi =$ constant and only as long as $J_-(\zeta)$ is a constant.
Recall that [see Eq. (2.23)]

$$\frac{d^2T}{dF^2} \leq 0.$$

This condition gives

$$\frac{dC}{dF} \leq 0. \tag{2.115}$$

Consequently, when nonlinear-elastic materials are compressed, the material wave velocity increases as F decreases. Combining this result with Eq. (2.114), we obtain

$$\frac{dC}{dv} \geq 0. \tag{2.116}$$

For a simple wave where $J_-(\zeta)$ is constant, the material wave velocity increases as v increases. In contrast, if we consider another simple wave where $J_-(\zeta)$ varies and $J_+(\xi)$ is constant, then C will increase as v decreases (see Exercise 2.32).

A similar situation exists in the spatial coordinates x and t. In this case, we determine the shape of the contours by solving Eq. (2.105),

$$\frac{dx}{dt} = v(\xi^i) + c(\xi^i).$$

We find that these contours are the straight lines

$$\xi = t - \frac{x}{v(\xi) + c(\xi)},$$

and the Riemann integral solutions are

$$v - v^0 = \frac{1}{2} J_+[t - x/(v + c)],$$

$$\int_1^F C(\bar{F})\, d\bar{F} = -\frac{1}{2} J_+[t - x/(v + c)].$$

From Eq. (2.114) and $c = FC$, we obtain

$$\frac{d(v + c)}{dF} = F \frac{dC}{dF}.$$

Because $F > 0$, from Eq. (2.115), we obtain

$$\frac{d(v + c)}{dF} \leq 0. \tag{2.117}$$

Thus $v + c$ increases for decreasing values of F and increasing values of v. For a simple wave traveling in the opposite direction, similar relationships exist (see Exercise 2.32). Notice that Eq. (2.115) applies to all structured waves, whereas Eqs. (2.116)–(2.117) are restricted to simple waves.

T–v Diagram of a Simple Wave

Recall our discussion of the T–v diagram for linear simple waves (see Section 2.4.3). We discovered that this diagram is a convenient method for tracking the changes in stress and velocity caused by these waves. In this section we extend this method to nonlinear simple waves.

The Riemann integrals, Eqs. (2.99), are

$$J_\pm = v - v^0 \mp \int_{T^0}^{T} \frac{d\bar{T}}{z(\bar{T})}.$$

Because J_- is constant for a simple wave propagating in the positive-x direction and J_+ is constant for a simple wave propagating in the negative-x direction, we first differentiate the expression for J_+ and then the expression for J_- to obtain the following two relationships for simple waves:

$$\frac{d(-T)}{dv} = \pm z(T). \tag{2.118}$$

The wave impedance z is a function of the stress. It is determined from the constitutive model of the material.

Consider a material point $\mathbf{X}$ in a body where the velocity and stress are known. It is represented by the point I in the T–v diagram (see Figure 2.38). We can integrate

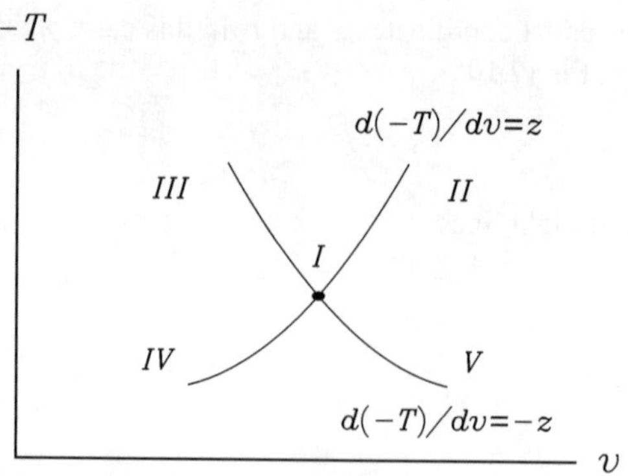

Figure 2.38 The T–v diagram for the material point **X**.

Eqs. (2.118) to obtain two curves that pass through the point I. The material at **X** follows the positive-sloping response path when it encounters a nonlinear simple wave propagating in the positive-x direction. It follows the negative-sloping response path when it encounters a simple wave propagating in the negative-x direction. Because $z = \rho_0 C$, Eq. (2.116) requires that the positive-sloping response path must steepen as v increases. We can also show that the negative-sloping response path must steepen as v decreases.

Linear-Elastic Diagram If the material is linear elastic, then we find that $z = z_0$, and we can integrate Eqs. (2.118) to obtain the two straight lines

$$-(T - T^0) = \pm z_0(v - v_0) \quad \text{for } z = z_0.$$

These results are identical to those we obtained directly from the linear wave equation [see Eq. (2.74)].

Adiabatic-Gas Diagram For an adiabatic gas, it is more convenient to use the pressure $p = -T$, so that

$$\frac{p}{p_0} = F^{-\gamma}.$$

From Eq. (2.108) the material wave velocity is

$$C = c_0 F^{-\frac{\gamma+1}{2}}.$$

By combining these two equations, we can write the wave impedance $z = \rho_0 C$ as

$$z = z_0 \left(\frac{p}{p_0}\right)^{\frac{\gamma+1}{2\gamma}}.$$

We substitute this expression into Eqs. (2.118) to obtain

$$\left(\frac{p}{p_0}\right)^{-\frac{\gamma+1}{2\gamma}} d\left(\frac{p}{p_0}\right) = \pm\frac{\gamma}{c_0} dv,$$

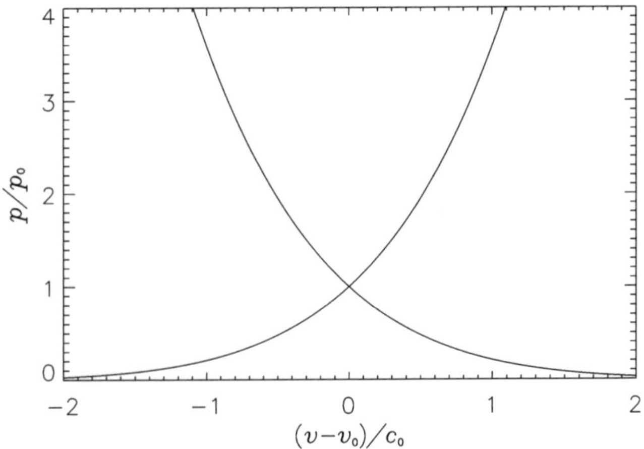

Figure 2.39 The p–v diagram for an adiabatic gas, $\gamma = 1.4$.

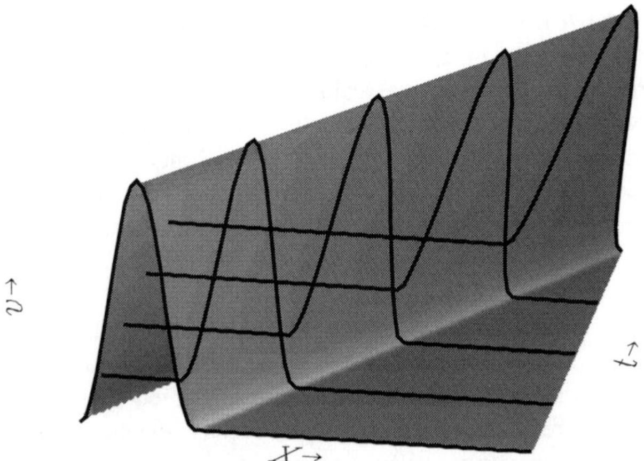

Figure 2.40 Distortion of a simple wave represented in the three-dimensional space defined by v, X, and t. The solid curves represent profiles of the wave at five different times. This solution must be discarded when the slope of the wave passes vertical.

where we have noticed that $z_0/p_0 = \gamma/c_0$. When we integrate this result, subject to the requirement that $p = p_0$ when $v = v^0$, we find that

$$\left(\frac{p}{p_0}\right)^{\frac{\gamma-1}{2\gamma}} - 1 = \pm\frac{\gamma-1}{2c_0}(v - v^0). \tag{2.119}$$

These are the two response paths for an adiabatic gas. We illustrate them in Figure 2.39 for $\gamma = 1.4$, which is the value of γ for air.

Distortion of a Nonlinear Simple Wave

Consider a nonlinear simple wave propagating in the positive-X direction. We show the solution for the velocity $v(X, t)$ of this wave in the three-dimension space defined by the coordinates v, X, and t (see Figure 2.40). The solid curves indicate profiles of the wave at

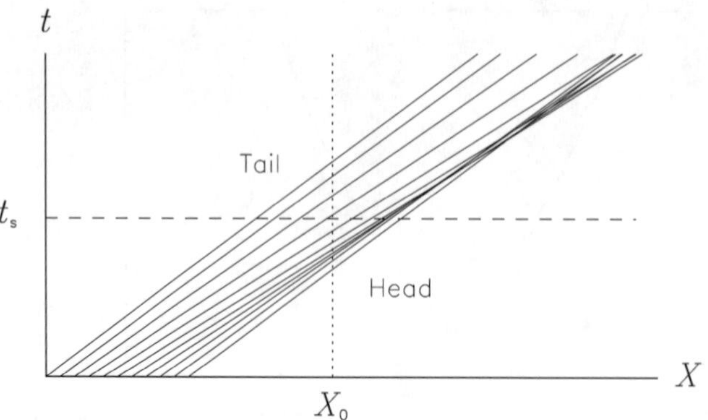

Figure 2.41 The X–t diagram of the simple wave.

five different times. Notice that each of the five wave profiles has a unique shape. As time increases, the leading portion of the wave steepens, and the trailing portion of the wave flattens. We call this process *distortion* of the wave. The wave distorts because even though the characteristic contours are straight, they are not parallel. We show the X–t diagram of the characteristic contours in Figure 2.41. Because velocity is a function of ξ only, the characteristic contours are also lines of constant velocity v. The portions of the wave where v is small propagate at the velocity of sound c_0, and because C increases with v, all other portions of the wave travel faster. The peak of the wave, where the velocity is maximum, has the highest propagation velocity. The regions of the material in front of and behind the wave do not move. The head and tail contours of the wave separate the interior region of the wave from these regions of zero velocity.

The trouble with the image in Figure 2.40 is that it has the appearance of a water wave cresting upon a beach, and Figure 2.41 shows that some of the characteristic contours intersect the head contour of the wave. There are similarities between the simple wave and the water wave. Water waves are nonlinear and they crest because the peak of the wave has the greatest wave velocity. However, water waves exist in the three-dimensional space X_k, and Figure 2.40 represents a nonlinear simple wave in the one-dimensional space $X = X_1$. The implication of this cresting is that certain material points $\mathbf{X}$ have multiple and distinct values of velocity v. Because we require that a material point $\mathbf{X}$ must have a single unique velocity, a nonlinear simple wave with intersecting contours is an unacceptable solution to our problem. We shall resolve this issue in the next section when we discuss shock waves. For the present, we notice that our solution for the nonlinear simple wave is valid only as long as the characteristic contours of the wave do not intersect. In Figure 2.41, we conclude that the solution for the simple wave is valid only while $t \leq t_s$.

Compression and Rarefaction Waves

If we observe the particular simple wave shown in Figure 2.40 as it passes some point X^0 in the body, we notice that the material is first compressed and then expanded by this simple wave. Some simple waves only cause compression, and they are called *compression waves*. Other simple waves only cause expansion, and they are called *rarefaction waves*. Let us examine these two types of waves. In Figure 2.42 we illustrate example histories of the deformation gradient for each type of wave. These histories are graphs of the deformation

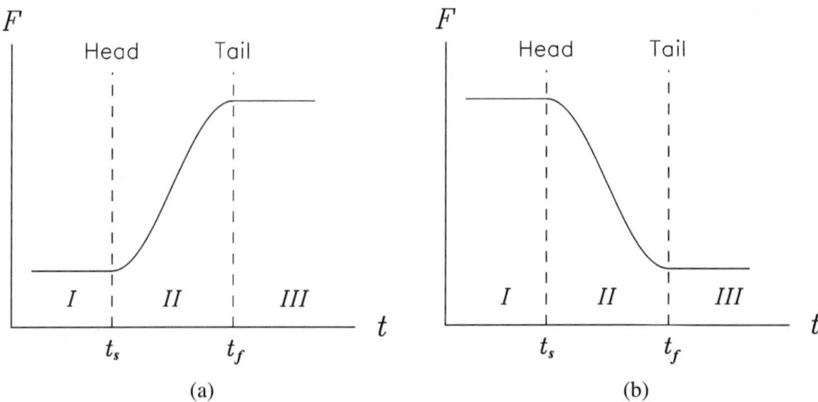

Figure 2.42 The history of the deformation gradient measured at the material point X^0 for (a) a rarefaction wave and (b) a compression wave.

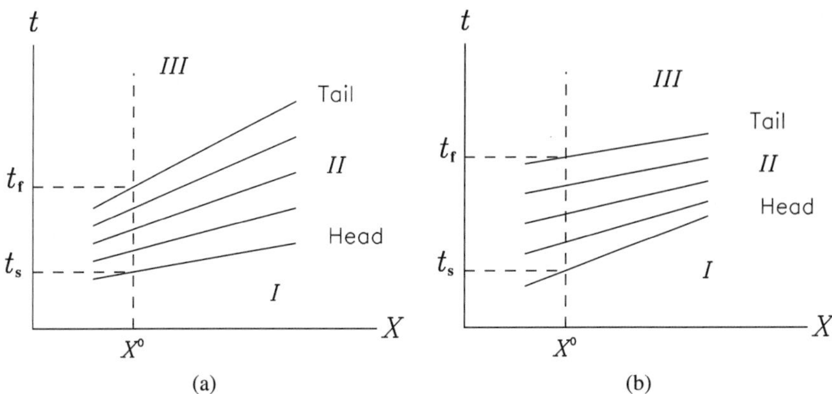

Figure 2.43 The characteristic contours for (a) a rarefaction wave and (b) a compression wave.

gradient at a material point X^0. To assist in the interpretation of these figures, we recall that [see Eq. (1.134) in Section 1.8.1]

$$F = \frac{dV}{dV_0}.$$

Thus the change of the volume of the material is represented by F. As a rarefaction wave passes the point X^0, the volume of the material increases, and as a compression wave passes, the volume decreases.

As a consequence of Eq. (2.115), $dC/dF \leq 0$, we find that the material wave velocity C decreases as a rarefaction wave passes point X^0, and it increases as a compression wave passes. Indeed, from these histories, we can easily construct the X–t diagrams for each type of wave. The results are shown in Figure 2.43. To understand how we obtain the X–t diagrams from the histories, you should examine the X–t diagram for the rarefaction wave. This diagram contains a vertical dashed line located at X^0. The values of F along this dashed line are given by the history of the rarefaction wave in Figure 2.42a. Because we know F along the dashed line, then we also know $C(F)$. Thus, at any time t along the dashed line, you can construct the characteristic contour by drawing a straight line with a slope of $dX/dt = C$.

Each X–t diagram contains a head contour and a tail contour. These contours separate each X–t diagram into three regions. In regions I and III, the solution is constant. In region II between the head and tail contours, the solution changes. Unlike the characteristic contours of the linear wave equation, these contours are not parallel. Thus, in an unbounded material, we conclude that the head and tail contours of the compression wave must eventually intersect and cause "cresting" of the wave. Because this leads to an unacceptable solution, the "life" of a compression wave in a nonlinear-elastic material is finite. The implication for the rarefaction wave is similar. If the contours of the rarefaction wave are extended into the past, they eventually intersect each other unless they first intersect a boundary or the $t = 0$ axis.

It is also possible to interpret the histories in Figure 2.42 as observations at the spatial point x^0 rather than the material point X^0. We can construct an x–t diagram by first using the history to evaluate F at some time t. The characteristic contours that intersect x^0 have slopes $dx/dt = v + c$, and because of Eq. (2.117), we obtain x–t diagrams that are similar to the previous X–t diagrams. That is, the characteristic contours for the rarefaction wave diverge as time increases, whereas those for the compression wave converge.

Simple Waves in an Adiabatic Gas

In this section we evaluate the solution for a simple wave propagating in the positive-x direction through an adiabatic gas. Noticing Eq. (2.111), we see that the simple wave solution Eq. (2.113) becomes

$$v - v^0 = \frac{2}{\gamma - 1}(c - c_0). \tag{2.120}$$

From Eqs. (2.109) and (2.120), we can relate the deformation gradient F to the velocity v. The result is

$$F = \left[\frac{\gamma - 1}{2}\left(\frac{v - v^0}{c_0}\right) + 1\right]^{-\frac{2}{\gamma-1}}. \tag{2.121}$$

We can also express the wave velocities and the pressure in terms of the velocity v. By using the previous result with Eqs. (2.28), (2.108), and (2.110), we obtain

$$\begin{aligned}
\frac{c}{c_0} &= \left[\frac{\gamma - 1}{2}\left(\frac{v - v^0}{c_0}\right) + 1\right], \\
\frac{C}{c_0} &= \left[\frac{\gamma - 1}{2}\left(\frac{v - v^0}{c_0}\right) + 1\right]^{\frac{\gamma+1}{\gamma-1}}, \\
\frac{p}{p_0} &= \left[\frac{\gamma - 1}{2}\left(\frac{v - v^0}{c_0}\right) + 1\right]^{\frac{2\gamma}{\gamma-1}}.
\end{aligned} \tag{2.122}$$

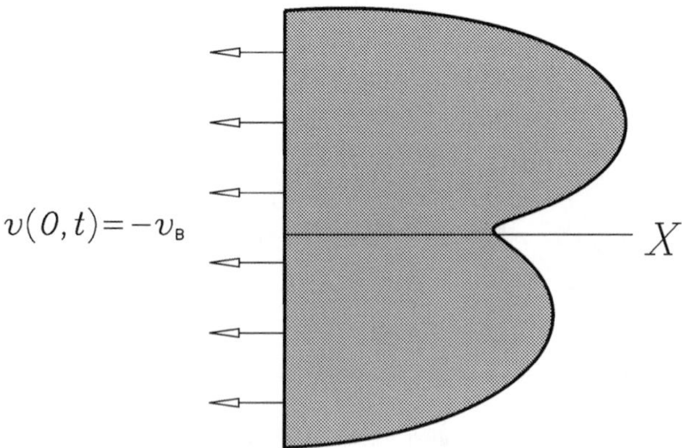

Figure 2.44 A half space with a velocity boundary condition.

2.6.2 *Centered Rarefaction Wave*

We now consider a special type of structured wave that we shall frequently encounter in our study of nonlinear waves. Consider a half space of nonlinear-elastic material occupying the region $X \geq 0$ (see Figure 2.44). The material is initially undisturbed, so that $\hat{v}(X, t) = 0$ and $F(X, t) = 1$ for $t \leq 0$. We subject the boundary at $X = 0$ to the velocity boundary condition

$$\hat{v}(0, t) = -v_B, \quad t > 0, \tag{2.123}$$

where v_B is a positive constant. At $X = 0$, the initial conditions and the boundary conditions require that

$$\hat{v}(0, t) = \begin{cases} 0, & t \leq 0, \\ -v_B, & t > 0. \end{cases}$$

Thus the velocity at the boundary is discontinuous at $t = 0$. However, we suppose that $\hat{v}(0, t)$ changes from 0 to $-v_B$ over a very small but finite amount of time. We shall now derive solutions for this problem in both the material and the spatial descriptions.

Material Solution We show the X–t diagram of the characteristic contours of this problem in Figure 2.45. Because of the initial conditions, J_+ and J_- are zero everywhere on the X axis, and $v^0 = 0$. Thus we find that $J_- = 0$ everywhere in the half space, and the solution is a nonlinear simple wave. The Riemann integral J_+ is nonzero only on the contours that intersect the t axis where

$$J_+(\xi) = 2\hat{v}(0, t).$$

For this simple wave, the variables v, F, c, and C are functions of ξ only. This allows us to express C as

$$C = C(v).$$

From the boundary condition, we know v on the t axis. Consequently, we also know C on the t axis. Because the change in the velocity of the boundary is confined to a very small region

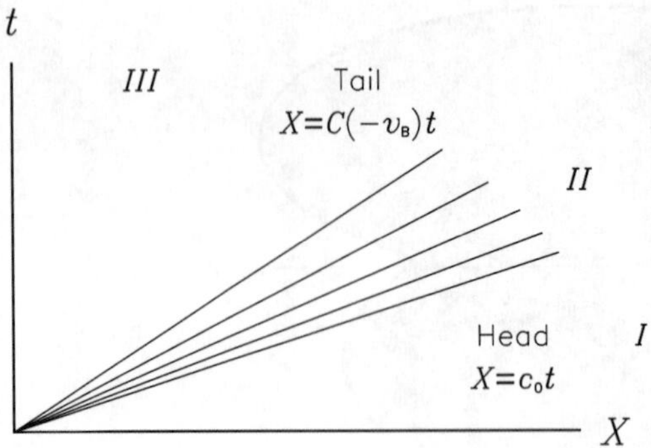

Figure 2.45 The X–t diagram for a half space with a centered rarefaction wave.

near the origin, the characteristic contours for different values of v give the appearance of a fan radiating from the origin. The contours diverge because the velocity $v(0, t)$ of the boundary decreases with time. The resulting wave is a rarefaction wave. Because the contours of the fan intersect at a common point, the wave is called a *centered rarefaction wave*.

The contours in the fan are described by the differential equation

$$\frac{dX}{dt} = C(v).$$

Because all of the contours in the fan intersect the origin, we can integrate this differential equation and obtain

$$X = C(v)t.$$

The fan has a head contour and a tail contour. On the head contour $v = 0$, and

$$X = c_0 t.$$

On the tail contour $v = -v_B$, and

$$X = C(-v_B)t.$$

Spatial Solution We show the x–t diagram of the characteristic contours of this problem in Figure 2.46. You will notice that we have prescribed the position of the boundary of the half space by the boundary condition, Eq. (2.123). We denote its position with a dashed line. The velocity on the boundary is $v = -v_B$, and because the solution is a simple wave, we know the value of $v + c = -v_B + c(-v_B)$ on the boundary.

The contours form a fan in the x–t diagram. They are described by the differential equation

$$\frac{dx}{dt} = v + c(v),$$

where each contour corresponds to a different value of v. Because all of the contours in the fan intersect the origin, we can integrate this expression to obtain

$$x = [v + c(v)]t.$$

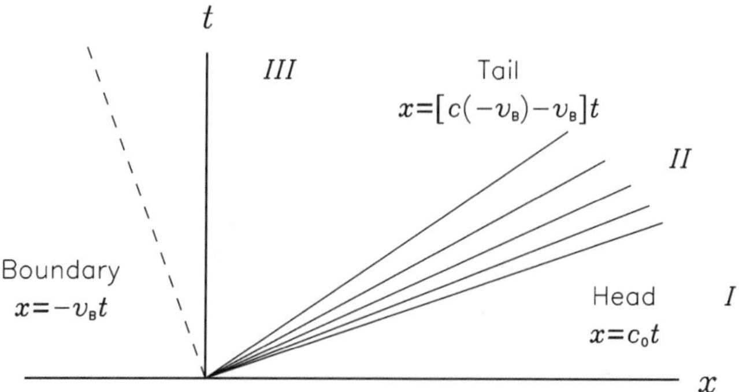

Figure 2.46 The x–t diagram for a half space with a centered rarefaction wave. The boundary of the half space is denoted as a dashed line.

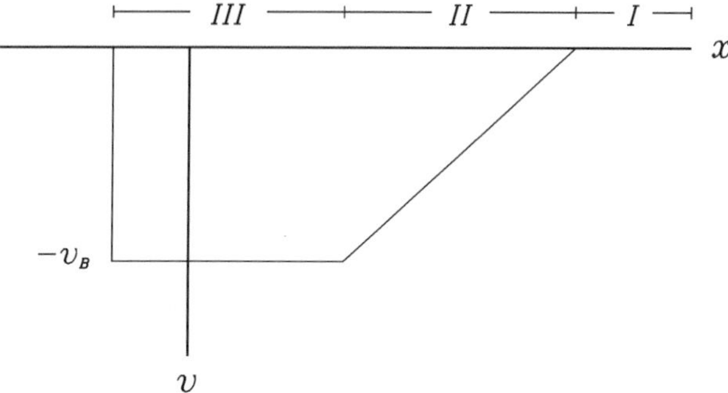

Figure 2.47 A typical velocity profile for a centered rarefaction wave.

The fan has a head contour and a tail contour. On the head contour $v = 0$, and

$$x = c_0 t.$$

On the tail contour $v = -v_B$, and

$$x = [-v_B + c(-v_B)]t.$$

Now consider a typical wave profile of this solution (see Figure 2.47). The discontinuous jump in the profile is located at the boundary of the half space. The ramp results from the centered rarefaction wave, although not all rarefaction waves are ramps. The labels *I*, *II*, and *III* indicate regions in the x–t diagram.

Escape Velocity of an Adiabatic Gas Suppose that the half space in this problem is an adiabatic gas. The expressions for C and c for a simple wave in an adiabatic gas are contained in Eqs. (2.122). By letting $v^0 = 0$, we substitute these expressions into the solutions for the contours in the fan to obtain

$$
\begin{aligned}
X &= c_0 t\left[\frac{\gamma-1}{2}\left(\frac{v}{c_0}\right)+1\right]^{\frac{\gamma+1}{\gamma-1}}, \\
x &= c_0 t\left[\frac{\gamma+1}{2}\left(\frac{v}{c_0}\right)+1\right].
\end{aligned}
\tag{2.124}
$$

We obtain the expression for the tail contours in each diagram by letting $v = -v_B$:

$$
\begin{aligned}
X &= c_0 t\left[-\frac{\gamma-1}{2}\left(\frac{v_B}{c_0}\right)+1\right]^{\frac{\gamma+1}{\gamma-1}}, \\
x &= c_0 t\left[-\frac{\gamma+1}{2}\left(\frac{v_B}{c_0}\right)+1\right].
\end{aligned}
$$

These results only yield real values for X at the tail contour when the prescribed boundary velocity obeys $v_B \leq v_E$ where

$$
v_E = \frac{2c_0}{\gamma - 1}. \tag{2.125}
$$

When we prescribe the boundary condition $v_B = v_E$, the relationships for the tail contour yield

$$
\begin{aligned}
X &= 0, \\
x &= -v_E t.
\end{aligned}
$$

This means the tail contour of the centered rarefaction wave is coincident with the boundary of the half space. The value v_E is called the *escape velocity* of the gas. By substituting the escape velocity into Eq. (2.121), we find that $F = \infty$ at the boundary. Consequently, when $v_B = v_E$, the density and pressure in the gas are zero, and from Eq. (2.109), $c = C = 0$. The greatest velocity the gas can attain is v_E, and if $v_B \geq v_E$, the gas will separate from the boundary. This phenomenon is called *cavitation*.

The condition $v_B = v_E$ represents the solution to another problem. In this problem a half space of gas is initially at rest with pressure p_0. At $t = 0$, we suddenly drop the boundary pressure to zero. The resulting expansion of the gas is described by the solution for the centered rarefaction wave with $v_B = v_E$. For example, air has a value of $\gamma = 1.4$, and from this solution, we see that the escape velocity for a half space of air is

$$
v_E = 5c_0.
$$

When we let $v_B = v_E$, the slopes of the head and tail contours have opposite signs in the x–t diagram. On the head contour $v = 0$, and $v + c = c_0$ is positive. On the tail contour $c = 0$ and $v = -v_E$. Thus $v + c = -v_E$ is negative. You may have a tendency to assume that in both the x–t and X–t diagrams, the ξ contours always have positive slopes while the ζ contours always have negative slopes. Although this is true for the X–t diagram, you should realize that it is not necessarily true for the x–t diagram.

Material Trajectories In the x–t diagram, the path followed by a material point $\mathbf{X}$ is called the *material trajectory* of $\mathbf{X}$. For the centered rarefaction wave in an adiabatic gas,

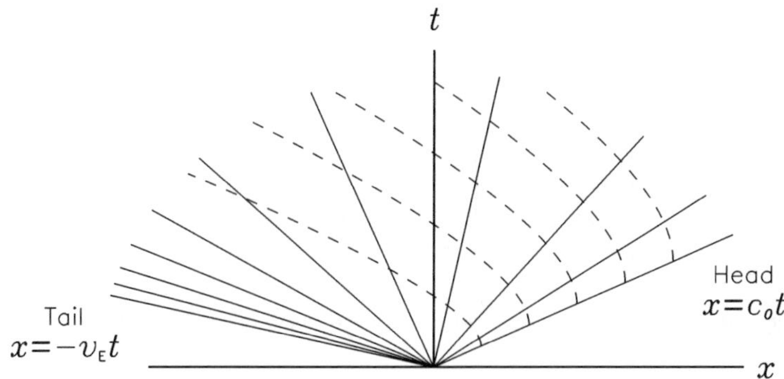

Figure 2.48 The material trajectories (– –) for the centered rarefaction wave where $v = -2c_0/(\gamma - 1)$ at the boundary of the half space.

we can determine this path by combining the first expression in Eqs. (2.124) with the simple wave expression Eq. (2.120). We obtain

$$X = c_0 t \left(\frac{c}{c_0}\right)^{\frac{\gamma+1}{\gamma-1}}.$$

Along the material trajectory, X is constant. Let us examine the material trajectory where

$$X = c_0 t_0.$$

By selecting the arbitrary constant t_0, we prescribe the fixed value of X. These two expressions yield

$$t = t_0 \left[\frac{\gamma - 1}{2}\left(\frac{v}{c_0}\right) + 1\right]^{-\frac{\gamma+1}{\gamma-1}}.$$

From this result and Eqs. (2.124), we can determine the material trajectory of $\mathbf{X}$. In Figure 2.48 we show the trajectories for the centered rarefaction wave for $v_B = v_E$. You can calculate them by first selecting a point x_0 on the head contour. This gives $t_0 = x_0/c_0$. You can then determine subsequent values of t and x on the material trajectory by letting the velocity decrease from $v = 0$ to $v = -2c_0/(\gamma - 1)$.

2.6.3 *Reflection at a Boundary*

In this section we analyze the reflection of a structured wave at a rigid boundary and a free boundary. Consider a half space of a nonlinear-elastic material in the region $X \leq 0$ (see Figure 2.49). The boundary of this material is at $X = 0$. Suppose a simple wave approaches the boundary. It is called the incident wave. The Riemann solution for $X \leq 0$ is [see Eqs. (2.101)]

$$v = \frac{1}{2}[J_+(\xi) + J_-(\zeta)],$$

$$\int_1^F C(\bar{F})\,d\bar{F} = \frac{1}{2}[-J_+(\xi) + J_-(\zeta)],$$

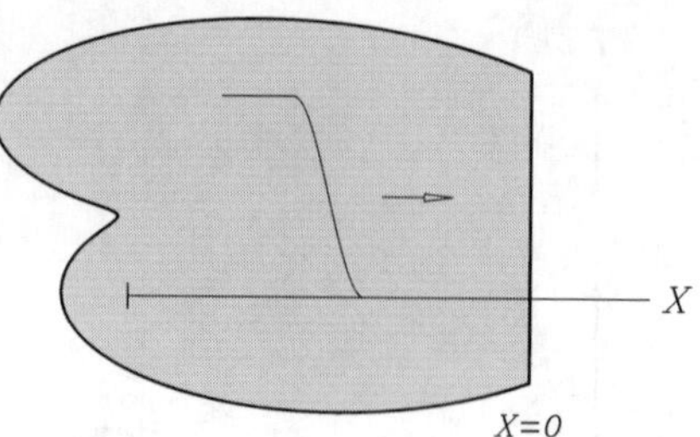

Figure 2.49 A structured wave incident on the boundary of a half space at $X = 0$.

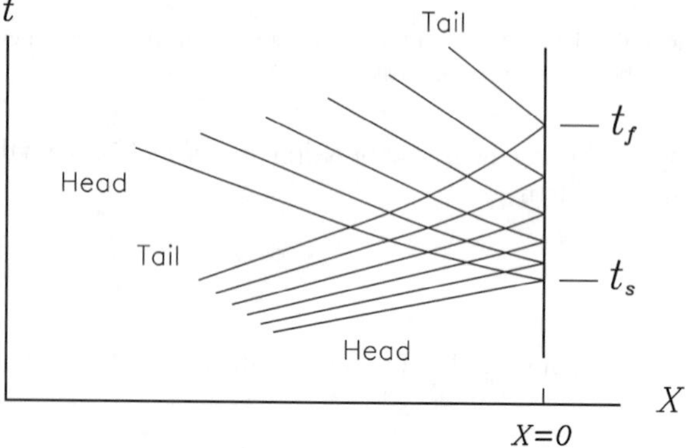

Figure 2.50 The reflection of a simple wave from the boundary of a half space.

where, initially, the body is stationary and $v^0 = 0$. We illustrate the characteristic contours for this solution in Figure 2.50. The head contour of the incident wave contacts the boundary at $t = t_s$. This results in the head contour of the reflected wave. Subsequent contours of the incident wave give rise to additional contours of the reflected wave until time $t = t_f$. At this time, the tail contour of the incident wave makes contact with the boundary. During the period of time $t_s \leq t \leq t_f$ when the incident wave is in contact with the boundary, neither the incident nor the reflected waves are simple waves. In this region of the diagram the characteristic contours are curved. Outside of this region the contours are straight.

Reflection at a Rigid Boundary

We next discuss the reflection of a structured wave at a rigid boundary. We assume that the solution for the structured wave does not become multivalued during the time that it is in contact with the boundary. The velocity at $X = 0$ is $v = 0$ for all $t \geq 0$, and the current position of a rigid boundary is always $x = 0$. We shall use the reference configuration and the X–t diagram in this section (see Figure 2.51), but because the boundary obeys $X = x = 0$, the x–t diagram has a similar qualitative appearance.

Although we have illustrated the reflection of a rarefaction wave in Figure 2.51, this discussion also applies to compression waves. The incident and reflected waves separate the diagram into three regions. In region I the initial conditions give $v = 0$ and $F = 1$.

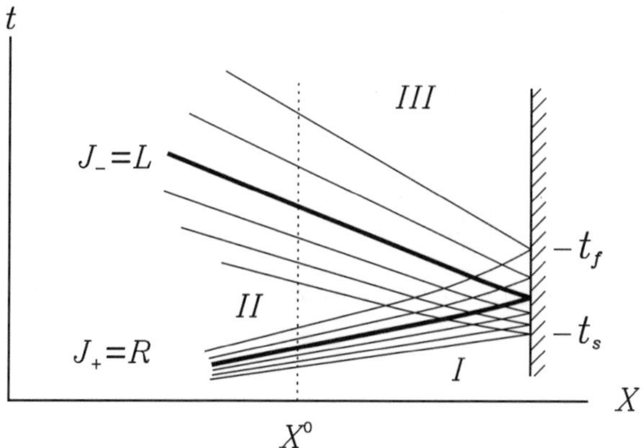

Figure 2.51 The reflection of a simple wave from a rigid boundary.

Thus

$$\left.\begin{aligned} J_+ &= 0 \\ J_- &= 0 \end{aligned}\right\} \text{ region } I.$$

In region *II* we prescribe the velocity $v = v_{II}$. Because J_- is constant across the incident wave, the Riemann solution yields

$$\left.\begin{aligned} J_+ &= 2v_{II} \\ J_- &= 0 \end{aligned}\right\} \text{ region } II.$$

In region *III* the boundary condition requires $v = 0$. Because J_+ is constant across the reflected wave, the Riemann solution yields

$$\left.\begin{aligned} J_+ &= 2v_{II} \\ J_- &= -2v_{II} \end{aligned}\right\} \text{ region } III.$$

Consider the contour of the incident wave represented by the bold line. On this contour, we prescribe $J_+ = R$. At the rigid boundary, this contour gives rise to a contour of the reflected wave where $J_- = L$. At the rigid boundary, $v = 0$. Substitution of this boundary condition into the Riemann solution gives

$$R = -L.$$

When $t < t_s$, we denote the velocity and deformation gradient on the contour of the incident wave as v_R and F_R. Because $J_- = 0$ in region *I* and across this portion of the incident wave, the Riemann solution yields

$$v_R = \frac{1}{2}R,$$

$$\int_1^{F_R} C(\bar{F})\, d\bar{F} = -v_R.$$

When $t > t_f$, we denote the velocity and deformation gradient on the contour of the reflected wave as v_L and F_L. Because $J_+ = 2v_{II}$ in region *II* and across this portion of the reflected wave, and $J_- = L = -R = -2v_R$, the Riemann solution yields

$$\left.\begin{aligned} v_L &= v_{II} - v_R \\ \int_1^{F_L} C(\bar{F})\,d\bar{F} &= -v_{II} - v_R \end{aligned}\right\} \quad t > t_f.$$

For the special case of a structured wave where we prescribe $v_{II} = 0$, we obtain

$$\left.\begin{aligned} v_L &= -v_R \\ F_L &= F_R \end{aligned}\right\} \quad t > t_f.$$

We also find that a compression wave incident on a rigid boundary is reflected as a compression wave, and a rarefaction wave is reflected as a rarefaction wave.

***T–v* Diagram** We have used the Riemann integrals to demonstrate how simple waves reflect from a rigid boundary. The T–v diagram offers us an alternative graphical method for analyzing this reflection. In the X–t diagram, the incident and reflected waves separate the diagram into three regions. Consider a material point X^0 that intersects the regions *I*, *II*, and *III* (see Figure 2.51). Initially, the material point X^0 is in region *I*. From the initial conditions, we know the stress and velocity are zero. As time advances, the material point at X^0 encounters the incident wave and moves into region *II*. By using the T–v diagram we can see how the values of stress and velocity move along the response path as the incident wave passes X^0 (see Figure 2.52). Because the incident wave travels in the positive-x direction, it moves the material along a response path with a positive slope. If the incident wave is a rarefaction wave as shown in the X–t diagram, the material moves to point *II*. If the wave is a compression wave, it moves to point *II′*. The material remains in region *II* until it encounters the reflected wave returning from the rigid boundary. This wave causes the material to move into region *III* where the boundary condition requires that $v = 0$. The response path for the reflected wave has a negative slope. The two possible conditions for region *III* are denoted by *III* and *III′*.

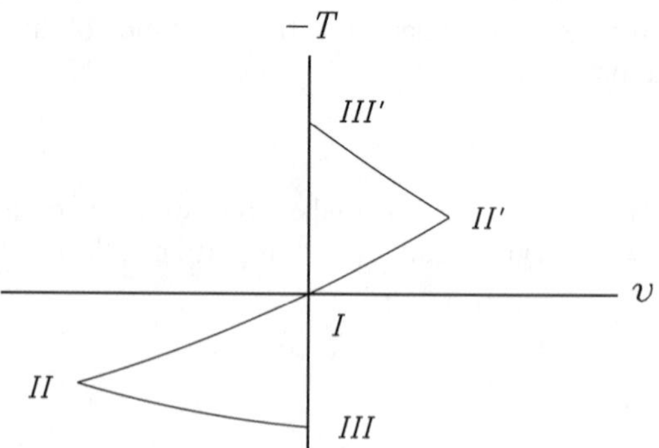

Figure 2.52 The T–v diagram for the material at X^0 during the reflection of a simple wave from a rigid boundary.

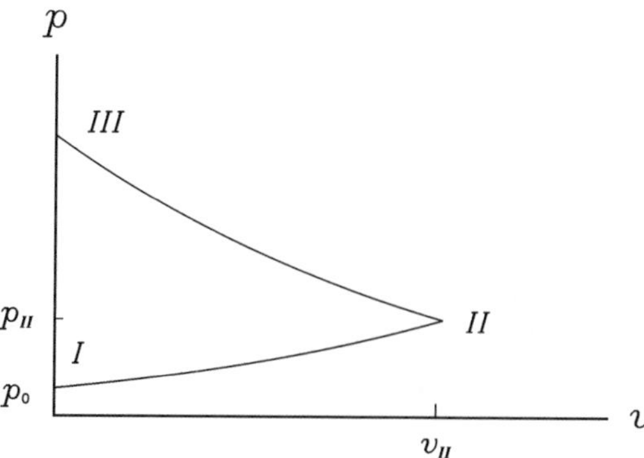

Figure 2.53 The p–v diagram for the material at X^0 during the reflection of a compression wave from the rigid boundary of a half space of adiabatic gas.

Reflection at a Rigid Boundary of an Adiabatic Gas Suppose the half space illustrated in Figure 2.51 contains an adiabatic gas. The boundary at $X = 0$ is rigid. Initially, the pressure in the undisturbed gas is p_0. A simple wave approaches the boundary. Behind the tail contour of the incident wave, we prescribe the pressure to be p_{II}. Let us determine the pressure at the boundary resulting from the reflection of this wave. We consider an incident compression wave first, so that $p_{II} > p_0$.

In Figure 2.53 we show the p–v diagram for the material at X^0. The equations for the response paths of an adiabatic gas are given by Eq. (2.119),

$$\left(\frac{p}{p_0}\right)^{\frac{\gamma-1}{2\gamma}} = 1 \pm \frac{\gamma-1}{2c_0}(v - v^0).$$

We obtain the response path for the incident wave, which connects points *I* and *II*, from this expression by selecting the positive sign and letting $v^0 = 0$:

$$\left(\frac{p}{p_0}\right)^{\frac{\gamma-1}{2\gamma}} = 1 + \frac{\gamma-1}{2c_0}v, \quad \text{incident wave.} \tag{2.126}$$

We obtain the response path for the reflected wave, which connects points *II* and *III*, by selecting the negative sign and letting $v^0 = 2v_{II}$:

$$\left(\frac{p}{p_0}\right)^{\frac{\gamma-1}{2\gamma}} = 1 - \frac{\gamma-1}{2c_0}(v - 2v_{II}), \quad \text{reflected wave,} \tag{2.127}$$

where v_{II} is the velocity in region *II*. When $v = v_{II}$ both response paths yield the same pressure p_{II}:

$$\left(\frac{p_{II}}{p_0}\right)^{\frac{\gamma-1}{2\gamma}} = 1 + \frac{\gamma-1}{2c_0}v_{II}. \tag{2.128}$$

We determine the pressure in region *III* by setting $v = 0$ in Eq. (2.127) and then substituting Eq. (2.128):

$$\left(\frac{p_{III}}{p_0}\right)^{(\gamma-1)/2\gamma} = -1 + 2\left(\frac{p_{II}}{p_0}\right)^{(\gamma-1)/2\gamma}. \tag{2.129}$$

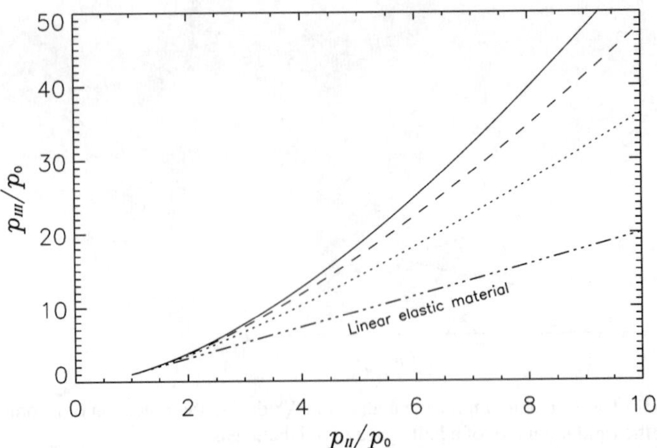

Figure 2.54 The pressure ratio p_{III}/p_0 at the rigid boundary of a gas caused by an incident compression wave of pressure ratio p_{II}/p_0: $\gamma = 3$ (···), $\gamma = 7/5$ (—), and $\gamma = 5/3$ (– –).

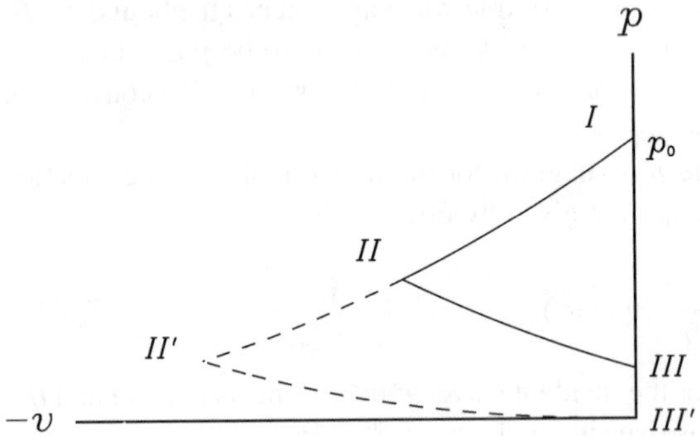

Figure 2.55 The p–v diagram for the material at X^0 during the reflection of a rarefaction wave from the rigid boundary of the half space of adiabatic gas.

This is the relationship between the prescribed pressure p_{II} behind the incident compression wave and the pressure p_{III} behind the reflected wave. We show a graph of p_{III}/p_0 versus p_{II}/p_0 in Figure 2.54. Curves for three values of γ are displayed. Recall that the reflection of a compression wave in a linear-elastic material causes the pressure to double. We see that significantly greater pressure increases are possible in an adiabatic gas.

When the incident wave is a rarefaction wave, and $p_{II} < p_0$, we obtain the p–v diagram in Figure 2.55. Let us focus our attention on the two solid curves first. They represent the response paths for the incident and reflected waves. The response path for the incident wave, which connects points I and II, is given by Eq. (2.126):

$$\left(\frac{p}{p_0}\right)^{\frac{\gamma-1}{2\gamma}} = 1 + \frac{\gamma - 1}{2c_0}v, \quad \text{incident wave.}$$

We leave it as an exercise to show that the response path for the reflected wave, which

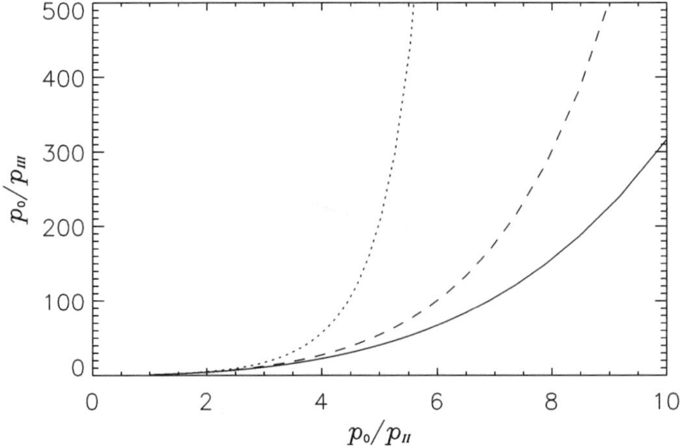

Figure 2.56 The pressure ratio p_0/p_{III} at the rigid boundary of a gas versus the pressure ratio of the incident rarefaction wave p_0/p_{II}: $\gamma = 3$ (···), $\gamma = 7/5$ (—), and $\gamma = 5/3$ (– –).

connects points *II* and *III*, is [see Exercise 2.42]

$$\left(\frac{p}{p_0}\right)^{\frac{\gamma-1}{2\gamma}} - 2\left(\frac{p_{II}}{p_0}\right)^{\frac{\gamma-1}{2\gamma}} + 1 = -\frac{\gamma - 1}{2c_0}v, \quad \text{reflected wave.} \tag{2.130}$$

When we set $v = 0$ in this expression, we obtain the pressure in region *III* in terms of the prescribed pressure in region *II*:

$$\left(\frac{p_{III}}{p_0}\right)^{\frac{\gamma-1}{2\gamma}} = -1 + 2\left(\frac{p_{II}}{p_0}\right)^{\frac{\gamma-1}{2\gamma}}. \tag{2.131}$$

We graph the inverted ratios p_0/p_{III} versus p_0/p_{II} in Figure 2.56. Notice that when

$$\frac{p_{II}}{p_0} = \left(\frac{1}{2}\right)^{2\gamma/(\gamma-1)}, \tag{2.132}$$

then $p_{III} = 0$. We use the dashed lines and points II' and III' to illustrate this case in the p–v diagram. If p_{II} is less than the value in Eq. (2.132), then p_{III} is negative. However, the constitutive model for an adiabatic gas does not admit negative values for the pressure. Because it is possible to produce rarefaction waves in which the pressure p_{II} behind the incident wave is less than that given by Eq. (2.132), we must conclude that such rarefaction waves can never completely reach the rigid boundary. Indeed, as the pressure approaches zero, the deformation gradient F approaches infinity. Consequently, the material wave velocity C also approaches zero [see Eq. (2.108)], and the characteristic contours become vertical and never reach $X = 0$.

Reflection at a Free Boundary

The stress is zero at a free boundary. In a nonlinear-elastic material, we require $F = 1$ at $X = 0$. Figure 2.57 shows the X–t diagram for the reflection of a compression wave from a free boundary. Because the boundary is free to move, its motion gives the x–t diagram an

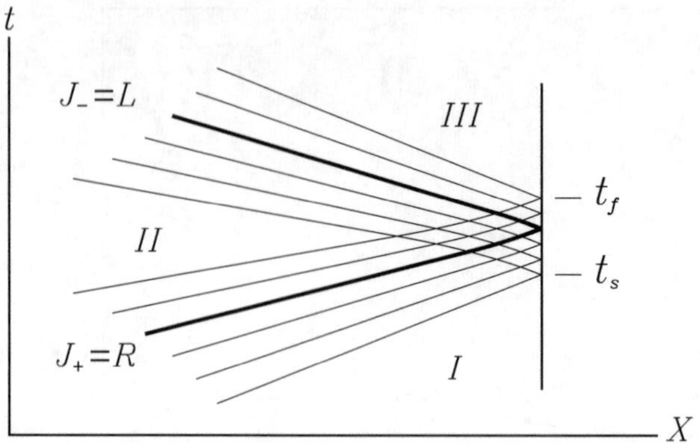

Figure 2.57 The X–t diagram for the reflection of a simple wave from a free boundary.

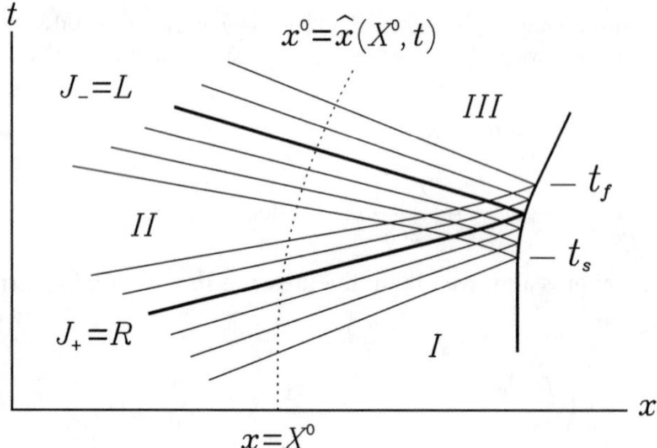

Figure 2.58 The x–t diagram for the reflection of a simple wave from a free boundary.

appearance that differs from the X–t diagram (see Figure 2.58). Comparison of Figures 2.57 and 2.58 illustrates a fundamental difference between spatial and material formulations. In the material formulation, we can state the boundary condition as $F = 1$ at $X = 0$. In the spatial formulation, the boundary condition is $F = 1$ at $x = \hat{x}(0, t)$. In the spatial description, we must take the additional step of evaluating the motion of the material point at $X = 0$.

We proceed just as we did for the reflection of a simple wave from the fixed boundary and separate these diagrams into three regions. In region I the initial conditions are $v = 0$ and $F = 1$. Thus,

$$\left.\begin{aligned} J_+ &= 0 \\ J_- &= 0 \end{aligned}\right\} \quad \text{region } I.$$

In region II

$$\left.\begin{aligned} J_+ &= 2v_{II} \\ J_- &= 0 \end{aligned}\right\} \quad \text{region } II.$$

In region *III* the boundary condition requires that $F = 1$. The Riemann solution yields

$$\left.\begin{aligned} J_+ &= 2v_{II} \\ J_- &= 2v_{II} \end{aligned}\right\} \quad \text{region } III.$$

We find that the velocity in region III, and consequently the final velocity of the free boundary, is double the velocity behind the incident wave:

$$v_{III} = 2v_{II}.$$

Consider the contour of the incident wave represented by the bold line. On this contour, we prescribe $J_+ = R$. At the rigid boundary, this contour gives rise to a contour of the reflected wave where $J_- = L$. At the free boundary $F = 1$; thus

$$R = L.$$

When $t < t_s$, we denote the velocity and deformation gradient on the contour of the incident wave by v_R and F_R, where

$$v_R = \frac{1}{2}R,$$

$$\int_1^{F_R} C(\bar{F})\, d\bar{F} = -v_R.$$

When $t > t_f$, we denote the velocity and deformation gradient on the contour of the reflected wave by v_L and F_L. Because $J_+ = 2v_{II}$ in region *II* and across this portion of the reflected wave, and $J_- = L = R = 2v_R$, the Riemann solution yields

$$v_L = v_{II} + v_R,$$

$$\int_1^{F_L} C(\bar{F})\, d\bar{F} = -v_{II} + v_R.$$

For a structured wave where $v_{II} = 0$, we find that

$$v_R = v_L,$$

$$\int_1^{F_R} C(\bar{F})\, d\bar{F} = -\int_1^{F_L} C(\bar{F})\, d\bar{F}.$$

When a compression wave is incident on a free boundary, it is reflected as a rarefaction wave. Similarly, a rarefaction wave is reflected as a compression wave. We leave it as an exercise to show that the velocity of the free boundary always equals $2v_R$ [see Exercise 2.40].

***T–v* Diagram** We can construct the T–v diagram for reflection from a free boundary in a nearly identical fashion to the way we constructed it for the reflection from a fixed boundary. Consider the material at X^0. In the x–t diagram, we use the dotted line to represent the motion of X^0, which is given by $x^0 = \hat{x}(X^0, t)$. In Figure 2.59 we show the T–v diagram for the material point at x^0. Because the incident wave travels in the positive-x direction, its response curve has a positive slope. If the incident wave is a compression wave like the one we have shown in the x–t diagram, the material moves to point *II*. If the wave is a rarefaction wave, it moves to point *II*′. The material remains in region *II* until it encounters

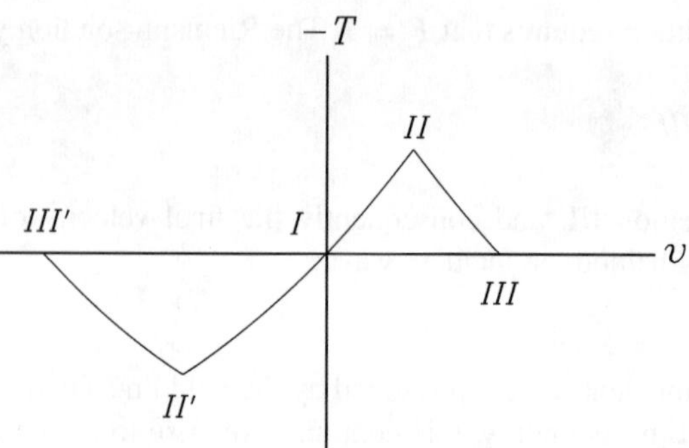

Figure 2.59 The T–v diagram for the material at x^0 during the reflection of a simple wave from a free boundary.

the reflected wave returning from the free boundary. This wave causes the material to move into region *III* where the boundary condition requires that $T = 0$.

2.6.4 *Exercises*

2.32 Consider an infinite nonlinear-elastic material. A simple wave propagates in the negative-x direction. For this wave $J_+(\xi) = 0$.

(a) Show that, in the material description, the solution for this wave is

$$v - v^0 = \int_1^F C(\bar{F})\, d\bar{F} = \frac{1}{2} J_-(t + X/C).$$

(b) Show that, in the spatial description, the solution for this wave is

$$v - v^0 = \int_1^F C(\bar{F})\, d\bar{F} = \frac{1}{2} J_-(t + x/(c - v)).$$

(c) Show that

$$\frac{dv}{dF} = C, \qquad \frac{dC}{dv} \le 0, \qquad \frac{d(c - v)}{dF} \le 0.$$

2.33 Use the relationship

$$J_\pm = v - v^0 \mp \int_{T^0}^{T} \frac{d\bar{T}}{z(\bar{T})},$$

where z is written a function of T alone, to verify that

$$\frac{d(-T)}{dv} = \pm z.$$

2.34 Suppose, at $t = 0$, the velocity profile of a simple wave propagating in the positive-x direction is

$$v = \begin{cases} 1 + \cos[\pi(X - 1)], & 0 \le X \le 2, \\ 0, & \text{otherwise.} \end{cases}$$

For this simple wave the relationship between the material wave velocity C and the velocity of the material v is

$$C(v) = 1 + \frac{1}{8}v^2.$$

(a) Graph the velocity profiles of this wave at $t = 0, 0.5, 1.0, 1.5$, and 2.0.
(b) Draw the X–t diagram for this wave.
(c) Graph the history of this wave at $X = 2$.

2.35 Consider a simple wave propagating in the negative-x direction in an adiabatic gas. Show that

$$v - v_0 = -\frac{2}{\gamma - 1}(c - c_0),$$

$$F = \left[-\frac{\gamma - 1}{2}\left(\frac{v - v^0}{c_0}\right) + 1\right]^{-\frac{2}{\gamma-1}},$$

$$\frac{c}{c_0} = \left[-\frac{\gamma - 1}{2}\left(\frac{v - v^0}{c_0}\right) + 1\right],$$

$$\frac{C}{c_0} = \left[-\frac{\gamma - 1}{2}\left(\frac{v - v^0}{c_0}\right) + 1\right]^{\frac{\gamma+1}{\gamma-1}},$$

$$\frac{p}{p_0} = \left[-\frac{\gamma - 1}{2}\left(\frac{v - v^0}{c_0}\right) + 1\right]^{\frac{2\gamma}{\gamma-1}}.$$

2.36 Consider a stress-free, nonlinear-elastic material at rest, which obeys the constitutive model

$$T = \rho_0 c_0^2 \ln F.$$

(a) For a simple wave propagating in the positive-x direction, verify that

$$C = \frac{2c_0^2}{2c_0 - v}, \qquad c = c_0 - \frac{v}{2}.$$

(b) Substitute these results into our solution for the centered rarefaction wave in the half space $X \geq 0$. If we suddenly apply a constant negative velocity to the boundary of the half space, for what value of the applied boundary velocity does the tail contour of the rarefaction wave remain coincident with the boundary?
(c) Is this centered rarefaction wave solution valid for applied boundary velocities that are positive? Why?

2.37 Suppose a centered rarefaction wave propagates in the positive-x direction within the material described in Exercise 2.36. This wave is centered at $X = 0$ and $t = 0$.
(a) Show that the motion within the rarefaction fan is

$$x = c_0 t(2 - c_0 t/X).$$

(b) Determine the relationship between x and X on the head contour of this wave.
(c) If the velocity on the tail contour of the rarefaction wave is $v = -2c_0$, determine the material and spatial descriptions of the tail contour.

2.38 The escape velocity for an adiabatic gas is

$$v_E = \frac{2c_0}{\gamma - 1}.$$

(a) Derive this result directly from Eqs. (2.122).
(b) Compare this result to that of Exercise 2.31.

2.39 In Section 2.6.2 we give the material trajectory resulting from a centered rarefaction wave in an adiabatic gas. Consider a gas in which $c_0 = 1$ and $\gamma = 3$. Suppose the left boundary of the gas moves to the left at the escape velocity. Plot the trajectory of the material at $X = 1$ for $0 \le t \le 4$.

2.40 Consider a half space of nonlinear-elastic material that occupies the region $X \le 0$. A simple wave is incident on the boundary. For a particular characteristic contour of this simple wave, the velocity and deformation gradient are v_I and F_I. At the point where this contour intersects the boundary, the velocity and deformation gradient are denoted by v_B and F_B.
(a) If the boundary is rigid, show that

$$\int_1^{F_B} C(\bar{F})\, d\bar{F} = 2 \int_1^{F_I} C(\bar{F})\, d\bar{F}.$$

(b) If the boundary is free, show that

$$v_B = 2v_I.$$

2.41 The pressure resulting from the reflection of a simple wave at a rigid boundary of an adiabatic gas is given by Eq. (2.129). Assume that the pressures in this expression are

$$p_{II} = p_0 + \Delta p_{II}, \qquad p_{III} = p_0 + \Delta p_{II},$$

where Δp_{II} and Δp_{III} are assumed to be infinitesimal. Show that

$$\Delta p_{III} = 2\Delta p_{II}.$$

2.42 Verify that Eq. (2.130) is the response path for a wave that is reflected from the rigid boundary of an adiabatic gas.

2.7 Shock Waves

In our discussion of structured waves we showed that compression waves have converging characteristic contours that eventually intersect. These contours converge because both the stiffness and the wave velocity of a nonlinear-elastic material increase as it is compressed. This is formally expressed by the condition [see Eq. (2.23)]

$$\frac{d^2T}{dF^2} \le 0.$$

If we extend the solution for a simple wave beyond the point of convergence, the contours cross and the solution becomes multivalued. In such a solution, several discrete values of velocity exist at the same position and time. Because the velocity v is an inherently single valued function of position, such multivalued solutions are impossible from a physical viewpoint, and we must search for an alternative solution. We consider two possibilities. The first approach is to replace the multivalued solution by a single contour. Then the solution must be discontinuous across the new contour. We call this discontinuity a *shock wave*. The second approach is to abandon the constitutive model for a nonlinear-elastic material and replace it with a constitutive model that does not allow the characteristic contours to intersect. In this section we discuss the first approach. We discuss the second approach later.

2.7.1 *Formation of a Shock Wave*

In a nonlinear-elastic material, the characteristic contours of a compression wave converge and intersect. Up to the time the contours begin to intersect, the compression wave is a structured wave. Beyond this time, we replace the smooth solution of the simple wave by a discontinuous solution. We represent this discontinuity with a single contour called a shock wave. We begin our study of shock waves with an example in which the characteristic contours of a simple wave converge to form a discontinuous shock wave. The example we use is the propagation of a compression ramp in an adiabatic gas. It is an interesting example because all of the characteristic contours of the compression ramp simultaneously converge to a single point in the gas. Thus, unlike other types of simple waves, which gradually transform into a shock wave, this compression ramp suddenly transforms into a shock wave.

We show the profile of a ramp wave in Figure 2.60. The equation for this profile is

$$v = -\alpha x, \qquad 0 \le v \le \alpha h_0. \tag{2.133}$$

The constant α is the initial slope of the ramp wave, and the constant h_0 is the initial thickness of the ramp wave. The head of the ramp wave, where $v = 0$, is located at $x = 0$. The tail of the ramp wave, where $v = \alpha h_0$, is located at $x = -h_0$. Using the head and tail of the ramp wave, we divide this illustration into three regions labeled, *I*, *II*, and *III*. Consider an adiabatic gas, which contains a ramp wave propagating in the positive-x direction. At time $t = 0$, the profile of this wave is given by Eq. (2.133). The characteristic contours of this simple wave are straight lines, and the Riemann solution for the ramp wave is

$$v = \alpha[(v + c)t - x], \qquad 0 \le v \le \alpha h_0.$$

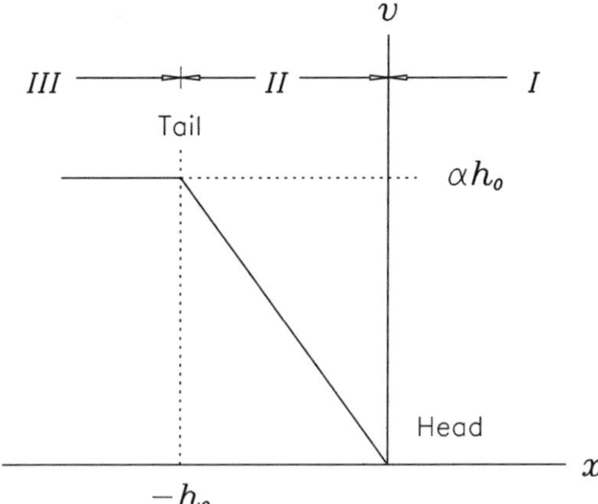

Figure 2.60 The profile of a ramp wave at time $t = 0$.

Substituting Eq. (2.122), we obtain

$$v = \alpha\left[\left(\frac{\gamma+1}{2}v + c_0\right)t - x\right], \qquad 0 \le v \le \alpha h_0.$$

When we solve for the velocity we get

$$v = \frac{\alpha(c_0 t - x)}{\left(1 - \frac{\gamma+1}{2}\alpha t\right)}, \qquad 0 \le v \le \alpha h_0.$$

Notice that at time t, the head of the wave, where $v = 0$, moves to $x = c_0 t$. Simultaneously, the tail of the wave, where $v = \alpha h_0$, moves to $x = c_0 t - h$, where

$$h(t) \equiv h_0\left(1 - \frac{\gamma+1}{2}\alpha t\right).$$

We call $h(t)$ the thickness of the wave. We can write the solution in terms of the thickness as

$$v = \frac{\alpha h_0}{h(t)}(c_0 t - x), \qquad 0 \le v \le \alpha h_0. \tag{2.134}$$

From this solution, we see that the original ramp wave at $t = 0$ continues to propagate as a ramp wave. However, the slope of the ramp wave changes with time. The change in the slope depends upon the sign of α.

When we prescribe a negative value for α, the ramp wave results in negative velocity. This is a rarefaction wave, and from the definition of $h(t)$, we see that the thickness of the wave increases with time, $h(t) > h_0$. When we prescribe a positive value for α, the ramp wave results in positive velocity. This is a compression wave, and its thickness decreases with time, $h(t) < h_0$. We show the head and tail characteristic contours for both of these cases in Figure 2.61. The three regions *I*, *II*, and *III* in Figure 2.60 are also illustrated here. Notice that the thickness $h(t)$ of the compression wave is zero at $t = t_s$ where

$$t_s = \frac{2}{(\gamma+1)\alpha}.$$

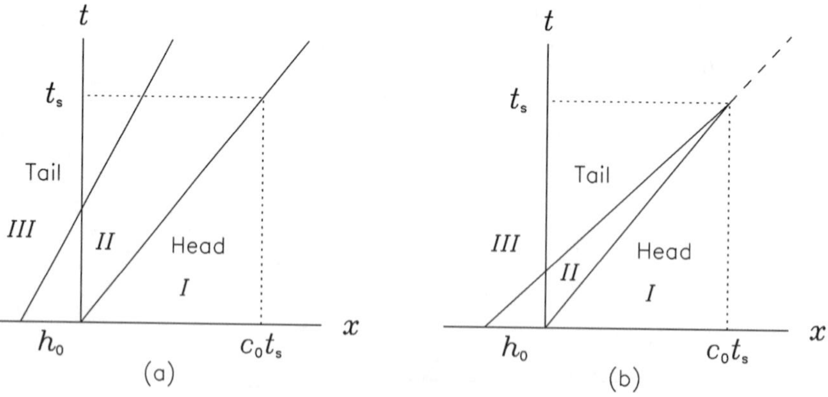

Figure 2.61 The head and tail characteristic contours for (a) the rarefaction wave, $\alpha < 0$, and (b) the compression wave, $\alpha > 0$.

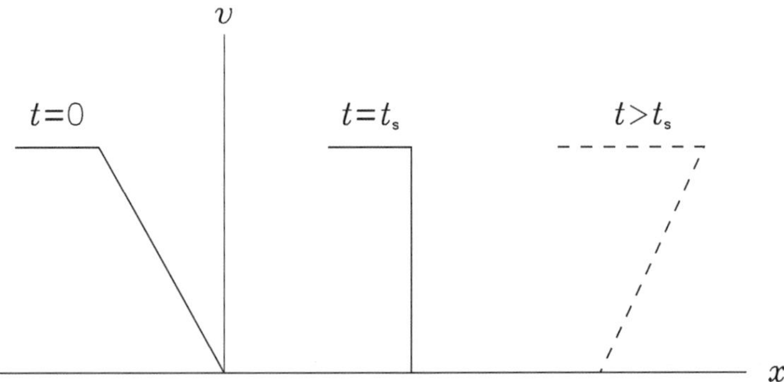

Figure 2.62 The evolution of a compression ramp to a shock wave.

At time t_s, region *II* disappears and the compression wave becomes a discontinuous step (see Figure 2.62). For values of time $t > t_s$, the solution for the compression ramp yields negative values of thickness and becomes multivalued. You will notice that over a particular range of x the dashed line shows three distinct values of velocity in the region between the head and tail contours (remember to include $v = 0$). Thus the simple wave solution is invalid for $t > t_s$ and an alternative solution must be found. Because the simple wave solution is discontinuous at $t = t_s$, suppose we require the solution to continue to propagate as a discontinuity for $t \geq t_s$. We shall represent this new solution with a single contour [see the dashed line in Figure 2.61b]. Across this new contour, we shall allow the value of the velocity to jump discontinuously. We shall call this new contour a shock wave. The question we must now answer is, *what is the slope of the contour for the shock wave?*

2.7.2 *Jump Conditions: One-Dimensional Motion*

In the previous section we motivated the need for a discontinuous solution to describe a shock wave. In this section we introduce this solution. It is called a *singular surface* because in three dimensions the location of the discontinuity is determined by a moving surface. In a one-dimensional theory this surface becomes a moving point. You will recall that we first derived the laws of motion for continuous solutions in one spatial dimension. Then we rederived them for three dimensions. We shall follow the same procedure here. However, unlike the previous derivations where the laws of motion yielded differential equations that govern continuous solutions such as simple waves, here we shall find that the same laws of motion yield discrete *jump conditions* for discontinuous solutions describing shock waves.

Singular Surfaces

Consider an arbitrary function $\hat{\phi}(X, t)$ of position X and time t. By means of the motion $\hat{x}(X, t)$, we can also express this function as $\phi(x, t)$. Let ϕ represent any dependent variable in the nonlinear-elastic equations. We shall assume that ϕ is continuous everywhere except at $x = x_S$, where it is discontinuous. The position $x = x_S$ is a singular surface. We shall describe a shock wave as a propagating singular surface. To do this, we must first derive some general mathematical relationships for the singular surface.

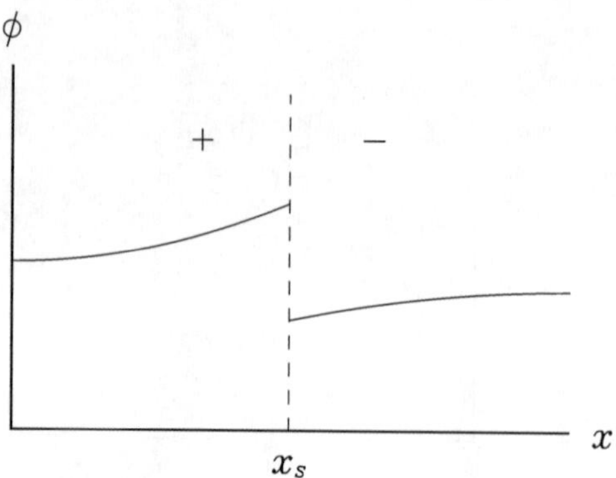

Figure 2.63 A discontinuous function.

Jumps and Kinematic Compatibility First we consider the scalar function ϕ when it is expressed in the spatial description as $\phi(x, t)$. Suppose that at time t, it is discontinuous at the point x_S where (see Figure 2.63)

$$x_S = x_S(t).$$

We denote the regions to the left and right of x_S by $+$ and $-$, and we define ϕ_+ and ϕ_- to be the limiting values of $\phi(x, t)$ as x approaches x_S from the $+$ and $-$ sides:

$$\phi_+ = \lim_{(x \to x_S)_+} \phi(x, t), \qquad \phi_- = \lim_{(x \to x_S)_-} \phi(x, t),$$

where, of course, we assume that these limits exist. We define the *jump* of $\phi(x, t)$ to be

$$[\![\phi]\!] \equiv \phi_+ - \phi_-. \tag{2.135}$$

Next we consider the same scalar function ϕ when it is expressed in the material description as $\hat{\phi}(X, t)$. It is discontinuous at the point X_S where

$$X_S = X_S(t).$$

The material and spatial descriptions of the singular surface are interrelated through the motion as

$$x_S = \hat{x}[X_S(t), t]. \tag{2.136}$$

Because x_S and X_S represent the same physical point in space, we find that in the material description the values of the limits

$$\phi_+ = \lim_{(X \to X_S)_+} \hat{\phi}(X, t), \qquad \phi_- = \lim_{(X \to X_S)_-} \hat{\phi}(X, t)$$

are identical to those in the spatial description.

We assume the motion is continuous across the singular surface; however, we allow the derivatives of the motion to be discontinuous. On the $+$ side of the singular surface, we differentiate Eq. (2.136) to obtain

$$\frac{dx_S}{dt} = \left(\frac{\partial \hat{x}}{\partial X}\right)_+ \frac{dX_S}{dt} + \left(\frac{\partial \hat{x}}{\partial t}\right)_+.$$

Now we define the material shock velocity C_S and the spatial shock velocity c_S to be

$$C_S \equiv \frac{dX_S}{dt}$$

and

$$c_S \equiv \frac{dx_S}{dt}.$$

We combine the previous three expressions to obtain

$$c_S = F_+ C_S + v_+. \tag{2.137}$$

Using the same procedure, we obtain a similar relationship on the $-$ side of the singular surface,

$$c_S = F_- C_S + v_-. \tag{2.138}$$

We call the two previous relationships the *kinematic compatibility conditions* of the singular surface. Using our definition of the jump of a function, Eq. (2.135), you can subtract the compatibility conditions to obtain

$$[\![F]\!] C_S + [\![v]\!] = 0. \tag{2.139}$$

Notice that as a singular surface moves into a material at rest, the velocity and deformation gradient are $v_- = 0$ and $F_- = 1$. Under these conditions, we find that $c_S = C_S$.

Comments on Wave Velocities The material wave velocity C and the spatial wave velocity c represent the velocity of propagation of a structured wave in the reference and current configurations of the body. We can write the relationships for the characteristic contours of a structured wave as [see Eqs. (2.91) and (2.105)]

$$\frac{dX}{dt} = \pm C, \qquad \frac{dx}{dt} = v \pm c.$$

Recall that we define the wave velocities C and c to be positive, and we use the $\pm$ symbol to denote the direction of propagation of the wave.

We describe the propagation of a singular surface as

$$\frac{dX_S}{dt} = \pm U_S, \qquad \frac{dx_S}{dt} = c_S,$$

where for convenience we have defined the *shock velocity*

$$U_S \equiv |C_S|.$$

By definition, c_S and C_S can be either positive or negative. Notice that it is even possible for the material shock velocity C_S and the spatial shock velocity c_S to have opposite signs.

Derivatives in the Spatial Description In the spatial description, during the interval of time from t to $t+dt$, the point x_S moves a distance $c_S\, dt$ (see Figure 2.64). We approximate the value of ϕ_+ at time $t + dt$ in terms of its value at time t as

$$\phi_+(t + dt) \approx \phi_+(t) + \left(\frac{\partial \phi}{\partial t}\right)_+ dt + \left(\frac{\partial \phi}{\partial x}\right)_+ c_S\, dt,$$

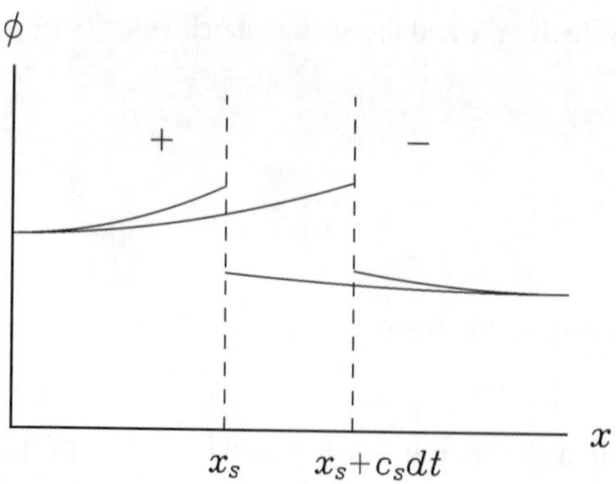

Figure 2.64 The scalar function $\phi(x, t)$ at time t and at time $t + dt$.

where the derivatives are evaluated at time t. From this expression, we see that the rate of change of ϕ_+ can be expressed as

$$\frac{d}{dt}\phi_+ = \left(\frac{\partial \phi}{\partial t}\right)_+ + \left(\frac{\partial \phi}{\partial x}\right)_+ c_S. \tag{2.140}$$

We can derive a corresponding expression for the rate of change of ϕ_-:

$$\frac{d}{dt}\phi_- = \left(\frac{\partial \phi}{\partial t}\right)_- + \left(\frac{\partial \phi}{\partial x}\right)_- c_S.$$

Subtracting this equation from Eq. (2.140), we obtain an equation for the rate of change of the jump of ϕ:

$$\frac{d}{dt}[\![\phi]\!] = \left[\!\!\left[\frac{\partial \phi}{\partial t}\right]\!\!\right] + \left[\!\!\left[\frac{\partial \phi}{\partial x}\right]\!\!\right] c_S.$$

For the special case when ϕ is continuous at x_S, $[\![\phi]\!] = 0$ and

$$\left[\!\!\left[\frac{\partial \phi}{\partial t}\right]\!\!\right] + \left[\!\!\left[\frac{\partial \phi}{\partial x}\right]\!\!\right] c_S = 0 \quad \text{(when } \phi \text{ is continuous)}. \tag{2.141}$$

Derivatives in the Material Description In the material description, during the interval of time from t to $t + dt$, the point X_S moves a distance $C_S\, dt$. We approximate the value of ϕ_+ at time $t + dt$ in terms of its value at time t as

$$\phi_+(t + dt) \approx \phi_+(t) + \left(\frac{\partial \hat{\phi}}{\partial t}\right)_+ dt + \left(\frac{\partial \hat{\phi}}{\partial X}\right)_+ C_S\, dt,$$

where the derivatives are evaluated at time t. By following the process of the previous paragraph, we obtain the following results:

$$\begin{aligned} \frac{d}{dt}\hat{\phi}_\pm &= \left(\frac{\partial \hat{\phi}}{\partial t}\right)_\pm + \left(\frac{\partial \hat{\phi}}{\partial X}\right)_\pm C_S, \\ \frac{d}{dt}[\![\hat{\phi}]\!] &= \left[\!\!\left[\frac{\partial \hat{\phi}}{\partial t}\right]\!\!\right] + \left[\!\!\left[\frac{\partial \hat{\phi}}{\partial X}\right]\!\!\right] C_S. \end{aligned} \tag{2.142}$$

For the special case when ϕ is continuous at X_S, $[\![\phi]\!] = 0$ and

$$\left[\!\left[\frac{\partial \hat{\phi}}{\partial t}\right]\!\right] + \left[\!\left[\frac{\partial \hat{\phi}}{\partial X}\right]\!\right] C_S = 0 \quad \text{(when } \phi \text{ is continuous).}$$

If we substitute $\phi = \hat{x}(X, t)$ into this expression, we recover Eq. (2.139).

An Integral Expression from Calculus Consider the following integral:

$$G(t) = \int_{a(t)}^{b(t)} \phi(\zeta, t)\, d\zeta.$$

Notice that the kernel ϕ of the integral as well as the limits a and b are functions of t. From the Leibniz theorem for differentiation of an integral, we obtain

$$\frac{d}{dt} G(t) = \int_{a(t)}^{b(t)} \frac{\partial \phi}{\partial t} d\zeta + \phi(b, t) \frac{db}{dt} - \phi(a, t) \frac{da}{dt}. \tag{2.143}$$

A Useful Integral Expression in the Material Description Consider the following integral of the function $\hat{\phi}(X, t)$:

$$G(t) = \int_{X_L}^{X_R} \hat{\phi}(X, t)\, dX.$$

The limits of this integral denote two material points that enclose a singular surface at X_S (see Figure 2.65). We divide the integral into the following two parts:

$$G(t) = \int_{X_L}^{X_S(t)} \hat{\phi}(X, t)\, dX + \int_{X_S(t)}^{X_R} \hat{\phi}(X, t)\, dX.$$

Now we evaluate the time derivative of this expression using Eq. (2.143). Because X_L and X_R are constants,

$$\frac{d}{dt} G(t) = \int_{X_L}^{X_S(t)} \frac{\partial \hat{\phi}}{\partial t} dX + C_S \phi_+ + \int_{X_S(t)}^{X_R} \frac{\partial \hat{\phi}}{\partial t} dX - C_S \phi_-.$$

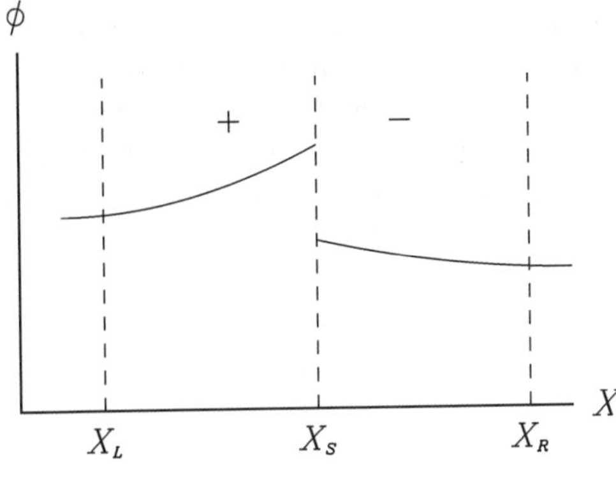

Figure 2.65 The interval of integration in the material description.

In the limit as $(X_L \to X_S)_+$ and $(X_R \to X_S)_-$, we obtain

$$\frac{d}{dt}G(t) = [\![\phi]\!]C_S \quad \text{when} \begin{cases} (X_L \to X_S)_+ \\ (X_R \to X_S)_- \end{cases}. \tag{2.144}$$

We shall use this expression to derive the laws of motion for the singular surface.

A Useful Integral Expression in the Spatial Description Consider the following integral of the function $\phi(x, t)$:

$$G(t) = \int_{x_L(t)}^{x_R(t)} \phi(x, t)\, dx.$$

This integral is evaluated between the two material points X_L and X_R,

$$x_L(t) = \hat{x}(X_L, t), \qquad x_R(t) = \hat{x}(X_R, t),$$

that enclose a singular surface at x_S. We divide this integral into two parts:

$$G(t) = \int_{x_L(t)}^{x_S(t)} \phi(x, t)\, dx + \int_{x_S(t)}^{x_R(t)} \phi(x, t)\, dx.$$

Using Eq. (2.143), we differentiate $G(t)$ to obtain

$$\frac{d}{dt}G(t) = \int_{x_L(t)}^{x_S(t)} \frac{\partial \phi}{\partial t} dx + c_S\phi_+ - v(x_L, t)\phi(x_L, t)$$
$$+ \int_{x_S(t)}^{x_R(t)} \frac{\partial \phi}{\partial t} dx - c_S\phi_- + v(x_R, t)\phi(x_R, t).$$

In the limit as $(x_L \to x_S)_+$ and $(x_R \to x_S)_-$, we obtain

$$\frac{d}{dt}G(t) = [\![\phi(c_S - v)]\!] \quad \text{when} \begin{cases} (x_L \to x_S)_+ \\ (x_R \to x_S)_- \end{cases}. \tag{2.145}$$

We shall also use this result to derive the laws of motion for the singular surface.

Derivation of the Jump Conditions

The laws of motion are composed of the conservation of mass and the balance of linear momentum. By assuming the smooth and continuous solution of a structured wave, we have derived differential relationships for these laws. Now we shall apply the laws of motion to a singular surface. The resulting expressions are called *jump conditions*. We derive jump conditions in both the spatial and material descriptions using the mathematical relationships for a singular surface presented in the previous section.

Material Description Consider the material element illustrated in Figure 2.66. The cross-sectional area of this element is unity. A singular surface is located at the position X_S. The mass of material in this element is

$$m = \int_{X_L}^{X_R} \hat{\rho}_0\, dX,$$

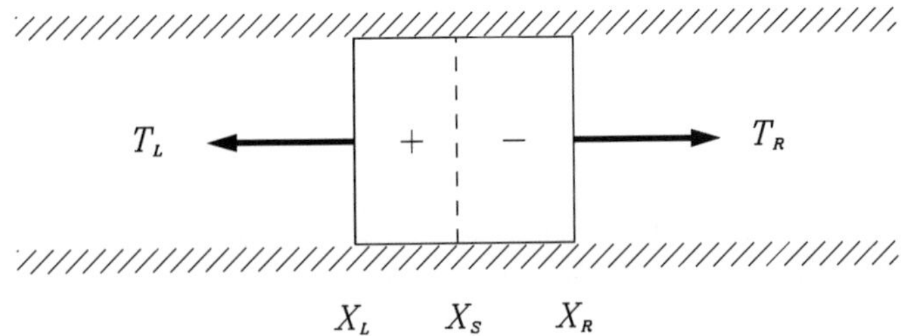

Figure 2.66 The material element with a singular surface (— —) in the material description.

where we notice that ρ_0 can at most be a function of X. Using the conservation of mass, we require that

$$\frac{dm}{dt} = 0.$$

This condition must hold in the limit as $(X_L \to X_S)_+$ and $(X_R \to X_S)_-$. By letting $\phi = \rho_0$, we can apply Eq. (2.144). We shall use a continuous function for ρ_0. Therefore $[\![\rho_0]\!] = 0$, and the conservation of mass is satisfied automatically.

In the same material element, we require the rate of change of linear momentum to be equal to the sum of the forces acting on the material element (see Figure 2.66),

$$T_R - T_L = \frac{d}{dt}\int_{X_L}^{X_R} \hat{\rho}_0 \hat{v}\, dX.$$

This condition must also hold in the limit as $(X_L \to X_S)_+$ and $(X_R \to X_S)_-$. By letting $\hat{\phi} = \hat{\rho}_0 \hat{v}$, we can apply Eq. (2.144). The result is

$$[\![T]\!] = -\rho_0 C_S [\![v]\!]. \tag{2.146}$$

This result is a basic tool in the analysis of shock wave experiments. To illustrate why, consider a material that is initially at rest. We assume that the stress and density of the material are known. Now suppose we observe the passage of a shock wave through this material. The propagation speed of the shock wave and the velocity of the material behind the shock wave are relatively easy quantities to measure. Once we measure these quantities, this jump condition allows us to calculate the stress behind the shock wave. Because it is a law of motion, this jump condition is not limited to nonlinear-elastic materials. It is true for *any* material that supports a steady wave, regardless of the constitutive model. Substituting Eq. (2.139) into Eq. (2.146), we obtain another useful form of this jump condition:

$$C_S^2 = \frac{1}{\rho_0}\frac{[\![T]\!]}{[\![F]\!]}. \tag{2.147}$$

Spatial Description Consider the material volume in Figure 2.66. The current positions of the left and right faces of this material element are $x_L(t) = \hat{x}(X_L, t)$ and $x_R(t) = \hat{x}(X_R, t)$. The position of the singular surface within the element is $x_S(t) = \hat{x}(X_S, t)$. The mass of material in this element is

$$m = \int_{x_L(t)}^{x_R(t)} \hat{\rho}\, dx.$$

Conservation of mass requires that

$$\frac{dm}{dt} = 0.$$

We shall require this condition to hold in the limit as $(x_L \to x_S)_+$ and $(x_R \to x_S)_-$. Although we have prescribed ρ_0 to be a smooth function, we shall allow ρ to be discontinuous at X_S. If we let $\phi = \rho$, we can use Eq. (2.145) to obtain the jump condition for the conservation of mass,

$$[\![\rho(c_S - v)]\!] = 0. \tag{2.148}$$

In the same material element, we require the rate of change of linear momentum to be equal to the sum of the forces acting on the element,

$$T_R - T_L = \frac{d}{dt}\int_{x_L(t)}^{x_R(t)} \rho v \, dx.$$

We also require this condition to hold in the limit as $(x_L \to x_S)_+$ and $(x_R \to x_S)_-$. If we let $\phi = \rho v$, we can use Eq. (2.145) to obtain

$$[\![T + \rho v(c_S - v)]\!] = 0. \tag{2.149}$$

This is the spatial description of the jump condition for the balance of linear momentum.

2.7.3 *Jump Conditions: Three-Dimensional Motion*

Now that we have derived the one-dimensional jump conditions for the conservation of mass and the balance of linear momentum, we shall present a derivation of the jump conditions for three-dimensional motion. With respect to the methods we have used to obtain the one-dimensional laws of motion, you may have noticed only a weak similarity between the method of derivation of the differential equations for continuous solutions and the method of derivation of the jump conditions for discontinuous solutions. In contrast, you shall see a very strong connection in the methods used for the three-dimensional derivations of the differential equations and the jump conditions.

Modified Transport Theorem

In Chapter 1, we used the transport theorem [see Eq. (1.149)],

$$\frac{d}{dt}\int_V \phi \, dV = \int_V \left[\frac{\partial \phi}{\partial t} + \frac{\partial(\phi v_k)}{\partial x_k}\right] dV,$$

to derive the three-dimensional differential equations for the conservation of mass and balance of linear momentum. This theorem is based upon the Leibniz rule for differentiation of an integral and the Gauss theorem. The transport theorem only applies to a material element V, which contains a continuous and differentiable solution. We cannot apply the transport theorem to a material element that contains a singular surface. Such a situation is shown in Figure 2.67 where the singular surface is illustrated with a dashed curve. However, we can modify the transport theorem to accommodate the presence of this singular surface.

We shall use the symbol Σ to denote the portion of the singular surface contained in the volume V. At the singular surface Σ, we split the material element V into two volumes $\mathcal{V}_+$ and $\mathcal{V}_-$ (see Figure 2.68). The unit vector $\mathbf{N}$ is the normal to the surface Σ. This unit vector

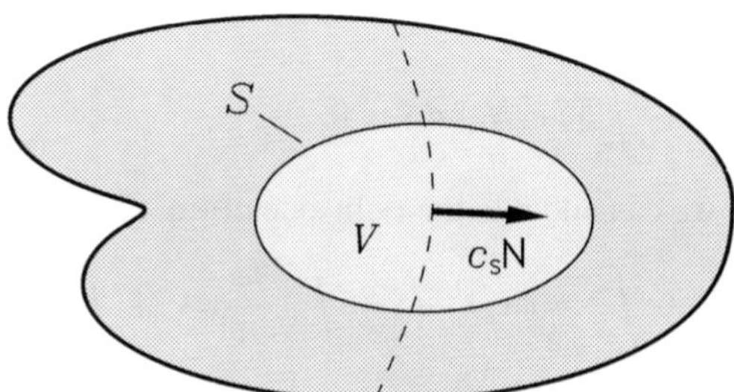

Figure 2.67 A material element V within a body. The singular surface (— —) passes through V. Its magnitude and direction of motion are represented by the velocity of propagation c_S and the unit normal **N**.

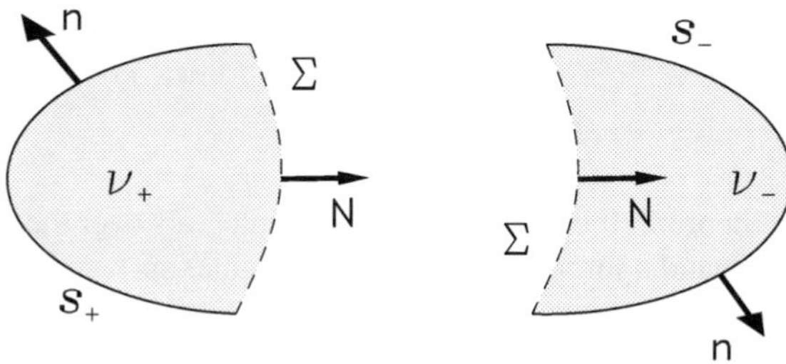

Figure 2.68 A material element V after it is split into two volumes $\mathcal{V}_+$ and $\mathcal{V}_-$.

points in the direction of motion of Σ. Thus, it points into the interior of $\mathcal{V}_-$. The unit vector $\mathbf{n}$ points outward from $\mathcal{V}_+$ and $\mathcal{V}_-$. Where the volume $\mathcal{V}_+$ intersects the surface Σ, we see that $\mathbf{n} = \mathbf{N}$, and where the volume $\mathcal{V}_-$ intersects the surface Σ, we see that $\mathbf{n} = -\mathbf{N}$. By splitting V into two parts, we obtain

$$\frac{d}{dt}\int_V \phi\, dV = \frac{d}{dt}\int_{\mathcal{V}_+} \phi\, dV + \frac{d}{dt}\int_{\mathcal{V}_-} \phi\, dV. \tag{2.150}$$

The components of the velocity of Σ are $c_S N_k$. Because $c_S N_k \neq v_k$, we see that $\mathcal{V}_+$ and $\mathcal{V}_-$ are not material volumes. This means that even though ϕ is continuous in both $\mathcal{V}_+$ and $\mathcal{V}_-$, we still cannot apply the transport theorem to these integrals; however, we can use Leibniz rule for differentiation of an integral [see Eq. (1.148)],

$$\frac{d}{dt}\int_{\mathcal{V}} \phi\, dV = \int_{\mathcal{V}} \frac{\partial \phi}{\partial t}\, dV + \int_{\mathcal{S}} \phi w_k n_k\, dS, \tag{2.151}$$

where w_k are the components of the velocity of the surface $\mathcal{S}$ that completely encloses the volume $\mathcal{V}$. To apply this rule, we notice that $\mathcal{V}_+$ is enclosed by $\mathcal{S}_+$ and Σ, and $\mathcal{V}_-$ is enclosed by $\mathcal{S}_-$ and Σ. Also, on $\mathcal{S}_+$ and $\mathcal{S}_-$, we see that $w_k n_k = v_k n_k$. Where the volume $\mathcal{V}_+$ intersects Σ, we have $w_k n_k = c_S$, and where the volume $\mathcal{V}_-$ intersects Σ, we have $w_k n_k = -c_S$. Thus, after applying Eqs. (2.151) to (2.150), we obtain

$$\begin{aligned}\frac{d}{dt}\int_V \phi\, dV = &\int_{\mathcal{V}_+} \frac{\partial \phi}{\partial t}\, dV + \int_{\mathcal{S}_+} \phi v_k n_k\, dS + \int_{\Sigma} \phi_+ c_S\, dS \\ &+ \int_{\mathcal{V}_-} \frac{\partial \phi}{\partial t}\, dV + \int_{\mathcal{S}_-} \phi v_k n_k\, dS - \int_{\Sigma} \phi_- c_S\, dS.\end{aligned} \tag{2.152}$$

Now we notice that

$$\int_{\Sigma}(\phi_+ v_k^+ n_k + \phi_- v_k^- n_k)\,dS - \int_{\Sigma} [\![\phi v_k N_k]\!]\,dS = 0.$$

Because this term is equal to zero, we can add it to the right side of Eq. (2.152) to obtain

$$\frac{d}{dt}\int_V \phi\,dV = \int_{\mathcal{V}_+} \frac{\partial \phi}{\partial t}\,dV + \int_{S_+ + \Sigma} \phi v_k n_k\,dS + \int_{\mathcal{V}_-} \frac{\partial \phi}{\partial t}\,dV + \int_{S_- + \Sigma} \phi v_k n_k\,dS$$

$$+ \int_{\Sigma} [\![\phi(c_S - v_k N_k)]\!]\,dS.$$

Now we apply the Gauss theorem to obtain

$$\begin{aligned}\frac{d}{dt}\int_V \phi\,dV = {} & \int_{\mathcal{V}_+}\left[\frac{\partial \phi}{\partial t} + \frac{\partial(\phi v_k)}{\partial x_k}\right]dV + \int_{\mathcal{V}_-}\left[\frac{\partial \phi}{\partial t} + \frac{\partial(\phi v_k)}{\partial x_k}\right]dV \\ & + \int_{\Sigma} [\![\phi(c_S - v_k N_k)]\!]\,dS.\end{aligned} \tag{2.153}$$

This is the modified transport theorem. It contains two volume integrals over the regions in which the solution is continuous and a jump condition at the singular surface.

Derivation of the Jump Conditions

We have just derived the modified transport theorem. Now we use it to determine the jump conditions at a singular surface. We also find that, in addition to these jump conditions, we again obtain the differential equations for the conservation of mass and balance of linear momentum.

Conservation of Mass Consider a material volume V. The law of conservation of mass for this volume is

$$\frac{d}{dt}\int_V \rho\,dV = 0.$$

When we apply the modified transport theorem, Eq. (2.153), to this relationship, we find that

$$\int_{\mathcal{V}_+}\left[\frac{\partial \rho}{\partial t} + \frac{\partial(\rho v_k)}{\partial x_k}\right]dV + \int_{\mathcal{V}_-}\left[\frac{\partial \rho}{\partial t} + \frac{\partial(\rho v_k)}{\partial x_k}\right]\partial V + \int_{\Sigma} [\![\rho(c_S - v_k N_k)]\!]\,dS = 0.$$

This equation must hold for every material volume V. It is satisfied only when the three integrands independently vanish at every point over their respective regions of integration. Thus, the volume integrals yield

$$\frac{\partial \rho}{\partial t} + \frac{\partial(\rho v_k)}{\partial x_k} = 0,$$

which is the differential equation for the conservation of mass in a region with a continuous and differentiable solution [see Eq. (1.150)]. In addition to this relationship, we obtain the jump condition for the conservation of mass at the singular surface:

$$[\![\rho(c_S - v_k N_k)]\!] = 0. \tag{2.154}$$

For one-dimensional motion, this result reduces to the jump condition for the conservation of mass in the one-dimensional, spatial description [Eq. (2.148)].

Balance of Linear Momentum The law of balance of linear momentum for the volume V is [see Eq. (1.152)]

$$\frac{d}{dt}\int_V \rho v_m \, dV = \int_S T_{km} n_k + \int_V \rho b_m \, dV dS.$$

By splitting the surface integral into two parts, we obtain

$$\frac{d}{dt}\int_V \rho v_m \partial V = \int_{S_+ + \Sigma} T_{km} n_k \, dS + \int_{S_- + \Sigma} T_{km} n_k \, dS$$
$$+ \int_V \rho b_m \, dV - \int_\Sigma [\![T_{km} N_k]\!] \, dS, \tag{2.155}$$

where

$$\int_\Sigma \left(T_{km}^+ n_k + T_{km}^- n_k \right) dS - \int_\Sigma [\![T_{km} N_k]\!] \, dS = 0.$$

We then apply the modified transport theorem to the left side of Eq. (2.155) and the Gauss theorem to the right side, obtaining

$$\int_{V_+} \left[\frac{\partial(\rho v_m)}{\partial t} + \frac{\partial(\rho v_m v_k)}{\partial x_k} - \frac{\partial T_{km}}{\partial x_k} - \rho b_m \right] dV$$
$$+ \int_{V_-} \left[\frac{\partial(\rho v_m)}{\partial t} + \frac{\partial(\rho v_m v_k)}{\partial x_k} - \frac{\partial T_{km}}{\partial x_k} - \rho b_m \right] dV$$
$$+ \int_\Sigma [\![\rho v_m (c_s - v_k N_k) + T_{km} N_k]\!] \, dS = 0.$$

This equation must hold for every material volume V. It is satisfied only when the three integrands independently vanish at every point over their respective regions of integration. The volume integrals yield

$$\frac{\partial(\rho v_m)}{\partial t} + \frac{\partial(\rho v_m v_k)}{\partial x_k} - \frac{\partial T_{km}}{\partial x_k} - \rho b_m = 0.$$

We have already demonstrated that this expression reduces to the differential equation for the balance of linear momentum (see Exercise 1.59 in Section 1.8.8). From the surface integral we obtain

$$[\![T_{km} N_k + \rho v_m (c_s - v_k N_k)]\!] = 0.$$

This is the three-dimensional jump condition for the balance of linear momentum. For one-dimensional motion, this expression reduces to the jump condition for the balance of linear momentum in the one-dimensional, spatial description [Eq. (2.149)].

2.7.4 *Comparison of Shock Waves and Simple Waves*

In this section we summarize and compare the relationships that describe simple waves and shock waves. We shall use these results in future sections to analyze the interactions of simple waves with shock waves.

By definition, J_- is constant for a simple wave propagating in the positive-x direction. Thus

$$\frac{dv}{dF} = -C, \qquad \frac{dT}{dv} = -\rho_0 C, \quad \text{positive-}x \text{ propagation.} \tag{2.156}$$

Remember that C is positive. Let us compare these results for a simple wave to the results for a shock wave propagating in the positive-x direction. We shall use the kinematic and linear-momentum jump conditions in Eqs. (2.139) and (2.146):

$$\frac{[\![v]\!]}{[\![F]\!]} = -U_S, \qquad \frac{[\![T]\!]}{[\![v]\!]} = -\rho_0 U_S, \quad \text{positive-}x \text{ propagation,} \tag{2.157}$$

where we recall that we have defined the shock velocity $U_S = |C_S|$ to be a positive quantity.

Similarly, for waves that propagate in the negative-x direction, by definition, J_+ is constant, and C_S is negative. Thus for the simple wave we have

$$\frac{dv}{dF} = C, \qquad \frac{dT}{dv} = \rho_0 C, \quad \text{negative-}x \text{ propagation,} \tag{2.158}$$

and for the shock wave

$$\frac{[\![v]\!]}{[\![F]\!]} = U_S, \qquad \frac{[\![T]\!]}{[\![v]\!]} = \rho_0 U_S, \quad \text{negative-}x \text{ propagation.} \tag{2.159}$$

If we consider shock waves with infinitesimal jumps in v, T, and F, it is clear that the jump conditions for a shock wave become the differential conditions for a simple wave. Under these conditions, we find that $U_S \to C$.

Previously, we defined the acoustic impedance to be $z_0 = \rho_0 c_0$ and the wave impedance to be $z = \rho_0 C = \rho c$. Let us define the *shock impedance* z_S to be

$$z_S \equiv \rho_0 U_S.$$

Using this definition, we can rewrite the second expressions in each of Eqs. (2.156)–(2.159) as

$$\frac{d(-T)}{dv} = \pm z, \tag{2.160}$$

$$\frac{[\![-T]\!]}{[\![v]\!]} = \pm z_S. \tag{2.161}$$

The positive signs are used for waves traveling in the positive-x direction, and the negative signs are used for waves traveling in the negative-x direction. Recall that with respect to the T–v diagram, the differential relationships Eq. (2.160) yield the response paths for simple waves. Similarly, the jump conditions, Eq. (2.161), represent the responses for shock waves.

Consider a body with known stress and velocity. The T–v diagram in Figure 2.69 represents the values of stress and velocity at the material point $\mathbf{X}$. The initial values of velocity and stress are given by point I. If the material at $\mathbf{X}$ encounters a compression wave, the stress and velocity will move along one of the two response paths shown by the solid lines.

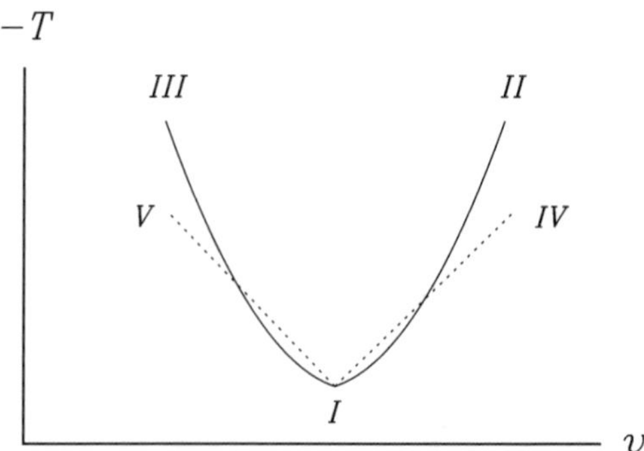

Figure 2.69 The T–v diagram for the material point **X**.

We obtain these paths by integration of the two differential relationships Eq. (2.160). When the compression wave propagates in the positive-x direction, the stress and velocity travel up the positive-sloping solid line from point *I* to point *II*, and when the compression wave propagates in the negative-x direction, the stress and velocity travel up the negative-sloping solid line from point *I* to point *III*.

Instead, let us now suppose that the material at point *I* encounters a shock wave. The stress and velocity will jump from point *I* to either point *IV* or *V*. The locations of the points *IV* and *V* are determined by the two jump conditions Eq. (2.161). When the shock propagates in the positive-x direction, the stress and velocity jump from point *I* to point *IV* along the positive-sloping dotted line. When the shock wave propagates in the negative-x direction, the stress and velocity jump from point *I* to point *V* along the negative-sloping dotted line. To indicate the difference between a continuous compression and a discontinuous jump, we connect points *V*, *I*, and *IV* with dotted lines. The dotted lines are called *Rayleigh lines*. The slopes of the Rayleigh lines are $\pm z_S$. In this chapter, we restrict our discussion to materials in which shock waves must cause compression. Discontinuities that cause expansion immediately become centered rarefaction waves (see Section 2.6.2). Thus we can only draw a Rayleigh line towards greater values of $-T$. Later we shall show that the Rayleigh line is much more than a convenient graphical notation for a jump.

Here we do not intend to preclude the possibility of rarefaction shock waves. They do exist but not in a nonlinear-elastic material as we have defined it. Rather, they exist in other materials. We shall return to this topic in Chapter 4.

2.7.5 *Linear-Elastic Discontinuity*

The sudden compression of a half space results in a propagating singular surface. Let us now examine how the simple wave solution, the d'Alembert solution, and the jump conditions all yield the same results if the half space in the region $X \geq 0$ is a linear-elastic material (see Figure 2.70).

Simple Wave First suppose that the half space is a nonlinear elastic material. If we attempt to construct a simple wave solution for this problem, we find that the head and tail characteristics of the simple wave solution form a fan, but because C on the tail characteristic is greater than C on the head characteristic, the tail characteristic must precede the head

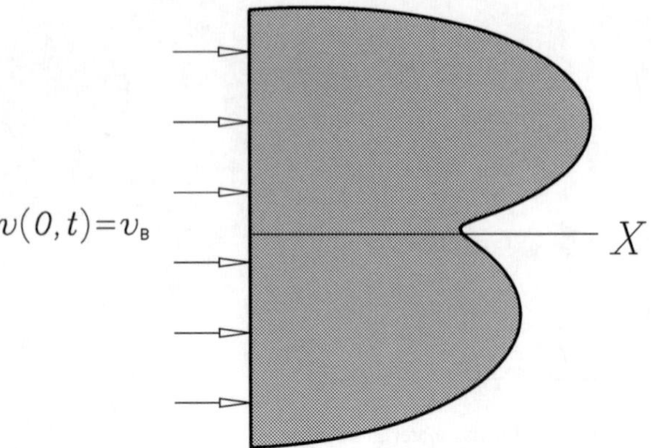

Figure 2.70 A half space subjected to a sudden compression.

characteristic. The resulting solution is multivalued and invalid. Suppose we now assume that the material in the half space is linear elastic. Then, both the head and tail contours are given by

$$x = c_0 t.$$

You can visualize the head and tail contours as having merged into a single contour with a slope determined by the velocity of sound c_0. All of the contours in this simple-wave solution are not only parallel, they are also coincident and represent a discontinuous solution. Even though this solution is discontinuous, we chose not to call it a shock wave. We reserve that term to describe a nonlinear wave.

D'Alembert Solution Now let us examine the d'Alembert solution. Suppose we represent the boundary condition as

$$p(t) = \begin{cases} 0, & t \leq 0, \\ v_B, & t > 0. \end{cases}$$

The d'Alembert solution for the velocity in the half space is [see Eq. (2.54)]

$$v(x, t) = p(t - x/c_0).$$

We see that the velocity jumps from $v = 0$ to $v = v_B$ across the single contour

$$x = c_0 t.$$

Thus both the simple wave solution and the d'Alembert solution yield jumps that propagate at the velocity of sound c_0.

Jump Conditions Next we turn to the jump conditions to evaluate this same solution. Recall that the constitutive equation for a nonlinear-elastic material is $T = T(F)$. For a linear-elastic material where the change in F is small, we can employ the Taylor's series of T about $F = 1$ to obtain the expansion

$$T = T^0 + \left(\frac{dT}{dF}\right)_{F=1} (F - 1) + \frac{1}{2}\left(\frac{d^2T}{dF^2}\right)_{F=1} (F - 1)^2 + \cdots.$$

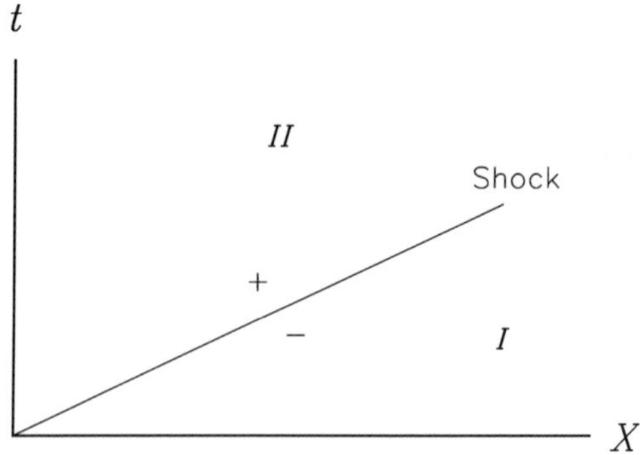

Figure 2.71 The X–t diagram of a shock wave solution for a half space subjected to a sudden compression.

We notice that

$$\rho_0 C^2 = \frac{dT}{dF}, \qquad 2\rho_0 C \frac{dC}{dF} = \frac{d^2T}{dF^2}.$$

Thus, because $C = c_0$ at $F = 1$,

$$T - T^0 = \rho_0 c_0^2 (F - 1) + \rho_0 c_0 \left(\frac{dC}{dF} \right)_{F=1} (F - 1)^2 + \cdots. \tag{2.162}$$

For a linear-elastic material, we shall ignore the terms containing products of $(F - 1)$. Then

$$T - T^0 = \rho_0 c_0^2 (F - 1),$$

which gives us the following expression for the jump in stress:

$$[\![T]\!] = \rho_0 c_0^2 [\![F]\!].$$

Comparing this result to the jump condition for the balance of linear momentum [Eq. (2.147)], we see that

$$|C_S| = U_S = c_0.$$

We represent this solution by the contour for a singular surface (see Figure 2.71),

$$X = c_0 t.$$

This shock contour separates the X–t diagram into two regions. The stress T and velocity v are uniform in each region. We conclude that when the half space contains a linear-elastic material, the simple wave solution, the d'Alembert solution, and the shock wave solution yield identical results.

2.7.6 *Weak Shock*

In the previous section we examined the sudden compression of a half space of linear-elastic material. Suppose we replace the linear-elastic material by a nonlinear-elastic material with a constitutive model represented by the truncated Taylor's series [see Eq. (2.162)]

$$T = T^0 + \rho_0 c_0^2 (F - 1) + \rho_0 c_0 \left(\frac{dC}{dF} \right)_{F=1} (F - 1)^2. \tag{2.163}$$

The material wave velocity is

$$C^2 = \frac{1}{\rho_0} \frac{dT}{dF} = c_0^2 + 2c_0 \left(\frac{dC}{dF} \right)_{F=1} (F - 1).$$

We approximate the square root of this result as

$$C \approx c_0 + \left(\frac{dC}{dF} \right)_{F=1} (F - 1). \tag{2.164}$$

Using this constitutive model, we shall now derive the relationships that govern both structured waves and shock waves.

Structured Waves Structured waves are described by the Riemann integrals

$$J_+(\xi) = v - v^0 - \int_1^F C(\bar{F})\, d\bar{F}\,,$$

$$J_-(\zeta) = v - v^0 + \int_1^F C(\bar{F})\, d\bar{F}\,.$$

After substituting Eq. (2.164), we can integrate these expressions to obtain

$$J_\pm = v - v^0 \mp \left[c_0 (F - 1) + \frac{1}{2} \left(\frac{dC}{dF} \right)_{F=1} (F - 1)^2 \right]. \tag{2.165}$$

Shock Waves We obtain the shock wave solution by substitution of Eq. (2.164) into Eq. (2.163):

$$T = T^0 + \rho_0 c_0^2 (F - 1) + \rho_0 c_0 (C - c_0)(F - 1).$$

This yields the following expression for the jump in stress at a singular surface:

$$[\![T]\!] = \rho_0 c_0^2 [\![F]\!] + \rho_0 c_0 [\![C]\!][\![F]\!].$$

From Eq. (2.147), we obtain

$$C_S^2 = c_0^2 + c_0 [\![C]\!] = c_0 C_+,$$

where C_+ is the material wave velocity behind the shock wave and $c_0 = C_-$ is the material wave velocity ahead of the shock wave. When terms containing $[\![C]\!]^2$ are neglected, it is possible to show that (see Exercise 2.47)

$$U_S = \frac{1}{2}(C_+ + c_0), \tag{2.166}$$

where $U_S = |C_S|$. We shall now evaluate the shock wave solution for the sudden compression of a half space of this material.

Sudden Compression of a Half Space We illustrate this shock wave solution in Figure 2.71. We use the shock wave contour to divide the diagram into regions *I* and *II*. The values of v, F, and T are constant in both regions. From the initial conditions, the velocity, deformation gradient, and stress in region *I* are $v_- = 0$, $F_- = 1$, and $T_- = T^0$. We prescribe the velocity in region *II* by using the boundary condition $v = v_+$. Then from the jump condition, Eq. (2.139), we obtain

$$F_+ = 1 - \frac{v_+}{U_S}. \tag{2.167}$$

Using the jump condition Eq. (2.146), we determine the stress in region *II*:

$$T_+ - T^0 = -\rho_0 U_S v_+. \tag{2.168}$$

From Eqs. (2.164) and (2.166), we find that

$$U_S = c_0 + \frac{1}{2}\left(\frac{dC}{dF}\right)_{F=1}(F_+ - 1). \tag{2.169}$$

Because F_+ is constant in region *II*, the shock velocity U_S is constant. Thus the shock wave contour is a straight line. We call this solution a *weak shock*. We use this terminology because the constitutive equation [Eq. (2.163)] of this material only contains the first two terms of a Taylor's series expansion. This implies that the wave is not strong enough to activate higher-order terms in this series.

A Weak Shock in an Adiabatic Gas

Consider a half space of an adiabatic gas. The material wave velocity for an adiabatic gas is given by Eq. (2.108):

$$C = c_0 F^{-\frac{\gamma+1}{2}}.$$

When we substitute this expression and Eq. (2.167) into the solution for a weak shock, Eq. (2.169), we obtain the following equation for U_S in terms of the boundary velocity v_+:

$$\frac{(U_S - c_0)^2}{c_0} + U_S = c_0 + \frac{\gamma + 1}{4} v_+. \tag{2.170}$$

For a weak shock, we assume the term $(U_S - c_0)^2/c_0$ is small with respect to U_S and can be ignored. Then

$$U_S = c_0 + \frac{\gamma + 1}{4} v_+. \tag{2.171}$$

This expression gives the shock velocity in terms of the velocity of sound and the material velocity v_+ behind the weak shock wave. When we subject a gas to a strong shock wave, this expression does not reliably predict the velocity of the shock wave. In fact, the analysis of the response of a gas to stronger shock waves is beyond the scope of this chapter. We shall return to this subject in Chapters 3 and 4. In Chapter 4 we shall determine the range of validity of Eq. (2.171).

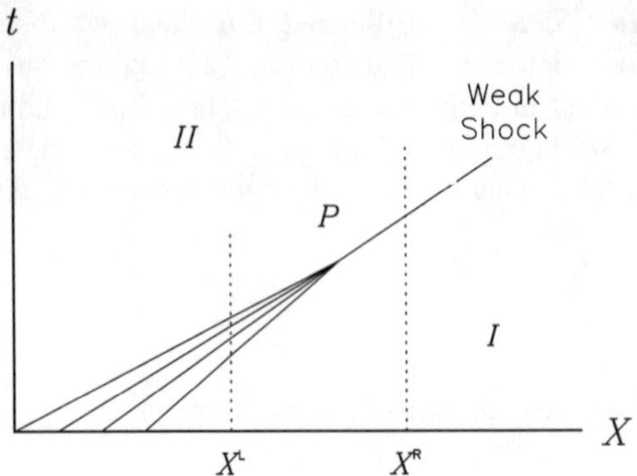

Figure 2.72 The X–t diagram for the transition of a compression wave into a weak shock.

Transition of a Compression Wave into a Weak Shock

Consider a nonlinear-elastic material in a reference configuration where $F = 1$ and $T = T^0$. This material obeys the constitutive relation Eq. (2.163):

$$T = T^0 + \rho_0 c_0^2 (F - 1) + \rho_0 c_0 \left(\frac{dC}{dF}\right)_{F=1} (F - 1)^2. \tag{2.172}$$

Suppose a compression wave propagates in the positive-x direction. This compression wave has characteristic contours that converge to a single point P (see Figure 2.72). For later times, a single contour representing the weak shock extends beyond the point P. These contours divide the diagram into regions *I* and *II*. In constructing this diagram, we have implicitly assumed that both the compression wave and the weak shock produce identical changes in v and F. Thus we assume that region *II* includes both the area behind the compression wave and the area behind the shock wave. If our assumption is wrong, then region *II* will have to be divided into two regions by additional contours that extend from point P into region *II* and separate the portion of region *II* behind the compression wave from the portion of region *II* behind the shock wave. To test our assumption that these additional contours are unnecessary, we shall calculate the changes in v and F at X^L and at X^R.

First we consider position X^L. The material at X^L encounters the compression wave. This material point begins in region *I*, where the initial conditions require that $v = 0$ and $F = 1$. Thus $J_\pm = 0$ in region *I*. The compression wave is a simple wave that propagates in the positive-x direction. Therefore $J_- = 0$ across the compression wave. In region *II* behind the compression wave, we prescribe that $F = F_{II}$. In region *II* the velocity is [see Eq. (2.165)]

$$v_{II} = -c_0(F_{II} - 1) - \frac{1}{2}\left(\frac{dC}{dF}\right)_{F=1} (F_{II} - 1)^2.$$

Noticing that the material wave velocity in region *II* is

$$C_{II} = c_0 + \left(\frac{dC}{dF}\right)_{F=1} (F_{II} - 1),$$

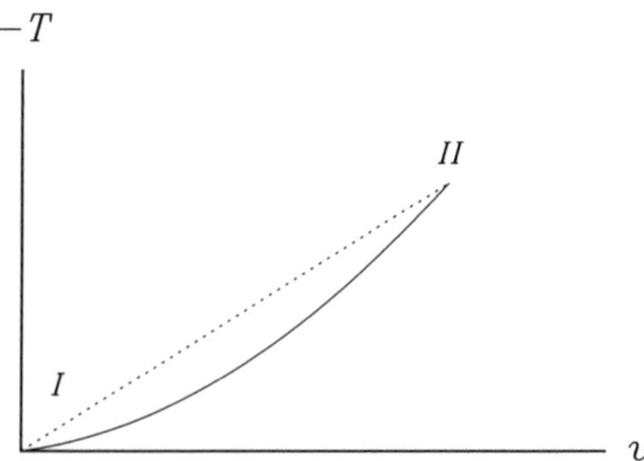

Figure 2.73 The transition of a compression wave into a weak shock. The material at X_L moves along the response path of the compression wave (—) and the material at X^R jumps between the two points connected by the Rayleigh line $(\cdots)$.

we obtain

$$v_{II} = -\frac{1}{2}(C_{II} + c_0)(F_{II} - 1). \tag{2.173}$$

Having determined the change in velocity v resulting from the compression wave at position X^L, we next determine the change in velocity resulting from the weak shock at position X^R. Because we wish to demonstrate that the solution behind the weak shock is equal to the solution behind the compression wave, let us assume that the deformation gradient at X^R behind the weak shock is equal to the deformation gradient at X^L behind the compression wave. By letting $F_+ = F_{II}$ in Eqs. (2.166) and (2.167), we obtain

$$v_+ = -U_S(F_{II} - 1),$$

where

$$U_S = \frac{1}{2}(C_{II} + c_0).$$

Comparison of these results to Eq. (2.173) shows that $v_+ = v_{II}$. We conclude that the weak shock and the compression wave produce the same changes in deformation gradient and velocity. This is illustrated in the T–v diagram in Figure 2.73. We compare the response of the material at X^L to the response of the material at X^R. The material at X^L encounters the compression wave and moves from point *I* to point *II* along the response path represented by the solid line. The material at X^R encounters the weak shock and jumps between points *I* and *II*, which are connected by the dotted Rayleigh line. This result is limited to weak shocks, which obey the constitutive relation Eq. (2.172). In general, shock waves and simple waves *do not* produce identical changes in the stress, deformation gradient, and velocity. We demonstrate this in the next section.

2.7.7 *Strong Shock*

Consider a volume of adiabatic gas that is initially at rest. For a weak shock propagating in the positive-x direction, we can express the propagation velocity of the shock wave as a linear function of the velocity v_+ of the gas behind the shock wave as [see Eq. (2.171)]

$$U_S = c_0 + \frac{\gamma + 1}{4} v_+. \tag{2.174}$$

We focus our attention on this expression because it suggests an empirical relationship for the response of solids and liquids to shock waves. For a shock wave propagating in the positive-x direction, this relationship is

$$U_S = c_0 + s[\![v]\!], \tag{2.175}$$

where s is an arbitrary constant. We call this type of expression a *Hugoniot*. We shall use the Hugoniot to determine the response of a material to a shock wave. In Chapter 4 we show how the Hugoniot can be derived from the constitutive model of the material. However, the opposite is not true. *If we know the Hugoniot of a material, we cannot uniquely determine the constitutive model of the material.*

The Hugoniot equation (2.175) is important for several reasons. First, unlike Eq. (2.174), which is limited to weak shocks in gases, Eq. (2.175) represents the behavior of a great number of solids and liquids to both weak and strong shocks. Here we define a *strong shock* to be a shock wave that causes changes in v and F that differ from the changes caused by a compression wave. Next, although this Hugoniot only yields partial information about the constitutive model of an unknown material, it is nevertheless significant information. Also, the Hugoniot contains quantities that we can easily measure in an experiment.

The Hugoniot Experiment

Consider the experiment shown in Figure 2.74. We show two identical plates of a solid material. The lateral dimensions of the plates are large in comparison to their thicknesses. The two plates move toward each other with equal and opposite velocities. Initially, the stress in each plate is zero. When the plates impact, two shock waves are generated at the impact plane. Initially, along the horizontal centerline of the experiment, the motion is one dimensional. As time advances, waves from the edges of the plates converge on the centerline, and the motion is no longer one dimensional. Typically, we confine our experimental measurements to the centerline, and we complete our measurements during the period of time in which the motion is one dimensional on the centerline.

Such an experiment is called a *Hugoniot experiment*. Indeed, this experiment is similar to the flyer-plate experiment that we discussed in Chapter 0. We show the X–t diagram for a Hugoniot experiment in Figure 2.75. The diagram does not include the interactions of the shock waves with the two free boundaries of the plates. We shall discuss this later in this chapter. Here we show the contours only up to the time that they reach the free boundaries.

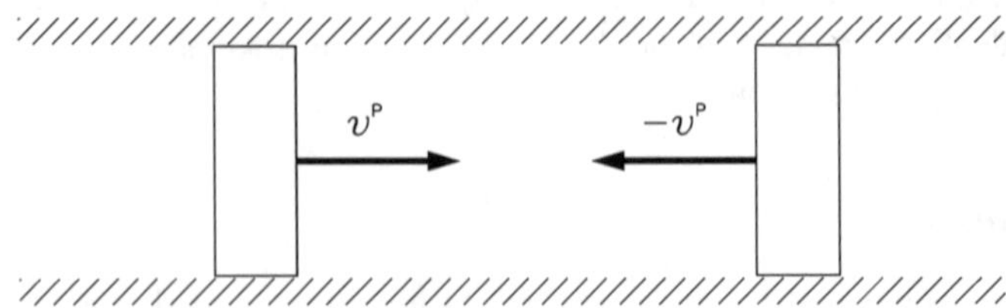

Figure 2.74 The Hugoniot experiment.

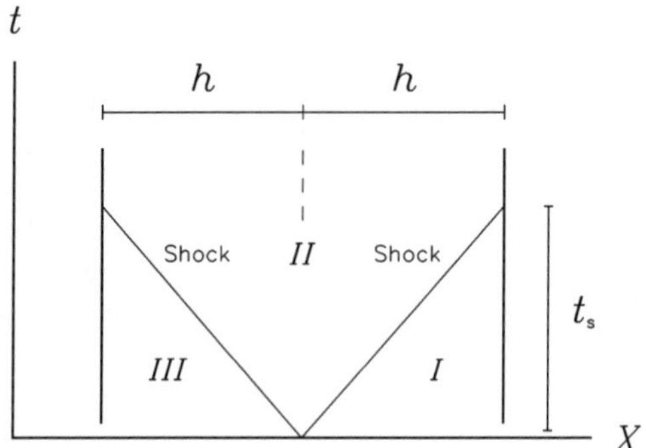

Figure 2.75 The X–t diagram for the Hugoniot experiment.

Because of the symmetry of the experiment, each shock wave propagates with the same velocity U_S.

We measure two quantities in this experiment: the velocity of the plates v^P and the time t_s between impact and the arrival of the shock waves at the free boundaries. If the thickness of each plate is h, then the material wave velocity of each shock wave is

$$U_S = h/t_s.$$

For the left plate $C_S = -U_S$, and for the right plate $C_S = U_S$. Using the shock contours, we divide the diagram into three regions. From the initial conditions, we know the velocity in region *I* is $-v^P$ and the velocity in region *III* is v^P. From the symmetry of the impact, we know the velocity in region *II* is zero. The jump in stress across each shock contour is [see Eq. (2.146)]

$$[\![T]\!] = -\rho_0 C_S [\![v]\!].$$

We apply this relationship to the two shock contours to obtain

$$T_{II} - 0 = \begin{cases} -\rho_0(-U_S)(0 - v^P), & \text{left contour,} \\ -\rho_0(U_S)(0 + v^P), & \text{right contour.} \end{cases}$$

Thus the stress in region *II* is

$$T_{II} = -\rho_0 U_S v^P.$$

Notice that this value of stress only depends upon the measured quantities v^P and U_S plus the jump condition for linear momentum. It does not depend upon a constitutive model.

By repeating the Hugoniot experiment over a range of impact velocities v^P, we can determine the Hugoniot $U_S = c_0 + s[\![v]\!]$ as a fit to the measured data. For many solids and liquids, the data from these experiments are surprisingly well fit by this linear relationship. In Appendix B, we list values of c_0 and s for a large variety of materials. We emphasize that the parameters c_0 and s in this relation have been chosen to empirically fit the data from Hugoniot experiments. As a consequence, these values of c_0 are not necessarily equal to the velocity of sound.

The Linear-Hugoniot Model

In this section we discuss a special type of nonlinear-elastic material that yields the linear Hugoniot $U_S = c_0 + s[\![v]\!]$. Remember that other constitutive models, which we have not yet discussed, can also yield a linear Hugoniot. We shall discuss this in greater detail in Chapter 4. However, even with its limitations, the nonlinear-elastic model described here has widespread utility. To determine this nonlinear-elastic model, consider a shock wave propagating in the positive-x direction into a material that is initially at rest. Notice that both $U_S = C_S$ and v are positive. The jumps in velocity and deformation gradient across this shock wave are related by

$$[\![F]\!]U_S + [\![v]\!] = 0.$$

Because $F_- = 1$ and $v_- = 0$, we obtain

$$F_+ = 1 - \frac{v_+}{U_S}.$$

Substituting the Hugoniot equation (2.175) gives

$$v_+ = \frac{c_0(1 - F_+)}{1 - s(1 - F_+)}. \tag{2.176}$$

When this result and the Hugoniot are substituted into the jump condition Eq. (2.146),

$$[\![T]\!] = -\rho_0 U_S [\![v]\!],$$

we obtain

$$T_+ = -\frac{\rho_0 c_0^2(1 - F_+)}{[1 - s(1 - F_+)]^2} \tag{2.177}$$

and

$$U_S = \frac{c_0}{1 - s(1 - F_+)}. \tag{2.178}$$

These expressions relate the jump in stress to the jump in deformation gradient across the shock wave. Using any two of the three quantities F_+, T_+, or v_+, we can draw a *Hugoniot curve*. In Figure 2.76 we show the T–v diagram of a Hugoniot curve and a Rayleigh line. Our use of the Rayleigh line is meant to emphasize that *a shock wave does not cause the*

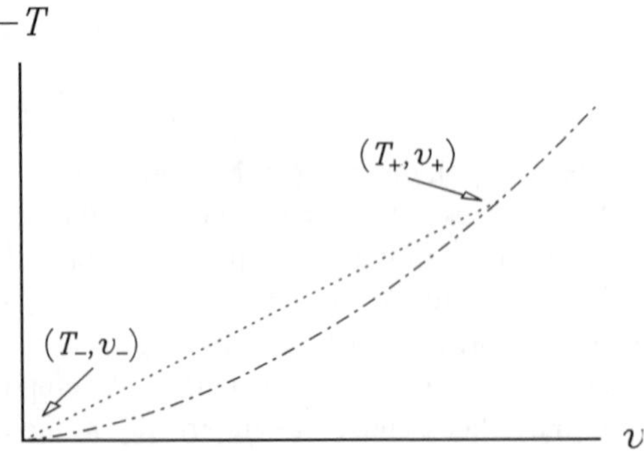

Figure 2.76 The Hugoniot curve (– · –) and the Rayleigh line (· · ·).

material to move along the Hugoniot. Instead, the shock solution causes the material to jump between two points on the Hugoniot curve. These points are located at the intersections of the Hugoniot curve and the Rayleigh line. The Hugoniot curve defines the response of the material to discontinuous shock waves. Let us next determine the response of the material to structured waves.

We represent the measured response of the material to a shock wave by Eq. (2.177). Although you might be tempted to call this expression a constitutive model, it is not. It is solely an artifact of the empirical relationship $U_S = c_0 + s[\![v]\!]$ and the jump conditions. However, we can use this expression as motivation for the following constitutive model:

$$T = -\frac{\rho_0 c_0^2(1-F)}{[1-s(1-F)]^2}. \tag{2.179}$$

This is a nonlinear-elastic material that yields the jump condition, Eq. (2.177), for the Hugoniot experiment. We shall assume that this expression for the stress can be used to predict the response of the material to structured waves. The material wave velocity for this constitutive model is given by

$$C^2 = \frac{1}{\rho_0}\frac{dT}{dF}.$$

Substituting Eq. (2.179) yields

$$C = c_0\left\{\frac{1+s(1-F)}{[1-s(1-F)]^3}\right\}^{1/2}, \tag{2.180}$$

and by substituting this result into the Riemann integrals

$$J_\pm = v - v_0 \mp \int_1^F C(\bar{F})\,d\bar{F},$$

we obtain

$$J_\pm = v - v_0 \pm 2\frac{c_0}{s}\left\{\sqrt{\frac{1+s(1-F)}{1-s(1-F)}} - \tan^{-1}\sqrt{\frac{1+s(1-F)}{1-s(1-F)}} + \frac{\pi}{4} - 1\right\}. \tag{2.181}$$

You can verify this expression by differentiating the terms containing F to recover the expression for C in Eq. (2.180).

We recall that the Hugoniot equation (2.178) only determines U_S for a shock wave that moves the material from $F = 1$ to $F = F_+$. However, with the constitutive relation Eq. (2.179), we can evaluate the shock velocity for a jump between any two values of F_- and F_+. We leave it as an exercise to show that (see Exercise 2.49)

$$U_S = c_0\frac{[1-s^2(1-F_-)(1-F_+)]^{1/2}}{[1-s(1-F_+)][1-s(1-F_-)]}. \tag{2.182}$$

We emphasize a very important difference between this Hugoniot and the one in Eq. (2.178). We obtain the Hugoniot in Eq. (2.178) solely from empirical observation – it does not depend upon a constitutive model. We obtain the data that yield that relationship solely from experiments on material with an initial deformation gradient $F = 1$. In contrast, our derivation of Eq. (2.182) depends upon the constitutive model Eq. (2.179). This distinction becomes even more apparent when we consider another experiment. In this experiment, we first compress the material from $F = 1$ to $F = 0.9$. Then we subject it to a shock wave. The value of the shock velocity that we measure in this experiment may or may not be correctly predicted by Eq. (2.182). If it is not, we must question the validity of our constitutive model.

To further emphasize this point, notice that the constitutive model Eq. (2.179) has an unusual property: When

$$F = 1 - \frac{1}{s},$$

both $T \to \infty$ and $C \to \infty$. Independently of how we load the material, it becomes rigid at this value of F and exhibits a *compression limit*. Most materials exhibit a compression limit in a Hugoniot experiment. This means that a shock wave cannot compress the material beyond $F = 1 - 1/s$, regardless of the impact velocity. However, although we usually observe a compression limit in a Hugoniot experiment, it is not observed in other types of experiments. When we slowly compress a material, its volume can be reduced below the compression limit. At this point, we can only take these observations as a sign that something is missing from our constitutive model. In later chapters we shall show how thermal effects can be introduced into constitutive models to account for this behavior. For the present, we notice that these deficiencies in this particular constitutive model limit, but do not eliminate, its usefulness.

Transition of a Compression Wave into a Strong Shock

Consider a material that obeys the constitutive model Eq. (2.179). Suppose a compression wave propagates in the positive-x direction. Let the characteristic contours of this compression wave converge to a single point. At this time, the simple wave transforms to a strong shock. Let us construct the X–t diagram for this solution (see Figure 2.77). The fan of characteristic contours including the head and tail contours of the compression wave converge to a point P. A contour representing the strong shock extends to the right beyond the point P. An additional fan of contours extends up and to the left from the point P. We show two wave profiles from this X–t diagram to illustrate how the compression wave transforms into a shock wave (see Figure 2.78). With the dashed line, we show the compression wave at the time $t = t_B$, before it transforms into a shock wave. We assume that it is a ramp wave propagating in the positive-x direction. With the solid line, we illustrate the wave

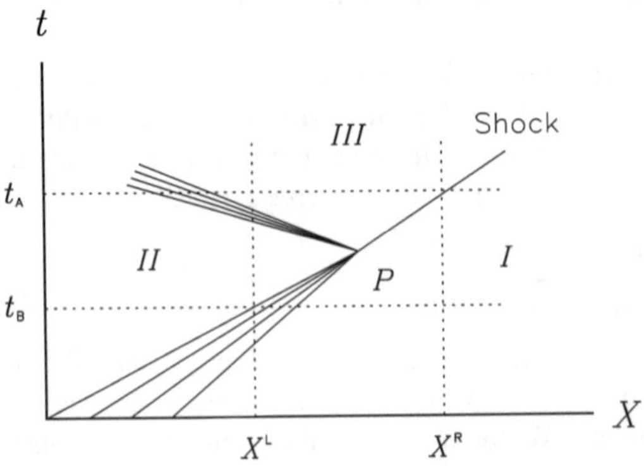

Figure 2.77 The X–t diagram for the transition of a compression wave into a strong shock.

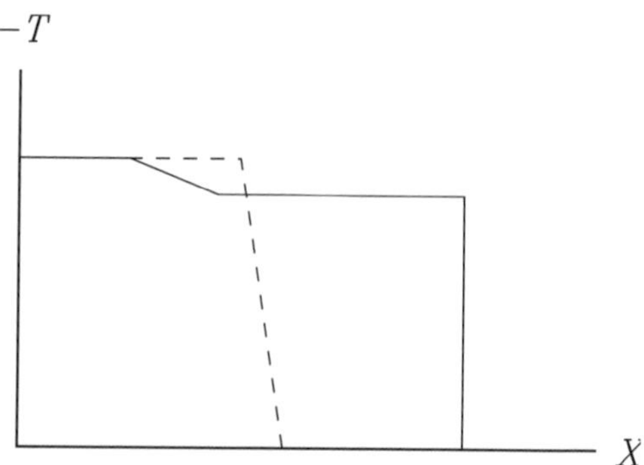

Figure 2.78 Two wave profiles for a strong shock at $t = t_B$ (– –) and $t = t_A$ (—).

at the time $t = t_A$, after it has transformed into a shock. This profile illustrates a shock wave propagating in the positive-x direction and the rarefaction wave propagating in the negative-x direction.

We use the characteristic contours to separate the X–t diagram into regions *I*, *II*, and *III*. In constructing this diagram, we have assumed that the compression wave and the strong shock cause different changes in the solution. By separating regions *II* and *III* with a rarefaction fan, we have assumed that the strong shock compresses the material less than the compression wave.

Our objective is to verify that these are indeed the correct X–t diagram and wave profiles for the transition of a compression wave into a strong shock. We shall test our assumptions by calculating the response of the material at X^L and X^R (see Figure 2.77). The material at X^L encounters the compression and rarefaction waves; the material at X^R encounters only the strong shock.

Consider the material at X^L. In region *I*, we prescribe initial conditions $T = 0$, $v = 0$, and $F = 1$. This material encounters the compression wave. We determine its response using the Riemann integrals. From the initial conditions, $J_{\pm} = 0$ in region *I*. Because the compression wave is a simple wave, $J_- = 0$ across the compression wave. Behind the compression wave, we prescribe that $F = F_{II}$. From Eq. (2.181), we obtain

$$v_{II} = 2\frac{c_0}{s}\left\{\sqrt{\frac{1+s(1-F_{II})}{1-s(1-F_{II})}} - \tan^{-1}\sqrt{\frac{1+s(1-F_{II})}{1-s(1-F_{II})}} + \frac{\pi}{4} - 1\right\}. \qquad (2.183)$$

We shall now determine that the changes at X^L due to the compression wave are not equal to the changes at X^R due to the shock wave. We shall do this by showing that even if F has equal values in regions *II* and *III*, the velocity v must be different in each region. Indeed, we can use Eq. (2.183) to evaluate v behind the compression wave in region *II*, and we can use Eq. (2.176) to evaluate v behind the shock wave in region *III*. We illustrate this comparison in Figure 2.79 for four values of s. For the smaller values of s and v, the compression wave and the shock wave produce approximately the same changes in v, and the shock wave is a weak shock. For these conditions, the fan of contours separating regions *II* and *III* is unnecessary. For the larger values of s when the velocities v behind the waves approach c_0, we observe significant differences between the solutions for the compression wave and the shock wave. This defines the region where the shock wave is a strong shock.

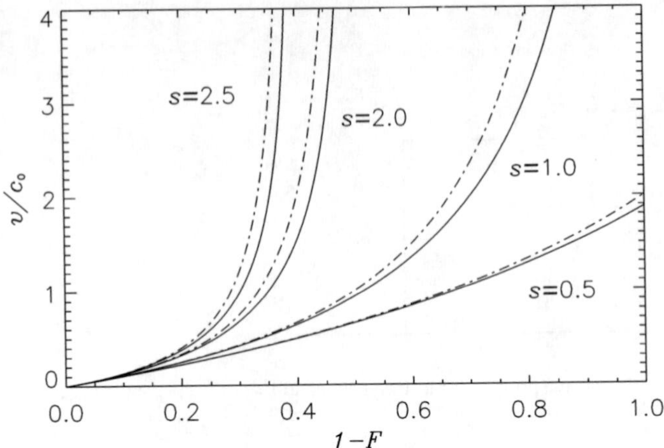

Figure 2.79 The velocity v and deformation gradient F behind a compression wave (—) and a strong shock (– · · –) for selected values of s.

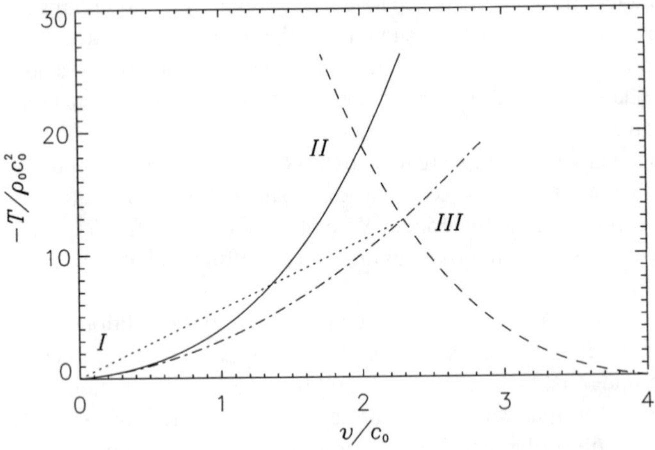

Figure 2.80 The T–v diagram for the transition of a compression wave into a strong shock when $s = 2$. The material at X^L first encounters a compression wave (—) and then a rarefaction wave (– –). The material at X^R encounters a shock wave represented by the Hugoniot curve (– · · –) and the Rayleigh line (· · ·).

***T–v* Diagram** We can also analyze the transition of a compression wave into a strong shock with a T–v diagram. We show the T–v diagram for the material at X^L and X^R in Figure 2.80. First consider the material at X^L. Initially, it is in region *I* of the X–t diagram (see Figure 2.77). It encounters the compression wave and moves into region *II*. Because the compression wave propagates in the positive-x direction, the material at X^L is compressed along the positive-sloping compression path (—) from point *I* to point *II* in the T–v diagram. We determine this path with the equations for stress [Eq. (2.179)] and velocity [Eq. (2.183)]. Next, the material at X_L encounters the rarefaction wave and moves from region *II* to region *III* in the X–t diagram. Because the rarefaction wave propagates in the negative-x direction, the material expands along the negative-sloping rarefaction path (– –) from point *II* to point *III* in the T–v diagram.

During the same period of time, the material at X^R starts in region *I* of the X–t diagram. This material encounters the shock wave and jumps into region *III*. In the T–v diagram, the material at X^R jumps between points *I* and *III*. Point *III* is at the intersection of the rarefaction path (– –) and the Hugoniot curve (– · –). We determine the Hugoniot curve with the jump conditions for stress, Eq. (2.179), and velocity, Eq. (2.176). The *Rayleigh line* (· · ·) connects point *I* to point *III*. We determine the shock velocity U_S from the slope of the Rayleigh line.

When we compare the T–v diagram for the strong shock in Figure 2.80 to the T–v diagram for the weak shock in Figure 2.73, we observe that the Hugoniot curve of the weak shock coincides with the response path for the compression wave, while the Hugoniot curve for the strong shock does not. When analyzing shock waves in solids and liquids we shall often find it convenient to approximate compression *and* rarefaction paths with the Hugoniot curve. We call this the *weak-shock assumption*. The linear-Hugoniot model is representative of the shock response of many solids and liquids, and the results in Figures 2.80 and 2.79 show that the weak-shock assumption is valid until shock waves with very large jumps in stress are encountered. (For example, $\rho_0 c_0^2 = 270$ GPa for a typical structural steel.)

2.7.8 *Exercises*

2.43 Consider the ramp wave with the profile

$$v = -\alpha x, \quad 0 \leq v \leq \alpha h_0,$$

at $t = 0$. This wave is propagating in the positive-x direction in an adiabatic gas.

(a) For $0 \geq x \geq -h_0$, show that the inverse motion at $t = 0$ is

$$X = \frac{2c_0}{(\gamma+1)\alpha}\left\{1 - \left[1 - \frac{(\gamma-1)\alpha x}{2c_0}\right]^{\frac{\gamma+1}{\gamma-1}}\right\}.$$

(b) Determine the inverse motion for $x > 0$ and $x < -h_0$.

(c) Determine the motion at $t = 0$.

(d) Determine the thickness of the ramp at $t = 0$ in the material description.

(e) Assume that $c_0 = 1$, $\gamma = 1.4$, $\alpha = 1$, and $h_0 = 1$. Using a single set of coordinates, graph the profile of the ramp wave at $t = 0$ in both the spatial and material descriptions.

2.44 Suppose a ramp wave propagates through a volume of air that is initially at rest. The velocity of the air behind the wave is $v = 50$ m/s. The initial thickness of the ramp is 0.1 m. Assume that $\gamma = 1.4$ and $c_0 = 345$ m/s.

(a) Determine the distance traveled by the head of the wave before a shock wave forms.

(b) Using the solution for a weak shock in an adiabatic gas, determine U_S.

(c) Construct the x–t diagram for the solution.

2.45 The motion of a singular surface is given by

$$x_S = \hat{x}[X_S(t), t].$$

Use this relationship to verify that [see Eq. (2.138)]

$$c_S = F_- C_S + v_-.$$

2.46 Consider a shock wave propagating in the positive-x direction in a nonlinear-elastic material. The velocity of propagation of this shock wave is

$$\frac{[\![v]\!]}{[\![F]\!]} = -U_S.$$

(a) Demonstrate that when the jumps $[\![v]\!]$ and $[\![F]\!]$ are small, we obtain the relationship for a simple wave

$$\frac{dv}{dF} = -C.$$

(b) Consider a simple wave propagating in the positive-x direction. How might we use the jump condition $[\![v]\!] = -U_S[\![F]\!]$ to approximate the behavior of the simple wave?

2.47 The velocity of a weak shock is given by Eq. (2.166):

$$U_S = \frac{1}{2}(c_0 + C_+).$$

Verify this result.

2.48 The velocity of a weak shock in an adiabatic gas is given by Eq. (2.170):

$$\frac{(U_S - c_0)^2}{c_0} + U_S = c_0 + \frac{\gamma + 1}{4} v_+.$$

Verify this result.

2.49 Consider a strong shock propagating in a material that exhibits the linear Hugoniot $U_S = c_0 + s[\![v]\!]$. The deformation gradients are F_- in front of the shock and F_+ behind the shock. Verify that [see Eq. (2.182)]

$$U_S = c_0 \frac{\left[1 - s^2(1 - F_-)(1 - F_+)\right]^{1/2}}{[1 - s(1 - F_+)][1 - s(1 - F_-)]}.$$

2.50 The Riemann integrals for a shock wave in a material that exhibits the linear Hugoniot $U_S = c_0 + s[\![v]\!]$ are contained in Eq. (2.181). Verify this result by demonstrating that

$$C = -2\frac{c_0}{s}\frac{d}{dF}\left\{\sqrt{\frac{1 + s(1 - F)}{1 - s(1 - F)}} - \tan^{-1}\sqrt{\frac{1 + s(1 - F)}{1 - s(1 - F)}} + \frac{\pi}{4} - 1\right\}.$$

2.51 A block of stainless steel is initially at rest. The linear Hugoniot of this material is

$$U_S = 4.58 + 1.49 v_+,$$

where U_S and v_+ are expressed in km/s. The density is $\rho_0 = 7.89$ Mg/m^3. A shock wave causes the deformation gradient to jump from $F_- = 1$ to $F = 0.7$.

(a) Determine v_+.
(b) Determine T_+.
(c) Determine C_- and C_+.

2.52 Consider a shock wave propagating in a nonlinear-elastic material. The velocity of the shock wave is U_S. The speed of sound in front of the shock wave is $C_- = c_0$, and the speed of sound behind the shock wave is C_+.

(a) Use the weak-shock relations to demonstrate that $C_+ \geq U_S \geq C_-$.
(b) Use the strong-shock relations for the linear Hugoniot to demonstrate that $C_+ \geq U_S \geq C_-$.
(c) For a linear-elastic material, demonstrate that $C_+ = U_S = C_-$.

2.53 The Hugoniot experiment illustrated in Figure 2.74 employs two identical plates. Suppose that the plate on the right is stationary and the plate on the left has a velocity of $v = 2v^P$.

(a) Using the jump conditions, demonstrate that the velocities of the plates at the plane of impact are equal to v^P.

(b) Use a T–v diagram to verify the previous result.

(c) Suppose two aluminum plates ($U_S = 5.37 + 1.29v_+$ in km/s and $\rho_0 = 2.784$ Mg/m^3) impact at a relative velocity of $2v^P = 1$ km/s. Determine the stress immediately after impact.

2.8 Wave–Wave Interactions

When two *linear waves* encounter each other, they do not interact. Throughout the region of penetration, the characteristic contours of both waves remain straight and parallel. The solution to the wave equation in the region of penetration is the sum of the solutions for each wave, and each wave behaves as if the other wave did not exist. When two *nonlinear waves* encounter each other, the situation is entirely different. In the region of penetration, the characteristic contours curve, and waves reflect off waves. We have also seen how boundaries and interfaces cause waves to reflect and interact. By virtue of the obvious fact that any two nonlinear waves traveling towards each other will interact, *wave–wave interactions* are an important part of the analysis of nonlinear waves. This importance is compounded by the not-so-obvious fact that any two nonlinear waves traveling in the *same* direction, with the exception of two rarefaction waves, will also eventually interact.

In this section we examine a variety of solutions for the interaction of shock waves and structured waves. Our analysis is presented primarily in graphical form with the aid of the X–t, x–t, and T–v diagrams. These diagrams are qualitative representations of solutions for any nonlinear-elastic material. They also represent solutions to many of the constitutive models that we shall discuss in Chapter 4.

2.8.1 *Reflection of a Shock Wave*

We first describe the interaction of a shock wave first with a rigid boundary and then with a free boundary. Next we consider the reflection and transmission of a shock wave at an interface. We shall construct our solutions in the context of the linear-Hugoniot model, but the resulting diagrams apply to any nonlinear-elastic material.

Reflection of a Shock Wave at a Rigid Boundary

Consider a half space of material in the region where $X \leq 0$ (see Figure 2.81). The boundary at $X = 0$ is rigid. The material obeys the linear-Hugoniot model. A shock wave approaches the boundary from the left. Behind the shock wave, we prescribe the velocity to be $v = v_{II}$. We show the X–t diagram for the reflection of the shock wave in Figure 2.82. We use the incident and reflected waves to separate the X–t diagram into three regions. Because of the rigid boundary condition, $v = 0$ in regions *I* and *III*. In region *II* the velocity is $v = v_{II}$.

Consider the material at X^0. We show the T–v diagram for X^0 in Figure 2.83. Initially, the material is in region *I* of the X–t diagram. As the material at X^0 encounters the incident wave, it moves into region *II*. In the T–v diagram, because the incident wave propagates in the positive-x direction, it moves the material from point *I* to point *II* on the positive-sloping Hugoniot curve. To calculate the jumps in F and T across the incident shock wave, we notice that in region *I* ahead of the shock wave $v_- = 0$ and $F_- = 1$, and in region *II*

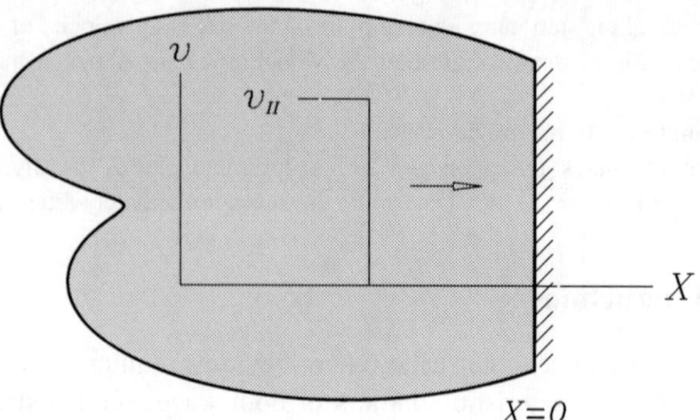

Figure 2.81 A shock wave incident on a rigid boundary at $X = 0$.

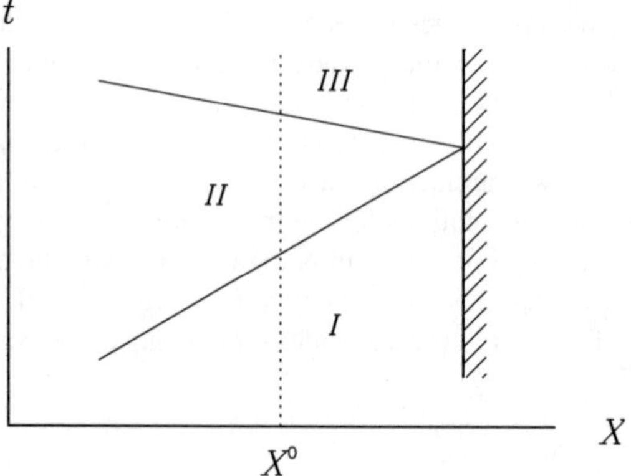

Figure 2.82 The X–t diagram of a shock wave reflected at a rigid boundary.

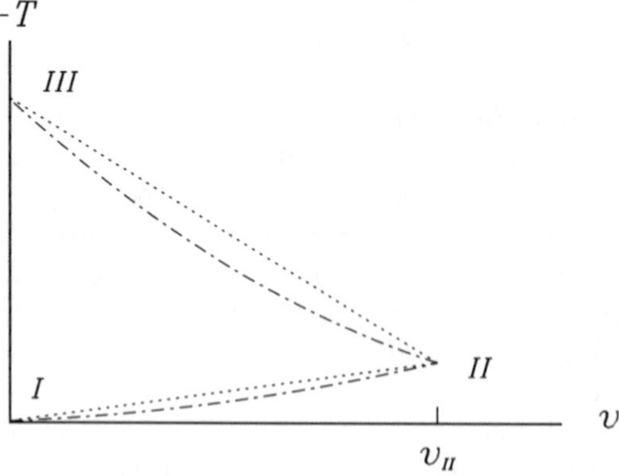

Figure 2.83 The T–v diagram of the material at X^0 due to the reflection of a shock wave from a rigid boundary.

behind the shock wave, $v_+ = v_{II}$. Using the jump condition Eq. (2.139), we determine the deformation gradient at point *II* to be

$$F_{II} = 1 - \frac{v_{II}}{U_S}.$$

The shock velocity of the incident wave is

$$U_S = \frac{c_0}{1 - s(1 - F_{II})},$$

and the stress is [see (Eq. 2.179)]

$$T_{II} = -\frac{\rho_0 c_0^2 (1 - F_{II})}{[1 - s(1 - F_{II})]^2}.$$

After passage of the incident shock wave, the material at X^0 next encounters the reflected wave. As the reflected wave passes, the material at X^0 moves from region *II* to region *III* in the X–t diagram. Because of the boundary condition, we know that $v = 0$ at point *III* of the T–v diagram. Because the reflected wave propagates in the negative-x direction, we must connect points *II* and *III* with a negative-sloping curve, which means the reflected wave causes additional compression and is a shock wave. We use the negative-sloping Hugoniot curve and Rayleigh line to connect points *II* and *III*. When we substitute $v_- = v_{II}$, $v_+ = 0$, and $F_- = F_{II}$ into the jump conditions, we obtain the deformation gradient,

$$F_{III} = F_{II} - \frac{v_{II}}{U_S},$$

the shock velocity [see Eq. (2.182)],

$$U_S = c_0 \frac{[1 - s^2(1 - F_{II})(1 - F_{III})]^{1/2}}{[1 - s(1 - F_{III})][1 - s(1 - F_{II})]},$$

and the stress,

$$T_{III} = -\frac{\rho_0 c_0^2 (1 - F_{III})}{[1 - s(1 - F_{III})]^2}.$$

From the T–v diagram, we find that the reflected shock wave has a greater shock velocity and jump in stress than the incident shock wave.

Reflection of a Shock Wave at a Free Boundary

Consider a half space of material in the region where $X \leq 0$. The boundary at $X = 0$ is free of stress. Suppose the material obeys the linear-Hugoniot model. A shock wave approaches the boundary from the left. We prescribe the velocity behind the shock wave to be $v = v_{II}$. In Figure 2.84 we show the X–t diagram of the reflection of the shock wave at the free boundary. We use the incident and reflected waves to separate the X–t diagram into three regions. We show the T–v diagram for the material at X^0 in Figure 2.85. From the boundary condition we know that the stress is zero at point *III*. Thus we see that the stress increases between points *II* and *III*, which means the reflected wave is a rarefaction wave. Because we have required the concave side of this response path to face up, the material wave velocity decreases as the stress increases and a fan of contours is produced.

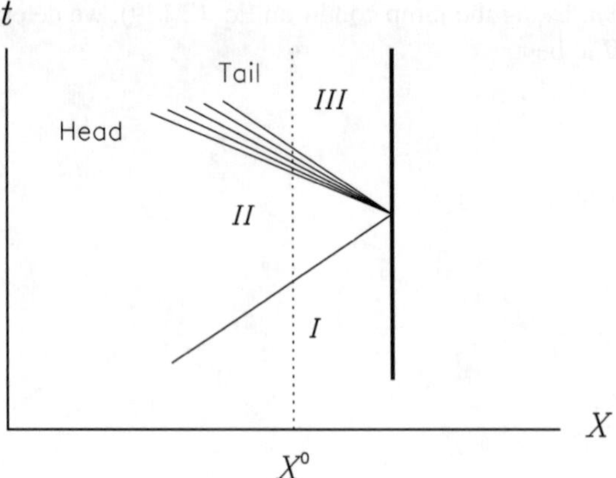

Figure 2.84 The X–t diagram of a shock wave reflected by a free boundary.

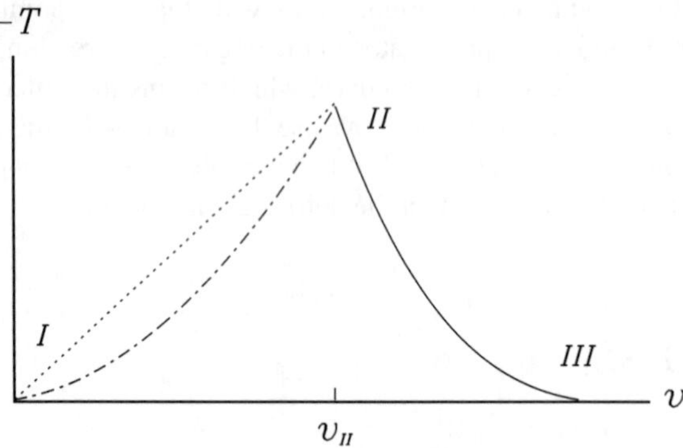

Figure 2.85 The T–v diagram for the material at X^0 due to the reflection of a shock wave by a free boundary.

The response path for this rarefaction wave is [see Eq. (2.181)]

$$v_{III} = v_{II} + 2\frac{c_0}{s}\left\{\sqrt{\frac{1+s(1-F_{II})}{1-s(1-F_{II})}} - \tan^{-1}\sqrt{\frac{1+s(1-F_{II})}{1-s(1-F_{II})}} + \frac{\pi}{4} - 1\right\}. \tag{2.184}$$

We leave the derivation of this result as an exercise (see Exercise 2.54). We determine the head contour of the reflected wave with the differential equation [see Eq. (2.180)]

$$\frac{dX}{dt} = -C(F_{II})$$

and the tail contour with the differential equation

$$\frac{dX}{dt} = -c_0.$$

We illustrate the x–t diagram for the reflection of the shock wave at a free boundary in Figure 2.86. When the shock wave contacts the free boundary, the velocity of the boundary

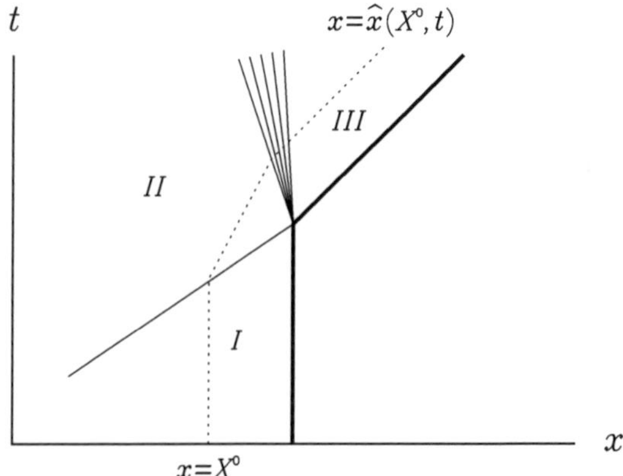

Figure 2.86 The x–t diagram of a shock wave reflection at a free boundary.

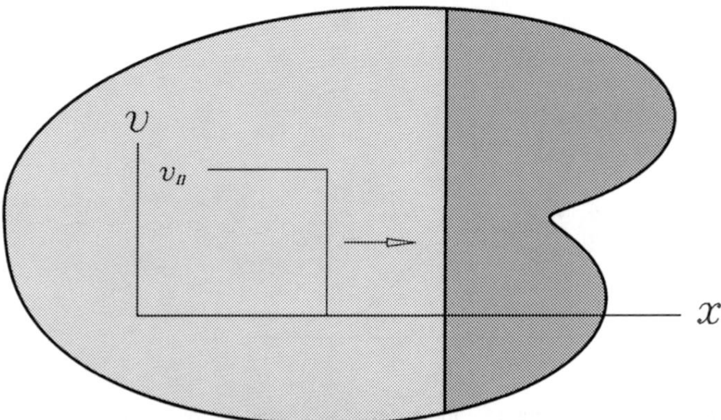

Figure 2.87 A shock wave incident on an interface.

changes abruptly. Because $v = 0$ and $F = 1$ in region *I*, the velocity of the shock wave in the spatial description is equal to the velocity of the shock wave in the material description: $c_s = C_S = U_S$ (see Section 2.7.2). However, the slopes of the contours of the reflected rarefaction wave are different in the x–t and X–t diagrams. In the x–t diagram, we determine the contours of the reflected wave by integrating $dx/dt = v - c$. In situations where v is greater than c, the slopes of the contours become positive. The dotted line represents the motion of the material point X^0. The slope of this line in region *II* is $1/v^{II}$. The slope in region *III* is the inverse of the velocity at point *III* in the T–v diagram. You can determine the motion of X^0 across the centered rarefaction fan from our analysis of the material trajectories in Section 2.6.2.

Reflection of a Shock Wave at an Interface

Suppose that two half spaces of nonlinear-elastic material are bonded together. The half spaces are initially at rest. A shock wave approaches the interface from the left (see Figure 2.87). We prescribe the velocity behind the shock wave to be v_{II}. We show the X–t diagram for the interaction of the shock with the interface in Figure 2.88. The vertical

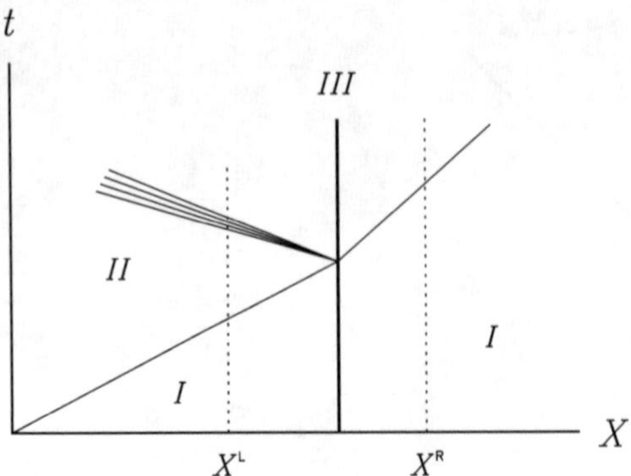

Figure 2.88 The X–t diagram of a shock wave incident on an interface where the left half space has a higher shock impedance than the right.

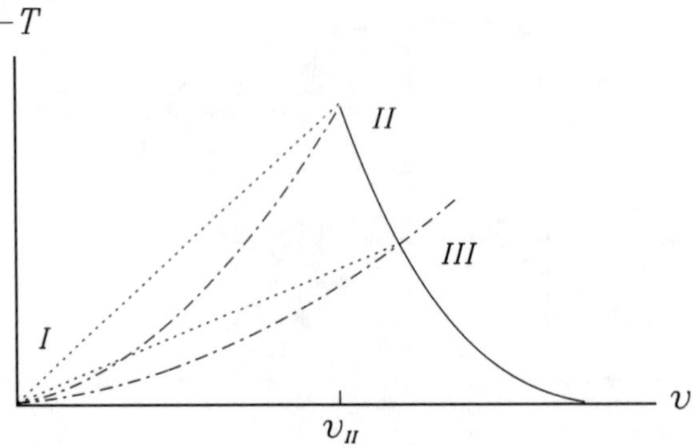

Figure 2.89 The T–v diagram for the material at X^L and X^R due to the reflection of a shock wave at an interface where the left half space has a higher shock impedance than the right.

bold line is the interface between the two materials. The interaction of the shock wave with the interface gives rise to a reflected wave and a transmitted wave. The reflected wave propagates in the negative-x direction, and the transmitted wave propagates in the positive-x direction. We use these waves to separate the diagram into three regions. Because T and v are continuous across the interface, we include the portions of the left and right materials under the incident and transmitted shock waves in region *I*, and we include the portions of both materials above the reflected rarefaction and transmitted shock waves in region *III*.

Consider the two material points at X^L to the left of the interface and at X^R to the right of the interface. We show the T–v diagram for these two points in Figure 2.89. We have drawn these diagrams to represent two materials where the shock impedance $z_S = \rho_0 U_S$ of the right half space is less than the shock impedance of the left half space. We conclude from the T–v diagram that the reflected wave is a rarefaction wave and the transmitted wave is a shock wave. Notice the similarity between this T–v diagram and the diagram for the reflection of a shock wave at a free boundary (Figure 2.85). When we imagine the shock impedance of the right half space approaching zero, the diagram for the reflection at an interface becomes the diagram for the reflection at a free boundary.

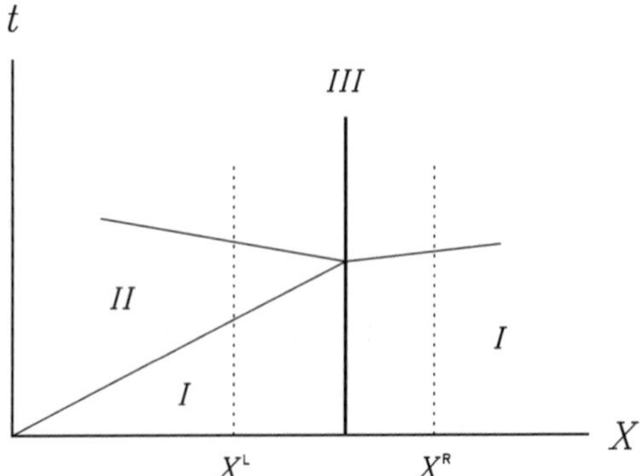

Figure 2.90 The X–t diagram of a shock wave reflected at an interface where the left half space has a lower shock impedance than the right.

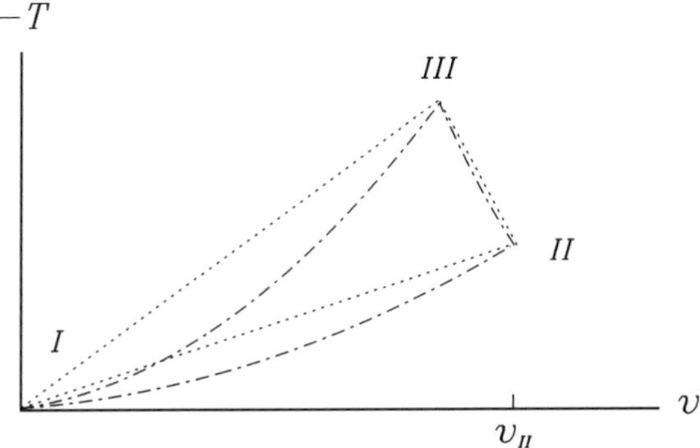

Figure 2.91 The T–v diagram for the material at X^L and X^R due to the reflection of a shock wave at an interface where the left half space has a lower shock impedance than the right.

Now we suppose that the shock impedance of the right half space is greater than the shock impedance of the left half space. We show the X–t diagram and the T–v diagram for this case in Figures 2.90 and 2.91. The reflected and transmitted waves are shock waves. By comparing these results to Figure 2.83, you will see the similarity to the solution for the reflection of a shock wave from a rigid boundary.

2.8.2 *Interaction of Two Shock Waves*

Here we discuss two problems where shock waves propagate in a homogeneous infinite medium. First, we consider two shock waves propagating in the same direction. We find that the trailing shock wave always overtakes the leading shock wave. Next, we consider two shock waves approaching each other.

Shock Wave Overtaking a Shock Wave

Consider a nonlinear-elastic material that is infinite in extent. Two shock waves propagate in the positive-x direction (see Figure 2.92). Initially, the velocity and stress in the material

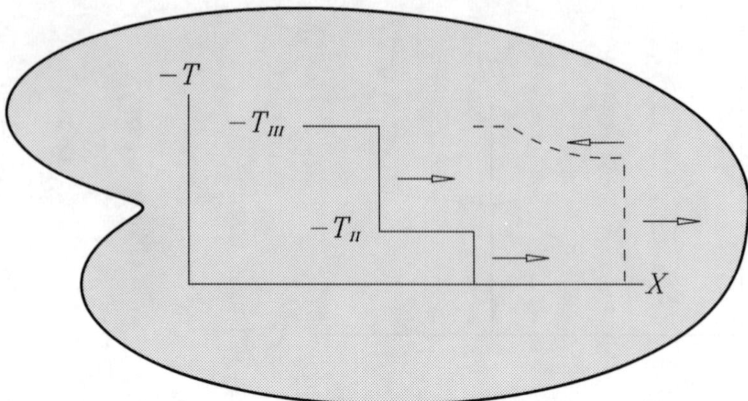

Figure 2.92 A shock wave overtaking a shock wave. Profiles are shown before (—) and after (– –) the wave interaction.

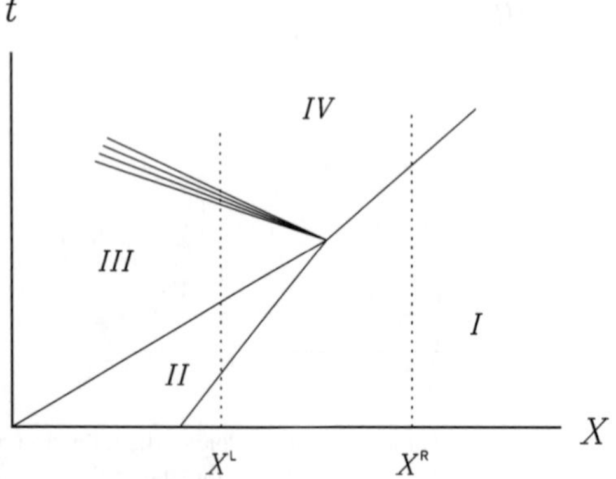

Figure 2.93 The X–t diagram for a shock wave overtaking a shock wave.

are zero. We prescribe the stresses behind the leading and trailing shock waves to be T_{II} and T_{III}, respectively. We choose negative values for both T_{II} and T_{III} so that each wave causes compression of the material. Clearly, we must select the magnitudes of these stresses so that $|T_{III}| > |T_{II}|$. We use a solid line to illustrate the profile of the two shock waves before the interaction. We use a dashed line to illustrate the profile of the waves after the interaction. By using X–t and T–v diagrams, we shall now show why the interaction of the two shock waves results in a shock wave propagating in the positive-x direction and the rarefaction wave propagating in the negative-x direction.

We show the X–t diagram in Figure 2.93. As the trailing shock wave overtakes the leading shock wave, the contours of the two shock waves converge to a point. Let us examine the T–v diagram for the material at the two positions X^L and X^R to the left and right of the point of convergence of the shock waves (see Figure 2.94). Point I represents the initial conditions in the material before the shock waves arrive. The material at position X^L encounters the leading shock wave first. We can determine the Hugoniot curve for this shock wave by letting $F_- = 1$ and $v_- = 0$. Then we evaluate Eqs. (2.139), (2.179), and (2.182) over the range of deformation gradients $1 - 1/s < F_+ < 1$. We graph this curve with the dot-dashed

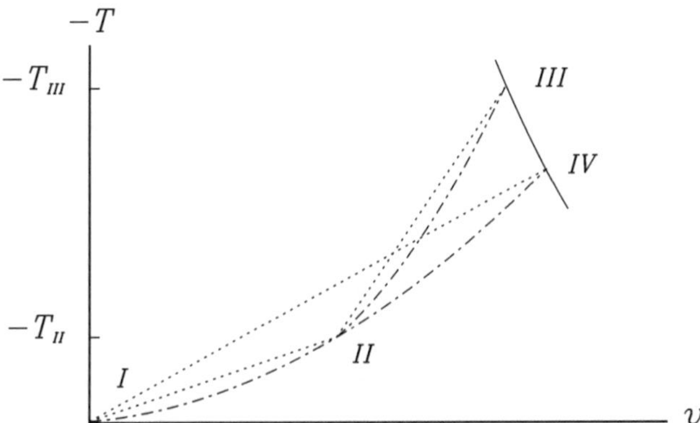

Figure 2.94 The T–v diagram for the material at X^L and X^R for a shock wave overtaking a shock wave.

line through point *I* in the T–v diagram. We place point *II* at the intersection of this curve with our prescribed value of stress T_{II} for the leading shock wave. Notice that we now know the velocity v_{II} and deformation gradient F_{II} at point *II*. The material at position X^L next encounters the trailing shock wave. Letting $F_- = F_{II}$ and $v_- = v_{II}$, we evaluate the Hugoniot curve for the trailing shock wave. We represent it with the steeper dot-dash curve through point *II*. Point *III* on this Hugoniot curve represents the stress and velocity behind the trailing shock wave. We place point *III* at the intersection of the second Hugoniot curve with our prescribed value of stress T_{III}. Notice that the Hugoniot curve, which connects point *II* with point *III*, *does not* intersect point *I*. The Hugoniot curve of the trailing shock wave is always steeper than the Hugoniot curve of the leading shock wave. Consequently, the trailing shock wave will always overtake the leading shock wave.

Next consider the material at position X^R. This material encounters only one shock wave, which separates region *I* from region *IV*. Point *IV* in the T–v diagram lies on the Hugoniot curve that passes through point *I*. We have placed point *IV* at the intersection of this Hugoniot curve and the negative-sloping response path for the rarefaction wave that separates regions *III* and *IV*.

You can gain additional insight into this wave interaction by comparing the T–v diagrams in Figures 2.94 and 2.89. Notice that points *II*, *III*, and *IV* in Figure 2.94 form a pattern that is quite similar to points *I*, *II*, and *III* in Figure 2.89. This implies that a shock wave overtaking a slower shock wave behaves like a shock wave striking a material interface with a sudden drop in shock impedance. Indeed, a shock wave is like a propagating material interface, because the material behind a shock wave has a higher shock impedance than the material in front of a shock wave. Thus when a trailing shock wave overtakes and tries to pass a leading shock wave, it encounters a sudden drop in shock impedance. We find that in both Figures 2.89 and 2.94 the transmitted wave is a shock wave and the reflected wave is a rarefaction wave.

Collision of Two Shock Waves

Consider a nonlinear-elastic material that is infinite in extent. Two shock waves approach each other (see Figure 2.95). Initially, the velocity and stress in the material are zero. Behind the shock wave propagating in the positive-x direction, we prescribe the stress to

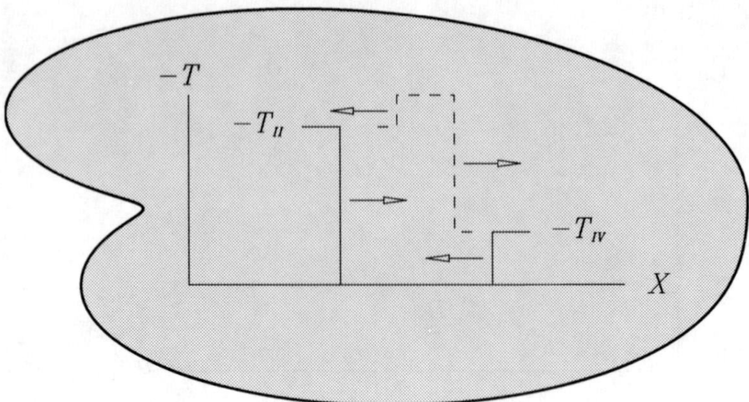

Figure 2.95 Collision of two shock waves.

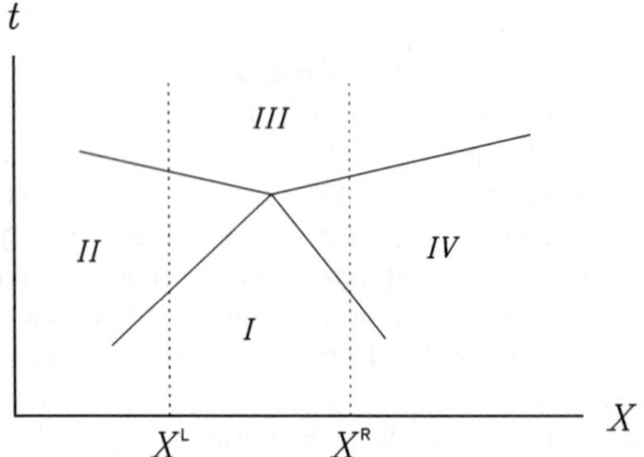

Figure 2.96 The X–t diagram for the collision of two shock waves.

be T_{II}. Behind the shock wave propagating in the negative-x direction, we prescribe the stress to be T_{IV}. The solid line represents the wave profiles before they interact; the dashed line represents the wave profiles after they interact. By using X–t and T–v diagrams, we shall now show why the interaction of the two shock waves results in two shock waves propagating in opposite directions.

We show the X–t diagram in Figure 2.96. Two material points are selected at X^L and X^R. We show the T–v diagrams for the material at these locations in Figure 2.97. These diagrams reveal that the collision of the two shock waves results in two shock waves. Both shock velocities after the collision are greater than either of the shock velocities before the collision. As in the previous example, we can argue that as one shock wave contacts another shock wave, it encounters a material with a higher shock impedance. This causes a reflected shock wave with a higher magnitude of stress and a higher propagation velocity.

2.8.3 *Interaction of Shock Waves with Structured Waves*

We have seen that two shock waves propagating in the same direction will interact. The trailing shock wave always overtakes the leading shock wave. It is also true that structured waves will overtake shock waves, and shock waves will overtake structured waves.

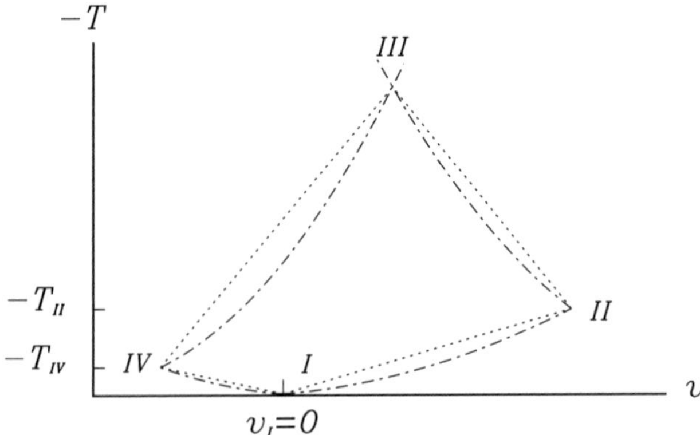

Figure 2.97 The T–v diagram for the material at X^L and X^R during the collision of two shock waves.

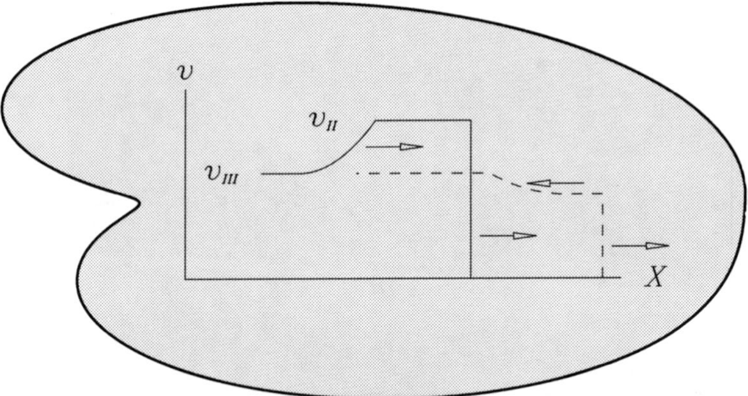

Figure 2.98 A rarefaction wave overtaking a shock wave.

First we examine the two cases of a rarefaction wave and a compression wave overtaking a shock wave. Then we examine a shock wave overtaking both a rarefaction wave and a compression wave.

Rarefaction Wave Overtaking a Shock Wave

Consider an infinite homogeneous nonlinear-elastic material. Initially, the material is at rest (see Figure 2.98). The solid line represents the profiles of the waves before the interaction. A shock wave propagates in the positive-x direction. Let the velocity behind the shock wave be $v = v_{II}$. A rarefaction wave follows behind the shock wave. Let the velocity behind the rarefaction wave be $v = v_{III}$. We prescribe the velocities so that $v_{II} > v_{III} > 0$. Suppose that $v_{II} - v_{III}$ is a fraction of v_{II} as shown in Figure 2.98. The dashed line illustrates the wave profiles after the interaction. We shall now use the X–t and T–v diagrams to examine just how the interaction produces a shock wave of diminished amplitude and a reflected compression wave.

We show the X–t diagram for the interaction of the rarefaction wave and the shock wave in Figure 2.99. A bold solid line is used to illustrate the leading shock wave. We illustrate the

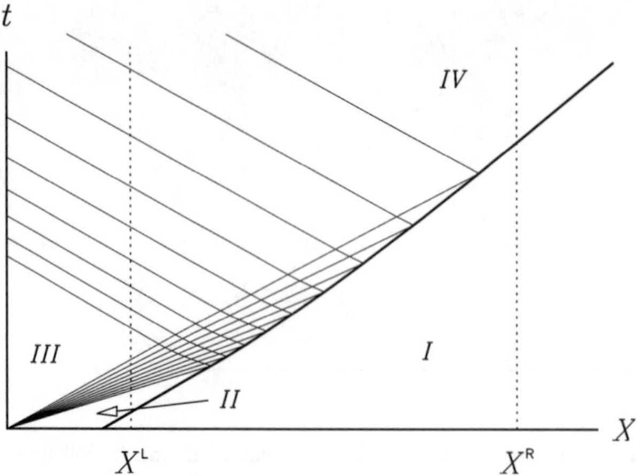

Figure 2.99 The X–t diagram for a rarefaction wave overtaking a shock wave (bold line).

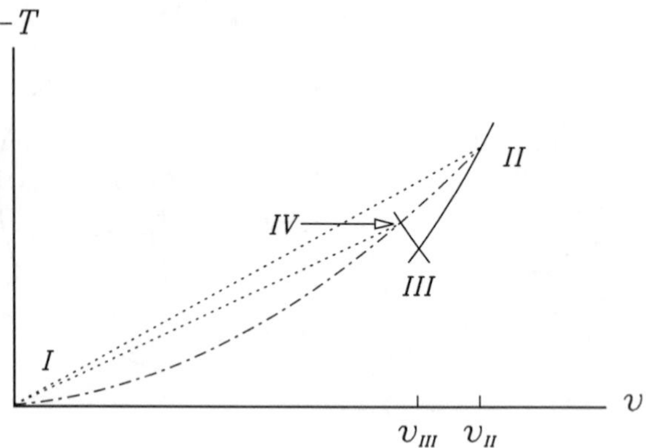

Figure 2.100 The T–v diagram for the material at X^L and X^R for a rarefaction wave overtaking a shock wave.

trailing rarefaction wave with a fan of diverging contours. The contours separate region *II* from region *III*. As the contours in the fan overtake the shock wave, we see that the velocity of propagation of the shock wave decreases. The interaction of the rarefaction wave and the shock wave produces a reflected wave, which we represent with the contours that separate region *III* from region IV. Two material points at X^L and X^R are shown. The material at X^L encounters the shock wave and the rarefaction wave before they interact. The material at X^R encounters only the shock wave after the interaction. We show the T–v diagram for X^L and X^R in Figure 2.100. You will recall that we have prescribed the velocities in regions I, *II*, and *III* to be $v = 0$, $v = v_{II}$, and $v = v_{III}$, respectively. We see that the rarefaction wave reflects off the shock wave and produces a compression wave propagating in the negative-x direction. This reflection weakens the shock wave. After the interaction, the shock wave has smaller jumps in velocity and stress, and it propagates at a smaller shock velocity. Notice that the response curve connecting regions *III* and IV is nearly straight. Thus the contours of the reflected compression wave are nearly parallel. These contours are also distributed over a large distance. Consequently, you may realize that we have taken some artistic liberty with our illustration of the reflected wave in Figure 2.98. This compression wave is quite spread out, and the actual wave profile after the interaction gives an appearance that is more like a single shock wave propagating in the positive-x direction. Again we see that the shock wave produces a sudden change in the wave impedance. Thus when the incident rarefaction wave encounters the shock wave, it also encounters a sudden drop in wave impedance. This change in impedance is what produces a reflected compression wave.

Case of $T_{III}=0$ Suppose we prescribe the stress behind the rarefaction wave to be zero (see Figure 2.101). The solid and dashed lines illustrate the wave profiles before and *during* the wave interaction. We show the X–t diagram for the interaction of the rarefaction wave and the shock wave in Figure 2.102. Notice that region IV is absent from the diagram. Region IV first appears at the completion of the interaction where the tail contour of the rarefaction wave intersects the shock wave. The tail contour of the rarefaction wave

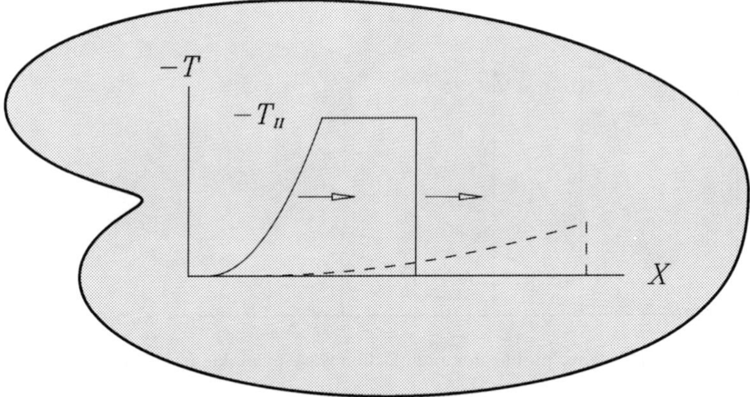

Figure 2.101 A rarefaction wave overtaking a shock wave, $T_{III}=0$.

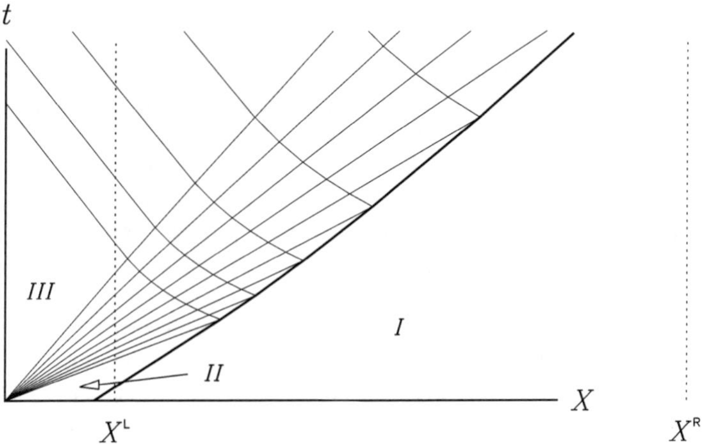

Figure 2.102 The X–t diagram for a rarefaction wave overtaking a shock wave (bold line).

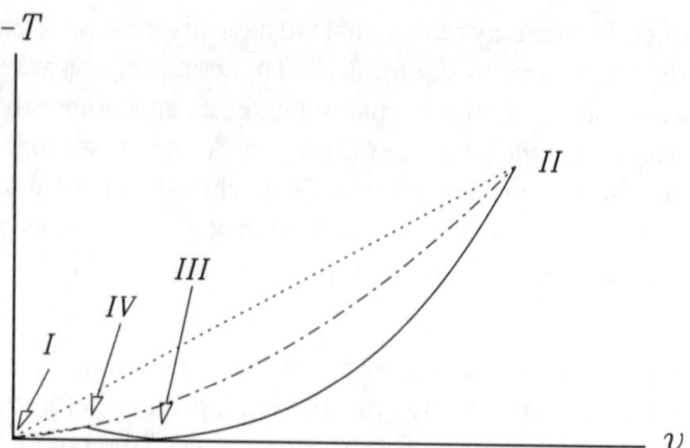

Figure 2.103 The T–v diagram for the material at X^L and X^R for a rarefaction wave overtaking a shock wave. The stress behind the rarefaction wave returns to zero, but the velocity does not.

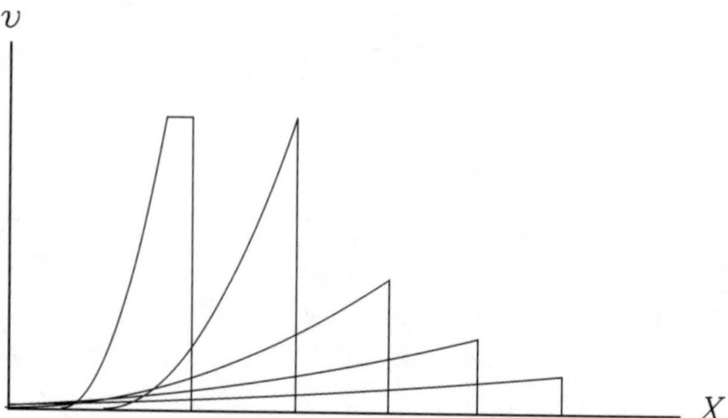

Figure 2.104 Wave profiles for a rarefaction wave overtaking a shock wave.

propagates at the velocity of sound c_0. Because $U_S \geq c_0$, the tail contour of the rarefaction wave never overtakes the shock wave, and region IV does not exist. However, the conditions at X^L and X^R for late times approach values of uniform velocity and stress that approximate region IV, and we can draw the T–v diagram for X^L and X^R as if region IV actually exists (see Figure 2.103). The results show that the rarefaction wave weakens the shock wave but does not eliminate it. Even though $T = T_{III} = 0$ behind the rarefaction wave, the stress behind the shock wave never reaches zero because $T = T_{IV} < 0$. There is always a small but finite shock wave propagating in the positive-x direction even in the limit as $t \to \infty$.

Using the wave profiles in Figure 2.104, we illustrate the evolution of the interaction of the rarefaction wave and the shock wave. In the left-most profile, the rarefaction wave has not made contact with the shock wave. The next profile shows the two waves at the instant of contact. In subsequent profiles, the rarefaction wave overtakes the shock wave. This causes the jump in velocity at the shock wave to decrease. The rarefaction wave continues to spread as time advances, and the tail of the rarefaction wave never catches the shock wave. At late times the tail of the rarefaction wave even appears to reverse its direction of propagation. This effect is caused by the reflected compression wave. At very late times the material behind the shock wave approaches a state of uniform velocity. From the T–v diagram, we find that the final velocity of the material is v_{IV}. This result is unexpected. We might reason that if the rarefaction wave causes the stress to return to zero, it should also cause the velocity to return to zero. Instead, we see that as the shock wave and the rarefaction wave pass a point in the material they leave behind a small residual velocity. This residual velocity exists because we have used the strong-shock solution in which the Hugoniot curve for the shock wave and the response path of the rarefaction wave are not coincident. In contrast, in the weak-shock solution these two curves converge to a single curve, and points I, III, and IV all reside at the origin of the T–v diagram. Thus the residual velocity behind a weak-shock interaction is zero. Our treatment of kinetic and internal energy in the next chapter will contribute to our understanding of the significance of the residual velocity. Indeed, the strong-shock solution with its residual velocity represents the true response of the nonlinear-elastic material, whereas the weak-shock solution is an approximation. The absence of a residual velocity in the weak-shock solution is an artifact of the weak-shock assumption.

Compression Wave Overtaking a Shock Wave

Consider an infinite material at rest where $T = v = 0$ (see Figure 2.105). The solid and dashed lines represent the wave profiles before and after the wave interaction. A shock wave propagates in the positive-x direction. We prescribe the velocity behind the shock wave to be v_{II}. A compression wave follows the shock wave. We prescribe the velocity behind the compression wave to be v_{III}. We shall now discuss how this wave interaction between a compression wave and a shock wave gives rise to a stronger shock wave and a reflected rarefaction wave.

We show the X–t diagram in Figure 2.106. We use a bold solid line to illustrate the leading shock wave. We illustrate the incident compression wave with a fan of converging contours, which separates region II from region III. As these contours overtake the shock wave, the velocity of propagation of the shock wave increases. The interaction of the compression wave with the shock wave results in a reflected wave, which we represent by the contours that separate region III from region IV. We also show two material points at X^L and X^R. The material at X^L encounters the shock wave, the compression wave, and the reflected wave. The material at X^R encounters only the shock wave. We illustrate the T–v diagram for the material at these locations in Figure 2.107. Because only one shock wave is present in the X–t diagram, only one Hugoniot curve is present in the T–v diagram. Points I, II,

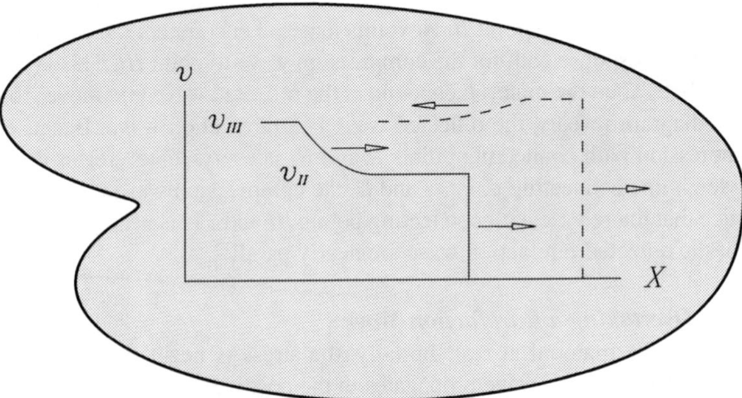

Figure 2.105 A compression wave overtaking a shock wave.

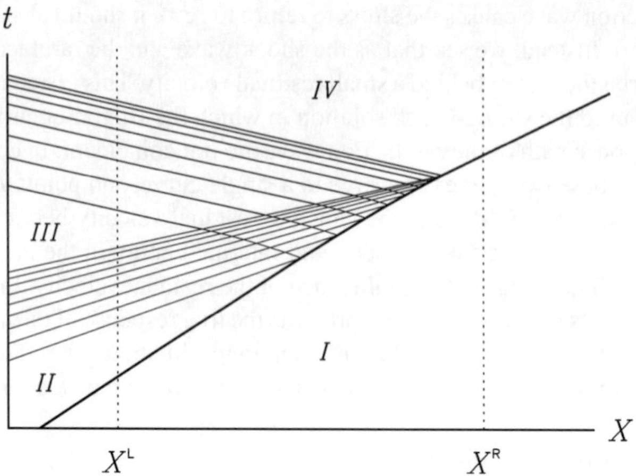

Figure 2.106 The X–t diagram for a compression wave overtaking a shock wave (bold line).

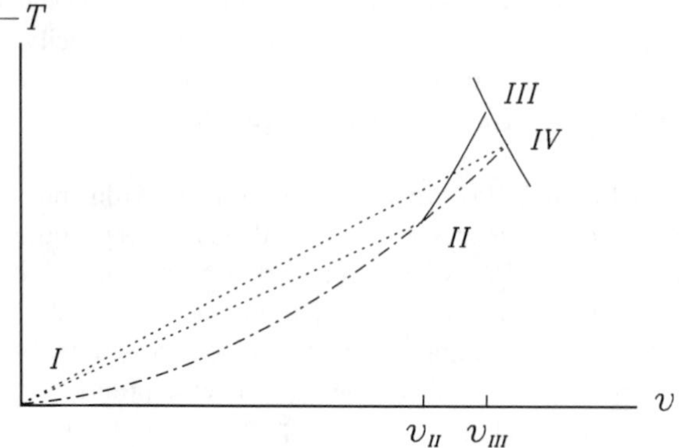

Figure 2.107 The T–v diagram for the material at X^L and X^R for a compression wave overtaking a shock wave.

and IV all lie on this Hugoniot curve. The material at X^L is initially at point *I*. It encounters the shock wave and jumps to point *II*. Next this material encounters the compression wave and moves up the response path for the compression wave to point *III*. This is the positive-sloping solid line. Then the material encounters the reflected wave and moves to point IV. As the T–v diagram reveals, the reflected wave is a rarefaction wave. Because the slope of the compression path connecting points *II* and *III* is everywhere greater than the slope of the Rayleigh line connecting points *I* and *II*, the compression wave overtakes the shock wave. Notice that the release path connecting points *III* and IV is nearly straight. Thus the contours of the reflected rarefaction wave are nearly parallel.

Shock Wave Overtaking a Rarefaction Wave

Consider an infinite material at rest. Initially, the stress is negative, $T = T_I < 0$ (see Figure 2.108). A rarefaction wave propagates in the positive-x direction. We prescribe the stress behind the rarefaction wave to be $T_{II} = 0$. A shock wave follows the rarefaction wave.

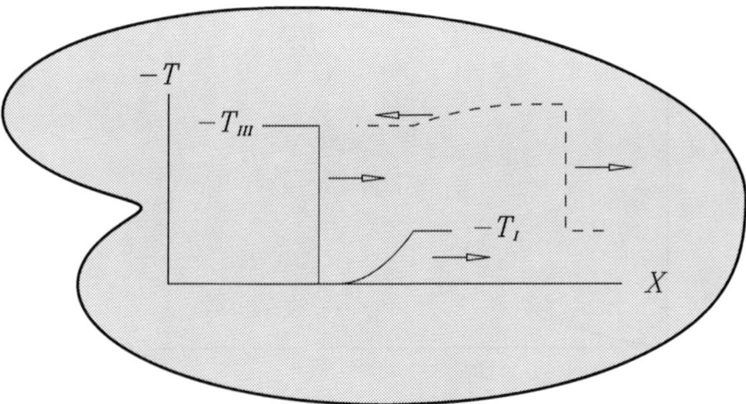

Figure 2.108 A shock wave overtaking a weaker rarefaction wave.

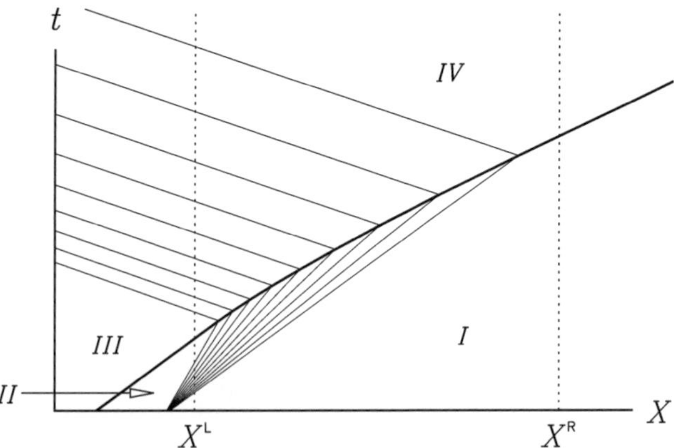

Figure 2.109 The X–t diagram for a shock wave (bold line) overtaking a weaker rarefaction wave.

We prescribe the stress behind the shock wave to be T_{III}. The solid and dashed lines represent the wave profiles before and after the interaction of the rarefaction and shock waves.

Stronger Shock Wave Suppose we prescribe the shock wave to be much stronger than the rarefaction wave, $T_{III} < T_I < 0$. In the X–t diagram in Figure 2.109 we illustrate the interaction of these two waves. We use a bold solid line to represent the shock wave. We divide the diagram into four regions. We also show the T–v diagram for the two material points X^L and X^R in Figure 2.110. The point X^L is initially in region *I*. When it encounters the rarefaction wave it moves down the response path (solid line) from point *I* to point *II*. Next it encounters the shock wave, and jumps from point *II* to point *III* along the Hugoniot curve (dot-dashed line). In the X–t diagram, we show a reflected wave resulting from the interaction of the shock wave with the leading rarefaction wave. The reflected wave is a compression wave. As the material at X^L encounters the reflected wave it moves from point *III* to point IV in the T–v diagram. Over the same interval of time, the material

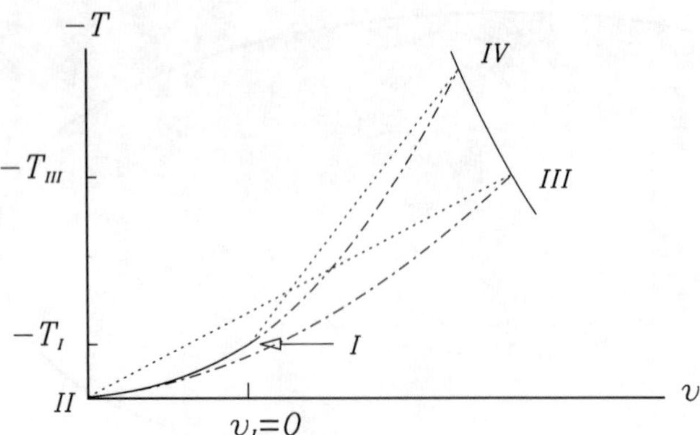

Figure 2.110 The T-v diagram for the material at X^L and X^R for a shock wave overtaking a weaker rarefaction wave.

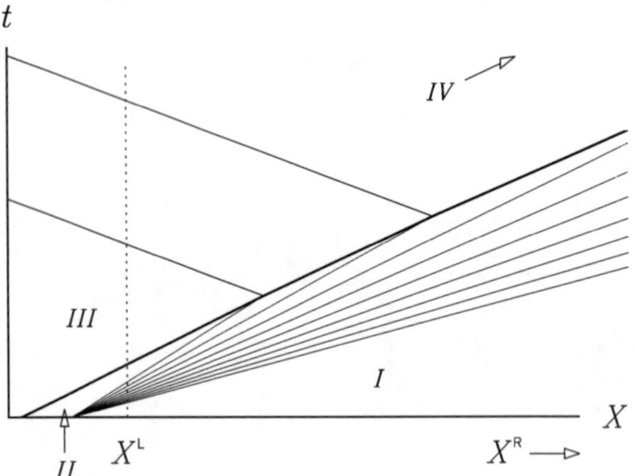

Figure 2.111 The X–t diagram for a shock wave (bold line) overtaking a stronger rarefaction wave.

at X^R encounters the shock wave after it has completely overtaken the leading rarefaction wave. This causes the material at X^R to jump directly from point I to point IV in the T–v diagram. Notice that the propagation velocity of the shock wave increases as it interacts with the rarefaction wave.

Stronger Rarefaction Wave We have just considered a stronger shock wave overtaking a weaker rarefaction wave. Let us instead consider a weaker shock wave overtaking a stronger rarefaction wave. In this case, we prescribe $T_I < T_{III} < 0$. We show the X–t diagram and the T–v diagram in Figures 2.111 and 2.112. The contours of the reflected wave are spread over a large distance and time, and region IV lies outside the boundaries of the illustration. Consequently, X^R is also beyond the right boundary of the illustration. The velocity of the shock wave increases as it interacts with the rarefaction wave. The interaction reduces the strength of the shock wave and results in a reflected compression wave. The

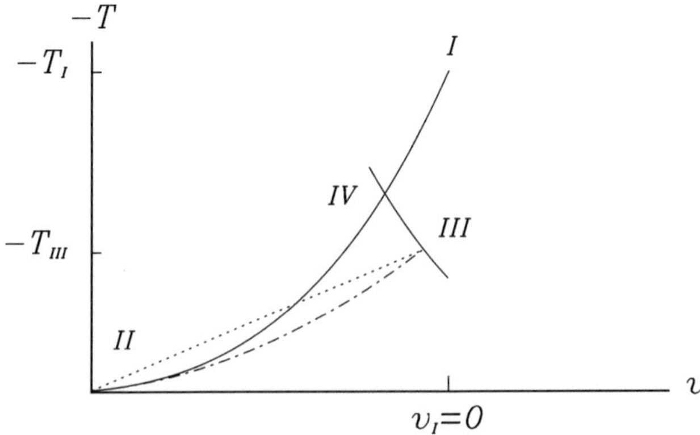

Figure 2.112 The T–v diagram for the material at X^L and X^R for a shock wave overtaking a stronger rarefaction wave.

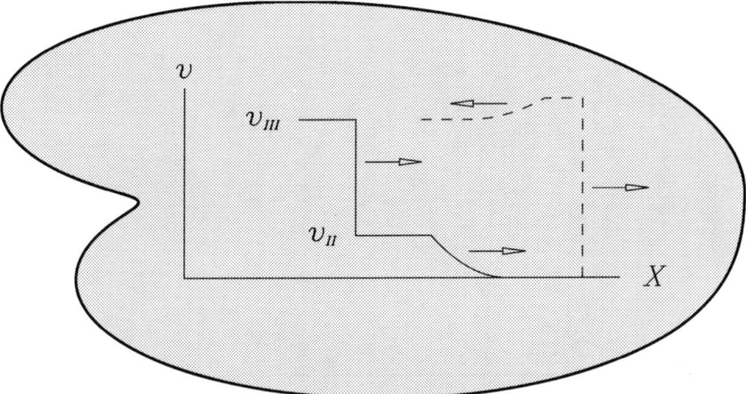

Figure 2.113 A shock wave overtaking a compression wave.

material at X_R only encounters a rarefaction wave of diminished amplitude, which causes the material to move from point I to point IV.

Shock Wave Overtaking a Compression Wave

Consider an infinite material at rest. A compression wave propagates in the positive-x direction (see Figure 2.113). We prescribe the velocity behind the compression wave to be v_{II}. A shock wave follows the compression wave. We prescribe the velocity behind the shock wave to be v_{III}. The solid line represents the profile of the waves before the interaction, and the dashed line represents the profile after the interaction. Using the X–t and T–v diagrams we shall now illustrate how this wave interaction results in a shock wave and a reflected rarefaction wave.

In the X–t diagram in Figure 2.114, we show the shock wave (bold line) overtaking the compression wave. We separate the diagram into four regions. We show the T–v diagram for the material points X^L and X^R in Figure 2.115. Initially, the material at X^L is at rest. It then encounters the compression wave and moves from point I to point II in the T–v diagram.

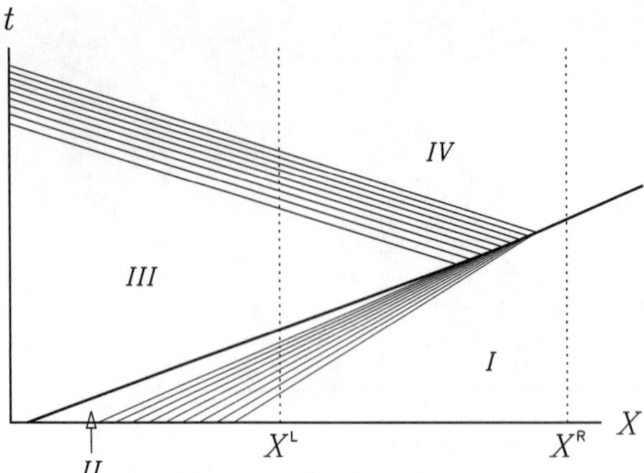

Figure 2.114 The X–t diagram for a shock wave (bold line) overtaking a compression wave.

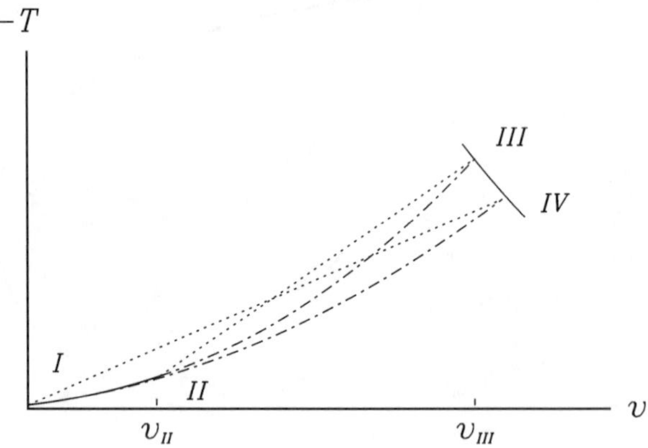

Figure 2.115 The T–v diagram for the material at X^L and X^R for a shock wave overtaking a compression wave.

Next it encounters the shock wave and jumps from point *II* to point *III*. The interaction of the shock with the compression wave results in a reflected rarefaction wave. When the material at X^L encounters this reflected wave, it moves to point IV in the T–v diagram. Over the same interval of time, the material at X^R encounters the shock wave after the interaction and jumps directly from point *I* to point IV. Notice that this interaction causes the shock velocity to decrease. Simultaneously, it also causes the velocity v behind the shock wave to increase.

2.8.4 *Flyer-Plate Experiment*

In Chapter 0 we introduced the flyer-plate experiment. We illustrate it again in Figure 2.116. We show a thin plate called a *flyer plate* traveling towards a half space of material called the *target plate*. The impact of the flyer plate produces two shock waves. One shock wave propagates in the negative-x direction into the flyer plate; the other shock wave propagates in the positive-x direction into the target plate. For purposes of analysis, we consider the lateral dimensions of both the flyer and the target to be infinite. It is not

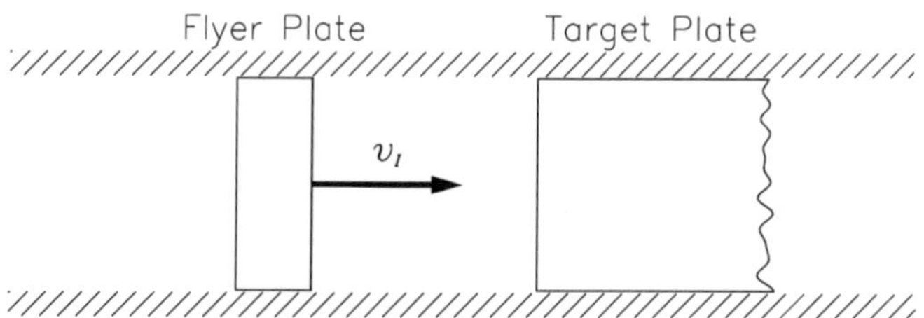

Figure 2.116 The flyer-plate experiment.

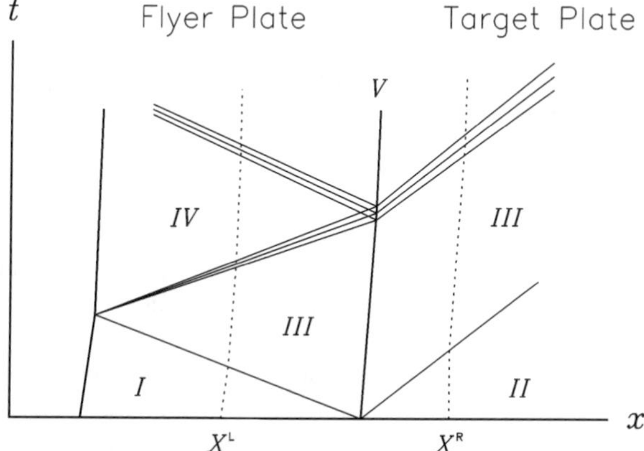

Figure 2.117 The x–t diagram for a flyer plate hitting a half space with lower shock impedance.

always necessary for the target plate to be thicker than the flyer plate, and sometimes the flyer plate and the target plate are identical. The Hugoniot experiment in Figure 2.74 is such an example.

We consider three example experiments in this section. In the first example, we consider a flyer plate with a shock impedance that is greater than that of the target plate. We shall see that this flyer plate sticks to the target plate. In the second example, we consider a flyer plate with a shock impedance that is lower than that of the target plate. Here the flyer plate bounces off the target plate. In the third example, we replace the half space by a target plate that is identical to the flyer plate. In this case, we show that the flyer plate transfers its motion to the target plate and then comes to a stop. To simplify the T–v diagram, we use the weak-shock assumption. Thus we assume the response path of a structured wave is also the Hugoniot curve.

Shock Impedance of Flyer Plate is Greater Than Target

Suppose the flyer plate has a greater shock impedance than the target plate. We show the x–t diagram of the impact in Figure 2.117. We use bold solid lines, which are nearly vertical, to represent both the left free boundary of the flyer plate and the interface between the flyer plate and the target plate. The impact produces two shock waves. The shock wave in the flyer plate reflects off the left free surface as a rarefaction wave. When this rarefaction wave returns to the interface, it produces a reflected compression wave and a transmitted rarefaction wave. We use these waves to separate the diagram into five regions. We show

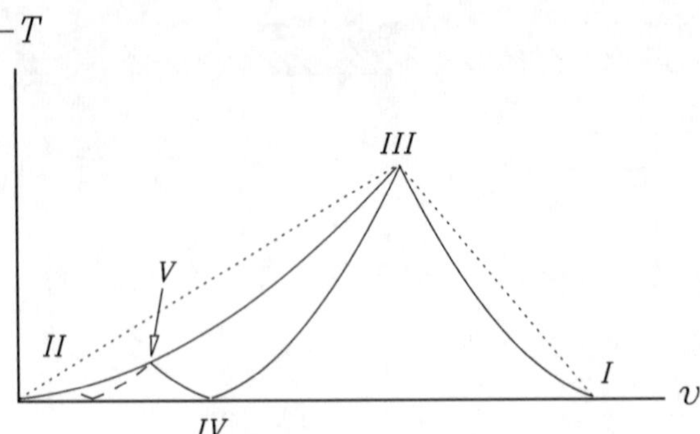

Figure 2.118 The T–v diagram of the material at X^L and X^R due to a flyer plate hitting a half space with lower shock impedance.

two material points X^L and X^R. The material at X^L is in the flyer plate, and the material at X^R is in the target plate. In the x–t diagram, these points move. We show the T–v diagram for X^L and X^R in Figure 2.118. Because the material obeys the weak-shock assumption, the response paths of the rarefaction and compression waves are coincident with the Hugoniot curves. Initially, the material at X^L in the flyer plate is in region *I* where we prescribe $T = 0$ and $v = v_I$. The material at X^R in the target plate is in region *II* where we prescribe $T = v = 0$. Impact creates two shock waves. The shock wave in the flyer plate propagates in the negative-x direction. This means the material at X^L jumps up the negative-sloping Hugoniot for the flyer plate. Similarly, the material at X^R jumps up the positive-sloping Hugoniot for the target plate. Both materials jump to point *III*, which is at the intersection of the two Hugoniot curves. Because we have prescribed that the shock impedance of the flyer plate is greater than that of the target plate, the velocity after impact v_{III} is closer to $v = v_I$ than to $v = 0$. In the flyer plate, the reflection of the shock wave from the left free surface results in a rarefaction wave. When the material X^L encounters this rarefaction wave, it moves down the positive-sloping response path for the flyer plate to point IV. Next, the interaction of the rarefaction wave with the interface of the two plates causes both X^L and X^R to move to region *V* in the x–t diagram. At point *V* in the T–v diagram, both the target and the flyer plates have the same stress and velocity. This means the two plates are still in contact at this time. For later times, the compression wave, which separates regions IV and V, will reflect off the left free boundary again and return to the interface between the flyer and target plates. We use dashed lines to illustrate these additional interactions in the T–v diagram. We often use the term *ring down* to describe these additional wave interactions. We leave it as an exercise for you to show that the plates do not separate during ring down (see Exercise 2.61).

Shock Impedance of Flyer Plate is Less Than Target

Suppose the flyer plate has a smaller shock impedance than that of the target plate. We show the x–t diagram of this impact in Figure 2.119. We also separate this diagram into five regions and use the same two material points at X^L and X^R. We illustrate the T–v diagram for these points in Figure 2.120. As in the previous case, impact causes the material at X^L and X^R to move to point *III*, but this time the velocity after impact $v = v_{III}$ is closer to $v = 0$ than to $v = v_I$. As a result, when the material at X^L encounters the rarefaction wave

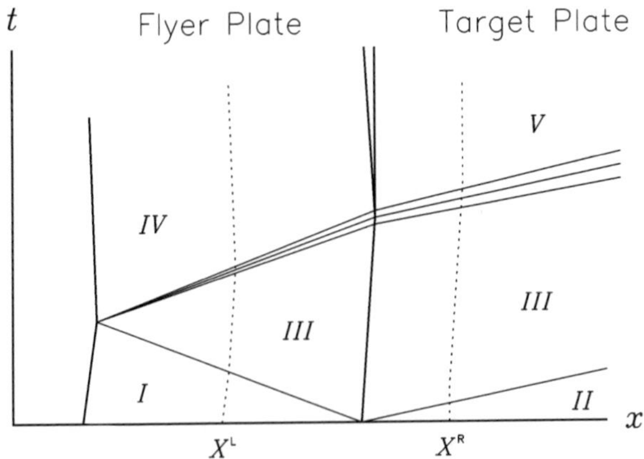

Figure 2.119 The x–t diagram for a flyer plate hitting a half space with higher shock impedance.

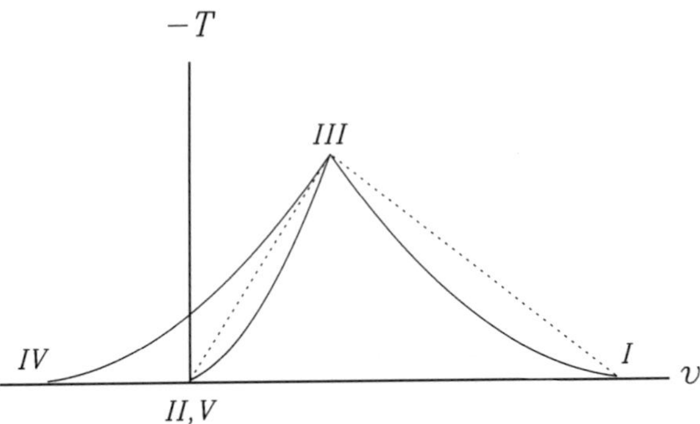

Figure 2.120 The T–v diagram of the material at X^L and X^R due to a flyer plate hitting a half space with higher shock impedance.

returning from the left free boundary of the flyer plate, the velocity changes sign as the flyer plate moves into region IV. When the rarefaction wave reaches the interface, the two plates separate. The flyer plate bounces off the target plate. At later times, the flyer plate has a uniform velocity $v = v_{IV}$ and a uniform stress $T = 0$. When the material at X^R encounters the transmitted rarefaction wave, it moves into region V where the stress and velocity are both equal to zero. Thus the target plate comes to rest after being displaced a small distance to the right.

Identical Flyer and Target

Suppose we replace the half space with a target plate that is identical to the flyer plate. We show the x–t diagram of the impact in Figure 2.121. We include the left and right free boundaries of the flyer and target plates as well as the interface between the flyer plate and the target plate. We illustrate the T–v diagram for the material points at X^L and X^R in Figure 2.122. Notice that the symmetry of the impact yields $v_{III} = v_I/2$. Initially, the flyer plate has a velocity $v = v_I$ and the target plate is stationary. After impact and separation of the plates, the flyer plate is stationary and the velocity of the target plate is $v = v_V = v_I$.

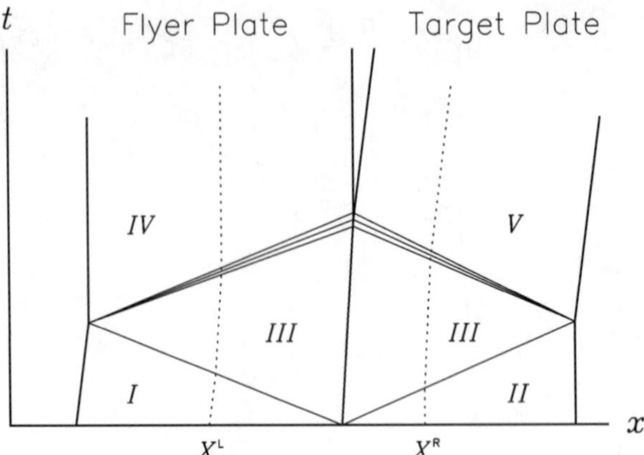

Figure 2.121 The x–t diagram for a flyer plate hitting an identical target plate.

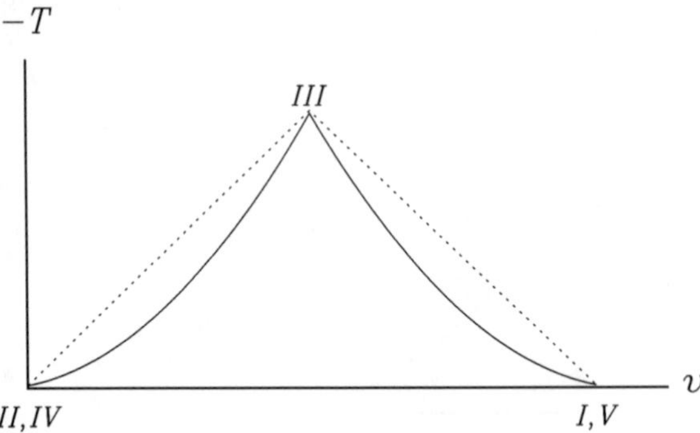

Figure 2.122 The T–v diagram of the material at X^L and X^R due to a flyer plate hitting an identical target plate.

Thus the two plates exchange positions in the T–v diagram. (This is a commonly observed phenomena in the everyday life of a billiard player.) You should recognize that the complete transfer of motion from the flyer plate to the target plate is an artifact of the weak-shock assumption. If we use a strong-shock solution, the response paths for the rarefaction waves will not be coincident with the Hugoniot curves. Consequently, the final velocity of the target plate will be less than $v = v_I$, and the flyer plate will retain a positive residual velocity (see Exercise 2.63).

2.8.5 *Exercises*

2.54 Consider a half space of material with a linear Hugoniot. Suppose a shock wave reflects off the free boundary of this half space. Verify that a reflected rarefaction wave results in the following response path [see Eq. (2.184)]:

$$v_{III} = v_{II} + 2\frac{c_0}{s}\left\{\sqrt{\frac{1+s(1-F_{II})}{1-s(1-F_{II})}} - \tan^{-1}\sqrt{\frac{1+s(1-F_{II})}{1-s(1-F_{II})}} + \frac{\pi}{4} - 1\right\}.$$

2.55 A half space of steel occupies the region $X \leq 0$. Assume that steel has a linear Hugoniot ($\rho_0 = 7{,}890\,\mathrm{kg/m^3}$, $c_0 = 4.58$ km/s, and $s = 1.49$). A shock wave approaches the stress-free boundary of the half space. The stress behind the shock wave is $T = -2$ GPa. Construct the X–t and T–v diagrams for the reflection of this wave.

2.56 Consider two half spaces that are initially free of stress. One half space is composed of the plastic Melmac ($\rho_0 = 1{,}453$ kg/m^3, $c_0 = 3.52$ km/s, and $s = 1.21$). The other half space is Neoprene ($\rho_0 = 1{,}439$ kg/m^3, $c_0 = 2.79$ km/s, and $s = 1.42$). A shock wave approaches the interface. When it contacts the interface, it does not produce a reflected wave. Determine the stress T behind the shock wave.

2.57 A shock wave propagates in an infinite block of brass ($\rho_0 = 8{,}450$ kg/m^3, $c_0 = 3.74$ km/s, and $s = 1.43$). The stress and velocity in front of the shock wave are both zero. The deformation gradient behind the shock wave is $F_+ = 0.9$. A rarefaction wave follows behind the shock wave. The two waves are not in contact. The stress behind the rarefaction wave is zero.

(a) Determine the stress T and the velocity v between the shock wave and the rarefaction wave.
(b) Determine the velocity behind the rarefaction wave.

2.58 Why is the case of a rarefaction wave overtaking a rarefaction wave not discussed in this section? Use the x–t and T–v diagrams to defend your answer.

2.59 Explain why the shock wave in Figure 2.99 slows down. What condition have we imposed on the linear-elastic material that causes this to happen?

2.60 Throughout this section, we have illustrated before-and-after profiles of wave–wave interactions.

(a) In Figures 2.98, 2.105, and 2.113, we illustrate profiles of the velocity v. Sketch the profiles of the stress T for these three cases.
(b) In Figures 2.92, 2.95, 2.101, and 2.108, we illustrate profiles of the stress T. Sketch the profiles of the velocity v for these four cases.

2.61 When a flyer plate of higher shock impedance strikes a half space of lower shock impedance, a process called ring down occurs. We have illustrated the initiation of ring down in Figures 2.117 and 2.118. Redraw these diagrams to include times up to the arrival of the second rarefaction wave at the left free surface of the flyer plate. Demonstrate that the flyer plate remains in contact with the half space throughout this period of time. Discuss why additional ringing of the flyer plate will not cause separation of the plates.

2.62 We have illustrated the X–t diagram for the Hugoniot experiment in Figure 2.75. This diagram does not include the reflections of the shock waves at the free boundaries.

(a) Extend this diagram to include times just beyond the return of the rarefaction waves to the impact plane.
(b) Draw the T–v diagram.
(c) Determine the velocities and stresses in each plate after the interaction of the rarefaction waves at the impact plane is complete.

2.63 We show the x–t and T–v diagrams for the impact of identical flyer and target plates in Figures 2.121 and 2.122. Because these diagrams are based upon the weak-shock assumption, the final velocity of the flyer plate is zero. Suppose both plates are aluminum ($U_S = 5.37 + 1.29v_+$ in km/s and $\rho_0 = 2{,}784$ kg/m^3) and the velocity of the flyer plate is $v_I = 10$ km/s. Determine the residual velocity in the flyer plate after separation.

2.64 All material will fail under sufficiently large positive stress T. A classic experiment, known as a *spall test*, is often used to measure the strength of material during dynamic loading.

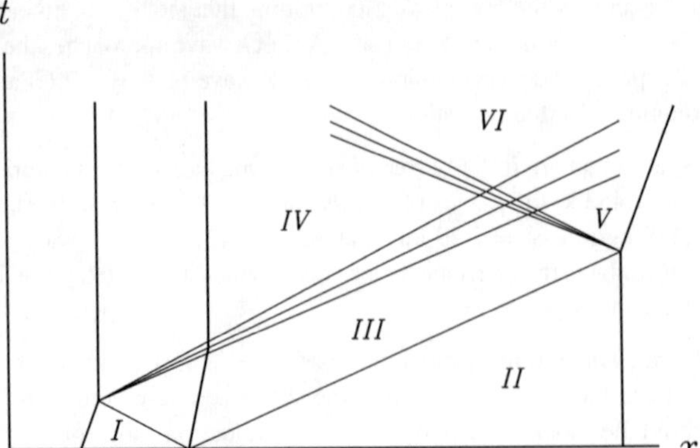

Figure 2.123 Exercise 2.64.

One implementation of this experiment is illustrated in Figure 2.123. Here two plates of aluminum are used ($U_S = 5.37 + 1.29v_+$ in km/s and $\rho_0 = 2{,}784$ kg/m^3). The flyer plate is 30-mm thick and the target plate is 120-mm thick. In constructing the x–t diagram, we have assumed that the shocks are weak and that aluminum can support unlimited tension.

(a) Draw a T–v diagram that includes all of the regions shown in the x–t diagram. In which region is the material likely to fail?

(b) Assume that the target plate fails when T exceeds 60 MPa. Sketch another x–t diagram that illustrates the target plate splitting into two plates. The plane at which the target separates is called the *spall plane*.

(c) If we assume that aluminum cannot support any tension, what are the final velocities and thicknesses of the flyer plate, the target plate to the left of the spall plane, and the *spall fragment* to the right of the spall plane?

2.9 Steady Waves

To analyze waves we have considered a special material, the elastic material. The constitutive model for an elastic material is $T = T(F)$. We have placed two constraints on this model. They are

$$\frac{dT}{dF} > 0, \qquad \frac{d^2T}{dF^2} \leq 0.$$

This constitutive model can be either linear or nonlinear. The first constraint requires that the elastic material has a positive stiffness. Because the wave velocities depend upon the square root of the first derivative of the stress, this constraint ensures that these velocities are real. The second constraint requires that nonlinear-elastic materials become stiffer as they are compressed. This means that the solutions for structured compression waves have characteristic contours that converge and eventually intersect. At the point of intersection, these solutions are multivalued. From a physical point of view, this is an unacceptable result. To avoid this result, we replace the overlapping contours of the multivalued solution by a single contour that represents a discontinuous solution. We call this discontinuous solution a shock wave. The jump conditions tell us how the shock wave changes the deformation gradient, stress, and velocity. The concept of the shock wave is an exceedingly useful tool that has allowed us to analyze a great variety of wave phenomena. However, the shock wave is an

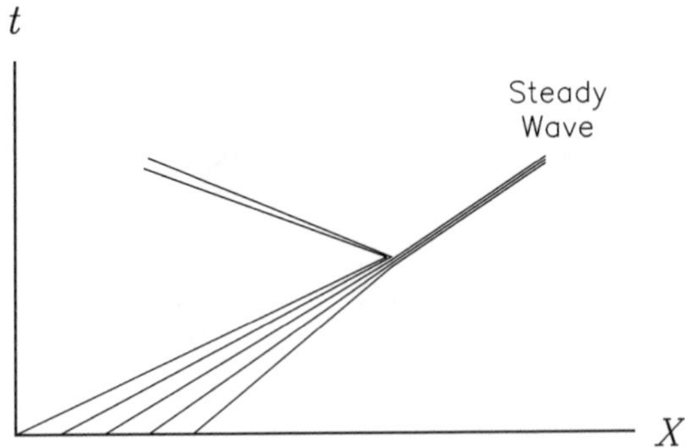

Figure 2.124 The X–t diagram for the transition of a compression wave into a steady wave.

approximation. In reality, shock waves do not exist! Given sufficiently precise experimental apparatus, it is possible to measure structure in waves that appear to be discontinuous. This means that a compression wave does not change into a discontinuous wave. Rather, it remains a structured wave in which the contours stop converging and become parallel (see Figure 2.124). We call this wave a *steady wave*. For many materials, the contours of the steady wave are very close together. For metals, the head and tail contours of the steady compression wave may only be separated by a few nanoseconds.

Because the constitutive model for a nonlinear-elastic material results in the shock wave, experimental observations of steady waves imply that real materials are not elastic. This does not mean that results based on the elastic material are worthless. Indeed, the nonlinear-elastic material is a very useful approximation. However, it does mean that if we wish to understand nonlinear waves in greater detail, we must at some point abandon the nonlinear-elastic material and turn to other constitutive models that predict the response of materials to steady waves. In this book, this section represents such a turning point. In this section, we analyze the steady wave. *Our analysis does not depend upon the constitutive model for the nonlinear-elastic material.* In this section the constitutive equation for the elastic material is replaced by another equation, which states that steady waves exist. Our analysis of the steady wave for one-dimensional motion is presented in both the material and the spatial descriptions. The analysis for three-dimensional motion is left as an exercise (see Exercises 2.66 and 2.67).

2.9.1 *Material Description*

Consider a steady wave propagating in an infinite material (see Figure 2.125). We prescribe the velocity, stress, and deformation gradient in front of the steady wave to be v_-, T_-, and F_-. For a steady wave to exist and propagate with constant shape in the material description, we require the solution of the equations of conservation of mass and balance of linear momentum in the material description to be a function of the variable $\hat{\varphi}$,

$$\hat{\varphi} = X - C_S t, \tag{2.185}$$

where C_S is the material shock velocity. Notice that the steady wave is similar to a simple wave with straight and parallel contours. When $C_S > 0$, the steady wave propagates in the positive-x direction, and when $C_S < 0$, the steady wave propagates in the negative-x direction.

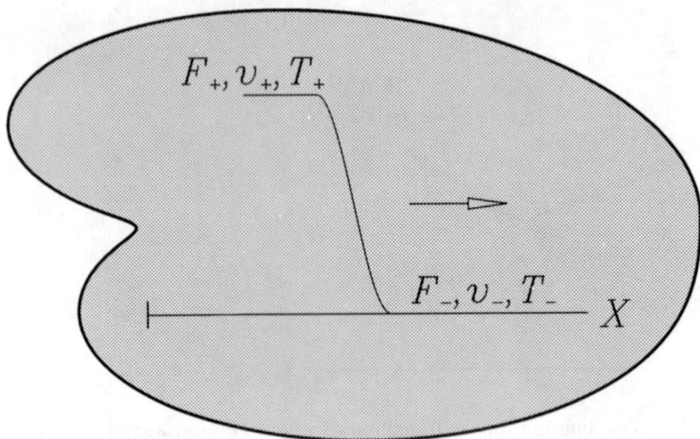

Figure 2.125 A steady wave.

You will recall that the velocity and the deformation gradient are related by [see Eq. (2.33)]

$$\frac{\partial \hat{v}}{\partial X} = \frac{\partial \hat{F}}{\partial t}.$$

By assuming that F and v are only functions of $\hat{\varphi}$, we apply the chain rule to this expression to obtain

$$\frac{dv}{d\hat{\varphi}} = -C_S \frac{dF}{d\hat{\varphi}}.$$

Because $F = F_-$ and $v = v_-$ ahead of the wave, we integrate this equation to obtain

$$v - v_- = -C_S(F - F_-). \tag{2.186}$$

Evaluating this result at the tail contour of the steady wave where $F = F_+$ and $v = v_+$, we find that

$$[\![F]\!]C_S + [\![v]\!] = 0, \tag{2.187}$$

which is equivalent to the compatibility condition for a discontinuous wave [see Eq. (2.139)].

The balance of linear momentum in the material description is [see Eq. (2.31)]

$$\rho_0 \frac{\partial \hat{v}}{\partial t} = \frac{\partial \hat{T}}{\partial X}.$$

By requiring that T and v are functions of $\hat{\varphi}$ alone, we obtain

$$-\rho_0 C_S \frac{dv}{d\hat{\varphi}} = \frac{dT}{d\hat{\varphi}}.$$

Upon integration, we find that

$$T - T_- = -\rho_0 C_S(v - v_-), \tag{2.188}$$

where T_- represents the stress ahead of the wave. We substitute Eq. (2.186) to obtain

$$T - T_- = \rho_0 C_S^2(F - F_-). \tag{2.189}$$

When we evaluate the previous results at the tail contour of the steady wave where T_+ and v_+, we obtain

$$[\![T]\!] = -\rho_0 C_S [\![v]\!], \qquad [\![T]\!] = \rho_0 C_S^2 [\![F]\!],$$

which are equivalent to the jump conditions for a discontinuous wave [see Eqs. (2.146)–(2.147)]. Thus we find that the steady wave propagates at the same velocity as a shock wave. The steady wave also produces the same changes in stress, velocity, and deformation gradient.

2.9.2 *Spatial Description*

For a steady wave to exist and propagate with constant shape in the spatial description, we require that the solution to the equations of conservation of mass and balance of linear momentum in the spatial description must be a function of the variable φ,

$$\varphi = x - c_S t, \tag{2.190}$$

where c_S is the spatial shock velocity.

The equation of conservation of mass of the material is [see Eq. (2.34)]

$$\frac{\partial \rho}{\partial t} + \frac{\partial(\rho v)}{\partial x} = 0. \tag{2.191}$$

By assuming that ρ and v are only functions of φ, and by using the chain rule, we obtain

$$\frac{d}{d\varphi}[\rho(c_S - v)] = 0. \tag{2.192}$$

Integration yields

$$[\rho(c_S - v)] = [\rho_-(c_S - v_-)], \tag{2.193}$$

where ρ_- and v_- are the density and velocity ahead of the steady wave. When we evaluate this result at the tail contour of the steady wave where $\rho = \rho_+$ and $v = v_+$, we obtain

$$[\![\rho(c_S - v)]\!] = 0,$$

which is equivalent to the jump condition for a discontinuous wave in the spatial description [see Eq. (2.148)].

The equation of balance of momentum in the spatial description is [see Eq. (2.35)]

$$\frac{\partial(\rho v)}{\partial t} = \frac{\partial}{\partial x}(T - \rho v^2).$$

By assuming that ρ, v, and T are only functions of φ, and by using the chain rule, we obtain

$$\frac{d}{d\varphi}[\rho v(c_s - v) + T] = 0.$$

Integration yields

$$\rho v(c_s - v) + T = \rho_- v_-(c_s - v_-) + T_-. \tag{2.194}$$

When we evaluate this result at the tail contour of the steady wave, we obtain

$$[\![T + \rho v(c_S - v)]\!] = 0,$$

which is equivalent to the jump condition for a discontinuous wave in the spatial description [see Eq. (2.149)]. We find that the spatial descriptions of the steady wave and the shock wave have the same changes in stress, density, and velocity.

2.9.3 *Equation of the Rayleigh line*

As a material point **X** encounters a steady wave, the stress and velocity change. This change is described by Eq. (2.188). This is the equation for a straight line,

$$\frac{T - T_-}{v - v_-} = -\rho_0 C_S. \tag{2.195}$$

Let us compare this equation to the jump condition that defines the Rayleigh line [see Eq. (2.146)]:

$$\frac{[\![T]\!]}{[\![v]\!]} = -\rho_0 C_S.$$

We see that this steady wave result is the equation of the Rayleigh line. *Thus the Rayleigh line is the response path for the steady wave.* Whereas the shock wave solution causes the material to *jump* between two points on the Hugoniot curve, the steady wave solution causes the material to *move* along the Rayleigh line between the same two points on the Hugoniot curve. It is important for you to notice that neither the steady wave solution nor the jump conditions for a shock wave depends upon the constitutive model for an elastic material. In particular, the steady wave solution depends only on the laws of motion and the existence of the steady-wave variable φ. Indeed, we might suppose that C_S and c_s in Eqs. (2.185) and (2.190) are in no way related to our previous definitions of the material and spatial wave velocities. In fact, this section on steady waves could just as easily be moved to the end of Chapter 1 where we had not yet introduced any constitutive model.

2.9.4 *A Little History*

It was the Reverend S. Earnshaw† in 1859 who first integrated the equations of conservation of mass and balance of linear momentum subject to the steady wave conditions. He obtained the result

$$[\![T]\!] = \rho_0 C_S^2 [\![F]\!],$$

which is our Eq. (2.147). Earnshaw presented his result in a purely mechanical context. Because he assumed that T was only a function F, he was left no choice but to interpret this equation as a constitutive model rather than a description of a response path. He also knew that measurements of the pressure of an ideal gas showed that

$$pF = p_0,$$

when the temperature of the gas is held constant. This is called *Boyle's law*. These two equations are incompatible. Earnshaw incorrectly concluded that steady waves could not

† S. Earnshaw, "On the mathematical theory of sound," *Transactions of the Royal Society of London* **150**, 133–148 (1860).

exist in an ideal gas because the constitutive model of an ideal gas contradicted his so-called *constitutive model* of a steady wave. Just ten years later, W. J. M. Rankine† analyzed the steady wave and concluded that "there must be both a change of temperature and a conduction of heat" in order that steady waves might exist. Thus Rankine recognized that Boyle's law was not wrong, but it only told part of the story. He found the next chapter of the story in how the pressure changed as the temperature changed. In doing so, Rankine made the fundamental transition from a purely mechanical theory to a thermodynamical theory. He recognized that steady waves not only cause the deformation gradient to change, they also cause the temperature to change. Hugoniot later showed how steady waves can exist even in the absense of heat conduction.‡ Thus although Earnshaw incorrectly concluded that steady waves are unusual and only exist in materials with a very special constitutive model, Rankine and Hugoniot showed that steady waves can exist in an ideal gas. Indeed, today we know that they exist in virtually every material.

In this section, with the benefit of the hindsight of history, we abandoned the constitutive model of an elastic material that contains Boyle's law as a special case and therefore avoided the conflict faced by Earnshaw. Thus we have an incomplete theory of waves. The next two chapters are devoted to constructing constitutive descriptions of materials that allow the existence of the steady wave.

2.9.5 *Exercises*

2.65 A nonlinear-elastic material, where $T = T(F)$, cannot support a steady wave unless T is proportional to F. A simple modification of this constitutive model that does yield a steady wave solution is obtained by letting the stress depend upon the rate of change of the deformation gradient, $T = T(F, \partial\hat{F}/\partial t)$. As an example, consider the simple constitutive model

$$T = E_0\left(1 - \frac{1}{F}\right) + \frac{b}{F}\frac{\partial \hat{F}}{\partial t},$$

where E_0 and b are prescribed positive constants.

(a) Substitute the steady wave condition, Eq. (2.185), into this expression. Then substitute the result into Eq. (2.195) and show that

$$\frac{dF}{(F - E_0/E_s)(1 - F)} = \frac{E_s}{bC_S}d\hat{\varphi},$$

where $E_s = \rho_0 C_s^2$, the initial velocity is zero, and the initial deformation gradient is one.

(b) Integrate this equation.

(c) The initial conditions require $F_- = 1$. Determine the values of C_S that ensure a bounded solution.

(d) Let $E_0 = 1$, $E_s = 2$, and $b = 1$. Obtain a profile for this wave by graphing $-T(F)$ versus $\hat{\varphi}(F)$ for $0.501 < F < 0.999$.

2.66 The solution for a steady wave with three-dimensional motion is solely a function of the variable $\varphi = x_k N_k - c_S t$, where the steady wave propagates in the direction prescribed by

† W. J. M. Rankine, "On the thermodynamic theory of waves of finite longitudinal disturbance," *Transactions of the Royal Society of London* **160**, 277–288 (1870).

‡ H. Hugoniot, "Sur la propagation du mouvement dans les corps et spécialement dans les gaz parfaits," *Journal de l'École Polytechnique* **58**, 1–125 (1889).

the unit normal with components N_k. Using the equation for the conservation of mass,

$$\frac{\partial \rho}{\partial t} + \frac{\partial(\rho v_k)}{\partial x_k} = 0,$$

require ρ, and v to be functions of φ alone. Integrate the resulting ordinary differential equation to obtain

$$\rho(c_s - N_k v_k) = [\rho(c_s - N_k v_k)]_-. \tag{2.196}$$

This is the balance of mass for a steady wave in three dimensions.

2.67 The solution for a steady wave with three-dimensional motion is solely a function of the variable $\varphi = x_k N_k - c_S t$, where the steady wave propagates in the direction prescribed by the unit normal with components N_k. Use the equation for the balance of linear momentum,

$$\frac{\partial(\rho v_m)}{\partial t} + \frac{\partial(\rho v_m v_k)}{\partial x_k} - \frac{\partial T_{km}}{\partial x_k} = 0$$

and require T_{km}, ρ, and v to be functions of φ alone. Integrate the resulting ordinary differential equation to obtain

$$T_{km} N_k + \rho v_m (c_s - v_k N_k) = [T_{km} N_k + \rho v_m (c_s - v_k N_k)]_-. \tag{2.197}$$

This is the balance of linear momentum for a steady wave in three dimensions.

CHAPTER 3

Thermomechanics

In the late seventeenth century, Isaac Newton presented the fundamental concepts of motion, force, and momentum. With these concepts, he made many successful predictions of physical phenomena; however, when he attempted to predict the speed of sound in air, he failed. It was not until a century later, when the distinction between temperature and heat was understood, that Laplace was able to achieve an accurate prediction of this wave velocity.

As in Newton's day, we often do not appreciate that even weak elastic waves cause changes not only in stress and strain but also temperature. Because these temperature changes are small and reversible, they are not in our common perception. In contrast, strong shock waves cause irreversible changes in temperature. Most solids when subjected to repeated hammer blows become noticeably hotter to the touch. To describe the distinctions between these types of waves, we must introduce the fundamentals of thermodynamics, including the concepts of heat, temperature, and entropy.

Why must we study thermodynamics? In Chapter 2 we studied a wide range of wave phenomena using the mechanical concepts of force and motion. Using only the constitutive model for an elastic material, we developed detailed analyses of the nonlinear structured wave, the transition of a compression wave into a shock wave, and the decay of a shock wave as it is overtaken by a rarefaction wave. We have analyzed wave motion in solids, liquids, and adiabatic gases. But now we must turn to thermodynamics and the concepts of temperature, entropy, and heat for several important reasons. Let us briefly look at two of these reasons.

Newton's famous dilemma over the prediction of the speed of sound in air was only resolved after it was understood that temperature and heat as well as force and motion govern the constitutive model of air. In Newton's day, people presumed that the constitutive model for air was

$$p = p_0 \frac{\rho}{\rho_0},$$

which is called *Boyle's law*. Using this result, Newton concluded that the velocity of sound is

$$c_0 = \left(\frac{p_0}{\rho_0} \right)^{1/2}.$$

By noting that Boyle's law is represented by an adiabatic gas where $\gamma = 1$, we can obtain the same results by substituting Eq. (2.29) in Section 2.1.4 into Eq. (2.38) in Section 2.3.1.

However, in Chapter 2, we model air as an adiabatic gas with $\gamma = 1.4$,

$$p = p_0\left(\frac{\rho}{\rho_0}\right)^{1.4}.$$

This leads to the following velocity of sound:

$$c_0 = \left(\frac{1.4p_0}{\rho_0}\right)^{1/2}.$$

This last result is a reasonably accurate estimation of the measured speed of sound in air. Does this then mean that Boyle's law is wrong? No, it doesn't. Boyle's law accurately predicts the response of air to pressure and density changes, when we hold the temperature of the air constant. The key to understanding Newton's dilemma is that sound waves cause temperature changes. To understand this, we need to understand the constitutive model for the ideal gas. To understand the ideal gas we need to understand thermodynamics.

We can find another motivation for studying thermodynamics in a simple experimental observation. Shock discontinuities do not exist! What at first appears to be a discontinuity is actually a structured wave. Given measurements of sufficient accuracy, we can always resolve the smooth profile of a steady wave. There seems to be no obvious way to construct a constitutive model that explains this observation unless we allow the material to absorb some of the energy of the wave. When we combine this fact with the additional observation that shock loading a material often causes dramatic increases in temperature, we must conclude that the absorbed energy is converted into heat. In turn, this heating can result in thermal expansion of the material. Often we see that strong shock waves cause large amounts of heating and thermal expansion. When this happens a compression limit is reached. At this limit, stronger shock waves will not cause additional compression, only additional heating.

Thermodynamics is a useful theory with broad application to virtually every technical pursuit. Thermodynamists tend to have opinions about how their version of thermodynamics should be tailored and specialized to the task at hand. I am no exception to this rule. For this reason our version of thermodynamics is called *thermomechanics*. This simply means that we are going to emphasize the connections between mechanical and thermal effects and mostly ignore aspects of the theory that relate to other fields such as chemistry and electromagnetism. We begin our discussion by introducing the concept of energy. Energy is the unifying concept that links force and motion to temperature and entropy. Force and motion describe work, and temperature and entropy describe heat. Both work and heat have the same physical dimensions and can be stored in a material as energy.

3.1 Balance of Energy

It is a fundamental premise of thermodynamics that energy can be stored in a material either by mechanical working or by thermal heating. This implies that a material can either transform heat into work or work into heat. The *first law of thermodynamics* states that the total amount of energy cannot change during such a transformation. The *second law of thermodynamics* states that work transforms into heat more readily than heat transforms into work. We use the physical dimension of joules to represent energy. Thus, both work and heat have dimensions of joules. When we work or heat an element of a body, we change its energy. We can store energy in a body in one of two ways, either by changing its *kinetic energy* $\mathcal{K}$ or by changing its *internal energy* $\mathcal{E}$. Kinetic energy is determined solely by the motion and mass of the material. However, to determine the internal energy, we must know the constitutive model of the material. In the next several sections, we first restrict our discussion to one-dimensional motion. We describe how working and heating are performed on an element of a body. Then we derive a mathematical relationship for the first law of thermodynamics. After the one-dimensional derivation, we present a three-dimensional derivation of the same relationships.

3.1.1 *Work*

Consider an element of a body undergoing one-dimensional motion. We subject this element to triaxial stress. We show the tractions resulting from this stress configuration in Figure 3.1. In the material description, the thickness of the element is dX, and the stress components at the center of the element are $T(X, t)$ and $\tilde{T}(X, t)$. We use a Taylor's series expansion of $T(X, t)$ to evaluate the tractions on two of the faces of this element. When a force acts through a distance, it does work. Because we restrict motion to the x direction, only the tractions $T + \frac{1}{2}\frac{\partial \hat{T}}{\partial X} dX$ and $T - \frac{1}{2}\frac{\partial \hat{T}}{\partial X} dX$ that act in the x direction do work.

Let us consider a two-dimensional view of this element in which we show only those forces that do work (see Figure 3.2). These forces include the two tractions, which act in the

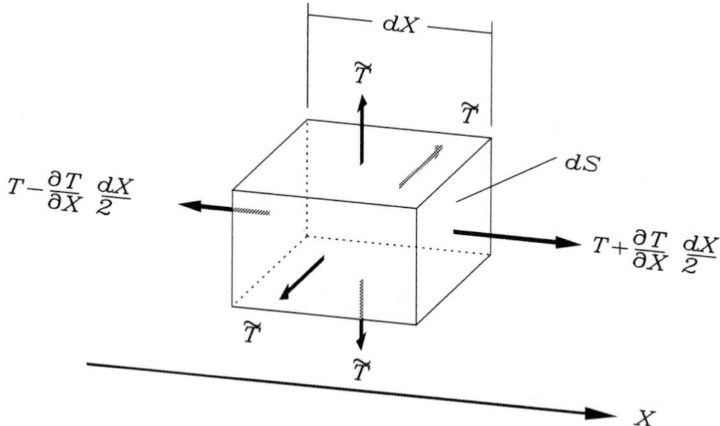

Figure 3.1 Triaxial stress acting on a volume element.

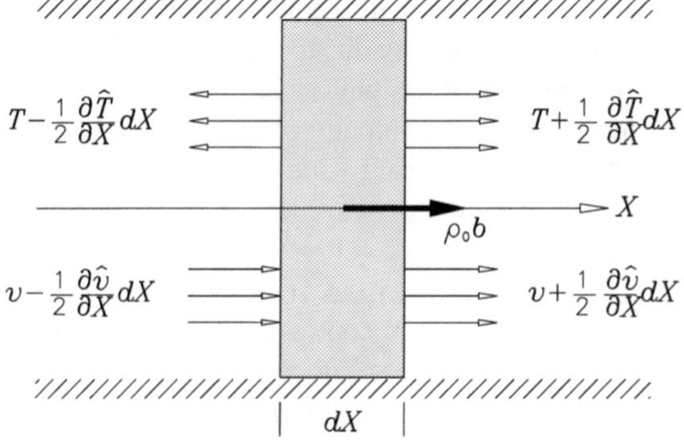

Figure 3.2 Free-body diagram of an element of the material.

x direction, and a body force $\rho_0 b\, dS\, dX$. We omitted the body force from Figure 3.1 in order to simplify that diagram. We recall that b is the body force per unit mass, dS is the area of each vertical face, and $dSdX$ is the volume of the element. During an infinitesimal interval of time Δt, the total work W done on the element by the body force and the stresses is

$$W = \left(T + \frac{1}{2}\frac{\partial \hat{T}}{\partial X}dX\right)\left(v + \frac{1}{2}\frac{\partial \hat{v}}{\partial X}dX\right)\Delta t\, dS$$
$$- \left(T - \frac{1}{2}\frac{\partial \hat{T}}{\partial X}dX\right)\left(v - \frac{1}{2}\frac{\partial \hat{v}}{\partial X}dX\right)\Delta t\, dS + \rho_0 b\, dX\, dS v \Delta t.$$

By convention, work is positive when stress and velocity act in the same direction to do work on the element. When they act in opposite directions, the work term is negative. We use the physical dimension of joules for work. Let $\mathcal{P}$ represent the rate of work per unit mass. Then

$$\mathcal{P} \equiv \frac{W}{\rho_0\, dS\, dX \Delta t}.$$

We combine the last two expressions to obtain

$$\rho_0 \mathcal{P} = v\left(\frac{\partial \hat{T}}{\partial X} + \rho_0 b\right) + T\frac{\partial \hat{v}}{\partial X}. \tag{3.1}$$

Recall that the material form of the balance of linear momentum is [see Eq. (1.144)]

$$\rho_0 \frac{\partial \hat{v}}{\partial t} = \frac{\partial \hat{T}}{\partial X} + \rho_0 b$$

and that the gradient of the velocity is [see Eq. (1.132)]

$$\frac{\partial \hat{F}}{\partial t} = \frac{\partial \hat{v}}{\partial X}.$$

We substitute these expressions into Eq. (3.1) to obtain

$$\rho_0 \mathcal{P} = \rho_0 \frac{\partial \hat{\mathcal{K}}}{\partial t} + T\frac{\partial \hat{F}}{\partial t}, \tag{3.2}$$

where we have defined

$$\mathcal{K} \equiv \frac{1}{2}v^2. \tag{3.3}$$

Notice that we have divided the work expression on the right-hand side of Eq. (3.2) into two terms. The first term represents the change in energy due to the acceleration of the element. We call $\mathcal{K}$ the *specific kinetic energy,* because it has physical dimensions of energy per unit mass (J/kg $=$ m^2/s^2) and it depends solely on the velocity v. The second term on the right-hand side of Eq. (3.2) represents the change in energy due to deformation of the element. We call $T\partial\hat{F}/\partial t$ the *stress power.* This division of work into two separate expressions is quite important. Let us see why.

Notice that the kinetic energy depends only on the velocity and density. It does not depend on the constitutive model of the material. In contrast, we cannot evaluate the stress power unless we know the constitutive model of the material. Suppose for a moment that the material is elastic. The constitutive model is $T = T(F)$, and substitution of this expression into the stress power $T\partial\hat{F}/\partial t$ allows us to integrate the stress power over an interval of the deformation gradient F. Thus we can characterize kinetic energy as external energy; that is, it is energy that *does not* depend upon the internal or constitutive properties of the material. Energy resulting from the stress power is internal energy; that is, it is energy that *does* depend upon the internal or constitutive properties of the material.

The Conjugate Pair T and F

Notice a very important symmetry between the elastic constitutive model and the stress power expression. Both expressions contain the same pair of functions: the stress T and the deformation gradient F. We call the functions T and F a *conjugate pair.* We describe the stress power with the product of a conjugate pair, which contains a *force term* T and a *flux term* $\partial \hat{F}/\partial t$.

In thermodynamics, the notion of the conjugate pair is fundamental to the concept of energy. Work, which results in a change of internal energy, is described by one conjugate pair of functions. In the next section, we shall see that heating, which also results in a change in internal energy, is described by another conjugate pair of functions. Even though we do not discuss them in this text, conjugate pairs of functions exist that describe internal energy arising from other fields such as the electromagnetic field. Conjugate pairs for phase transformations and chemical reactions also exist. We shall return to this topic in Chapter 4.

3.1.2 *Heat*

We have just considered how work is done on an element of volume of a body. We can also heat this element. We can deliver heat to the element from external sources, and we can also generate heat internally.

External Heating We first consider heating the element from its surroundings. This can occur in two ways: We can conduct heat to the element from neighboring elements and we can radiate heat to the element from remote sources. We describe the conduction of heat with the *heat flux vector* $\mathbf{q}$ and the radiation of heat with the *specific external heat supply* r. We use the physical dimensions of W/m^2 to represent the heat flux vector, and we use the physical dimensions of W/kg to represent the specific external heat supply. In the case of one-dimensional motion, we shall assume that heat only flows in the X direction. We describe the flow of heat with the axial component of the heat flux $q = \hat{q}_1(X, t)$ (see Figure 3.3). We use a Taylor's series to evaluate the heat flux at the left and right faces of the element. *By convention, heating is positive when heat is delivered to the element.* During an infinitesimal interval of time Δt, the total external heating H delivered to the element is

$$H = \left(q - \frac{1}{2}\frac{\partial \hat{q}}{\partial X} dX\right) dS\,\Delta t - \left(q + \frac{1}{2}\frac{\partial \hat{q}}{\partial X} dX\right) dS\Delta t + \rho_0 r\, dX\, dS\,\Delta t.$$

We use the physical dimension of joules for H. Let $\mathcal{Q}$ represent the rate of external heating per unit mass of the element. Then

$$\mathcal{Q} \equiv \frac{H}{\rho_0\, dX\, dS\,\Delta t}.$$

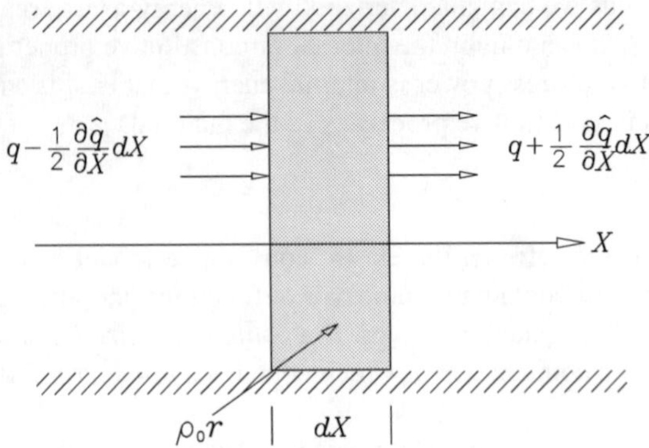

Figure 3.3 Heat supplies to an element of the material.

We combine these two equations to obtain

$$\rho_0 \mathcal{Q} = -\frac{\partial \hat{q}}{\partial X} + \rho_0 r. \tag{3.4}$$

Internal Heating As illustrated in Figure 3.2, the process of stretching or compressing our element of material involves work. For example, suppose our element of material is initially free of stress. Let us compress this element and do work on it. Next we allow the element to expand and return to its initial configuration. During this expansion process, the element will do work on us. Ideally, the work we do on the element to compress it will be equal to the work the element does on us when it expands. Any material that exhibits this type of behavior is said to have undergone a *reversible process*. However, in all real materials, a portion of the work we put into the material is retained as heat and not returned. Real materials are said to undergo *irreversible processes*. The portion of the work that is converted to heat is called the *dissipation* $\mathcal{D}$. We assign physical dimensions of W/kg to the dissipation.

The Conjugate Pair ϑ and η

Remember that we describe the change in internal energy due to work by the product of a conjugate pair called stress power. We shall describe the change in internal energy due to heating by using the product of another conjugate pair. This conjugate pair is composed of the temperature ϑ and the entropy η. Both are scalar quantities. The physical dimensions that we assign to ϑ and η are the kelvin degree K and the joule per kilogram per kelvin degree J/kg/K. This pair of variables has a force term ϑ and a flux term $\partial\hat{\eta}/\partial t$. The product $\vartheta\,\partial\hat{\eta}/\partial t$ is called the *heating power*. At first, you might view this representation of heating power as a simple convenience or possibly a purely mathematical convention. However, it represents the essence of thermodynamics. Indeed, the twentieth-century, Nobel physicist Richard Feynman called it, "the center of the universe of thermodynamics."† Both temperature and entropy are essential to the description of internal energy.

† See pp. 44–49 in *The Feynman Lectures on Physics* by Feynman, R. P., Leighton, R. B., and Sands, M. (Addison-Wesley, 1963).

Absolute Temperature The force term of the heating power is the temperature ϑ. We shall only consider materials that exist at positive temperature. Often, we use the term *absolute temperature* to emphasize this assumption. The relationship between the temperature and the amount of heat in the material depends upon the constitutive model of the material. When we bring two bodies with different temperatures into contact and leave them undisturbed, heat will flow from the body with higher temperature to the body with lower temperature until the temperatures of the two bodies become equal.

Entropy The concept of temperature seems obvious because everyone has used a thermometer to measure temperature. In contrast, the concept of the scalar quantity entropy seems obscure. No one seems to have used an entropy meter. However, it is entropy that provides us with the link between temperature and heat. The origin of the concept of entropy is found in the work of the engineer Nicholas Leonard Sadi Carnot. In 1824, Carnot published a work entitled *The Motive Power of Heat*. In this work Carnot analyzed the efficiency of heat engines and showed *there can be no transformation of heat into work without a change of temperature.* Carnot arrived at his conclusions by studying a "perfect" engine – the Carnot engine. Let us briefly consider the operation and analysis of the Carnot engine because it demonstrates how entropy lies at the heart of the theory of thermodynamics.

By using entropy, we can easily see why Carnot's engine is the most efficient heat engine possible, and we can also see why a change of temperature is required to operate a heat engine. In Figure 3.4a we represent a Carnot engine as a simple piston and cylinder. We place a heat source of temperature ϑ_{in} above the cylinder and a heat sink of temperature ϑ_{out}

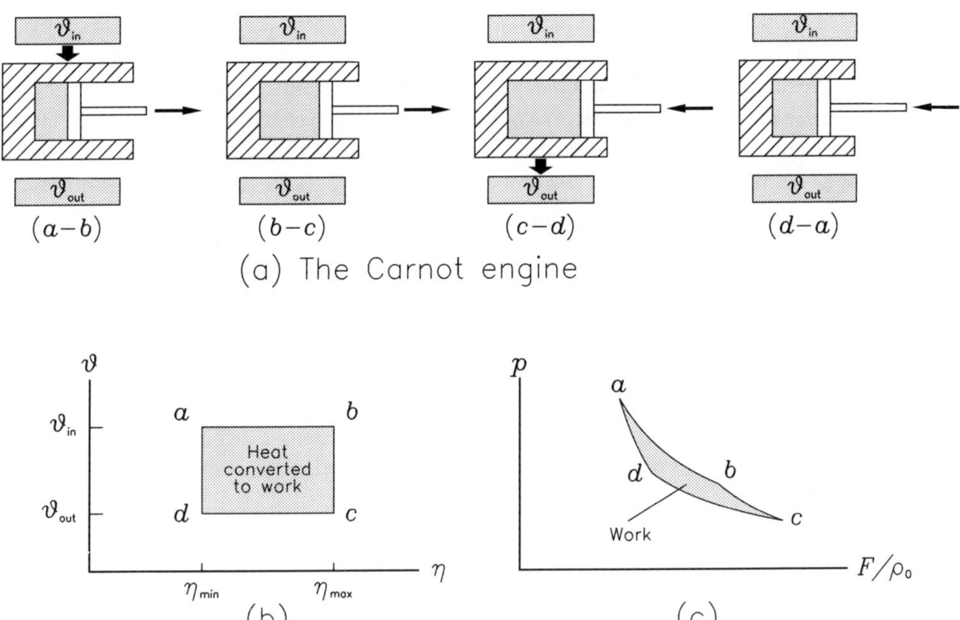

Figure 3.4 The Carnot engine and the Carnot cycle.

below the cylinder. A gas is trapped inside the cylindrical chamber. We assume that this gas does not dissipate work and only undergoes reversible processes. To describe the gas, we shall replace the stress T by the pressure p. Thus the gas is described by values of the two conjugate pairs (p, F) and (ϑ, η). We use four illustrations to represent one complete cycle of the Carnot engine. Initially, the temperature of the gas is ϑ_{in}. During the first part of the cycle, shown in the illustration labeled (*a–b*), we allow the gas to push the piston to the right. Normally, when a gas expands, it cools. However, in this case, we add heat from the heat source above the cylinder. Enough heat is added to keep the temperature constant at ϑ_{in}. During the second part of the cycle, labeled (*b–c*), we disconnect the heat source and continue to allow the gas to push the piston to the right until the temperature of the gas drops to ϑ_{out}. This is the temperature of the heat sink below the cylinder. During the third part of the cycle, shown in the illustration labeled (*c–d*), we connect the cylinder to the heat sink and push the piston to the left. Normally, as a gas is compressed it will heat up, but by using the heat sink, we draw enough heat out of the gas to keep its temperature at ϑ_{out}. During the final part of the cycle, labeled (*d–a*), we disconnect the heat sink and continue to compress the gas until its temperature returns to ϑ_{in}.

In Figures 3.4b and 3.4c we illustrate what happens to the values of the two conjugate pairs that describe the gas. At the beginning of the cycle, the gas is at point a. Point a in Figure 3.4b represents the initial values of ϑ and η, and point a in Figure 3.4c represents the initial values of p and F. During the operation of this engine, the gas in the chamber moves to points b, c, and d. Finally, it returns to point a to complete a *cycle*. During the (*a–b*) portion of the cycle, the gas moves from point a to point b. Because the gas is connected to the heat source during this portion of the cycle, the entropy of the gas increases. Because the temperature is constant, we can easily integrate the heating power $\vartheta \partial\hat{\eta}/\partial t$ to obtain the amount of heat that we must add to the gas. We must add $\vartheta_{in}(\eta_{max} - \eta_{min})$ to each unit of mass of gas in the cylinder. As we move the gas from point b to point c, we do not add heat; therefore, the entropy remains constant while the temperature of the gas drops to ϑ_{out}. Following these two expansion processes, we push the piston to the left and compress the gas. Simultaneously, we remove heat from the gas to keep the temperature at ϑ_{out}. This causes the gas to move to point d. The amount of heat per unit mass that we must remove during this part of the cycle is $\vartheta_{out}(\eta_{max} - \eta_{min})$. To complete the cycle, we continue to compress the gas from point d to point a. We hold the entropy constant by not adding heat to the gas.

The difference between the heat that we put into the gas when moving from a to b and the heat that we removed from the gas when moving from c to d is

$$\text{heat consumed per unit mass of gas} = (\vartheta_{in} - \vartheta_{out})(\eta_{max} - \eta_{min}).$$

This quantity of heat is equal to the shaded area in Figure 3.4b. Furthermore, the rate of work performed by each unit of mass of gas in the engine during the cycle can be evaluated from Eq. (3.2):

$$\mathcal{P} = -\frac{p}{\rho_0}\frac{\partial \hat{F}}{\partial t}. \tag{3.5}$$

Here we have assumed that the kinetic energy of the gas is negligible. We can integrate this relationship with respect to time to obtain work. This integral is equal to the areas under the p–F/ρ_0 curves in Figure 3.4c. For example, we see that work is performed by the engine as the gas moves from a to b and then to c. The work we obtain from each unit of mass of gas in the engine during this portion of the cycle is the area under the curve connecting points a, b, and c. But we must do work on the engine to compress the gas from c to d and then to a. This is the area under the curve connecting points c, d, and a. Clearly, we get more work out of the engine then we put back into it. The work per unit mass of gas produced by the engine during this cycle is equal to the shaded area in Figure 3.4c. The first law of thermodynamics states that the shaded areas in Figures 3.4b and 3.4c are equal. Now it is clear that if $\vartheta_{in} = \vartheta_{out}$, both shaded areas will vanish. Thus we conclude, as Carnot did, that a temperature difference is necessary to make the engine work.†

Next let us compare the Carnot engine to other types of heat engines. By definition, the Carnot engine has a rectangular ϑ–η diagram and the other engines do not. Suppose we require that all of these engines must operate between the prescribed temperatures ϑ_{in} and ϑ_{out}. We also require them to operate with a prescribed input of heat $\vartheta_{in}(\eta_{max} - \eta_{min})$. We conclude that the maximum amount of heat that can be converted into work is $(\vartheta_{in} - \vartheta_{out})(\eta_{max} - \eta_{min})$. We see that the Carnot engine converts the largest amount of heat into work. Other types of engines produce less work because their ϑ–η diagrams are not rectangular.

Despite its demonstrated usefulness, the topic of entropy still remains an elusive quantity to many students of thermodynamics. Innumerable books offer examples designed to clarify this topic. In particular, one of the many successes of the kinetic theory of gases is the quantitative analysis of entropy for a gas. You will find reading some of this work to be quite rewarding. However, here we shall follow Carnot's dictum, *speak little of what you know, and not at all of what you do not know*, and continue with our discussion of the theory of thermomechanics.

Production of Entropy

Let us next derive an expression for the production of entropy in an element of volume of a body in one dimension. If this element is heated by its surroundings, the entropy will change at a rate equal to the flow of heat divided by the temperature of the element. The element can also be heated by its internal dissipation of work. The dissipation $\mathcal{D}$ causes entropy to change at the rate $\mathcal{D}/\vartheta$. We illustrate each of the contributions to the production of entropy in Figure 3.5. Thus the production of entropy in the element is

$$\rho_0 \frac{\partial \hat{\eta}}{\partial t} dX\, dS = \left[\frac{q}{\vartheta} - \frac{1}{2} \frac{\partial}{\partial X} \widehat{\left(\frac{q}{\vartheta} \right)} dX \right] dS - \left[\frac{q}{\vartheta} + \frac{1}{2} \frac{\partial}{\partial X} \widehat{\left(\frac{q}{\vartheta} \right)} dX \right] dS + \frac{\rho_0 r}{\vartheta} dX dS + \frac{\rho_0 \mathcal{D}}{\vartheta}\, dX\, dS.$$

Simplification and division by $dX\, dS$ yields

$$\rho_0 \frac{\partial \hat{\eta}}{\partial t} = -\frac{\partial}{\partial X} \widehat{\left(\frac{q}{\vartheta} \right)} + \frac{\rho_0 r}{\vartheta} + \frac{\rho_0 \mathcal{D}}{\vartheta}, \tag{3.6}$$

which can also be written as

$$\rho \frac{\partial \hat{\eta}}{\partial t} = -\frac{\partial}{\partial x} \left(\frac{q}{\vartheta} \right) + \frac{\rho r}{\vartheta} + \frac{\rho \mathcal{D}}{\vartheta}. \tag{3.7}$$

† Actually, Carnot used quite different arguments to reach this conclusion because amazingly he did not know about the first law of thermodynamics.

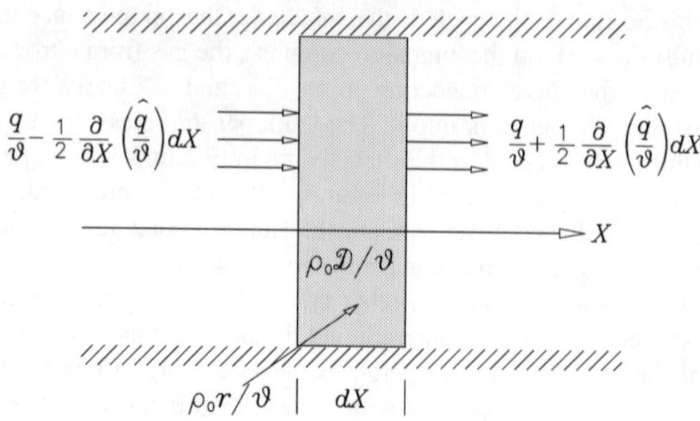

Figure 3.5 The contributions to the production of entropy.

These are the desired expressions that determine the production of entropy in terms of the heat flux, the external heat supply, and the dissipation. The word entropy means *transformability*, and as we proceed it will become apparent that η determines both the sign and the magnitude of $\mathcal{D}$. Thus η controls the amount of work that is dissipated as heat, and, as a result, it also controls the amount of internal energy remaining in the material that is available to do mechanical work. Thus from a purely pragmatic viewpoint, entropy is a variable we invent to describe heating and to control the transformation of energy within the material.

3.1.3 *First Law of Thermodynamics*

Let us review some of the quantities we have defined in the previous sections. A material can store energy in two ways, either as internal energy $\mathcal{E}$ or as kinetic energy $\mathcal{K}$. Both types of energy are represented by *specific* functions, meaning both represent energy per unit mass. We assign to them physical dimensions of J/kg. Thus $\mathcal{E}$ is often called the specific internal energy. In addition to storing energy, an element of material also exchanges energy with its surroundings. Energy can be exchanged either as work or as heat. The rate of exchange of work per unit mass is given by $\mathcal{P}$ in Eq. (3.2), and the rate of exchange of heat per unit mass is given by $\mathcal{Q}$ in Eq. (3.4). We have also introduced the dissipation $\mathcal{D}$. This variable does not represent an exchange of energy with the surroundings of the element. Rather, it represents the rate of dissipation of energy within the element.

Our objective now is to derive an equation connecting these quantities. We shall use the first law of thermodynamics to do so. This law states that energy can be neither created nor destroyed in the element. We shall derive our equation for this law, first in the material description and then in the spatial description.

Material Description

Consider an element of volume in the material description. The total energy of this element changes as it exchanges work and heat with its surroundings. The first law of thermodynamics states that the rate of change of the total energy within an element of the body is equal to the rate of exchange of work and heat with the surroundings:

$$\rho_0 \frac{\partial}{\partial t}(\hat{\mathcal{E}} + \hat{\mathcal{K}}) = \rho_0 \mathcal{P} + \rho_0 \mathcal{Q}. \tag{3.8}$$

Substitution of Eqs. (3.1) and (3.4) yields

$$\rho_0 \frac{\partial}{\partial t}(\hat{\mathcal{E}} + \hat{\mathcal{K}}) = \frac{\partial}{\partial X}(\hat{T}\hat{v}) - \frac{\partial \hat{q}}{\partial X} + \rho_0 b v + \rho_0 r. \tag{3.9}$$

This result, which includes both the internal and kinetic energies, is called the material description of the *balance of total energy*.

When Eqs. (3.2) and (3.4) are substituted into Eq. (3.8), we obtain

$$\rho_0 \frac{\partial \hat{\mathcal{E}}}{\partial t} = T \frac{\partial \hat{F}}{\partial t} - \frac{\partial \hat{q}}{\partial X} + \rho_0 r. \tag{3.10}$$

This expression does not include the kinetic energy. It is called the material description of the *balance of energy*.

Spatial Description

We leave it as an exercise to show that the balance of total energy in the spatial description is (see Exercise 3.5)

$$\frac{\partial}{\partial t}[\rho(\mathcal{E} + \mathcal{K})] + \frac{\partial}{\partial x}[\rho v(\mathcal{E} + \mathcal{K})] = \frac{\partial}{\partial x}(T v) - \frac{\partial q}{\partial x} + \rho b v + \rho r. \tag{3.11}$$

If the equations of conservation of mass and balance of linear momentum are applied to this result, the kinetic energy terms can be removed to obtain the spatial description of the balance of energy (see Exercise 3.6).

We can also derive the equation for the balance of energy directly from Eq. (3.10). Let us see how. Using Eq. (1.135) in Section 1.8.1, we obtain

$$\frac{\partial \hat{F}}{\partial t} = F L = \frac{\rho_0}{\rho} L.$$

Moreover, application of the chain rule yields

$$\frac{\partial \hat{q}}{\partial X} = \frac{\partial \hat{x}}{\partial X}\frac{\partial q}{\partial x} = \frac{\rho_0}{\rho}\frac{\partial q}{\partial x}$$

and

$$\frac{\partial \hat{\mathcal{E}}}{\partial t} = \frac{\partial \mathcal{E}}{\partial t} + v \frac{\partial \mathcal{E}}{\partial x}.$$

Substituting these expressions into Eq. (3.10), we obtain the spatial description of the balance of energy,

$$\frac{\partial \mathcal{E}}{\partial t} + v \frac{\partial \mathcal{E}}{\partial x} = \frac{TL}{\rho} - \frac{1}{\rho}\frac{\partial q}{\partial x} + r. \tag{3.12}$$

We can also write this result as

$$\rho\left(\frac{\partial \mathcal{E}}{\partial t} + v \frac{\partial \mathcal{E}}{\partial x}\right) = TD - \frac{\partial q}{\partial x} + \rho r,$$

where we have noticed that the deformation rate D is equal to the velocity gradient L [see Eq. (1.108)].

3.1.4 *Balance of Energy: Three-Dimensional Motion*

In the previous section, we derived the equation of balance of energy for a material. By restricting our discussion to one-dimensional motion, we were able to present simple derivations in both the material and spatial descriptions. In this section we repeat the derivation of the equation for the balance of energy for general three-dimensional motion. This time we employ the Gauss and transport theorems. Recall that these theorems are stated for a material volume $\mathcal{V}$; that is, $\mathcal{V}$ is a volume that always contains the same material. Moreover, these theorems can only be applied to volumes that contain solutions that are continuous and differentiable. Thus, in this section, we shall exclude the possibility of a discontinuous solution resulting from a shock wave. However, we shall return to this topic later in this chapter. Here we shall derive the differential equation for balance of energy in three dimensions using a method similar to our one-dimensional derivation. That is, first we derive an expression for the rate of work done on the material volume, then we determine the heating of the volume, next we define the production of entropy, and finally we derive the equation for the balance of energy.

Work

Work is done on the material volume by the surface tractions $t_m(\mathbf{n})$ and the external body force b_k. The rate at which work is done on the material volume is denoted by W, where

$$\begin{aligned} W &\equiv \int_{\mathcal{S}} t_m(\mathbf{n}) v_m \, dS + \int_{\mathcal{V}} \rho b_k v_k \, dV \\ &= \int_{\mathcal{S}} T_{km} v_m n_k \, dS + \int_{\mathcal{V}} \rho b_k v_k \, dV. \end{aligned}$$

Using the Gauss theorem, we obtain [see Eq. (1.147)]

$$W = \int_{\mathcal{V}} \left(v_m \frac{\partial T_{km}}{\partial x_k} + T_{km} L_{mk} + \rho b_k v_k \right) dV. \tag{3.13}$$

This is the required expression for the rate of work done on the material volume. Notice the similarity of this expression to the definition of work for one-dimensional motion, Eq. (3.1).

Heat

Heat is added to the material volume by the heat flux q_k and the external heat supply r. The rate at which heat is added to the material volume is denoted by H, where

$$H \equiv \int_{\mathcal{V}} \rho r \, dV - \int_{\mathcal{S}} q_k n_k \, dS.$$

Application of the Gauss theorem yields

$$H = \int_{\mathcal{V}} \left(\rho r - \frac{\partial q_k}{\partial x_k} \right) dV. \tag{3.14}$$

This is the required expression for the heating of the material volume. Notice the similarity of this expression to the definition of heating for one-dimensional motion, Eq. (3.4).

Entropy
In the material volume, entropy is produced by the heat flux, the external heat supply, and the dissipation. Thus

$$\frac{d}{dt}\int_{\mathcal{V}} \rho\eta\, dV = -\int_{\mathcal{S}} \frac{q_k}{\vartheta} n_k\, dS + \int_{\mathcal{V}} \frac{\rho r}{\vartheta} dV + \int_{\mathcal{V}} \frac{\rho \mathcal{D}}{\vartheta} dV. \tag{3.15}$$

By applying the Gauss theorem, the transport theorem, and the conservation of mass, we can demonstrate that

$$\int_{\mathcal{V}} \left[\rho \frac{\partial \hat{\eta}}{\partial t} + \frac{\partial}{\partial x_k}\left(\frac{q_k}{\vartheta}\right) - \frac{\rho r}{\vartheta} - \frac{\rho \mathcal{D}}{\vartheta}\right] dV = 0. \tag{3.16}$$

This equation must hold for every material volume. Because the integrand is continuous, this equation can be satisfied only if the integrand vanishes at each point:

$$\rho \frac{\partial \hat{\eta}}{\partial t} = -\frac{\partial}{\partial x_k}\left(\frac{q_k}{\vartheta}\right) + \frac{\rho r}{\vartheta} + \frac{\rho \mathcal{D}}{\vartheta}. \tag{3.17}$$

This is the three-dimensional version of the equation for the production of entropy, Eq. (3.7).

Equation of Balance of Energy
The total energy in the material volume is

$$E(t) = \int_{\mathcal{V}} \left(\frac{1}{2}\rho v_k v_k + \rho \mathcal{E}\right) dV.$$

By applying the transport theorem [see Eq. (1.149)] to the first term in the integral, we obtain

$$\begin{aligned} \frac{d}{dt}\int_{\mathcal{V}} \frac{1}{2}\rho v_k v_k\, dV &= \int_{\mathcal{V}} \left\{\frac{1}{2} v_k v_k \left[\frac{\partial \rho}{\partial t} + \frac{\partial(\rho v_m)}{\partial x_m}\right] + \rho v_k \left[\frac{\partial v_k}{\partial t} + \frac{\partial v_k}{\partial x_m} v_m\right]\right\} dV \\ &= \int_{\mathcal{V}} \rho v_k a_k\, dV, \end{aligned}$$

where we have used the equation for the conservation of mass [see Eq. (1.150)]. Similarly, we can show that

$$\frac{d}{dt}\int_{\mathcal{V}} \rho \mathcal{E}\, dV = \int_{\mathcal{V}} \rho \frac{\partial \hat{\mathcal{E}}}{\partial t} dV.$$

Therefore, the rate of change of the total energy in the material volume is

$$\frac{dE(t)}{dt} = \int_{\mathcal{V}} \left(\rho v_k a_k + \rho \frac{\partial \hat{\mathcal{E}}}{\partial t}\right) dV. \tag{3.18}$$

The balance of energy requires that this expression must be equal to the work and heat added to the material volume:

$$\frac{d\,E(t)}{dt} = W + H. \tag{3.19}$$

Substitution of Eqs. (3.13), (3.14), and (3.18) yields

$$\int_{\mathcal{V}} \left(\rho \frac{\partial \hat{\mathcal{E}}}{\partial t} - T_{km} L_{mk} + \frac{\partial q_k}{\partial x_k} - \rho r\right) dV = 0,$$

where we have used the equation for the balance of linear momentum [see Eq. (1.115)]. This equation must hold for every material volume. Because the integrand is continuous, this equation can be satisfied only if the integrand vanishes at each point:

$$\rho \frac{\partial \hat{\mathcal{E}}}{\partial t} = T_{km} D_{mk} - \frac{\partial q_k}{\partial x_k} + \rho r. \tag{3.20}$$

Here we have used the symmetry of the stress tensor to replace L_{mk} by D_{mk}. This is the equation for the balance of energy in three dimensions.

3.1.5 *Exercises*

3.1 The human body is a heat engine that performs work by burning food. Dietitians determine the energy content of food by measuring the heat produced as food is burned. They use the physical dimension of a "kilogram calorie" when assembling dietary tables of common foods. Usually, they delete the prefix "kilogram." Thus we might see a table entry such as

one jelly donut = 250 calories.

One kilogram calorie is equal to 1000 "gram calories." The gram calorie was a commonly used, scientific unit of energy prior to the introduction of SI units. Physical scientists also dropped their prefix "gram," so confusion arose when physical scientists talked to dietitians. We use SI units in this book, and our physical measure of energy is the joule. There are 4.186 joules in one gram calorie. Determine the number of joules in a jelly donut.

3.2 An average automobile weighs about 15 kN.

(a) Determine the number of joules of energy that are required to lift this automobile a vertical distance of 10 m.

(b) Determine the number of jelly donuts we must burn to get this amount of energy (see Exercise 3.1).

(c) What does this imply about exercise and human weight loss?

3.3 Consider a Carnot engine that contains 1 kg of air ($\gamma = 1.4$). We show the p–F/ρ_0 diagram for this engine in Figure 3.6. This engine operates in the clockwise direction. During the (a–b) and (c–d) portions of the cycle, the air in the cylinder obeys Boyle's law:

$$pF = \text{constant}, \qquad \vartheta = \text{constant}.$$

During the (b–c) and (d–a) portions of the cycle, the air in the cylinder behaves as an adiabatic gas:

$$pF^{\gamma} = \text{constant}, \qquad \vartheta F^{\gamma-1} = \text{constant}.$$

At point a, we know that $p_a = 11.7$ MPa, $\vartheta_a = 400$ K, and the volume of the cylindrical chamber is equal to 0.01 m^3. We have designed the engine so that $F_a = 1$, $F_b = 2$, and $F_c = 3$.

(a) Determine the values of ϑ at points b, c, and d.

(b) Determine the value of F_d.

(c) Determine the value of the shaded area in this diagram. What does this area represent?

(d) Determine the change in entropy of the gas as it moves from point a to point b.

(e) Determine the amount of heat we must put into the gas during the (a–b) portion of the cycle and the amount of heat we must remove from the gas during the (c–d) portion of the cycle.

(f) Explain how this device can be operated to cool a refrigerator.

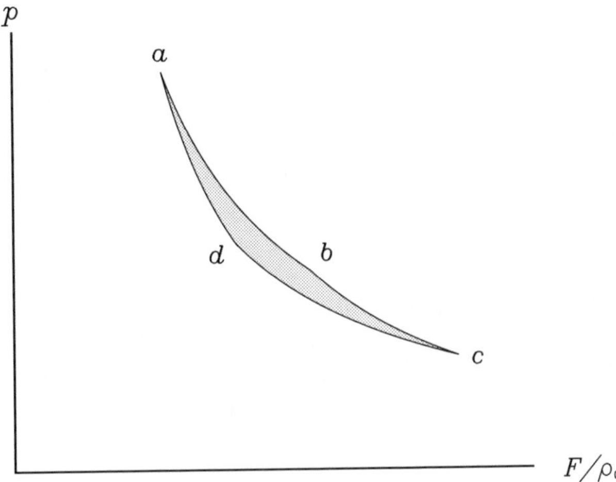

Figure 3.6 Exercise 3.3.

3.4 The production of entropy in an element of material is given by Eq. (3.6). This result is expressed in the material description. Verify that the production of entropy in the spatial description is given by Eq. (3.7).

3.5 In the spatial description, the differential equation for the balance of total energy is given by Eq. (3.11). Derive this equation by following the procedure we used to derive the one-dimensional, spatial descriptions of the equations for the conservation of mass and the balance of linear momentum (see Figure 1.37 in Section 1.8.2).

3.6 Use the equations for the conservation of mass and the balance of linear momentum to reduce the equation for the balance of total energy Eq. (3.11) to the equation for the balance of energy Eq. (3.12).

3.7 In Eq. (3.15), we postulate the relationship for the production of entropy for three dimensions. Use the Gauss theorem and the transport theorem to obtain Eq. (3.16) from this postulate.

3.8 Consider a material undergoing one-dimensional motion. Boundaries exist at x_L and x_R. Use Eq. (3.19) to show that

$$\frac{d}{dt}E(t) = T(x_R, t)v(x_R, t) - T(x_L, t)v(x_L, t)$$
$$- q(x_R, t) + q(x_L, t) + \int_{x_L}^{x_R} \rho(bv + r)\, dx. \tag{3.21}$$

3.2 Transformability of Energy

We have defined energy and shown how an element of material can exchange energy with its surroundings through working and heating. We described the change in internal energy of the element with two conjugate pairs. The conjugate pair (T, F) describes work, and the conjugate pair (ϑ, η) describes heat. The entropy of the element changes when heat is conducted or radiated to the element. The entropy of the element also changes when energy from work is dissipated. The dissipation $\mathcal{D}$ represents a transformation of energy within the element.

Individual materials transform energy between work and heat in different ways, and this difference must be reflected in the constitutive equations of the material. In this chapter we consider two mechanisms. The first mechanism is *viscosity*. By viscosity we simply mean that the stress depends upon the deformation rate D, where in one-dimensional motion

$$D = \frac{1}{F}\frac{\partial \hat{F}}{\partial t}.$$

We shall see that this results in a direct transformation of work into heat within an element of the material. The second mechanism is heat conduction. Because work results in temperature changes, heat conduction allows an element of material to exchange energy with neighboring elements. We shall use these two mechanisms to describe four general constitutive classes of materials. They are:

- *Thermoviscous*, where stress depends upon the deformation rate, and heat conduction is present.
- *Viscous*, where stress depends upon the deformation rate, and heat conduction is absent.
- *Thermoelastic*, where stress is independent of the deformation rate, but heat conduction is present.
- *Elastic*, where stress is independent of the deformation rate and entropy, and heat conduction is absent.

We considered the elastic material in detail in Chapter 2. You will recall that the constitutive model for an elastic material contains only one conjugate pair, (T, F). The constitutive models for the other three classes of material contain both conjugate pairs, (T, F) and (ϑ, η). In the next section we discuss the viscous material. We shall discuss the thermoelastic and thermoviscous materials later. However, for each constitutive model, we shall find that the forms of the constitutive equations are not arbitrary. Their forms are constrained by the *second law of thermodynamics*.

3.2.1 *Viscous Material*

With the possible exception of the nonlinear-elastic material, the viscous material is the constitutive model that is most commonly used to describe shock waves. This is because it is generally believed that heat conduction is not important to the analysis of a large majority of materials. To describe a viscous material, let us start by assuming the following functional dependence for the constitutive equations:

$$\mathcal{E} = \mathcal{E}(F, D, \eta), \tag{3.22}$$

$$T = T(F, D, \eta), \tag{3.23}$$

$$\vartheta = \vartheta(F, D, \eta), \tag{3.24}$$

$$q = 0. \tag{3.25}$$

Through the constitutive function $\mathcal{E}$ we prescribe the internal mechanism by which energy is stored in the material. With the functions T and ϑ, we describe the force terms of the conjugate pairs. Because this material cannot conduct heat, the heat flux q is zero.

Several important features about these assumptions should be noted. First, all of the dependent variables $\mathcal{E}$, ϑ, and T are functions of the same set of independent variables, F, D, and η. It is considered good practice to assume that each function depends on the same set of variables, unless you can prove otherwise. This practice has a name, *the principle of equipresence*. Next, we notice that the flux terms F and η of the two conjugate pairs appear as independent variables, whereas the force terms T and ϑ appear as dependent variables. Also, these constitutive assumptions are *rate sensitive*, meaning they depend upon the deformation rate D. Rate sensitivity is a crucial concept, which we use here to distinguish, for example, the behavior of materials subjected to steady waves from those subjected to structured waves. However, we emphasize that constitutive assumptions based upon a rate-sensitive model are not a unique choice. Often, people construct constitutive formulations by alternative means. For example, you might describe the stress in a material with the present value of the deformation gradient as well as past values of the deformation gradient. This will lead you to an alternate constitutive formulation of *materials with memory*.

3.2.2 *Admissible Thermodynamic Process*

Having defined an arbitrary set of constitutive functions for a viscous material, we must determine the constraints, if any, imposed on these functions by the second law of thermodynamics. The second law of thermodynamics requires the dissipation to be positive or zero:

$$\mathcal{D} \geq 0.$$

This means we must construct constitutive functions that ensure positive dissipation for any possible solution to the governing equations. We must consider all possible solutions without regard to the initial conditions, the boundary conditions, or the shape of the body. In this section we identify all of these possible solutions. Taken together, they are called the *admissible thermodynamic processes.*

Governing Equations

The viscous material is described by eight equations. They are the three constitutive equations,

$$\mathcal{E} = \mathcal{E}(F, D, \eta), \qquad T = T(F, D, \eta), \qquad \vartheta = \vartheta(F, D, \eta); \tag{3.26}$$

the definitions of velocity, deformation gradient, and deformation rate,

$$v = \frac{\partial \hat{x}}{\partial t}, \qquad F = \frac{\partial \hat{x}}{\partial X}, \qquad D = \frac{1}{F}\frac{\partial \hat{F}}{\partial t}; \tag{3.27}$$

the equation for the balance of linear momentum [see Eq. (1.144)],

$$\rho_0 \frac{\partial \hat{v}}{\partial t} = \frac{\partial \hat{T}}{\partial X} + b; \tag{3.28}$$

and the equation for the balance of energy Eq. (3.10),

$$\frac{\partial \hat{\mathcal{E}}}{\partial t} = \frac{TD}{\rho} + r. \tag{3.29}$$

States and Processes

There are eight unknowns, x, v, F, D, η, T, $\mathcal{E}$, and ϑ, in the governing equations. We can obtain solutions by prescribing the boundary conditions for the body, the initial conditions for the body, the body force b, and the external heat supply r. The solutions are functions of two independent variables X and t. The set of values obtained from these functions for a particular choice of X and t is called a *state*. When we watch the state at a fixed value of X, the evolution of the state values is called a *process*.

By varying our choices for b, r, the shape of the body, the initial conditions, and the boundary conditions, we can produce any one of an infinite number of processes at the point X. Most of these processes involve transformation of energy between work and heat. In the next section we shall see that some types of energy transformation are not allowed. To examine the transformability of energy within a material, we must first identify all possible processes. The only restriction we shall place on these processes is that they must satisfy the eight governing equations. Any process that meets these requirements is called an *admissible thermodynamic process.*

Defining a Process

We now show that any definition of the motion $\hat{x}(X,t)$ and the entropy $\hat{\eta}(X,t)$ throughout the volume of the material results in an admissible thermodynamic process. After we have chosen the functions $\hat{x}(X,t)$ and $\hat{\eta}(X,t)$, we satisfy the eight governing equations as follows:

- We determine the velocity v and deformation gradient F from their definitions.
- We evaluate the acceleration $\partial\hat{v}/\partial t$, deformation rate D, and entropy rate $\partial\hat{\eta}/\partial t$.
- We use the constitutive assumptions to evaluate the internal energy $\mathcal{E}$, temperature ϑ, and stress T.
- We evaluate $\partial\hat{\mathcal{E}}/\partial t$ and $\partial\hat{T}/\partial X$.
- We select the body force b so that the equation for the balance of linear momentum is satisfied.
- We select the external heat supply r so that the equation for the balance of energy is satisfied.

It is important to notice that we can independently choose the motion $\hat{x}(X,t)$ and the entropy $\hat{\eta}(X,t)$, and for each of our choices there is a corresponding admissible thermodynamic process.

3.2.3 *Second Law of Thermodynamics*

The second law of thermodynamics states that the dissipation must be positive or zero:

$$\mathcal{D} \geq 0. \tag{3.30}$$

Because the temperature is positive and the heat flux is zero, we can substitute Eq. (3.6) to obtain

$$\vartheta \frac{\partial\hat{\eta}}{\partial t} \geq r. \tag{3.31}$$

Thus entropy cannot decrease unless heat is removed from the material.

It was first proposed by Coleman and Noll† that we should view the second law of thermodynamics as a constraint on the constitutive assumptions. This means that we must construct our constitutive equations so that Eq. (3.31) *is never violated by any admissible thermodynamic process.* To determine these constraints, we first solve the balance of energy Eq. (3.10) for r and substitute the result into Eq. (3.31):

$$\mathcal{D} = -\frac{\partial \mathcal{E}}{\partial D}\frac{\partial \hat{D}}{\partial t} + \left(\vartheta - \frac{\partial \mathcal{E}}{\partial \eta}\right)\frac{\partial \hat{\eta}}{\partial t} + \left(\frac{T}{\rho} - F\frac{\partial \mathcal{E}}{\partial F}\right)D \geq 0. \tag{3.32}$$

Here $q = 0$, and the time derivative of the internal energy has been expanded by the chain rule.

Our objective now is to find a set of admissible thermodynamic processes that violate this inequality. Consider the processes defined by

$$x = \hat{x}(X, t) = A(t)X, \qquad \eta = \hat{\eta}(X, t) = B(t),$$

where we can arbitrarily choose the functions $A(t)$ and $B(t)$. For these processes

$$F = A, \qquad D = \frac{1}{A}\frac{dA}{dt}, \qquad \frac{\partial \hat{D}}{\partial t} = \frac{1}{A^2}\left[A\frac{d^2A}{dt^2} - \left(\frac{dA}{dt}\right)^2\right], \qquad \frac{\partial \hat{\eta}}{\partial t} = \frac{dB}{dt},$$

and Eq. (3.32) becomes

$$-\frac{1}{A^2}\frac{\partial \mathcal{E}}{\partial D}\left[A\frac{d^2A}{dt^2} - \left(\frac{dA}{dt}\right)^2\right] + \left[\vartheta - \frac{\partial \mathcal{E}}{\partial \eta}\right]\frac{dB}{dt} + \frac{1}{A}\left[\frac{T}{\rho} - F\frac{\partial \mathcal{E}}{\partial F}\right]\frac{dA}{dt} \geq 0.$$

At any given time t, we can independently choose the values of A, B, dA/dt, dB/dt, and d^2A/dt^2. We notice that d^2A/dt^2 explicitly appears in this inequality, but no other term depends upon d^2A/dt^2. Consequently, we can choose different processes in which d^2A/dt^2 assumes arbitrarily large positive or negative values without changing any other quantity in the expression. We see that some of these processes will violate the inequality unless we require that

$$\frac{\partial \mathcal{E}}{\partial D} = 0.$$

Thus $\mathcal{E}$ cannot be a function of D, and

$$\mathcal{E} = \mathcal{E}(F, \eta).$$

Similarly, we can select different processes in which dB/dt assumes arbitrarily large positive or negative values, and some of these processes will violate the inequality unless we require that

$$\vartheta = \frac{\partial \mathcal{E}}{\partial \eta}.$$

After we apply these two constraints to the constitutive equations, the remaining term in the inequality is

$$\left(\frac{T}{\rho} - F\frac{\partial \mathcal{E}}{\partial F}\right)D \geq 0.$$

† See "The thermodynamics of elastic materials with heat conduction and viscosity" by Coleman, B. D., and Noll, W. in *Archives of Rational Mechanics and Analysis* **13**, pp. 167–184 (1963).

Because T depends on D through its constitutive equation, we cannot apply the preceding arguments to this expression.

The Constrained Constitutive Assumptions

By examining a special set of processes, we have determined that our original assumptions about the constitutive equations Eqs. (3.26) are too general and must be replaced by the following equations:

$$\mathcal{E} = \mathcal{E}(F, \eta), \tag{3.33}$$

$$T = T(F, D, \eta), \tag{3.34}$$

$$\vartheta = \partial \mathcal{E} / \partial \eta. \tag{3.35}$$

When these relations are substituted into Eq. (3.32), we obtain the *reduced dissipation inequality*:

$$\mathcal{D} = \left(\frac{T}{\rho} - F \frac{\partial \mathcal{E}}{\partial F} \right) D \geq 0. \tag{3.36}$$

In a viscous material, only the stress can be a function of D, and it must be of a form that ensures that the reduced dissipation inequality is always satisfied.

Because $\mathcal{E}$ can only be a function of F and η, we can apply the chain rule to Eq. (3.29) to obtain the following equation for the balance of energy:

$$\vartheta \frac{\partial \hat{\eta}}{\partial t} - r = \left(\frac{T}{\rho} - F \frac{\partial \mathcal{E}}{\partial F} \right) D. \tag{3.37}$$

Stress for Nondissipative Material

Consider a material where the stress is *not* a function of D. When we examine the reduced dissipation inequality, Eq. (3.36), we find that D appears explicitly in this inequality while no other term in the inequality depends on D. Because the sign of D is arbitrary, the dissipation inequality will be violated unless we require that

$$T = \rho_0 \frac{\partial \mathcal{E}}{\partial F}, \qquad \text{when } T = T(F, \eta). \tag{3.38}$$

As a result of this constraint, the dissipation $\mathcal{D}$ is identically zero. For this reason, this material is called *nondissipative*. You will remember that we defined a reversible process in Section 3.1.2. In a reversible process, the dissipation $\mathcal{D}$ is always zero. Thus a nondissipative material always exhibits reversible processes.

Definition of Equilibrium

For a viscous material, we define an *equilibrium state* to be one in which $D = 0$. The stress at this state is

$$T = T(F, 0, \eta). \tag{3.39}$$

Reversible Process

A process in which F evolves very slowly so that D can be approximated as zero is often called a *quasistatic process*. During a quasistatic process, the stress in a viscous material is still given by the equilibrium relation Eq. (3.39). During this process, the reduced dissipation inequality for the viscous material, Eq. (3.36), becomes

$$\rho_0 \mathcal{D} = [T(F, 0, \eta) - T_e] F D \geq 0,$$

where

$$T_e \equiv \rho_0 \frac{\partial \mathcal{E}}{\partial F}. \tag{3.40}$$

Notice that D appears explicitly in this inequality while no other term in the inequality depends on D. Because the sign of D is arbitrary, a quasistatic process can violate this inequality unless

$$T = T(F, 0, \eta) = T_e. \tag{3.41}$$

We also conclude that $\mathcal{D} = 0$, and thus a quasistatic process in a viscous material is a reversible process. This expression for the stress in a viscous material during a reversible process is similar to the expression for the stress in a nondissipative material, Eq. (3.38). However, there is an important difference: In a nondissipative material, the stress is *always* equal to the *equilibrium stress*. In a viscous material, the stress is *only* equal to the equilibrium stress during a reversible process.

After substitution of Eq. (3.41) into the equation for the balance of energy, Eq. (3.37), we obtain the change of entropy for a reversible process,

$$\vartheta \frac{\partial \hat{\eta}}{\partial t} = r. \tag{3.42}$$

During a reversible process, entropy can change only when heat is supplied from an external source.

Viscous Stress

Because of Eq. (3.41), we can represent the stress T by the additive decomposition

$$T = T_e + \mathcal{T},$$

where

$$\mathcal{T} = \mathcal{T}(F, D, \eta)$$

is called the *viscous stress* and

$$\mathcal{T}(F, 0, \eta) = 0.$$

Energy and Dissipation Relations

In terms of the equilibrium and viscous stresses, the equation for the balance of energy, Eq. (3.37), becomes

$$\begin{aligned} \rho\left(\vartheta \frac{\partial \hat{\eta}}{\partial t} - r\right) &= (T - T_e) D \\ &= \mathcal{T} D, \end{aligned} \tag{3.43}$$

and the reduced dissipation inequality, Eq. (3.36), becomes

$$\begin{aligned} \rho \mathcal{D} &= (T - T_e) D \geq 0 \\ &= \mathcal{T} D \geq 0. \end{aligned} \tag{3.44}$$

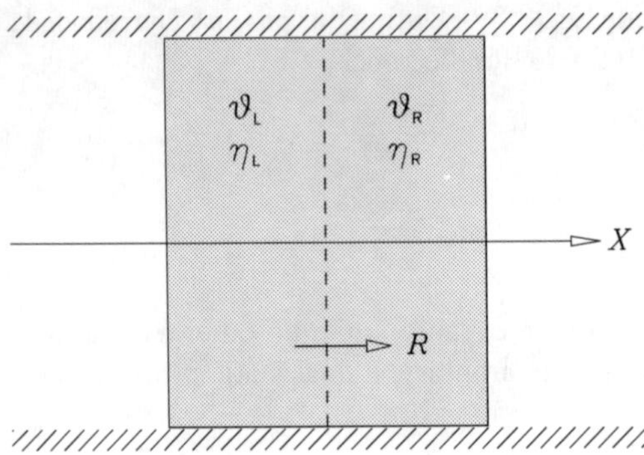

Figure 3.7 Heat flow between two halves of a volume element.

The Demon

The first and second laws of thermodynamics lead to an interesting conclusion about heat flow. Consider an element of material that has mass dm and does not exchange heat with its surroundings. The second law of thermodynamics, Eq. (3.31), yields the following inequality for the entropy of this element:

$$\frac{\partial \hat{\eta}}{\partial t} \geq 0. \tag{3.45}$$

Suppose that this element is divided into two halves, each with mass $dm/2$ (see Figure 3.7). Furthermore, let us assume that each half undergoes a reversible process to exchange heat between the left half and the right half. We use the symbol R to represent the rate of heat flow. We define R to be positive when heat flows from left to right. We denote the temperature and entropy of the left half by ϑ_L and η_L. For the right half, we use ϑ_R and η_R. From Eq. (3.42), the equations for the balance of energy for each half of the element are

$$\frac{\partial \hat{\eta}_L}{\partial t} = -\frac{R}{\vartheta_L}, \qquad \frac{\partial \hat{\eta}_R}{\partial t} = \frac{R}{\vartheta_R}. \tag{3.46}$$

Entropy is a specific quantity, which means it is measured per unit mass. Because the mass of each half of the element is $dm/2$, the total entropy η of the element is

$$\eta = \frac{1}{2}(\eta_L + \eta_R).$$

Combination of this result with Eqs. (3.45) and (3.46) yields

$$R\left(\frac{1}{\vartheta_R} - \frac{1}{\vartheta_L}\right) \geq 0.$$

If $\vartheta_L \geq \vartheta_R$, then $R \geq 0$, and if $\vartheta_L \leq \vartheta_R$, then $R \leq 0$. Simply said, this means that the heat must flow from the hotter half to the colder half and cause the temperatures of both halves to converge to a common value. Notice that, to obtain this result, we used Eq. (3.42), which is the equation for the balance of energy for a reversible process. Thus we have assumed that each half undergoes a reversible process. If we drop this assumption, this result cannot be obtained. This is important, because heat has been observed to flow from a cold body to a hot body during certain irreversible processes.

In the nineteenth century, the physicist and mathematician James Clerk Maxwell considered a similar problem. Referring to a gas in a vessel at a uniform temperature, he said,

> Now let us suppose that such a vessel is divided into two portions, A and B, by a division in which there is a small hole, and that a being, who can see the individual molecules, opens and closes this hole, so as to allow only the swifter molecules to pass from A to B, and only the slower ones to pass from B to A. He will thus, without expenditure of work, raise the temperature of B and lower that of A, in contradiction to the second law of thermodynamics.

Today, this "being" is called Maxwell's Demon (see Figure 3.8). Many have searched for this demon, but none have found him.

3.2.4 *Partial Derivatives in Thermodynamics*

You will recall that we use a special notation to distinguish between the time derivatives in the spatial and material descriptions. For example, the time derivative of the displacement in the material description is equal to the velocity. Thus

$$v = \frac{\partial \hat{u}}{\partial t} \neq \frac{\partial u}{\partial t}.$$

We use the circumflex to make the very important distinction between these two types of derivatives. We discuss this in detail in Section 1.2.2, and you may wish to read that section again.

Consider now that six variables $\mathcal{E}, T, \vartheta, F, D$, and η appear in the constitutive equations for a viscous material. This gives rise to many other possible combinations of these variables to produce other constitutive functions. Thus, just as the partial derivatives of the two functions $\hat{u}(X, t)$ and $u(x, t)$ yield different results, other constitutive functions for the same quantity, say $\mathcal{E}$, will also yield different partial derivatives. So far we have avoided this problem by limiting ourselves to only one possible formulation. In particular, we have assumed that $\mathcal{E} = \mathcal{E}(F, \eta)$, and from the dissipation inequality, we find that

$$T_e = \rho_0 \frac{\partial \mathcal{E}}{\partial F}, \qquad \vartheta = \frac{\partial \mathcal{E}}{\partial \eta}.$$

Because we know the functional dependence of $\mathcal{E}$, we know that η is constant for the first partial derivative and F is constant for the next partial derivative. Furthermore,

$$T_e = T_e(F, \eta), \qquad \vartheta = \vartheta(F, \eta). \tag{3.47}$$

Thus, if we define a new term C^η, where

$$C^\eta \equiv F \frac{\partial T_e}{\partial F},$$

we know that η is to be held constant for this partial derivative. So far, everything is clear because we have explicitly stated the functional dependence of $\mathcal{E}$ and stayed with this explicit form.

Figure 3.8 Maxwell's Demon. With a total lack of respect for the second law of thermodynamics, the demon separates fast moving molecules from slow moving molecules, causing the temperatures of two portions of a container to spontaneously diverge. Reprinted with kind permission of the artist Art Bailey and Tom Wright, Sandia National Laboratories.

However, T_e is also an implicit function of ϑ. You can see this by solving $\vartheta = \vartheta(F, \eta)$ for η and substituting the result into the first of Eq. (3.47). Then we obtain

$$T_e = T_e(F, \vartheta).$$

Now we see that for this stress function

$$C^\eta \neq F\frac{\partial T_e}{\partial F},$$

because $\partial T_e(F,\eta)/\partial F \neq \partial T_e(F,\vartheta)/\partial F$. Thus, if we include implicit functional dependences, pretty soon we will lose sight of which variable we must hold constant while taking a partial derivative. One way out of this difficulty is to use the notation

$$C^\eta = F\left(\frac{\partial T_e}{\partial F}\right)_\eta \neq F\left(\frac{\partial T_e}{\partial F}\right)_\vartheta.$$

Indeed, this is the notation adopted in most textbooks on thermodynamics. However, if the stress is a function of a large number of variables, this notation can get quite cumbersome; for example, in Chapter 4, we might have to write

$$C_s^\eta \equiv F_s\left(\frac{\partial T_s}{\partial F}\right)_{\eta_s,\eta_f,\bar{\rho}_f,\varphi_s^P,\varphi_s^E}$$

for a porous solid.

In this book, *we shall avoid this problem by excluding partial derivatives based on implicit functional dependences*. Thus, for example, when we discuss a material where the constitutive relationship for temperature is $\vartheta = \vartheta(F,\eta)$, then we shall use only two partial derivatives – $\partial\vartheta/\partial\eta$ with F held constant and $\partial\vartheta/\partial F$ with η held constant. We shall not use the partial derivatives $\partial\vartheta/\partial T_e$ and $\partial\eta/\partial\vartheta$ because ϑ is not an explicit function of T_e and because η is an independent variable. Therefore, throughout this work, to clarify the meaning of any partial derivative, you need only refer back to the relevant constitutive equations. We shall periodically do this to clarify the meaning of important partial derivatives.

3.2.5 *Exercises*

3.9 In Exercise 2.65 in Section 2.9.5 we considered the constitutive model

$$T = E_0\left(1 - \frac{1}{F}\right) + \frac{b}{F}\frac{\partial \hat{F}}{\partial t},$$

where E_0 and b are prescribed constants.

(a) Determine the equilibrium stress.

(b) Use the reduced dissipation inequality for a viscous material to determine the constraint on the value of b.

3.10 Show that the three-dimensional form of the second law of thermodynamics is

$$\rho\frac{\partial\hat{\eta}}{\partial t} \geq \frac{\rho r}{\vartheta} - \frac{\partial}{\partial x_k}\left(\frac{q_k}{\vartheta}\right).$$

3.11 Consider a viscous material with the following three-dimensional constitutive equations:

$$\mathcal{E} = \mathcal{E}(F_{ij},\eta), \qquad T_{mj} = T_{mj}(F_{ik}, D_{rs}, \eta),$$
$$\vartheta = \vartheta(F_{ij},\eta), \qquad q = 0.$$

Assume that the stress tensor is symmetric.

(a) Use the dissipation inequality from the previous exercise to show that

$$\vartheta = \frac{\partial\mathcal{E}}{\partial\eta}.$$

(b) Demonstrate that the equilibrium stress for this material is

$$T_{mj} = \rho F_{jr} \frac{\partial \mathcal{E}}{\partial F_{mr}}.$$

Compare this result to Eq. (2.14) in Section 2.1.3.

3.3 Equilibrium States and Processes

The constitutive equations for a viscous material are quite general and can represent solids, liquids, and gases. In Chapter 4 we shall study some specific examples that describe particular types of materials. However, for the present, we have more to learn about the general constitutive formulation of the viscous material. In this section we shall show how we can describe the internal energy with a curved surface in the space defined by coordinates $(\mathcal{E}, \eta, F)$. We shall also show that states are points on this surface and all processes are lines on this surface. Special types of processes have names, and we shall study each of these named processes.

In the initial constitutive formulation for the viscous material, Eqs. (3.22)–(3.24), we had assumed that the internal energy was a rate-sensitive function and that the temperature could be independently prescribed. Then we turned our attention to the second law of thermodynamics. We interpreted this law to mean all materials must be dissipative; that is, their constitutive equations must ensure positive dissipation for any possible process. This led us to a constrained set of constitutive equations, Eqs. (3.33)–(3.35). We found that the internal energy cannot be a function of D and that the temperature cannot be independently prescribed. It must be given by $\vartheta = \partial\mathcal{E}/\partial\eta$. Furthermore, for a reversible process, the stress cannot be independently prescribed. It must be given by $T = \rho_0 \partial\mathcal{E}/\partial F$. Thus to define an equilibrium state or a reversible process, *the only constitutive equation we need to know is the equation for the internal energy* $\mathcal{E}$. Sometimes the constitutive equation for the internal energy is called the *equation of state*. Sometimes all of the constitutive equations are called *equations of state*.

Clearly, the constitutive equation for the internal energy is fundamental to our understanding of the constitutive behavior of a viscous material. We assume the function $\mathcal{E}$ is continuous and sufficiently smooth to yield continuous and differentiable values of ϑ and T_e everywhere. It forms a surface in the three-dimensional space defined by the coordinates $(\mathcal{E}, \eta, F)$ (see Figure 3.9). As we shall see, it is possible to mathematically represent this same three-dimensional surface not only with the function $\mathcal{E}$, but with other mathematical descriptions. In this book, we shall use the terms constitutive equation and constitutive model to refer to the different mathematical descriptions. We shall reserve the term *state surface* to refer to the three-dimensional surface described by these different constitutive equations.

We can draw a variety of two-dimensional topographical diagrams of the state surface. We can construct these diagrams by drawing contours on the state surface and then projecting them to any one of the flat surfaces of a cube circumscribed about the state surface. For example, we show lines of constant entropy on the state surface in Figure 3.9 that are similar to elevation contours on a geographical map of a mountain. Then we form a diagram by projecting these contours onto the upper flat face of the cube. We call this an $\mathcal{E}$–F diagram. An $\mathcal{E}$–η diagram is obtained by projecting the contours to the left vertical face of the cube.

We represent a state of the material as a point both on the state surface and on the projected diagrams. Wave phenomena cause the state to move across the state surface. The

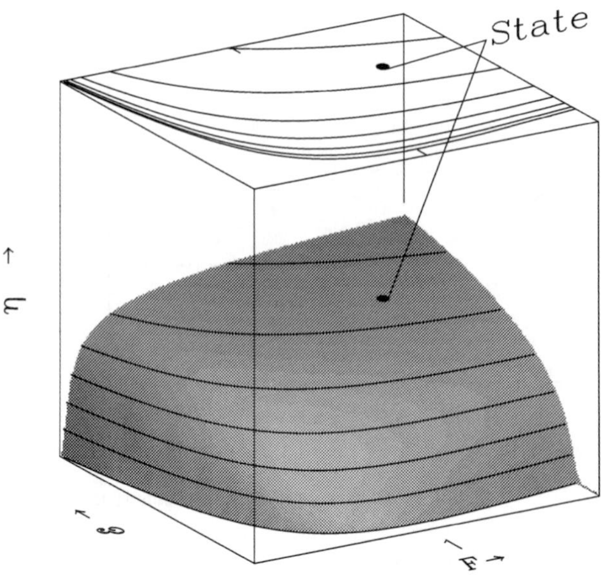

Figure 3.9 The state surface with isentropes (—).

motion of the state is described by the equation for the balance of energy, Eq. (3.10), and the path followed by the state is called a process. When the position of the state does not change with time, it is an equilibrium state. When the state moves sufficiently slowly so that $T = T_e$, the path it follows represents a reversible process. Because of Eq. (3.33), *all processes describe paths on the state surface*, and Eq. (3.35) requires that the temperature is always given by the partial derivative $\partial\mathcal{E}/\partial\eta$. This is true even for processes produced by strong shock waves.

In the following sections we study some important processes. We start by examining four reversible processes. Recall that we describe the material with two conjugate pairs, (T, F) and (ϑ, η). The four reversible processes are directly connected to the four variables in the conjugate pairs. The first reversible process is called an *isentropic process*. Here η is constant. This process describes the stiffness of the material when heat is not radiated and work is not dissipated. The second reversible process is called an *isochoric process*. Here F is constant. This process describes how the material absorbs heat when it is not allowed to expand. The third reversible process is called an *isothermal process*. Here ϑ is constant. This process describes the stiffness of the material when the temperature is held constant. The fourth reversible process is called an *isobaric process*. Here T is constant. This process describes how the material absorbs heat when it is allowed to expand against a constant stress.

3.3.1 *Isentropic Process*

For a reversible process, Eq. (3.41) gives

$$T = T_e = \rho_0 \frac{\partial\mathcal{E}}{\partial F}. \tag{3.48}$$

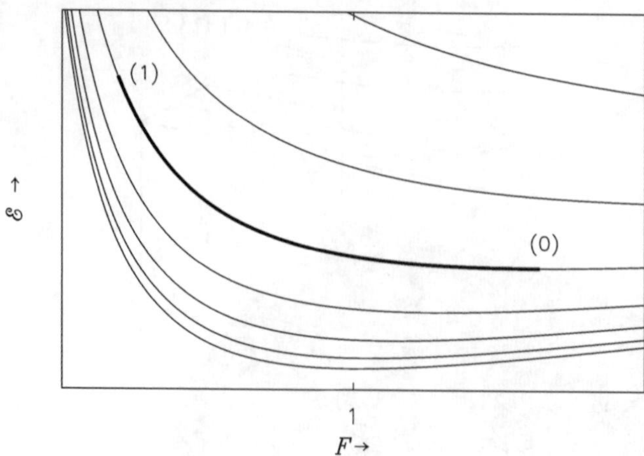

Figure 3.10 The $\mathcal{E}$–F diagram of an isentropic process connecting state (0) and state (1).

Thus the right side of the equation for the balance of energy, Eq. (3.37), is zero. When $r = 0$, we find that the equation for the balance of energy becomes

$$\frac{\partial \hat{\eta}}{\partial t} = 0.$$

We call this type of deformation an *isentropic process*, and we call the path it follows an *isentrope*. The contours illustrated in Figure 3.9 are isentropes. We show the $\mathcal{E} - F$ diagram of the isentropes with an isentropic process connecting state (0) with state (1) (see Figure 3.10). On an isentrope, the internal energy is a function of F alone, and from Eq. (3.48), the equilibrium stress is also a function of F alone. Recall that $T = T(F)$ is the constitutive equation for an elastic material. Thus elastic and isentropic processes are similar in that each occurs in the absence of heating and under a stress that only depends upon F.

Stiffness at Constant Entropy

As the material follows an isentropic process, we can use the chain-rule expansion of Eq. (3.48), $T = T_e(F, \eta)$, to obtain

$$\frac{\partial \hat{T}_e}{\partial t} = C^\eta D \text{ (isentropic process),} \tag{3.49}$$

where

$$C^\eta \equiv F\frac{\partial T_e}{\partial F} = \rho_0 F\frac{\partial^2 \mathcal{E}}{\partial F^2} \tag{3.50}$$

is called the *isentropic longitudinal stiffness*. The inverse of this quantity is called the *isentropic longitudinal compressibility*.

Geometry of the Isentrope

The plot of a portion of a single isentrope in the $\mathcal{E}$–F diagram reveals several interesting geometrical properties of the isentrope. In Figure 3.11 we plot the dimensionless quantity

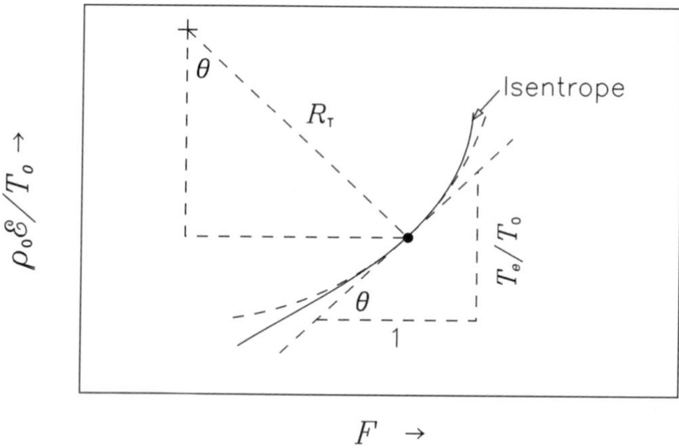

Figure 3.11 Geometry of an isentrope.

$\rho_0\mathcal{E}/T_0$ against F. We use an arbitrary reference stress T_0 to normalize the internal energy and to create a dimensionless graph. The slope of this isentrope at the state point is

$$\frac{\partial}{\partial F}\left(\frac{\rho_0\mathcal{E}}{T_0}\right) = \frac{T_e}{T_0}.$$

We use a right triangle with a hypotenuse that is tangent to the isentrope to illustrate this slope. The length of the two legs are equal to 1 and T_e/T_0, respectively. We denote the slope of the tangent with the angle θ, where

$$\theta \equiv \tan^{-1}(T_e/T_0).$$

We illustrate the curvature of the isentrope at the state point with a dashed circular arc. We use R_T to denote the radius of the arc. If s is a measure of length along the circular arc, then we find that

$$\frac{1}{R_T} \equiv \frac{d\theta}{ds} = \frac{dF}{ds}\frac{\partial\theta}{\partial F} = \cos\theta\left(\cos^2\theta\,\frac{\rho_0}{T_0}\frac{\partial^2\mathcal{E}}{\partial F^2}\right),$$

which gives

$$\frac{C^\eta}{T_0} = \frac{F}{R_T\cos^3\theta}. \tag{3.51}$$

Thus the slope of the isentrope gives the equilibrium stress in the material, and the slope and radius of curvature of the isentrope give the isentropic longitudinal stiffness. We see from this illustration that a straight isentrope indicates a state of zero stiffness $C^\eta = 0$, and a horizontal isentrope indicates a state of zero equilibrium stress, $T_e = 0$. Thus state (0) in Figure 3.10 is stress free, and state (1) is under compressive stress. Also, the stiffness at state (1) is greater than the stiffness at state (0). When the concave side of the isentrope faces the positive-$\mathcal{E}$ direction, the isentropic longitudinal stiffness C^η is positive; otherwise C^η is negative. All of the isentropes in Figure 3.10 have positive stiffness everywhere.

Temperature Change

Most processes cause the temperature to change. Let us determine the temperature change during an isentropic process. From Eq. (3.35), we see that

$$\vartheta = \vartheta(F, \eta).$$

The chain rule expansion of $\partial\hat{\vartheta}/\partial t$ along an isentrope yields

$$\frac{1}{\vartheta}\frac{\partial\hat{\vartheta}}{\partial t} = -\Gamma D, \tag{3.52}$$

where the *Grüneisen coefficient* is defined to be

$$\Gamma \equiv -\frac{F}{\vartheta}\frac{\partial\vartheta}{\partial F} = -\frac{F}{\vartheta}\frac{\partial^2\mathcal{E}}{\partial F\partial\eta}. \tag{3.53}$$

It is generally observed that temperature increases for $D < 0$. The use of a minus sign in this definition normally results in positive values for Γ. Although the Grüneisen coefficient yields a simple relationship for the change in temperature during an isentropic process, we shall see that it also describes temperature changes in other processes.

3.3.2 *Isochoric Process*

An isochor is a process that preserves volume. During uniaxial deformation, volume is constant when $D = 0$. We plot isochors on the state surface as solid lines (see Figure 3.12). We show the $\mathcal{E}$–η diagram of the isochors in Figure 3.13. Each isochor represents a different constant value of F. We produce an isochoric process by heating the material with the external supply r and moving the material from state (0) to state (1). The temperature of the material as it follows the isochoric process is given by $\partial\mathcal{E}/\partial\eta$, which is

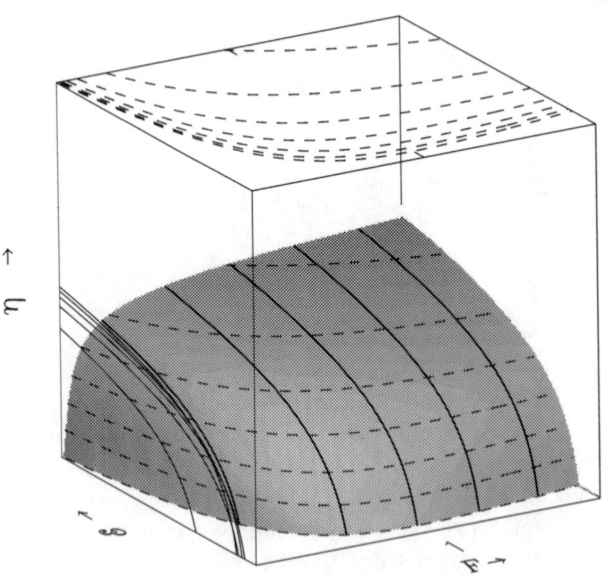

Figure 3.12 The state surface with isochors (—) and isentropes (– –).

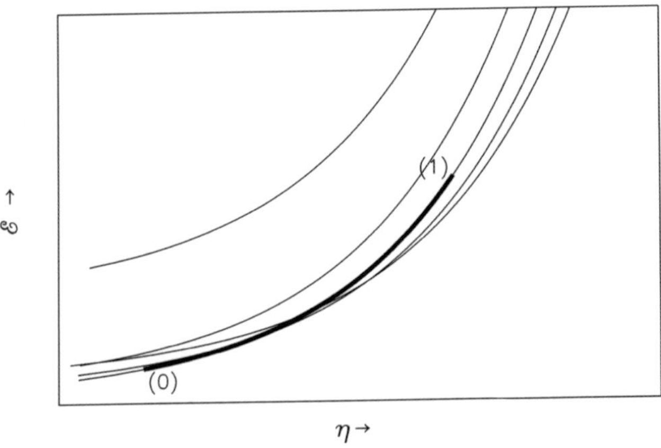

Figure 3.13 The $\mathcal{E}$–η diagram of an isochoric process connecting state (0) and state (1).

the slope of the isochor in the $\mathcal{E}$–η diagram. We see that as the material moves from state (0) to state (1) the temperature increases.

Specific Heat at Constant Volume

Along the isochoric process we can derive a simple relationship between the change in temperature ϑ and the supplied heat r. From the equation for the balance of energy, Eq. (3.37), we obtain

$$\vartheta \frac{\partial \hat{\eta}}{\partial t} = r. \tag{3.54}$$

Because $\vartheta = \vartheta(F, \eta)$, the chain-rule expansion of the temperature along an isochor is

$$\frac{\partial \hat{\vartheta}}{\partial t} = \frac{\partial \vartheta}{\partial \eta} \frac{\partial \hat{\eta}}{\partial t}.$$

If we define the *isochoric specific heat* c_v with

$$\frac{\vartheta}{c_v} \equiv \frac{\partial \vartheta}{\partial \eta} = \frac{\partial^2 \mathcal{E}}{\partial \eta^2}, \tag{3.55}$$

the equation for the balance of energy becomes

$$c_v \frac{\partial \hat{\vartheta}}{\partial t} = r \quad \text{(isochoric process)}. \tag{3.56}$$

Hence when we add heat to a body whose volume is held fixed, the isochoric specific heat c_v determines the change in temperature of the body.

Geometry of the Isochor

The plot of a portion of a single isochor in the $\mathcal{E}$–η diagram reveals several interesting geometrical properties (see Figure 3.14). We plot the quantity $\mathcal{E}/\vartheta_0$ against η. In this

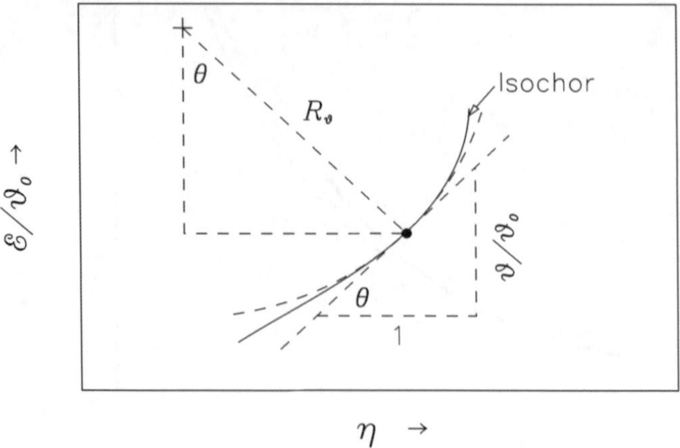

Figure 3.14 Geometry of an isochor.

plot, both the ordinate and abscissa have physical dimensions of entropy. The reference temperature ϑ_0 is an arbitrary quantity. The slope of this isochor at the state point is

$$\frac{\partial(\mathcal{E}/\vartheta_0)}{\partial\eta} = \frac{\vartheta}{\vartheta_0}.$$

We use a right triangle with a hypotenuse tangent to the isochor to illustrate this slope, and we denote the slope of the tangent with the angle θ. We illustrate the curvature of the isochor at the state point with a dashed circular arc, and we denote the radius of the arc as R_ϑ. By the procedure we used in the last section, we obtain

$$c_v = R_\vartheta \sin\theta \cos^2\theta. \tag{3.57}$$

Thus the slope of the isochor gives the temperature of the material, and the slope and radius of curvature of the isochor give the specific heat of the material. We see from this illustration that a straight isochor indicates a state of infinite specific heat, $c_v \Rightarrow \infty$, and a horizontal isochor indicates a state of zero temperature, $\vartheta = 0$. When the concave side of the isochor faces in the positive-$\mathcal{E}$ direction, the specific heat c_v is positive, else c_v is negative. All the isochors in Figure 3.13 have positive specific heat; consequently, the temperature increases when we heat this material.

Thermal Expansion

Let us combine the isochoric process in Figure 3.13 with the isentropes in the $\mathcal{E}$–F diagram (see Figure 3.15). In the $\mathcal{E}$–F diagram an isochor always appears as a straight line. The isentrope at state (0) has a zero slope; consequently, the material at state (0) is free of stress. Suppose we clamp this material to prevent deformation and then heat it with $r > 0$, so that the material moves along an isochoric process from state (0) to state (1). With each incremental increase in r, we move the material to an isentrope with a larger value of η. At state (1) the slope of the isentrope is negative, indicating that we must apply a compressive clamping load to maintain the isochoric process. After reaching state (1), we set the heat supply to zero ($r = 0$) and slowly remove the clamping load. The material expands along

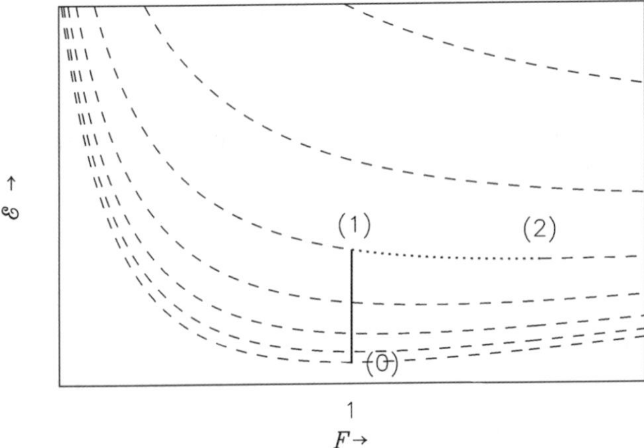

Figure 3.15 The $\mathcal{E}$–F diagram of the isentropes (– –) with an isochoric process (—) and an isentropic expansion (· · ·).

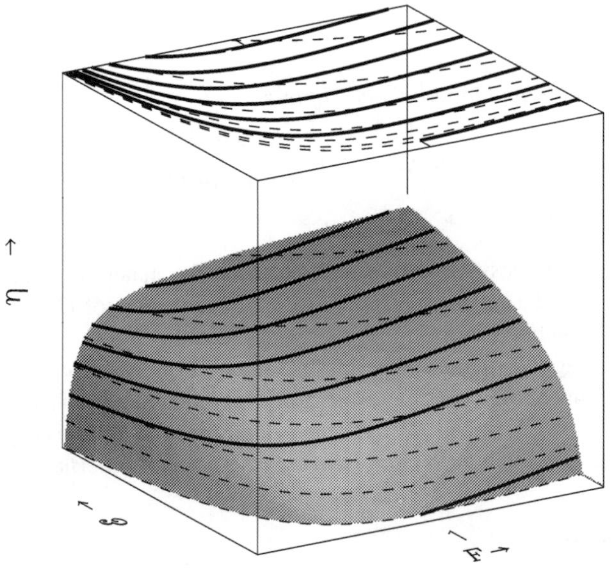

Figure 3.16 The state surface with isotherms (—) and isentropes (– –).

the isentrope to state (2), where the material is once again free of stress. This expansion of the material due to heating is called *thermal expansion*.

3.3.3 *Isothermal Process*

A process in which temperature does not change is called an *isotherm*. Here we plot the isotherms as solid lines on the state surface and then project them onto the $\mathcal{E}$–F diagram (see Figure 3.16). We have spaced them at uniform temperature intervals. Because

$\vartheta = \vartheta(F, \eta)$ is constant on an isotherm, the chain rule yields

$$\frac{\partial \hat{\vartheta}}{\partial t} = \frac{\partial \vartheta}{\partial \eta}\frac{\partial \hat{\eta}}{\partial t} + \frac{\partial \vartheta}{\partial F}\frac{\partial \hat{F}}{\partial t} = 0.$$

After substitution of Eq. (3.55), we obtain the following relationship between the deformation gradient and the entropy along an isotherm:

$$\frac{\partial \hat{\eta}}{\partial t} = -\frac{c_v}{\vartheta}\frac{\partial \vartheta}{\partial F}\frac{\partial \hat{F}}{\partial t}.$$

Stiffness at Constant Temperature

Recall that in a reversible process $T = T_e(F, \eta)$. The rate of change of stress is

$$\frac{\partial \hat{T}_e}{\partial t} = \frac{\partial T_e}{\partial F}\frac{\partial \hat{F}}{\partial t} + \frac{\partial T_e}{\partial \eta}\frac{\partial \hat{\eta}}{\partial t}.$$

When we simultaneously solve the two previous relationships and substitute Eqs. (3.48), (3.50), and (3.53), we obtain

$$\frac{\partial \hat{T}_e}{\partial t} = C^{\vartheta} D, \tag{3.58}$$

where C^{ϑ} is called the *isothermal longitudinal stiffness*,

$$C^{\vartheta} \equiv C^{\eta} - \rho_0 F \frac{c_v}{\vartheta}\left(\frac{\partial \vartheta}{\partial F}\right)^2 = C^{\eta} - \rho_0 F \frac{c_v}{\vartheta}\left(\frac{\partial^2 \mathcal{E}}{\partial F \partial \eta}\right)^2 \tag{3.59}$$

$$C^{\eta} = C^{\vartheta} + \rho c_v \vartheta \Gamma^2. \tag{3.60}$$

When the specific heat c_v is positive, $C^{\vartheta} \leq C^{\eta}$, meaning the material is softer during an isothermal process than during an isentropic process. The inverse of C^{ϑ} is called the *isothermal longitudinal compressibility.*

We show the $\mathcal{E}$–F diagram of the isotherms in Figure 3.17. We plot them as solid contours with higher temperatures residing at the top of the diagram. Earlier we noticed that elastic processes follow isentropes. *Because the isotherms are normally not parallel to the isentropes, elastic deformation follows an isentrope but not an isotherm.*

3.3.4 *Isobaric Process*

The conventional definition of an isobaric process is one in which a volume of material is held in a configuration of constant spherical stress. However, here we define it to be a process where the volume of material is held in a configuration of triaxial stress in which *only the axial stress T is held constant.* We notice that when the volume contains a fluid, holding T constant is equivalent to holding the pressure constant; however, when the volume contains a solid, the lateral stresses, and therefore the mean normal stress, will change. We shall discuss processes produced by spherical stress in the next section.

During an isobaric process the material follows an *isobar*. We plot isobars on the state surface as solid lines (see Figure 3.18). We space them at uniform stress intervals. At large values of compression where $F < 1$, the contours are very close together indicating a stiff material.

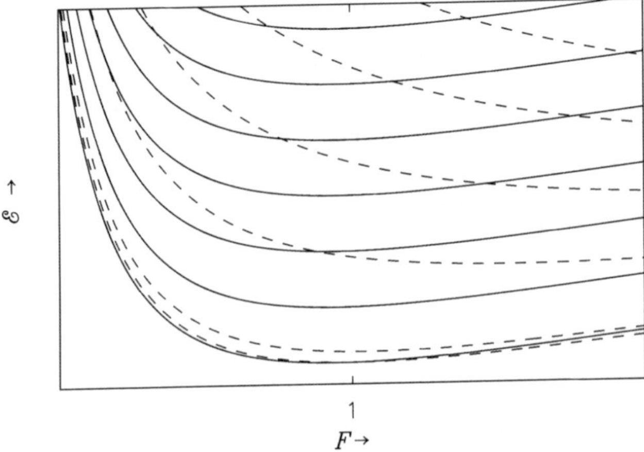

Figure 3.17 The $\mathcal{E}$-F diagram of isotherms (—) and isentropes (– –).

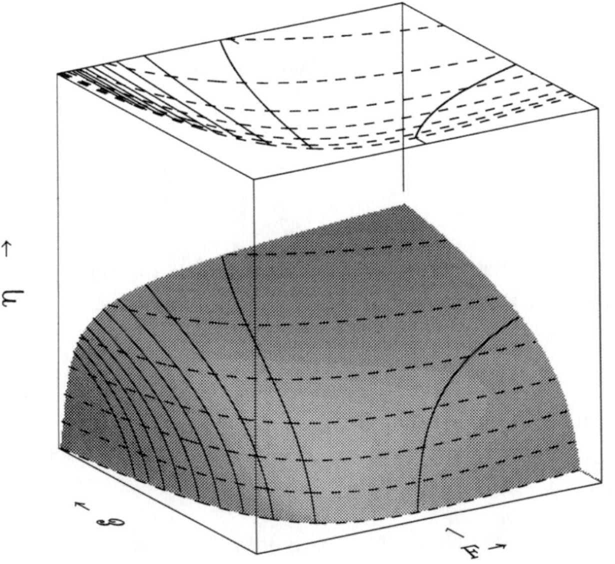

Figure 3.18 The state surface with isobars (—) and isentropes (– –).

Because the stress $T = T_e(F, \eta)$ is constant during an isobaric process,

$$\frac{1}{\rho_0}\frac{\partial \hat{T}_e}{\partial t} = \frac{\partial^2 \mathcal{E}}{\partial F^2}\frac{\partial \hat{F}}{\partial t} + \frac{\partial^2 \mathcal{E}}{\partial F \partial \eta}\frac{\partial \hat{\eta}}{\partial t} = \frac{C^\eta}{\rho_0 F}\frac{\partial \hat{F}}{\partial t} + \frac{\partial \vartheta}{\partial F}\frac{\partial \hat{\eta}}{\partial t} = 0.$$

The change of temperature $\vartheta = \vartheta(F, \eta)$ is

$$\frac{\partial \hat{\vartheta}}{\partial t} = \frac{\partial \vartheta}{\partial F}\frac{\partial \hat{F}}{\partial t} + \frac{\partial \vartheta}{\partial \eta}\frac{\partial \hat{\eta}}{\partial t} = \frac{\partial \vartheta}{\partial F}\frac{\partial \hat{F}}{\partial t} + \frac{\vartheta}{c_v}\frac{\partial \hat{\eta}}{\partial t},$$

and the change in internal energy $\mathcal{E} = \mathcal{E}(F, \eta)$ is

$$\frac{\partial \hat{\mathcal{E}}}{\partial t} = \frac{\partial \mathcal{E}}{\partial F}\frac{\partial \hat{F}}{\partial t} + \frac{\partial \mathcal{E}}{\partial \eta}\frac{\partial \hat{\eta}}{\partial t} = \frac{T_e}{\rho_0}\frac{\partial \hat{F}}{\partial t} + \vartheta\frac{\partial \hat{\eta}}{\partial t}.$$

When we simultaneously solve these three equations, we obtain

$$\frac{\partial \hat{\mathcal{E}}}{\partial t} = \frac{c_v}{C^\vartheta}\left(C^\eta - \frac{FT_e}{\vartheta}\frac{\partial \vartheta}{\partial F}\right)\frac{\partial \hat{\vartheta}}{\partial t}$$

and

$$D = -\frac{\rho_0 c_v}{\vartheta C^\vartheta}\frac{\partial \vartheta}{\partial F}\frac{\partial \hat{\vartheta}}{\partial t}. \tag{3.61}$$

Specific Heat at Constant Stress

When we substitute these results into the equation for the balance of energy, Eq. (3.10), we obtain the following relationship for the change in temperature along an isobar:

$$c_T\frac{\partial \hat{\vartheta}}{\partial t} = r \quad \text{(isobaric process)}, \tag{3.62}$$

where the *isobaric specific heat* c_T is given by the elegant result

$$\frac{c_T}{c_v} = \frac{C^\eta}{C^\vartheta}. \tag{3.63}$$

Thermal Expansion

Along an isobaric process, the relationship between the change in the deformation gradient and the change in temperature is given by Eq. (3.61),

$$D = \alpha\frac{\partial \hat{\vartheta}}{\partial t}, \tag{3.64}$$

where α is called the *longitudinal coefficient of thermal expansion*,

$$\alpha \equiv -\frac{\rho_0 c_v}{\vartheta C^\vartheta}\frac{\partial \vartheta}{\partial F}. \tag{3.65}$$

In general, $\partial\vartheta(F, \eta)/\partial F < 0$. You can see this in Figure 3.17, where by following any of the isentropes, you will find that ϑ decreases as F increases.

When we substitute Eq. (3.53), we obtain

$$\alpha = \rho c_v \Gamma / C^\vartheta. \tag{3.66}$$

When we substitute Eq. (3.65) into Eq. (3.59), we obtain

$$\frac{1}{C^\vartheta} = \frac{1}{C^\eta} + \frac{\vartheta\alpha^2}{\rho c_T}. \tag{3.67}$$

This relationship shows that thermal expansion accounts for the difference between the isentropic longitudinal stiffness and the isothermal longitudinal stiffness.

In Figure 3.19 we show the $\mathcal{E}$–F diagram of the isobars. We plot them as solid contours. The stress-free isobar is the solid contour on the right that intersects each isentrope at the

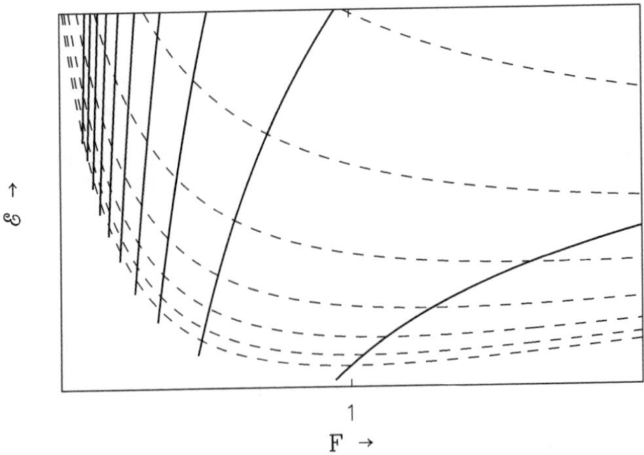

Figure 3.19 The $\mathcal{E}$-F diagram with isobars (—) and isentropes (– –).

point where the slope of the isentrope is zero. When we add heat to a stress-free body, it expands along this process.

3.3.5 *Spherical-Stress Processes*

We have restricted our study of nonlinear waves primarily to uniaxial deformation in which the only component of the deformation gradient that changes is F_{11}. We have used the processes resulting from this type of deformation to define the longitudinal stiffnesses, C^{η} and C^{ϑ}, the thermal expansion coefficient α, and the specific heats, c_T and c_v. These parameters provide a characterization of the state surface that is sufficient to describe one-dimensional nonlinear wave motion. However, most measurements of thermodynamic quantities are not made under conditions of uniaxial deformation. Rather, they are made under conditions of spherical stress. Therefore, in this section we derive the thermodynamic parameters for spherical stress and demonstrate how they are related to our results for unaxial deformation.

If we were to restrict our study to fluids, this section would not be necessary, because uniaxial deformation produces a configuration of spherical stress in a fluid. In contrast, an isotropic solid in equilibrium under uniaxial deformation supports a configuration of triaxial stress. When we assume that the internal energy is represented by

$$\mathcal{E} = \mathcal{E}(F_{ij}, \eta),$$

we find that the equilibrium stress $(T_{ij})_e$ must be derivable from the internal energy as (see Exercise 3.11 in Section 3.2.5)

$$(T_{km})_e = \rho F_{mj} \frac{\partial \mathcal{E}}{\partial F_{kj}}.$$

Notice the similarity between this result and Eq. (2.14) in Section 2.1.3. Notice also that

$$\frac{\partial (T_{km})_e}{\partial \eta} = \rho F_{mj} \frac{\partial \vartheta}{\partial F_{kj}}.$$

For an isotropic solid we assume the following symmetry conditions hold:

$$\frac{F_{ii}}{\vartheta}\frac{\partial \vartheta}{\partial F_{ii}} = -\Gamma, \qquad F_{jj}\frac{\partial (T_{ii})_e}{\partial F_{jj}} = \lambda + 2\mu\delta_{ij} \text{ (no summation)}, \tag{3.68}$$

where summation over the indices is *not* implied. Here the Grüneisen parameter Γ and the Lamé constants, λ and μ, are functions of η and F_{ij}. When we subject this material to the spherical deformation

$$[F_{km}] = \begin{bmatrix} F & 0 & 0 \\ 0 & F & 0 \\ 0 & 0 & F \end{bmatrix},$$

a configuration of spherical stress is produced in which the equilibrium mean normal stress is $\sigma_e = \frac{1}{3}(T_{kk})_e$.

To determine the thermodynamic parameters for spherical stress, we must evaluate the time derivatives of σ_e, ϑ, and $\mathcal{E}$. The derivative of the mean normal stress with respect to time is

$$\frac{\partial \hat{\sigma}_e}{\partial t} = \frac{1}{3}\frac{\partial (T_{kk})_e}{\partial F_{ij}}\frac{\partial F_{ij}}{\partial t} + \frac{1}{3}\frac{\partial (T_{kk})_e}{\partial \eta}\frac{\partial \hat{\eta}}{\partial t},$$

which after substitution of Eq. (3.68) yields

$$\frac{\partial \hat{\sigma}_e}{\partial t} = \frac{1}{3}[3(\lambda + 2\mu) + 6\lambda]D - \rho\vartheta\Gamma\frac{\partial \hat{\eta}}{\partial t}.$$

For spherical deformation $F^3 = \rho_0/\rho$; therefore,

$$\frac{\partial \hat{\sigma}_e}{\partial t} = -\frac{\left(\lambda + \frac{2}{3}\mu\right)}{\rho}\frac{\partial \hat{\rho}}{\partial t} - \rho\vartheta\Gamma\frac{\partial \hat{\eta}}{\partial t}. \tag{3.69}$$

The derivative of the temperature is

$$\frac{\partial \hat{\vartheta}}{\partial t} = \frac{\partial \vartheta}{\partial F_{ij}}\frac{\partial \hat{F}_{ij}}{\partial t} + \frac{\partial \vartheta}{\partial \eta}\frac{\partial \hat{\eta}}{\partial t},$$

which, after substituting Eq. (3.68) and using the definition of the isochoric specific heat, yields

$$\frac{\partial \hat{\vartheta}}{\partial t} = \frac{\vartheta\Gamma}{\rho}\frac{\partial \hat{\rho}}{\partial t} + \frac{\vartheta}{c_v}\frac{\partial \hat{\eta}}{\partial t}. \tag{3.70}$$

The derivative of the internal energy is

$$\frac{\partial \hat{\mathcal{E}}}{\partial t} = \frac{\partial \mathcal{E}}{\partial F_{ij}}\frac{\partial \hat{F}_{ij}}{\partial t} + \vartheta\frac{\partial \hat{\eta}}{\partial t},$$

which yields

$$\frac{\partial \hat{\mathcal{E}}}{\partial t} = -\frac{\sigma_e}{\rho^2}\frac{\partial \hat{\rho}}{\partial t} + \vartheta\frac{\partial \hat{\eta}}{\partial t}. \tag{3.71}$$

Stiffness at Constant Entropy For an isentropic process, we find that Eq. (3.69) yields

$$\frac{\partial \hat{\sigma}_e}{\partial t} = -\frac{\mathcal{K}^\eta}{\rho}\frac{\partial \hat{\rho}}{\partial t},$$

where

$$\mathcal{K}^\eta = \lambda + \frac{2}{3}\mu \tag{3.72}$$

is called the *isentropic bulk stiffness*. Here we have defined the bulk stiffness during a spherical-stress process. However, this definition is also compatible with our previous definition of the bulk stiffness during uniaxial deformation [see Eq. (2.26)]. You can verify this result by recognizing that $F = \rho_0/\rho$ during uniaxial deformation and then evaluating the quantity $(d\sigma/d\rho)_{\rho=\rho_0}$ using Eq. (2.26).

We can also compare the isentropic bulk stiffness $\mathcal{K}^\eta$ to the isentropic longitudinal stiffness C^η by substituting Eq. (3.68) into Eq. (3.50) to obtain

$$C^\eta = \lambda + 2\mu.$$

Notice that for a fluid, because $\mu \equiv 0$, the bulk stiffness and the longitudinal stiffness are equal: $C^\eta = \mathcal{K}^\eta$.

Stiffness at Constant Temperature During an isothermal process, the temperature is constant, and Eq. (3.70) yields

$$\frac{\partial \hat{\eta}}{\partial t} = -\frac{\Gamma c_v}{\rho}\frac{\partial \hat{\rho}}{\partial t}.$$

We substitute this result into Eq. (3.69) to obtain

$$\frac{\partial \hat{\sigma}_e}{\partial t} = -\frac{\mathcal{K}^\vartheta}{\rho}\frac{\partial \hat{\rho}}{\partial t},$$

where

$$\mathcal{K}^\vartheta = \mathcal{K}^\eta - \rho c_v \vartheta \Gamma^2$$

is called the *isothermal bulk stiffness*.

Thermal Expansion at Constant Mean Normal Stress When the mean normal stress is constant, simultaneous solution of Eqs. (3.69) and (3.70) yields

$$\frac{1}{\rho}\frac{\partial \hat{\rho}}{\partial t} = -\beta\frac{\partial \hat{\vartheta}}{\partial t}, \qquad \frac{\partial \hat{\eta}}{\partial t} = \frac{\mathcal{K}^\eta}{\mathcal{K}^\vartheta}\frac{c_v}{\vartheta}\frac{\partial \hat{\vartheta}}{\partial t}, \tag{3.73}$$

where

$$\beta = \frac{\rho c_v \Gamma}{\mathcal{K}^\vartheta} \tag{3.74}$$

is called the *volumetric coefficient of thermal expansion*. For a fluid, because $C^\vartheta = \mathcal{K}^\vartheta$, we find that the longitudinal coefficient of thermal expansion and the volumetric coefficient of thermal expansion are equal. For spherical deformation, we notice that

$$D = \frac{\beta}{3}\frac{\partial \hat{\vartheta}}{\partial t}.$$

The quantity $\beta/3$ is called the *linear coefficient of thermal expansion.*

Specific Heat at Constant Mean Normal Stress We can determine the change in internal energy during an isobaric process for constant mean normal stress by substituting Eq. (3.73) into Eq. (3.71) to obtain

$$\frac{\partial \hat{\mathcal{E}}}{\partial t} = \left(\frac{c_v \mathcal{K}^\eta}{\mathcal{K}^\vartheta} + \frac{\sigma_e \beta}{\rho}\right)\frac{\partial \hat{\vartheta}}{\partial t}.$$

Substitution of these results into the balance of energy Eq. (3.20) yields

$$c_p \frac{\partial \hat{\vartheta}}{\partial t} = r,$$

where the *specific heat for constant mean normal stress* c_p is

$$\frac{c_p}{c_v} = \frac{\mathcal{K}^\eta}{\mathcal{K}^\vartheta}.$$

By comparing this result to Eq. (3.63), we find that for fluids $c_p = c_T$. Consequently, c_p is also called the *specific heat for constant pressure.*

For most isotropic solids you will typically find published values for c_p, β, Young's Modulus E^0, and Poisson's ratio ν. Also, for many solids, such as metals, the values of E_0 and ν measured during an isentropic process do not differ significantly from those measured during an isothermal process. Using the table in Appendix B, you can determine the Lamé constants, λ and μ, from these published values. You can then use the results in this section to determine the other thermodynamic parameters, such as c_v and Γ.

3.3.6 Exercises

3.12 Following the procedure we used to derive Eq. (3.51), derive Eq. (3.57),

$$c_v = R_\vartheta \sin\theta \cos^2\theta.$$

3.13 We show the $\mathcal{E}$–η and $\mathcal{E}$–F diagrams for materials A (solid lines) and B (dashed lines) in Figure 3.20. Point P represents the same state in both diagrams. Assuming that both materials have the same density, at the state P in each material determine the following:

(a) What are the signs of the specific heats, and which material has the greater specific heat?

(b) Which material has a higher temperature?

(c) What are the signs of the stress for each material, and which material has the greatest magnitude of stress?

(d) What are the signs of the isentropic stiffnesses, and which material is stiffer?

3.14 At $\vartheta = 295$ K, the specific heat of water during an isobaric process is $c_p = 4.187$ kJ/kg/K, the coefficient of thermal expansion is $\alpha = 227 \times 10^{-6}\ \mathrm{K}^{-1}$, and the density is $\rho = 998\ \mathrm{kg/m^3}$. In Exercise 3.1, we found that burning a jelly donut releases about 1 MJ of heat. Assuming that c_p and α are constant, if this heat is absorbed by 10 kg of water, by how much will the temperature and the volume of the water increase?

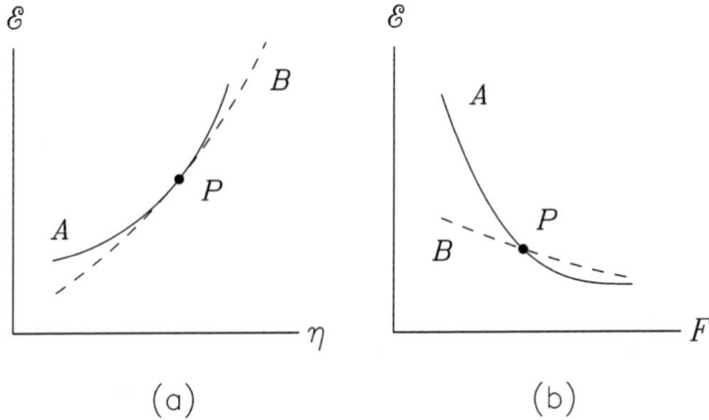

Figure 3.20 Exercise 3.13.

3.15 For fluids it is often preferable to express the internal energy as a function of density and entropy,

$$\mathcal{E} = \mathcal{E}(\rho, \eta).$$

Show that the equilibrium pressure is $p_e = -T_e$ and

$$p_e = \rho^2 \frac{\partial \mathcal{E}}{\partial \rho}, \qquad C^\eta = \rho \frac{\partial p_e}{\partial \rho}, \qquad \Gamma = \frac{\rho}{\vartheta} \frac{\partial \vartheta}{\partial \rho}.$$

3.16 The internal energy of a particular material has the following form:

$$\mathcal{E} = \mathcal{E}_0 \rho^{NR/c_v} e^{\eta/c_v}$$

where c_v, R, N, and $\mathcal{E}_0$ are prescribed constants.

(a) Derive the relationship for the temperature, by taking the partial derivative of this expression holding ρ constant.

(b) Derive the relationship for the equilibrium pressure $p_e = -T_e$, by taking the partial derivative of this expression holding η constant.

(c) Express the equilibrium pressure in terms of the density ρ and the temperature ϑ.

(d) What kind of material does this equation represent?

(e) Show that

$$C^\eta = \left(\frac{NR}{c_v} + 1 \right) p_e, \qquad C^\vartheta = p_e.$$

3.17 In the preceding exercise, you will notice that you can write the pressure p_e either as a function of ρ and η or as a function ρ and ϑ.

(a) Use these expressions for pressure to show that

$$C^\eta = \rho \frac{\partial p_e(\rho, \eta)}{\partial \rho}, \qquad C^\vartheta = \rho \frac{\partial p_e(\rho, \vartheta)}{\partial \rho}.$$

(b) Explain this result in the context of the isentropic and isothermal process.

3.18 Suppose we use the chain rule to write Eq. (3.64) as follows:

$$\frac{1}{\alpha F} \frac{\partial \hat{F}}{\partial t} = \frac{\partial \hat{\vartheta}}{\partial t} = \frac{\partial \vartheta}{\partial F} \frac{\partial \hat{F}}{\partial t}.$$

Then we conclude that

$$\frac{\partial \vartheta}{\partial F} = \frac{1}{\alpha F}.$$

Thus the partial derivative of ϑ with respect to F is positive if α is positive. But from Eq. (3.65), we know that

$$\frac{\partial \vartheta}{\partial F} = \frac{-\alpha \vartheta C^{\vartheta}}{\rho_0 c_v},$$

which is a negative number if c_v is positive. Clearly, these two partial derivatives represent different quantities. Explain this result. *Hint:* Although we have always used the explicit functional dependence $\vartheta = \vartheta(F, \eta)$, one of these relationships is based upon the implicit function $\vartheta = \vartheta(F, T)$.

3.19 The isentropic and isothermal stiffnesses of a material are related by Eqs. (3.59) and (3.60). Using the relationships we present for the isothermal process, verify these expressions.

3.20 The longitudinal stiffnesses and the specific heats are related by Eq. (3.63),

$$\frac{c_T}{c_v} = \frac{C^{\eta}}{C^{\vartheta}}.$$

Using the relationships we present for the isobaric process, verify this expression.

3.21 For many solids the difference between, for example, c_T and c_p is quite small. To illustrate this point, consider a block of aluminum with the following measured properties:

$$\begin{aligned} &E^0 = 75\,\text{GPa}, && \nu = 0.333, \\ &\rho = 2.7\,\text{Mg/m}^3, && \beta = 21.6 \times 10^{-6}\,\text{K}^{-1}, \\ &\vartheta = 293\,\text{K}, && c_p = 900\,\text{J/kg/K}. \end{aligned}$$

Assume that Young's modulus and Poisson's ratio are measured under isentropic conditions. Determine $\mathcal{K}^{\eta}$, C^{η}, Γ, c_v, c_T, $\mathcal{K}^{\vartheta}$, and C^{ϑ}.

3.4 Wave States and Processes

Now that we have completed our study of the basic reversible processes, we turn our attention to thermodynamical processes produced by waves. First, we consider elastic waves, which include linear-elastic waves and structured waves. We show that an elastic wave is an isentropic process, not an isothermal process. For example, as you speak, sound waves cause the pressure in air to oscillate, but surprisingly, they also cause the temperature to oscillate. Following our discussion of structured waves, we consider shock waves and steady waves, which cause the material to follow a process we have not yet discussed. It is called the *Rayleigh-line process.*

3.4.1 *Viscous Equations*

Here we provide both the material and spatial descriptions of the equations that govern wave motion in a viscous material. We call these the *viscous equations.*

Material Description

Recall that the material description of the governing equations for the viscous material are summarized in Section 3.2.2. Let us combine Eq. (3.27) to obtain

$$FD = \frac{\partial \hat{v}}{\partial X}, \qquad D = \frac{1}{F}\frac{\partial \hat{F}}{\partial t}.$$

This eliminates the motion $\hat{x}(X, t)$ from the unknown variables. Next, we assume that the body force $b = 0$ and that the external heat supply $r = 0$. The equations for the balance of linear momentum, Eq. (3.28), and the balance of energy, Eq. (3.29), become

$$\begin{aligned} \rho_0 \frac{\partial \hat{v}}{\partial t} &= \frac{\partial \hat{T}}{\partial X}, \\ \rho_0 \frac{\partial \hat{\mathcal{E}}}{\partial t} &= TFD. \end{aligned} \tag{3.75}$$

This system of equations is completed by the set of constrained constitutive equations, Eqs (3.33)–(3.35):

$$\mathcal{E} = \mathcal{E}(F, \eta), \tag{3.76}$$

$$T = T(F, D, \eta), \tag{3.77}$$

$$\vartheta = \partial\mathcal{E}/\partial\eta. \tag{3.78}$$

This system of seven equations has seven unknowns: $F, D, v, T, \mathcal{E}, \vartheta$, and η. We must write the constitutive equation for the stress in a form that guarantees that the reduced dissipation inequality, Eq. (3.36), is never violated.

Spatial Description

In the spatial description we have

$$F = \frac{\rho_0}{\rho}, \qquad D = \frac{1}{F}\frac{\partial \hat{F}}{\partial t}.$$

When b, r, and q are equal to zero, the equations for the conservation of mass, Eq. (2.34), the balance of linear momentum, Eq. (2.35), and balance of energy, Eq. (3.12), are

$$\begin{aligned} \frac{\partial \rho}{\partial t} &= -\frac{\partial(\rho v)}{\partial x}, \\ \frac{\partial(\rho v)}{\partial t} &= \frac{\partial}{\partial x}(T - \rho v^2), \\ \frac{\partial \mathcal{E}}{\partial t} + v\frac{\partial \mathcal{E}}{\partial x} &= \frac{T}{\rho}\frac{\partial v}{\partial x}. \end{aligned} \tag{3.79}$$

We complete this system of equations with the set of constrained constitutive equations, Eqs (3.33)–(3.35):

$$\begin{aligned} \mathcal{E} &= \mathcal{E}(F, \eta), \\ T &= T(F, D, \eta), \\ \vartheta &= \partial\mathcal{E}/\partial\eta. \end{aligned}$$

We have eight equations with eight unknowns: F, D, ρ, v, T, $\mathcal{E}$, ϑ, and η. We must write the constitutive equation for the stress in a form that guarantees that the reduced dissipation inequality, Eq. (3.36), is never violated.

3.4.2 *Elastic-Wave Process*

We have discussed elastic waves in Chapter 2. Elastic waves include linear-elastic waves and structured waves. Compression waves and rarefaction waves are special types of structured waves. These waves are solutions to the nonlinear-elastic equations. In this section we show that under special conditions they are also solutions to the viscous equations.

Consider the constitutive equation for stress in a viscous material,

$$T = T(F, D, \eta).$$

Suppose a wave results in values of D that are sufficiently close to zero so that

$$\begin{aligned} T &\approx T_e, \\ \mathcal{T} &\approx 0. \end{aligned} \tag{3.80}$$

When $r = 0$, the equation for the balance of energy, Eq. (3.43), becomes

$$\frac{\partial \eta}{\partial t} = 0.$$

We integrate this equation to obtain

$$\eta = \eta_0,$$

where η_0 is a constant. Thus we find that this wave causes the material to follow an isentrope. Because we have solved the balance of energy, we can remove it from the system of equations and replace it by the equation $\eta = \eta_0$.

The specific internal energy is

$$\mathcal{E} = \mathcal{E}(F, \eta_0),$$

and the stress is [see Eqs. (3.40) and (3.80)]

$$T = \rho_0 \frac{\partial \mathcal{E}}{\partial F}(F, \eta_0). \tag{3.81}$$

Because the entropy is constant, the stress is solely a function of F. Now let us recall the constitutive equation for an elastic material. In Eq. (2.19), we expressed the stress for an elastic material in terms of the strain-energy potential $\psi_F(F)$ as

$$T = \rho F \frac{\partial \psi_F}{\partial F}(F).$$

By comparing these two expressions for stress, we see that the internal energy and the strain-energy potential are related by

$$\frac{\partial \psi_F}{\partial F}(F) = \frac{\partial \mathcal{E}}{\partial F}(F, \eta_0). \tag{3.82}$$

Thus the isentropic response of the viscous material is equivalent to the response of an elastic material. For each value of η_0 that we assign to the isentrope, we obtain a unique elastic response.

Because entropy is constant during the elastic-wave process, the constitutive equations for the viscous material, Eqs. (3.76)–(3.78), become

$$T = T(F), \tag{3.83}$$

$$\vartheta = \vartheta(F). \tag{3.84}$$

With the exception of the constitutive equation for the temperature, we see that the viscous equations reduce to the nonlinear-elastic equations, Eqs. (2.30)–(2.33). We conclude that as long as we can ignore dissipation, the linear-elastic wave and the structured wave are solutions to both the nonlinear-elastic equations and the viscous equations.

Having established this connection to the nonlinear-elastic equations, let us evaluate the wave velocities for the viscous material in terms of the thermodynamic stiffness C^η. You will recall that the characteristic forms of the nonlinear-elastic equations are based upon the definition of the material wave velocity C [see Eq. (2.92)],

$$C = \left(\frac{1}{\rho_0}\frac{dT}{dF}\right)^{1/2}. \tag{3.85}$$

Remember that C determines the slopes of the characteristic contours in the X–t diagram. We derived this expression for the nonlinear-elastic material where T is solely a function of F. However, we have just shown that entropy is constant during an elastic-wave process in a viscous material. Thus we can apply Eq. (3.85) to the viscous material by replacing the total derivative with a partial derivative to obtain

$$C = \left(\frac{1}{\rho_0}\frac{\partial T}{\partial F}\right)^{1/2}. \tag{3.86}$$

Here T is a function of F, D, and η. Hence this definition of C depends only upon the existence of the characteristic contours and does not depend upon the precise constitutive equation for T. Indeed, in Chapter 4, we shall find that Eq. (3.86) is sufficient to define C in materials where T is not even an explicit function of F.

Now we can employ the constitutive equations for the viscous material to obtain additional useful relationships for the material and spatial wave velocities. We do this by noticing that the elastic wave is a reversible process where

$$T = T_e = \rho_0 \frac{\partial \mathcal{E}}{\partial F}.$$

Substituting this expression into Eq. (3.86), we obtain

$$C = \left[\frac{\partial^2 \mathcal{E}}{\partial F^2}(F, \eta_0)\right]^{1/2}. \tag{3.87}$$

Then we substitute Eq. (3.50) to obtain

$$C = \left(\frac{C^\eta}{\rho_0 F}\right)^{1/2}. \tag{3.88}$$

The spatial wave velocity is $c = FC$ [see Eq. (2.104) in Section 2.5.2]. Thus

$$c = \left(\frac{C^\eta}{\rho}\right)^{1/2}. \tag{3.89}$$

These are the desired expressions for the material and spatial wave velocities for an elastic wave in a viscous material.

Linear-Elastic Waves

In this section we show how the results we just derived for the elastic wave can be used to obtain the description of a linear-elastic wave. Suppose we let

$$F = 1 + \Delta F$$

and expand the internal energy in a Taylor's series about $F = 1$:

$$\mathcal{E}(F, \eta_0) = \mathcal{E}(1, \eta_0) + \frac{\partial \mathcal{E}}{\partial F}(1, \eta_0)\Delta F + \frac{1}{2}\frac{\partial^2 \mathcal{E}}{\partial F^2}(1, \eta_0)[\Delta F]^2 + \cdots .$$

If we ignore the higher-order terms in ΔF, we can substitute Eqs. (3.40) and (3.50) to obtain

$$\mathcal{E}(F, \eta_0) = \mathcal{E}(1, \eta_0) + \frac{T^0}{\rho_0}\Delta F + \frac{C_0^\eta}{2\rho_0}(\Delta F)^2, \tag{3.90}$$

where $T^0 = T_e(1, \eta_0)$ and $C_0^\eta = C^\eta(1, \eta_0)$.

By substituting this result into Eq. (3.81), we obtain the following expression for the stress:

$$T - T^0 = C_0^\eta(F - 1).$$

This is the constitutive model for a linear-elastic material [see Eq. (2.11)]. We find that

$$C_0^\eta = \lambda + 2\mu, \tag{3.91}$$

where λ and μ are the Lamé constants of a linear-elastic material. As a consequence, the velocity of sound is [see Eq. (2.38)]

$$c_0 = \left(\frac{C_0^\eta}{\rho_0}\right)^{1/2}. \tag{3.92}$$

Change in Temperature In Figure 3.21 we show the isentrope for the elastic process about an initial state, where $F = 1$ and $T^0 = 0$. The temperature and Grüneisen coefficient at this state are ϑ_0 and Γ_0. Notice that the isotherms are not parallel to the isentrope. Thus, because the wave causes the material to follow an isentrope, it causes the temperature to change. Let the change in temperature be denoted by $\Delta\vartheta$. Then from Eq. (3.52)

$$\Delta\vartheta = -\Gamma_0\vartheta_0\Delta F. \tag{3.93}$$

This relationship also holds at states where $T_0 \neq 0$. In a material where the Grüneisen coefficient Γ_0 is zero, the isentropes and the isotherms are parallel, and the temperature remains constant. Certain ceramic materials have very small values of Γ_0. However, for many materials including metals, the value of Γ_0 is on the order of one (see Exercise 3.22).

As a linear-elastic wave propagates, it changes the temperature of the material by the amount $\Delta\vartheta$. To do this, it must change the internal energy of the material by the amount $\Delta\mathcal{E} = \mathcal{E}(F, \eta_0) - \mathcal{E}(1, \eta_0)$. Suppose we wish to change the temperature by the same amount by adding heat to the material during an isochoric process. The amount of heat we must add to the material is $\Delta r = c_v\Delta\vartheta$. It is interesting to note that in general, as illustrated in Figure 3.21, $\Delta\mathcal{E}$ is considerably smaller than Δr [see Exercise 3.25].

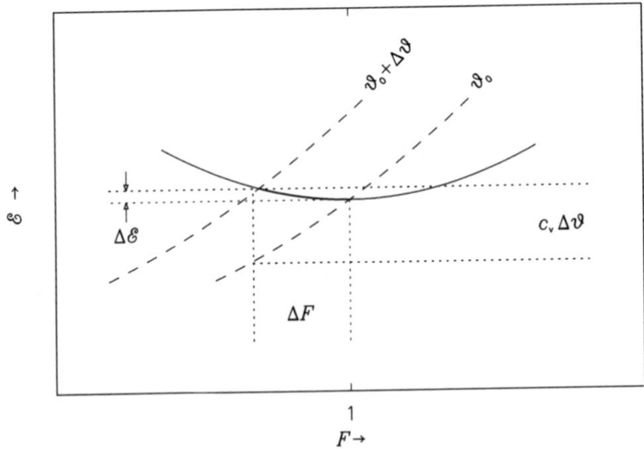

Figure 3.21 A linear elastic process (—) with isotherms (– –).

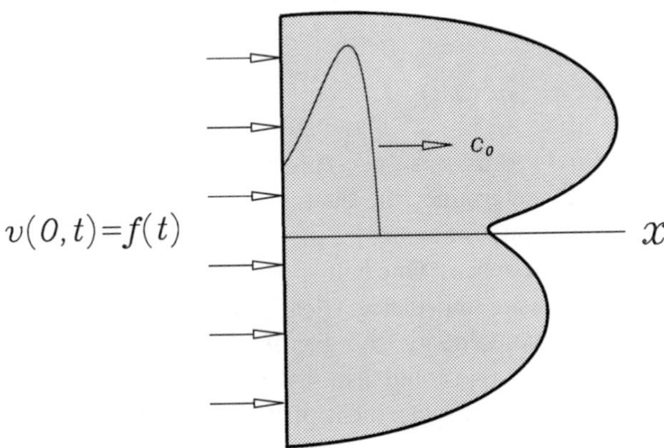

Figure 3.22 A half space with an elastic wave.

Linear-Elastic Wave in a Half Space Consider a linear-elastic half space initially at rest and free of stress. We subject the boundary of the half space to the boundary condition $v(0, t) = f(t)$ (see Figure 3.22). The solution for this boundary-value problem is a simple wave that propagates in the positive-x direction [see Eq. (2.77)]:

$$v(x, t) = f(t - x/c_0).$$

The stress and strain for a simple wave are [see Eqs. (2.72) and (2.73)]:

$$T = -z_0 v$$

and

$$\frac{\partial u}{\partial x} = -\frac{v}{c_0},$$

where we recall that z_0 is the acoustic impedance of the material. The displacement is

$$u = x - X.$$

In linear elasticity, derivatives of the displacement are assumed to be "small," meaning that we can neglect terms containing products of derivatives of the displacement. Recall that

$$F = 1 + \Delta F.$$

Using Eq. (1.131) in Section 1.8.1, we obtain

$$\Delta F = \frac{\partial \hat{u}}{\partial X} = \frac{\partial u}{\partial x} = -\frac{v}{c_0}.$$

Thus, from Eq. (3.93), we see that the simple wave causes the following change in temperature:

$$\Delta\vartheta = -\Gamma_0\vartheta_0\frac{\partial \hat{u}}{\partial X} = \Gamma_0\vartheta_0\frac{v}{c_0}. \tag{3.94}$$

The change in temperature is proportional to v. For waves where the velocity v returns to its initial value, the temperature will also return to its initial value. Temperature changes resulting from elastic waves are reversible.

The change of total energy is given by Eq. (3.9). We can substitute the simple wave results for stress and strain to obtain

$$\frac{\partial}{\partial t}\int_0^\infty \rho_0(\hat{\mathcal{E}} + \hat{\mathcal{K}})dX = z_0 v^2(0, t), \tag{3.95}$$

where we have assumed that $v \to 0$ as $x \to \infty$. This result illustrates why z_0 is called an "impedance." We see that z_0 determines the energy that can be introduced into the half space by the boundary condition. For example, suppose we apply the same boundary velocity $v(0, t)$ to two different half spaces. One half space has a large acoustic impedance and the other has a very small acoustic impedance. Clearly, we see that the half space with the larger impedance will accept more energy from the boundary condition. Noticing that the stress at the boundary is $z_0 v(0, t)$, we could also say that the half space with the small impedance offers little "resistance" to displacement of the boundary, and therefore little energy can enter the material.

3.4.3 *Newtonian Fluid*

Before we continue with our discussion of wave processes, let us consider a simple example of a constitutive equation for stress that always satisfies the reduced dissipation inequality, Eq. (3.36). It is called a *Newtonian fluid.* With this special example, we discuss how some waves cause dissipation whereas other waves do not cause dissipation.

Consider the following constitutive equation for the internal energy:

$$\mathcal{E} = \mathcal{E}(\rho, \eta). \tag{3.96}$$

Notice that $\mathcal{E}$ is a function of the density ρ and not the deformation gradient F. At first this appears to be the result of a simple substitution of the expression $F = \rho_0/\rho$ into $\mathcal{E} = \mathcal{E}(F, \eta)$. However, this is not the case. Let us see why. Remember that $F = F_{11}$ is a component of a tensor. The general relationship between density and the deformation gradient is given by [see Eq. (2.15)]

$$\det[F_{km}] = \frac{\rho_0}{\rho}.$$

Thus we see that density is related to a very special combination of the components of the deformation gradient that just happens to be $F = \rho_0/\rho$ for uniaxial deformation. For a moment, let us recall some results for the strain energy ψ_F. In Section 2.1.3, we saw that ψ_F can depend upon all of the components F_{ij}. When it does, shear stresses result, and the strain energy function represents a solid. However, when we restrict the strain energy to be a function of density alone, a spherical stress configuration results, shear stresses are absent, and the function represents a fluid. Because in the previous section we established the equivalence between the internal energy $\mathcal{E}$ and the strain energy ψ_F, we find that the same conclusions apply to the internal energy. Indeed, when we write $\mathcal{E} = \mathcal{E}(F, \eta)$, we are actually evaluating a more general function $\mathcal{E} = \mathcal{E}(F_{ij}, \eta)$. This function represents a solid and results in shear stresses. When we write $\mathcal{E} = \mathcal{E}(\rho, \eta)$, we are restricting this function to depend upon $\det[F_{km}]$ alone. As a result, the internal energy function represents a fluid.

Suppose we consider a body where $\mathcal{E} = \mathcal{E}(\rho, \eta)$. For a moment let us restrict our discussion to a reversible process. Because the internal energy is a function of density alone, the equilibrium stress is spherical. Using the definition of the equilibrium stress, Eq. (3.40), we obtain the following expression for the equilibrium stress $(T_{ij})_e$:

$$(T_{ij})_e = \rho_0 \frac{\partial \mathcal{E}}{\partial \rho} \frac{\partial \rho}{\partial F} \delta_{ij} = -p_e \delta_{ij},$$

where

$$p_e \equiv \rho^2 \frac{\partial \mathcal{E}}{\partial \rho}.$$

If we employ the equivalence between the strain energy and the internal energy, Eq. (3.82), we find that this is identical to the result for the nonlinear-elastic fluid [see Eq. (2.16)].

In a Newtonian fluid, the internal energy is given by Eq. (3.96). As we have shown, at equilibrium, this leads to a spherical stress configuration described by the pressure p_e. Recall that the stress is the sum of the equilibrium stress and the viscous stress. In a Newtonian fluid, we define the viscous stress to be

$$\mathcal{T}_{ij} \equiv \rho_0 \lambda' D_{kk} \delta_{ij} + 2\rho_0 \mu' D'_{ij},$$

where the deformation rate is [see Eq. (1.87) in Section 1.4.7]

$$D_{km} = \frac{1}{2}(L_{km} + L_{mk})$$

and the deviatoric deformation rate is

$$D'_{km} \equiv D_{km} - \frac{1}{3} D_{nn} \delta_{km}.$$

The constants λ' and μ' have physical dimensions of length times velocity. Because length and velocity are kinematic quantities, the coefficients λ' and μ' are called *kinematic viscosities*. The products $\rho_0\lambda'$ and $\rho_0\mu'$ are also called the *bulk* and *dynamic viscosities.*

During one-dimensional motion, the deformation rate is given by $FD = \partial \hat{F}/\partial t$. We substitute this equation into the axial component of the viscous stress to obtain

$$\mathcal{T} = \mathcal{T}_{11} = \rho_0 \left(\lambda' + \frac{4}{3} \mu' \right) D. \tag{3.97}$$

Using this constitutive equation, we find that the equation for the balance of energy, Eq. (3.38), becomes

$$\vartheta \frac{\partial \hat{\eta}}{\partial t} = \mathcal{D},$$

and the dissipation inequality, Eq. (3.44), becomes

$$\mathcal{D} \geq 0,$$

where the dissipation is

$$\mathcal{D} = \left(\lambda' + \frac{4}{3}\mu'\right) F D^2.$$

We see that the dissipation inequality is satisfied for any positive value of the coefficient $\lambda' + \frac{4}{3}\mu'$.

Clearly, for sufficiently small values of D, the resulting dissipation produces negligible changes in entropy. Suppose a compression wave of this type propagates in this material. Recall that as it propagates, its characteristic contours converge and the magnitude of D increases. Eventually, the wave reaches a state where the dissipation becomes significant. At this state, the compression wave transforms into a steady wave. If necessary, the quantity D in a steady wave will approach infinity and produce significant dissipation irrespective of how small $\lambda' + \frac{4}{3}\mu'$ happens to be. In the limit as $\lambda' + \frac{4}{3}\mu' \to 0$, we find that $D \to \infty$, and a shock discontinuity will form.

3.4.4 *Rayleigh-Line Process*

During a reversible process, $T = T_e$. Now we consider an irreversible process where $T \neq T_e$. It is called the *Rayleigh-line process*. In Chapter 2 we discussed the Rayleigh line in the context of the T–v diagram. We discovered that the stress in a steady wave must follow a response path defined by the Rayleigh line [see Eq. (2.188)],

$$T - T_- = -\rho_0 C_S (v - v_-), \tag{3.98}$$

where T_- and v_- are the stress and velocity ahead of the steady wave. This response path *is not* a constitutive equation for the stress. Rather, it is a condition, imposed by the balance of linear momentum, that the stress must satisfy for a steady wave to exist. Because the equilibrium stress T_e does not depend on velocity v, it is clear that if the stress T is given by Eq. (3.98), it cannot always be equal to T_e.

The existence of this response path is extremely important. To illustrate, let us consider it in the context of a flyer-plate experiment. Suppose the target plate in this experiment is constructed of an unknown material. In most materials, the impact of the flyer plate will produce a steady wave in the target plate. Because of the existence of the Rayleigh line, measurements of the velocity of the wave C_S and the velocity of the material v determine the stress T from Eq. (3.98). Thus, we can determine the stress during the Rayleigh-line process independently of any knowledge of the constitutive equations of the material. Hence we can use the steady wave as a tool to analyze unknown constitutive equations. It also means that we are able to present all of the thermodynamic results for the Rayleigh-line process without actually specifying a particular constitutive equation for T.

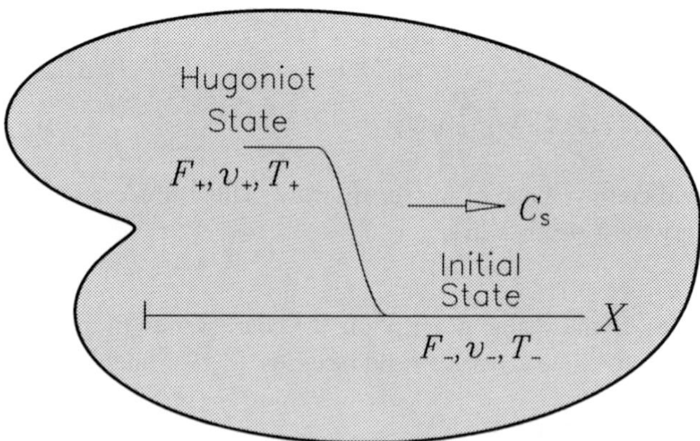

Figure 3.23 A steady wave propagating at the velocity C_S. The material ahead of the wave is at the initial state, and the material behind the wave is at the Hugoniot state.

Balance of Energy for Steady Waves

Consider a steady wave propagating in an infinite material (see Figure 3.23). In Section 2.9, we presented the analysis of a steady wave in a nonlinear-elastic material. Here, by inclusion of the equation for the balance of energy, we extend this analysis to a viscous material. We consider both the material and spatial descriptions for one-dimensional motion. The derivation for three-dimensional motion is left as an exercise (see Exercise 3.31). We then show that the steady wave causes the material to follow the Rayleigh-line process. During this process, the material moves from the initial state to the Hugoniot state.

Material Description Recall that, for a steady wave, the conservation of mass and the balance of linear momentum yield the following relationships [see Eqs. (2.186) and (2.189)]:

$$\begin{aligned} v - v_- &= -C_S(F - F_-), \\ T - T_- &= \rho_0 C_S^2 (F - F_-), \end{aligned} \tag{3.99}$$

where v_-, T_-, and F_- are the velocity, stress, and deformation gradient ahead of the steady wave. Let us derive a similar relationship for the balance of energy. The equation of the balance of energy in the material description is [see Eq. (3.75)]

$$\frac{\partial \hat{\mathcal{E}}}{\partial t} = \frac{T}{\rho_0}\frac{\partial \hat{F}}{\partial t}. \tag{3.100}$$

For a steady wave to exist, the solution to the equation for the balance of energy must only be a function of the variable $\hat{\varphi}$,

$$\hat{\varphi} = X - C_S t.$$

By assuming that $\mathcal{E}$, T, and F are only functions of $\hat{\varphi}$, we apply the chain rule to Eq. (3.100) to obtain

$$\rho_0 \frac{d\mathcal{E}}{d\hat{\varphi}} = T\frac{dF}{d\hat{\varphi}}.$$

Next we combine the stress relationship in Eq. (3.99) with this expression and integrate the result to obtain

$$\rho_0(\mathcal{E} - \mathcal{E}_-) = T_-(F - F_-) + \frac{1}{2}\rho_0 C_S^2 (F - F_-)^2, \tag{3.101}$$

where $\mathcal{E}_-$ is the internal energy ahead of the steady wave. This is the material description of the balance of energy for a steady wave.

Spatial Description In the spatial description, the conservation of mass and the balance of linear momentum for a steady wave yield [see Eqs. (2.193) and (2.194)]

$$\begin{aligned} \rho_0 C_S &= \rho_-(c_S - v_-) = \rho(c_S - v), \\ T &= T_- - \rho_0 C_S (v - v_-), \end{aligned} \tag{3.102}$$

where Eq. (2.138) has been used. Noticing that $F_- = \rho_0/\rho_-$ and $F = \rho_0/\rho$, we solve the first of these expressions to obtain

$$(F_- - F)C_S = (v - v_-). \tag{3.103}$$

Now we derive a relationship for the balance of energy for a steady wave. The equation for the balance of energy in the spatial description is [see Eq. (3.79)]

$$\frac{\partial \mathcal{E}}{\partial t} + v\frac{\partial \mathcal{E}}{\partial x} = \frac{T}{\rho}\frac{\partial v}{\partial x}.$$

For a steady wave to exist, the solution to this equation must be solely a function of the variable φ,

$$\varphi = x - c_S t.$$

By assuming that $\mathcal{E}$, ρ, T, and v are only functions of φ, we can apply the chain rule to obtain

$$-\rho(c_S - v)\frac{d\mathcal{E}}{d\varphi} = T\frac{dv}{d\varphi}.$$

Next we combine Eq. (3.102) with this expression and integrate the result to obtain

$$-\rho_0 C_S(\mathcal{E} - \mathcal{E}_-) = T_-(v - v_-) - \frac{1}{2}\rho_0 C_S (v - v_-)^2.$$

Substituting Eq. (3.103), we obtain the spatial description of the balance of energy for a steady wave. It is identical to the material description, Eq. (3.101).

Description of the Rayleigh-Line Process

We now have descriptions of the conservation of mass, balance of linear momentum, and balance of energy for a steady wave. Let us apply these relationships to analyze the steady wave in Figure 3.23. Consider a material point $X = X_0$. Before this material point encounters the steady wave, it is at an initial state where the deformation gradient, stress, and internal energy are F_-, T_-, and $\mathcal{E}_-$. As this material point encounters the steady wave, the change in the stress and internal energy are described by $T = T_r(U_S, F)$ and $\mathcal{E} = \mathcal{E}_r(U_S, F)$, where

$$\begin{aligned} T_r(U_S, F) &= \rho_0 U_S^2 (F - F_-) + T_-, \\ \mathcal{E}_r(U_S, F) &= \frac{1}{2} U_S^2 (F - F_-)^2 + \frac{T_-}{\rho_0}(F - F_-) + \mathcal{E}_-, \end{aligned} \tag{3.104}$$

and $U_S = |C_S|$ [see Eqs. (3.99) and (3.101)]. Thus the stress and velocity of the steady wave are

$$T_r = \rho_0 \frac{\partial \mathcal{E}_r}{\partial F} \tag{3.105}$$

and

$$U_S = \left(\frac{\partial^2 \mathcal{E}_r}{\partial F^2} \right)^{1/2}. \tag{3.106}$$

Notice the similarity between these expressions and the corresponding expressions for the equilibrium stress T_e, Eq. (3.48), and the material wave velocity C, Eq. (3.87).

The velocity of the steady wave U_S is a constant; therefore, the relationship $\mathcal{E}(\eta, F) = \mathcal{E}_r(U_S, F)$ defines a path on the state surface. We call this path the *Rayleigh-line process*. In Figure 3.24 we illustrate a Rayleigh-line process on a state surface. Here we have illustrated the initial state on an isentrope at $F_- = 1$ where $T_- = \rho_0 \partial \mathcal{E} / \partial F = 0$. We determine the Rayleigh line process by selecting a value of U_S and computing the energy with Eq. (3.104). This process ends at the Hugoniot state. Next we examine the conditions that cause the Rayleigh-line process to terminate at the particular isentrope shown in this figure.

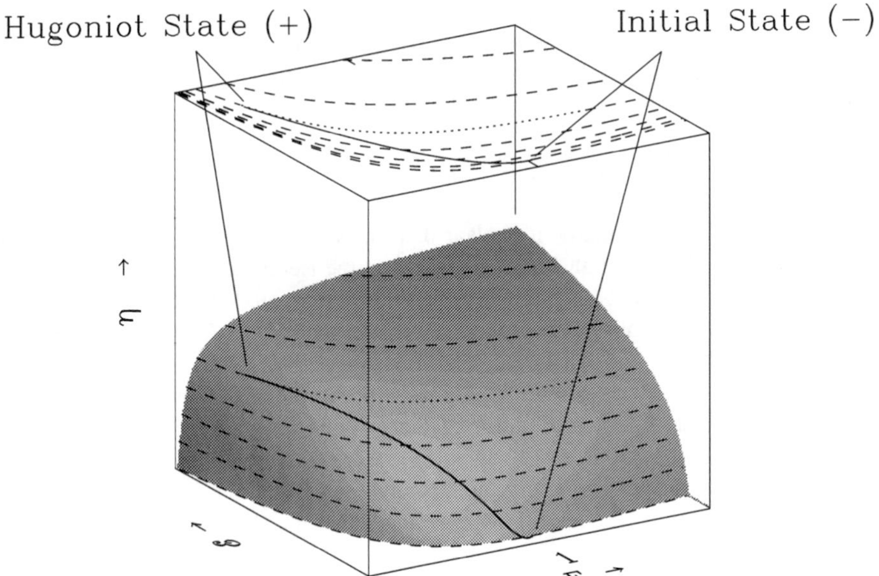

Figure 3.24 State surface with isentropes (– –) and a Rayleigh line process (—) followed by isentropic expansion (· · ·).

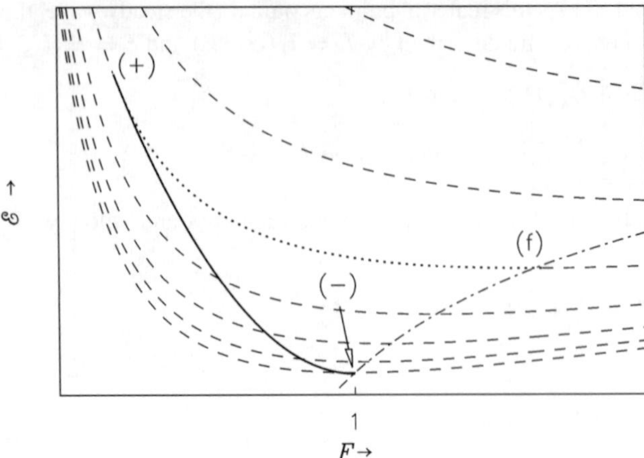

Figure 3.25 The $\mathcal{E}$–F diagram of the Rayleigh-line process (—), isentropic expansion (· · ·), stress-free isobar (− · −), and isentropes (− −).

Existence of the Rayleigh-Line Process

The Rayleigh-line process can only exist in a region where the dissipation $\mathcal{D}$ is positive. Steady waves cause compression, and during compression, $D \leq 0$. Thus, the reduced dissipation inequality, Eq. (3.44), becomes

$$T_r \leq T_e, \quad \text{for compression.}$$

To locate the region where this inequality is satisfied, we examine the isentropes and the Rayleigh line with the $\mathcal{E}$–F diagram (see Figure 3.25). You will notice that the stresses T_r and T_e are given respectively by the slopes of the isentropes and the Rayleigh line. For $F < 1$, these slopes are negative. Therefore, we rewrite the dissipation inequality in the following form:

$$-\frac{\partial \mathcal{E}_r}{\partial F} \geq -\frac{\partial \mathcal{E}}{\partial F}, \quad \text{for compression.} \tag{3.107}$$

With this form of the inequality, it is clear that the Rayleigh-line process can only exist in a region where the Rayleigh line is steeper than the isentropes. We see that the initial state and the final Hugoniot state, which bound this region, are located at states where the Rayleigh line and the isentrope are tangent. At these points, $T = T_r = T_e$ and $\mathcal{D} = 0$.

From Eq. (3.104), we find that the Rayleigh-line process has a parabolic form that projects up and across the isentropes. As the steady wave propagates past a point in the material, it decreases the value of the deformation gradient F and causes the material at this point to move away from the initial state (−) and follow the Rayleigh-line process. During this process, entropy increases until equilibrium is again reached at the Hugoniot state (+). At state (+), the slope of the Rayleigh-line process matches the slope of the isentrope, and T_r is equal to T_e.

Now suppose the steady wave is followed by a rarefaction wave that releases the stress and causes the material to move back to a stress-free state (see Figure 3.26). The rarefaction wave causes the material to unload by following an isentropic process. As a result, as the material at the Hugoniot state encounters the rarefaction wave, it moves from state (+) to

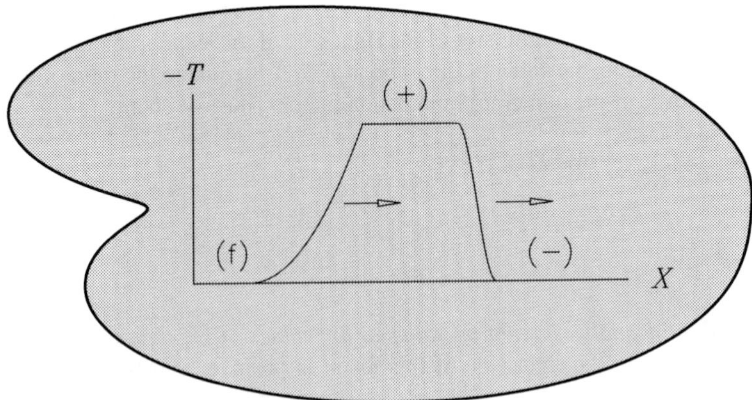

Figure 3.26 A steady wave followed by a rarefaction wave. The material ahead of the steady wave is at the initial state (−), the material between the steady wave and the rarefaction wave is at the Hugoniot state (+), and the material behind the rarefaction wave is at the final state (f).

state (f) (see Figure 3.25). State (f) resides at the stress-free state of the isentrope. Notice that both the initial state (−) and the final state (f) reside on the stress-free isobar. The final state (f) has a higher temperature than the initial state (−) and larger volume due to the thermal expansion of the material. It also has a greater internal energy and entropy. During loading along the Rayleigh-line process, the material dissipates some of the work produced by the steady wave. Consequently, the work put into the material by the steady wave cannot be completely recovered as work during the isentropic unloading process. The material retains some internal energy that can only be extracted as heat. Additionally, as we learned in Section 2.8, the material also retains some kinetic energy, which manifests itself as a small residual velocity at state (f).

Hugoniot Curve

Each value of U_S produces a unique Rayleigh-line process. In fact, we can use U_S as a parameter to describe all Rayleigh-line processes centered or starting at the initial state (−). When we decrease the value of U_S, Eq. (3.104) shows us that the resulting Rayleigh-line process is less steep, the Hugoniot state (+) moves closer to the initial state (−), and the region in which the Rayleigh-line process exists becomes smaller (see Figures 3.25 and 3.27). Recall that the Hugoniot curve is the curve traced out by the Hugoniot states for different values of U_S. The Hugoniot curve is *not* a steady-wave process. It is a locus of all of the possible Hugoniot states that can be reached by each of the Rayleigh-line processes. Each state on the Hugoniot curve is uniquely connected to a Rayleigh-line process.

Stability of the Rayleigh-Line Process

Here we show that a minimum value of U_S exists where the Rayleigh-line process collapses to a point and that the location of the Hugoniot state coincides with the initial state where $F_- = 1$. We evaluate this minimum value of U_S by substituting Eq. (3.104) into the dissipation inequality, Eq. (3.107), to obtain

$$U_S^2 \geq \frac{\partial \mathcal{E}/\partial F - T_-/\rho_0}{F - 1}, \quad \text{for compression,}$$

where we notice that division by $F-1$, which is a negative number, reverses the inequality. Both the numerator and denominator of the right side of this expression are zero at $F = 1$. From L'Hospital's rule, the value of this ratio at $F = 1$ is equal to the ratio of the values of the first derivatives of the numerator and denominator. Thus we obtain

$$U_S^2 \geq \frac{\partial^2 \mathcal{E}}{\partial F^2}, \quad \text{for } F = 1,$$

or

$$U_S^2 \geq \frac{C^\eta}{\rho_0}, \quad \text{for } F = 1.$$

The dissipation inequality cannot be satisfied for values of U_S that are smaller than this limit. When $C^\eta > 0$, the right side of this result is positive. Substituting Eq. (3.92), we obtain

$$U_S \geq c_0. \tag{3.108}$$

Recall that c_0 is the velocity of sound at the initial state ahead of the steady wave. We see that the steady wave propagates at a velocity that is greater than or equal to c_0. This is called the *supersonic condition.*

In Figure 3.27 we show that the Rayleigh line is tangent to the isentropes at the initial and Hugoniot states. On the Rayleigh line, we see that the entropy is monotonic, meaning it continually increases from the initial state to the Hugoniot state. At the Hugoniot state, the curvature of the isentrope is greater than or equal to the curvature of the Rayleigh line. Thus, at the Hugoniot state, we find that

$$\frac{\partial^2 \mathcal{E}}{\partial F^2} \geq \frac{\partial^2 \mathcal{E}_r}{\partial F^2}.$$

Substituting Eqs. (3.87) and (3.106), we obtain

$$C \geq U_S \quad \text{at the Hugoniot state.} \tag{3.109}$$

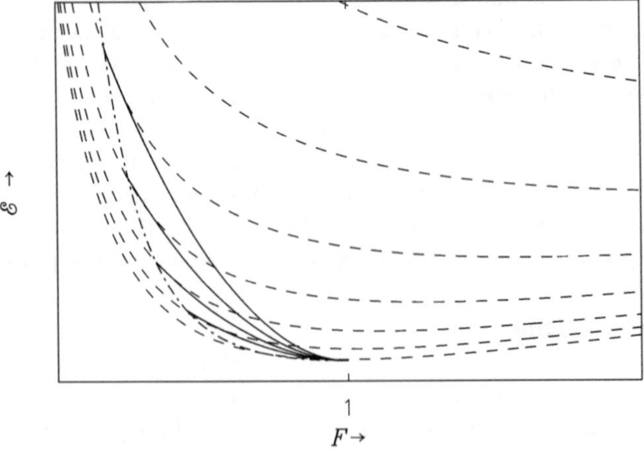

Figure 3.27 The $\mathcal{E}$–F diagram of four Rayleigh-line processes (—), the Hugoniot curve (– · –), and the isentropes (– –).

We find that velocity of the steady wave is less than or equal to the material wave velocity at the Hugoniot state. For a body in equilibrium at the Hugoniot state, the velocity of sound is C. Therefore, the velocity of a steady wave is less than or equal to the velocity of sound at the Hugoniot state. This is called the *subsonic condition*. Taken together, the supersonic and subsonic conditions mean that U_S must lie between the velocities of sound ahead of and behind the steady wave. These conditions ensure the stability of the steady wave because energy traveling within the wave cannot get ahead of the wave or fall behind it. Recall that mechanical waves also satisfy these conditions (see Exercise 2.52 in Section 2.7.8).

3.4.5 *Shock Discontinuities*

We have shown how the shock wave was born out of necessity to analyze the nonlinear-elastic material. In the absence of dissipation, the characteristic contours of compression waves eventually converge and intersect. The solution for the compression wave then acquires nonphysical attributes and must be abandoned. We replaced it with a discontinuous solution called a shock discontinuity or shock wave. Recall that shock waves cause the nonlinear-elastic material to jump between two points on the Hugoniot curve. These two points, which represent equilibrium states, are linked together by the Rayleigh line. In the strictest sense, the shock discontinuity does not cause the material to "follow" a path, and therefore, the Rayleigh line is not a process for the shock wave. As we discovered in Chapter 2, if a shock discontinuity exists in a nonlinear-elastic material, we can determine the solution with the jump conditions for the conservation of mass and the balance of linear momentum.

Shock discontinuities do not propagate in a viscous material. The viscous stress causes any discontinuity in the solution to quickly evolve into a steady wave. Thus, you might well question the value of the jump conditions when applied to viscous materials. However, we shall find that the jump conditions are consistent with the steady wave solution, and, as such, they are quite important analytic tools. To extend their application to the viscous material, we must supplement the jump conditions for mass and linear momentum with a third jump condition for the balance of energy. We shall derive the one-dimensional form of the jump condition for the balance of energy using both the material and spatial descriptions. We leave the derivation of the three-dimensional form of this jump condition as an exercise (see Exercise 3.32). The jump condition for the balance of energy was not required in Chapter 2 when we derived the solution for weak and strong shocks in a nonlinear-elastic material. We conclude this section with a discussion of why this is so.

Material Description

Consider the material element illustrated in Figure 3.28. It contains a nonlinear-elastic material. Recall that we analyzed a shock wave as a propagating singular surface. This element contains a singular surface that is located at the position X_S. No heat is added to this element, and no body force is present. Only the stresses on the left and right faces of the element do work. The balance of energy requires that the rate of change of the total energy in the element must equal the work done by these stresses:

$$T_R v_R - T_L v_L = \frac{d}{dt} \int_{X_L}^{X_R} \rho_0 (\mathcal{E} + \mathcal{K})\, dX.$$

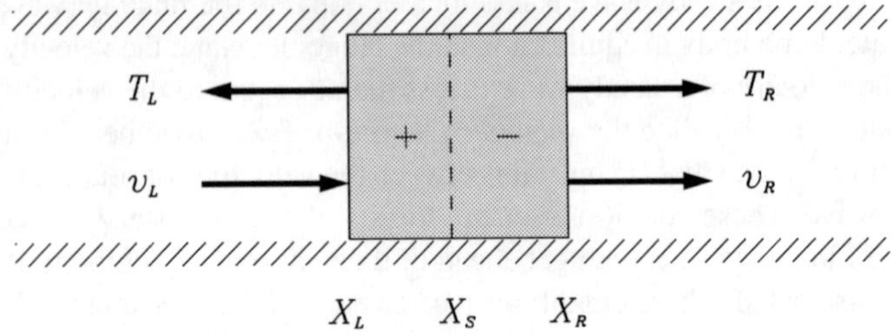

Figure 3.28 The volume element in the material description. A singular surface is located at the position X_S.

We require this condition to hold in the limit as $(X_L \to X_S)_+$ and $(X_R \to X_S)_-$. If we let $\phi = \rho_0(\mathcal{E} + \mathcal{K})$, we can employ Eq. (2.144) in Section 2.7.2. The result is

$$[\![Tv]\!] = -\rho_0 C_S [\![(\mathcal{E} + \mathcal{K})]\!]. \tag{3.110}$$

This is the jump condition for the balance of total energy. Recall that we were able to remove the kinetic energy term in the differential equation for the balance of total energy, Eq. (3.9). We can achieve a similar result with this jump condition. Let us rewrite this expression as follows:

$$\rho_0 C_S [\![\mathcal{E}]\!] = -[\![T]\!]v_+ - T_-[\![v]\!] + \frac{1}{2}\rho_0 C_S [\![v]\!]^2 - \rho_0 C_S [\![v]\!]v_+.$$

Next, we substitute the jump condition for the balance of linear momentum [see Eq. (2.146)],

$$[\![T]\!] = -\rho_0 C_S [\![v]\!],$$

and the compatibility condition [see Eq. (2.139)],

$$[\![F]\!]C_S + [\![v]\!] = 0,$$

to obtain

$$\rho_0 [\![\mathcal{E}]\!] = T_-[\![F]\!] + \frac{1}{2}\rho_0 C_S^2 [\![F]\!]^2, \tag{3.111}$$

which is the desired jump condition for the internal energy. This result looks similar to Eq. (3.101), but there is an important difference. We have just derived a jump condition for a nonlinear-elastic material, which links the initial state to the Hugoniot state. In Eq. (3.101), we derived the description of the entire Rayleigh-line process for a viscous material. However, when we use the balance of energy for a steady wave, Eq. (3.101), to evaluate the Hugoniot state, we obtain a result that is equivalent to this jump condition for the nonlinear-elastic material. *Thus the jump condition for a singular surface is equivalent to the jump condition for a smooth steady wave.*

We can obtain another useful form of this jump condition by expanding it as follows:

$$\rho_0 [\![\mathcal{E}]\!] = \frac{1}{2}T_-[\![F]\!] + \frac{1}{2}\{T_- + \rho_0 C_S^2 [\![F]\!]\}[\![F]\!],$$

which yields

$$\rho_0 [\![\mathcal{E}]\!] = \frac{1}{2}(T_+ + T_-)[\![F]\!]. \tag{3.112}$$

We find that the jump in internal energy is equal to the average stress times the jump in F, which is the area under the Rayleigh line in a T–F diagram. This relationship is called the *Rankine–Hugoniot equation.*

Spatial Description

Consider the material element in Figure 3.28. The current positions of the left and right faces of this element are $x_L(t) = \hat{x}(X_L, t)$ and $x_R(t) = \hat{x}(X_R, t)$. The current position of the singular surface within the element is $x_S(t) = \hat{X}(X_S, t)$. No heat is added to this element, and no body force is present. Only the stresses on the left and right faces of the element do work. Using the balance of energy, we require that the rate of change of the total energy in the element must equal the work done by these stresses:

$$T_R v_R - T_L v_L = \frac{d}{dt} \int_{x_L(t)}^{x_R(t)} \rho(\mathcal{E} + \mathcal{K})\, dx.$$

We require this condition to hold in the limit as $(x_L \to x_S)_+$ and $(x_R \to x_S)_-$. If we let $\phi = \rho(\mathcal{E} + \mathcal{K})$, we can employ Eq. (2.145) to obtain

$$-[\![Tv]\!] = [\![\rho(c_S - v)(\mathcal{E} + \mathcal{K})]\!]. \tag{3.113}$$

This is the spatial description of the jump condition for the balance of total energy.

To remove the kinetic energy term from this result, first notice that the jump condition for conservation of mass [see Eq. (2.148)] and the kinematic compatibility condition [see Eq. (2.138)] yield

$$\rho_0 C_S = \frac{\rho_0}{F_+}(c_S - v_+) = \frac{\rho_0}{F_-}(c_S - v_-). \tag{3.114}$$

We can substitute this result into Eq. (3.113) to obtain an expression that is identical to the material description of the jump condition for the balance of total energy, Eq. (3.110). Thus the spatial description of the jump condition for the balance of energy is identical to Eq. (3.112).

Hugoniot Entropy

The Hugoniot curve represents all of the Hugoniot states that we can attain either by a shock discontinuity or a steady wave. The initial state on the Hugoniot curve has internal energy, stress, entropy, and deformation gradient that are given by $\mathcal{E}_-$, T_-, η_-, and F_-. To determine the response of a material to either a shock discontinuity or a steady wave, we must determine five quantities: the four quantities $\mathcal{E}_+$, T_+, η_+, and F_+ at the Hugoniot state and the velocity C_S of the wave. We have four equations that can be used. They are the jump condition for the balance of linear momentum, the constitutive equation for the internal energy, the equation for the equilibrium stress, and the jump condition for the balance of energy:

$$\begin{aligned} &[\![T]\!] = \rho_0 C_S^2 [\![F]\!], && \mathcal{E} = \mathcal{E}(F, \eta), \\ &T = \rho_0 \frac{\partial \mathcal{E}}{\partial F}, && \rho_0 [\![\mathcal{E}]\!] = \frac{1}{2}(T_+ + T_-)[\![F]\!]. \end{aligned} \tag{3.115}$$

Let us use the first of these equations to determine the unknown velocity C_S. If we are given a value for F_+, then we can solve the remaining three equations for $\mathcal{E}_+$, T_+, and η_+. A plot of the solutions for a range of F_+ is called the Hugoniot curve (for example, see Figure 3.27).

Indeed, we can use F_+ as a parameter to identify a unique Hugoniot state on the Hugoniot curve. This means that we can represent the Hugoniot curve with the following relationship:

$$\eta = \check{\eta}(F). \tag{3.116}$$

Then we can represent the stress, internal energy, and temperature on the Hugoniot curve as

$$\begin{aligned} T &= T[F, \check{\eta}(F)] \equiv \check{T}(F), \\ \mathcal{E} &= \mathcal{E}[F, \check{\eta}(F)] \equiv \check{\mathcal{E}}(F), \\ \vartheta &= \frac{\partial \mathcal{E}}{\partial \eta} = \vartheta[F, \check{\eta}(F)] \equiv \check{\vartheta}(F). \end{aligned}$$

Although we use F as a parameter to define the particular Hugoniot state on the Hugoniot curve, you should recognize that we could use any of the five variables for this purpose. Also, you should remember that the Hugoniot curve is the locus of states that can be reached by either a shock discontinuity or a steady wave. It does not represent a process for these waves. The shock discontinuity jumps to the Hugoniot state and the steady wave follows the Rayleigh line process to the Hugoniot state. *Thus, although T on the Hugoniot curve is given by the expression for the equilibrium stress, this does not imply that the material maintains an equilibrium stress during the Rayleigh-line process that moves it from the initial state to the Hugoniot state.*

Our objective now is to determine how the entropy changes along the Hugoniot curve. Because we are using F as a parameter to determine the Hugoniot curve, we express the entropy of the Hugoniot state with the following Taylor's series expansion of Eq. (3.116):

$$[\![\eta]\!] = \left(\frac{d\check{\eta}}{dF}\right)_- [\![F]\!] + \frac{1}{2}\left(\frac{d^2\check{\eta}}{dF^2}\right)_- [\![F]\!]^2 + \frac{1}{6}\left(\frac{d^3\check{\eta}}{dF^3}\right)_- [\![F]\!]^3 + \cdots,$$

where the notation $(\cdot)_-$ indicates that the derivatives of $\check{\eta}$ are evaluated at the initial state. Neither the function $\check{\eta}(F)$ nor its Taylor's series expansion can be arbitrarily prescribed. In fact, they represent the solution to the Hugoniot relations in Eq. (3.115). Moreover, we can use the Hugoniot relations to evaluate the coefficients in this Taylor's series. We do this by taking successively higher derivatives of the jump condition for the balance of energy. We take the first derivative with respect to F and evaluate it at $F = F_+$ to obtain

$$\rho_0\left[\frac{\partial \mathcal{E}}{\partial \eta}\frac{d\check{\eta}}{dF} + \frac{\partial \mathcal{E}}{\partial F}\right]_+ = \frac{1}{2}\left(\frac{d\check{T}}{dF}\right)_+ (F_+ - F_-) + \frac{1}{2}(T_+ + T_-),$$

or

$$\rho_0\left(\vartheta\frac{d\check{\eta}}{dF}\right)_+ = \frac{1}{2}\left(\frac{d\check{T}}{dF}\right)_+ (F_+ - F_-) - \frac{1}{2}(T_+ - T_-). \tag{3.117}$$

Because the density and the temperature are positive, we can evaluate this expression at $F_+ = F_-$ and $T_+ = T_-$ and show that the coefficient in the first term of the Taylor's series is zero,

$$\left(\frac{d\check{\eta}}{dF}\right)_- = 0.$$

Next we take the derivative of Eq. (3.117) to obtain

$$\rho_0\left[\frac{d\breve{\vartheta}}{dF}\frac{d\breve{\eta}}{dF}+\vartheta\frac{d^2\breve{\eta}}{dF^2}\right]_+=\frac{1}{2}\left(\frac{d^2\breve{T}}{dF^2}\right)_+(F_+-F_-), \tag{3.118}$$

and we see that the coefficient in the second term of the Taylor's series is zero,

$$\left(\frac{d^2\eta}{dF^2}\right)_-=0.$$

Then we take the derivative of Eq. (3.118) to obtain

$$\rho_0\left[\frac{d^2\breve{\vartheta}}{dF^2}\frac{d\breve{\eta}}{\partial F}+2\frac{d\breve{\vartheta}}{dF}\frac{d^2\breve{\eta}}{dF^2}+\vartheta\frac{d^3\breve{\eta}}{dF^3}\right]_+=\frac{1}{2}\left(\frac{d^2\breve{T}}{dF^2}\right)_++\frac{1}{2}\left(\frac{d^3\breve{T}}{dF^3}\right)_+(F_+-F_-),$$

and we see that

$$\rho_0\left(\vartheta\frac{d^3\breve{\eta}}{dF^3}\right)_-=\frac{1}{2}\left(\frac{d^2\breve{T}}{dF^2}\right)_-.$$

Thus the Taylor's series becomes

$$[\![\eta]\!]=\frac{1}{12\rho_0}\left(\frac{1}{\vartheta}\frac{d^2\breve{T}}{dF^2}\right)_-[\![F]\!]^3+\cdots. \tag{3.119}$$

From this Taylor's series, we find that the jump in entropy across either a shock discontinuity or a steady wave is third order with respect to the jump in the deformation gradient.

Compression Shock To satisfy the second law of thermodynamics, the shock discontinuity must produce positive dissipation. Thus the jump in entropy must be positive, $[\![\eta]\!]>0$. For many liquids and solids, we can accurately determine the production of entropy by retaining only the third-order term in Eq. (3.119). For a shock wave to exist during compression of these materials, we find that

$$\left(\frac{d^2\breve{T}}{dF^2}\right)_-\leq 0,\quad [\![F]\!]<0. \tag{3.120}$$

The first of these inequalities is a constraint on the constitutive equation. In Chapter 2 we imposed a similar constraint on the constitutive equation for the nonlinear-elastic material [see Eq. (2.23)]. However, you should notice that

$$\frac{d^2\breve{T}}{dF^2}=\frac{\partial^2T}{\partial F^2}+2\frac{\partial^2T}{\partial F\partial\eta}\frac{d\breve{\eta}}{dF}+\frac{\partial^2T}{\partial\eta^2}\left(\frac{d\breve{\eta}}{dF}\right)^2+\frac{\partial T}{\partial\eta}\frac{d^2\breve{\eta}}{dF^2}.$$

Therefore, $d^2\breve{T}/dF^2$ can be negative when $\partial^2T/\partial F^2$ is positive. This is particularly true in a material, such as a porous solid, where a shock wave can produce a large increase in entropy. *It is important to remember that it is the shape of the Hugoniot curve, Eq. (3.120), and not the shape of the isentrope that determines if a compression shock can exist.*

The one exception to this rule is the nonlinear-elastic material where the stress is not a function of the entropy. Here, we see that $d^2\breve{T}/dF^2=d^2T/dF^2$. Thus Eq. (3.120) means that the compressed material must be stiffer than the uncompressed material. This ensures that characteristic contours of compression waves converged to form a shock discontinuity.

Rarefaction Shock Notice also that Eq. (3.119) yields a positive jump in entropy when

$$\left(\frac{d^2\check{T}}{dF^2}\right)_- \geq 0, \quad [\![F]\!] > 0.$$

In the particular case of a nonlinear-elastic material, this means that *rarefaction shocks can exist if the material becomes stiffer as it is stretched.* In general, most materials, but not all, violate this constraint. Even though rarefaction shocks are unusual, flyer-plate experiments have been employed to accurately measure them (see Chapter 0).

Weak Shock

Recall our discussion of the weak shock in Chapter 2. We showed that the Hugoniot for a weak shock can be approximated by the response path for a structured wave. We have found that the response path for a structured wave is an isentrope. This means that we can approximate the Hugoniot curve for a weak shock with an isentrope because the weak shock does not change entropy. Remember that we described the weak shock using the following constitutive equation for an elastic material [see Eq. (2.163)]:

$$T = T^0 + \rho_0 c_0^2 (F - 1) + \rho_0 c_0 \left(\frac{dC}{dF}\right)_- (F - 1)^2,$$

where $F - 1$ is the jump in the deformation gradient across the weak shock. In the analysis of the weak shock, we ignored terms containing $F - 1$ to third order. When we apply this condition to Eq. (3.119), we see that the weak shock assumption is equivalent to ignoring changes in entropy. This makes the Hugoniot coincident with the isentrope of the initial state.

Strong Shock

When the Hugoniot is not coincident with the isentrope of the initial state, we call the wave a strong shock. We can determine the Hugoniot state of a strong shock in a viscous material by solving Eq. (3.115). In general, we must solve these four equations simultaneously.

The procedure for a nonlinear-elastic material is different. In Chapter 2 we analyzed a strong shock for a nonlinear-elastic material. We used the constitutive equation, $T = T(F)$. This constitutive equation does not depend upon entropy. Recall that we evaluated the Hugoniot state for a strong shock using only the jump conditions for conservation of mass and balance of linear momentum. However, even though we did not use the jump condition for the balance of energy, it still applies to the strong shock in a nonlinear-elastic material.

To demonstrate this point, suppose for a moment that we insist that the constitutive equations for the nonlinear-elastic material are only a function of F. Hence the constitutive equation for the internal energy is only a function of F. Then we can obtain the equations that describe the Hugoniot of a nonlinear-elastic material from Eq. (3.115) after we remove the dependence on η. But this is an overdetermined system of equations because we have removed one unknown without removing an equation. The weak-shock approximation will still work because entropy is constant in that solution, but to make the strong-shock solution tractable, we must allow the internal energy to be a function of entropy. We can resolve this discrepancy by allowing the internal energy to be a function of F and η while still allowing the stress to only be a function of F.

Suppose the internal energy of a nonlinear-elastic material is

$$\mathcal{E} = \mathcal{E}_F(F) + \mathcal{E}_\eta(\eta). \tag{3.121}$$

We call $\mathcal{E}_F$ the *cold energetic*. It is a prescribed function of F. We call $\mathcal{E}_\eta$ the *thermal energetic*. It is a prescribed function of η. The equilibrium stress then becomes

$$T_e = \rho_0 \frac{d\mathcal{E}_F}{dF}.$$

Thus the equilibrium stress for this special material is only a function of F and is equivalent to the constitutive equation for the nonlinear-elastic material. Notice that because T_e is constant when F is constant, the isobars and isochors are coincident on the state surface of this material. The temperature is

$$\vartheta = \frac{d\mathcal{E}_\eta}{d\eta}. \tag{3.122}$$

Because ϑ is constant when η is constant, the isotherms and isentropes are also coincident. From Eqs. (3.53) and (3.65), the Grüneisen coefficient and the coefficient of thermal expansion are both zero. Consequently, linear-elastic waves and structured waves do not cause the temperature to change.

When we are given a material that is described by the internal energy Eq. (3.121), we can determine the Hugoniot state in two steps. First, we use Eq. (3.115) to determine the jumps in stress and velocity,

$$[\![T]\!] = -\rho_0 C_S [\![v]\!], \qquad T = \frac{d\mathcal{E}_F}{dF}.$$

This is identical to the procedure we used in Chapter 2. Once we obtain these solutions, we evaluate

$$\mathcal{E} = \mathcal{E}_F(F) + \mathcal{E}_\eta(\eta), \qquad \rho_0 [\![\mathcal{E}]\!] = \frac{1}{2}(T_+ + T_-)[\![F]\!],$$

for the jumps in internal energy and entropy.

3.4.6 *Exercises*

3.22 A block of aluminum at a temperature of 300 K has the following properties:

$$C^\vartheta = 70\,\text{GPa}, \qquad c_T = 900\,\text{J/kg/K},$$
$$\alpha = 75 \times 10^{-6}\,\text{K}^{-1}, \qquad \rho_0 = 2.7\,\text{Mg/m}^3.$$

(a) Determine the Grüneisen coefficient Γ, the isochoric specific heat c_v, and the isentropic stiffness C^η.

(b) Determine the speed of sound c_0 of an elastic wave.

(c) For a linear-elastic wave, determine the ratio of the temperature change $\Delta\vartheta$ to the change of the strain $\Delta F = \partial u/\partial X$.

(d) If this linear-elastic wave causes the stress to change by 0.1 MPa (a standard atmosphere), determine the change in temperature.

3.23 Repeat Exercise 3.22 using the following properties for water at $\vartheta = 295$ K:

$$C^\vartheta = 2.19\,\text{GPa}, \qquad c_p = 4.187\,\text{kJ/kg/K},$$
$$\alpha = 227 \times 10^{-6}\,\text{K}^{-1}, \qquad \rho_0 = 998\,\text{kg/m}^3.$$

3.24 The temperature ϑ of air during an isobaric process is given by

$$\rho\vartheta = \text{constant},$$

and the pressure p during an isothermal process is given by

$$p/\rho = \text{constant}.$$

At a pressure of $p = 0.1013$ MPa and a temperature of $\vartheta = 295$ K, the speed of sound and density of air are

$$c_0 = 345\,\text{m/s}, \qquad \rho_0 = 1.2\,\text{kg/m}^3.$$

(a) Determine the values of C^ϑ, C^η, and α. Hint: Use Eqs. (3.58) and (3.64).
(b) Determine the value of c_p/c_v.
(c) Determine the values of c_p and c_v.
(d) Determine the value of Γ.
(e) For a linear-elastic wave, determine the ratio of the change in the temperature to the change in the pressure.

3.25 The block of aluminum in Exercise 3.22 is initially at rest. A linear-elastic wave propagates into this block. Behind this wave $v/c_0 = 10^{-5}$.
(a) Determine the change of internal energy $\Delta\mathcal{E}$ resulting from the linear-elastic wave.
(b) Determine the change in kinetic energy $\Delta\mathcal{K}$.
(c) Determine the change in temperature $\Delta\vartheta$.
(d) For an isobaric process, determine the amount of heat required to change temperature by $\Delta\vartheta$.

3.26 Consider a linear-elastic half space, which is subjected to a stress $T(0, t)$ applied to the boundary at $X = 0$.
(a) Show that the energy introduced into the half space by the boundary condition is

$$\frac{\partial}{\partial t}\int_0^\infty \rho_o(\hat{\mathcal{E}} + \hat{\mathcal{K}})\,dX = T^2(0, t)/z_0.$$

(b) Determine the ratio of internal energy to kinetic energy.

3.27 Consider a material that is initially free of stress and at rest. Suppose a steady wave, which causes *expansion*, is observed in this material. Starting from the dissipation inequality

$$(T - T_e)D \geq 0,$$

(a) determine a condition similar to Eq. (3.107) that allows for the existence of this steady wave;
(b) determine the minimum propagation velocity of this wave if $F_- = 1$.
(c) What constraint does the subsonic condition place upon the curvatures of the isentropes and the Rayleigh-line in the $\mathcal{E}$–F diagram?

3.28 For a nonlinear-elastic material, derive the supersonic condition for a steady wave by using the known conditions on the curvatures of the Rayleigh-line and isentrope at the initial state in the $\mathcal{E}$–F diagram.

3.29 In Exercise 2.65 we considered a material in which

$$T = E_0\left(1 - \frac{1}{F}\right) + \frac{b}{F}\frac{\partial \hat{F}}{\partial t}.$$

(a) Determine the speed of sound c_0, the material wave velocity C, and the steady wave velocity C_S for $F_- = 1$.

(b) Determine the conditions on F that ensure the supersonic and subsonic conditions are satisfied, $c_0 \leq C_S \leq C$.

(c) Compare the answer for part (b) to the answer for part (c) of Exercise 2.65.

3.30 In Exercise 3.29, the stress of the material is not a function of η. We have shown that this implies that the temperature is not a function of F. However, we have also demonstrated that we can use F as a parameter to describe the Hugoniot curve so that $\eta = \breve{\eta}(F)$ [see Eq. (3.116)]. Therefore, along the Hugoniot curve for this material $\vartheta = \vartheta(\eta) = \breve{\vartheta}(F)$. For the Hugoniot curve of this material:

(a) Demonstrate that $(d\breve{\vartheta}/dF)_- = 0$.

(b) Letting ρ_0 and ϑ_0 represent the initial density and temperature of the material, show that

$$[\![\eta]\!] = \frac{E_0}{\rho_0 \vartheta_0}\left(-\frac{1}{6}[\![F]\!]^3 + \frac{1}{4}[\![F]\!]^4 + \cdots\right).$$

(c) Provided only the third- and fourth-order dependences of $[\![\eta]\!]$ on $[\![F]\!]$ are significant, determine the values of E_0 and b that allow for the existence of a compression shock.

(d) Under the constraints of Part (c), are there conditions that suggest a rarefaction shock wave might be possible?

3.31 Verify that the equation for the balance of total energy in the spatial description can be written as

$$\frac{\partial}{\partial t}\left[\rho\left(\mathcal{E} + \frac{1}{2}v_m v_m\right)\right] + \frac{\partial}{\partial x_k}\left[\rho v_k\left(\mathcal{E} + \frac{1}{2}v_m v_m\right)\right] = \frac{\partial}{\partial x_k}(T_{km} v_m),$$

where we assume that q, b, and r are zero. Assume that a steady wave solution exists in which all of the dependent variables in this equation are solely functions of $\varphi = x_k N_k - c_s t$. Here N_k is the direction of propagation and c_S is the spatial wave velocity of the steady wave. Obtain the following relationship for this steady wave by integrating the equation for the balance of total energy:

$$\rho(N_k v_k - c_S)\left(\mathcal{E} + \frac{1}{2}v_m v_m\right) - T_{km} v_m N_k = \text{constant}. \tag{3.123}$$

This is the three-dimensional form of the balance of energy for a steady wave.

3.32 The balance of energy requires that [see Eq. (3.19)]

$$\frac{dE(t)}{dt} = W + H.$$

Recall that, assuming a continuous and differentiable solution, we applied the transport and Gauss theorems to this equation to obtain the differential equation for the balance of energy in three dimensions. Now let us assume that a singular surface exists in a material volume. The spatial velocity of this singular surface is c_S, and its direction of propagation is N_k. Assuming b and r are equal to zero, apply the modified transport theorem, Eq. (2.153), and the Gauss theorem to this relationship. Show that the resulting jump condition is

$$\left[\!\!\left[\rho(N_k v_k - c_S)\left(\mathcal{E} + \frac{1}{2}v_m v_m\right)\right]\!\!\right] = [\![T_{km} v_m]\!] N_k.$$

3.33 Consider an infinite homogeneous body. The internal energy of this material is

$$\mathcal{E} = \frac{E_0}{\rho_0}(F - \ln F) + \mathcal{E}_\eta(\eta).$$

Initially the body is at rest. The initial values of the deformation gradient and entropy are $F_- = 1$ and $\eta = \eta_-$. We also require that $\mathcal{E}_\eta(\eta_-) = 0$. Now suppose that a steady wave propagates through this material. The propagation velocity of this wave is $U_S/c_0 = 5$.

(a) Show that the speed of sound at the initial state and the deformation gradient at the Hugoniot state are

$$c_0^2 = E_0/\rho_0, \qquad F_+ = (c_0/U_S)^2.$$

(b) Create an $\mathcal{E}$–F diagram for the Rayleigh-line process by plotting $\mathcal{E}/c_0^2$ versus F. Extend the process beyond the initial and Hugoniot states by letting $1/100 \le F \le 1.25$.

(c) In the same diagram, graph the isentrope of the initial state.

(d) Determine the value of $\mathcal{E}_\eta/c_0^2$ at the Hugoniot state and then graph the isentrope of the Hugoniot state.

(e) Use the $\mathcal{E}$–F diagram you have created to demonstrate that the Rayleigh-line is tangent to the isentropes of the initial and the Hugoniot states.

(f) Demonstrate that the subsonic and supersonic conditions are satisfied by comparing the appropriate curvatures of the three processes you have illustrated.

3.5 Heat Conduction

Thus far we have ignored heat conduction by requiring the heat flux to be zero ($q = 0$). Traditionally, heat conduction has been considered inappropriate to the description of waves. In part, this prejudice is based upon the analysis for a steady wave in an ideal gas first presented by Rayleigh (see Chapter 4). Additionally, it is also true that many constitutive models that omit heat conduction have been successful in describing experimental data. In this section, we include heat conduction in our constitutive description of the viscous material. This results in the *thermoviscous material.* We also consider a constitutive description where we only include heat conduction and ignore viscosity. Materials thus described are called *thermoelastic.*

In a material that obeys Fourier's law of heat conduction, the heat flux is

$$q = -K\frac{\partial \vartheta}{\partial x}. \tag{3.124}$$

This is a constitutive equation. Often, we assume that K is constant. In this constitutive equation, q is the dependent variable, and the gradient of the temperature $\partial\vartheta/\partial x$ is the independent variable. This suggests that if we wish to construct a constitutive model for a thermoviscous material, then we should require temperature to be an independent variable in *all* of the constitutive equations in the model. Typically, in a constitutive model, we select one member of a conjugate pair to be a dependent variable and the other to be an independent variable. In the constitutive model for a viscous material, we selected temperature as the dependent variable and entropy as the independent variable [see Eqs. (3.33)–(3.35)]. At the time, you may have considered this to be an arbitrary choice. However, in this section, we discover that there is a fundamental reason behind our selection of entropy as the independent variable. This does not obviate the necessity of making temperature the independent variable in the description of heat conduction, but it does add complexity to the formulation. For example, to make temperature an independent variable, you might be tempted to replace the internal energy function $\mathcal{E}(F, \eta)$ with the function $\mathcal{E}(F, \vartheta)$. However, we cannot arbitrarily alter this function in this way. To illustrate, we shall look at a difficulty that can arise with an even more basic alteration of this function.

The state surface describes the response of a material to deformation and heating. We have represented this surface by the constitutive equation for the internal energy

$$\mathcal{E} = \mathcal{E}(F, \eta).$$

This constitutive relationship for $\mathcal{E}$ is single valued, meaning it uniquely determines the value of internal energy for every choice of F and η. Inspection of Figure 3.9 shows that the state surface we have illustrated throughout this chapter meets this requirement.

Now let us consider another representation of the state surface where we express the deformation gradient in terms of the internal energy and the entropy

$$F = F(\mathcal{E}, \eta).$$

Because the isochors for the state surface overlap in the $\mathcal{E}$–F diagram, this expression is multivalued (see Figure 3.13). The overlapping of isochors occurs because the equilibrium stress T_e can have both positive and negative values. Because this function for F is multivalued, we cannot uniquely determine the state of the material from the values of $\mathcal{E}$ and η. In contrast, the temperature is always positive. Consequently, in the $\mathcal{E}$–F diagram, the isentropes never overlap (see Figure 3.10). The expression

$$\eta = \eta(F, \mathcal{E})$$

is single valued and uniquely determines the state. Therefore, we can use this function to describe the state surface. Other constitutive descriptions of the state surface are also possible.

3.5.1 *Helmholtz Energy*

Because temperature and entropy are a conjugate pair of variables, let us see if we can interchange the roles of temperature and entropy to obtain a constitutive equation for the state surface that is a function of the deformation gradient and the temperature. To do so, we must be able to write a function that uniquely defines each state of the material given the deformation gradient and the temperature. In general, this is possible only over limited regions of the state surface. For example, recall that the temperature of the material is equal to the slope of the tangent to the isochor in the $\mathcal{E}$–η diagram. In Figure 3.29 we illustrate an isochor on which two unique states have the same temperature. Notice that in the region between the two vertical dotted lines the concave side of the isochor faces downward. As discussed in Section 3.3.2, we find that the specific heat c_v is negative in this region, and adding heat to the material will cause its temperature to decrease. In this region of the state surface the material will exhibit unstable behavior, and even infinitesimally small perturbations in temperature and stress will result in very large changes in the state. The topic of stability of thermodynamic systems is outside the scope of this text; however, all materials undergoing a phase transformation, such as water transforming to steam, exhibit this type of behavior.† The representation of the state surface that we are about to discuss is limited to single-phase regions. The connection of representations of different phases across phase boundaries must be treated with additional techniques that are not discussed in this book.

† You will find a discussion of stability in *Thermodynamics* by H. B. Callen (Wiley, 1960).

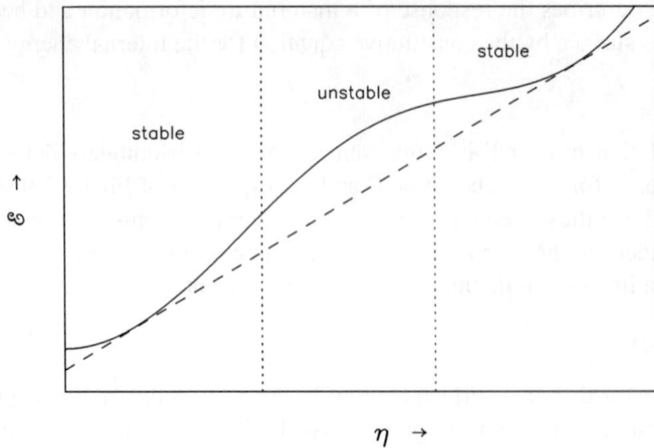

Figure 3.29 The $\mathcal{E}$–η diagram of an isochor (—) with unstable states, and a tangent line (– –) intersecting two states with equal temperature.

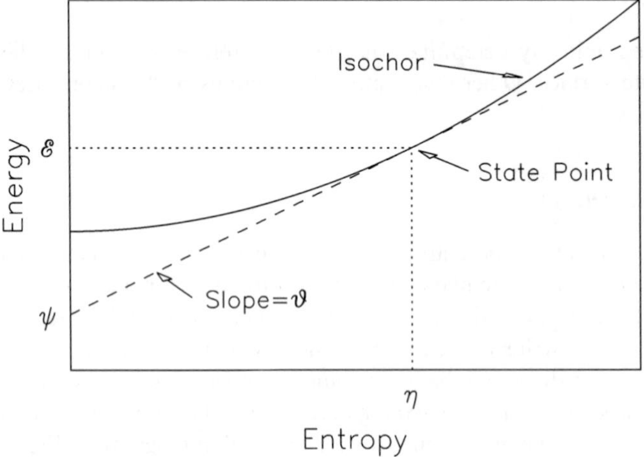

Figure 3.30 The $\mathcal{E}$–η diagram of an isochor.

In regions where the specific heat c_v is positive it is possible to uniquely determine the state with the temperature. We show this case in Figure 3.30, on an isochor that has stable states everywhere. Each state has a unique temperature represented by a tangent line that touches the state surface at one point. This tangent line intersects the vertical axis of the diagram at a point called the *Helmholtz energy* ψ of the state. We assign physical dimensions of J/kg to the Helmholtz energy. We find that each state point on this isochor has a unique value of ψ given by the function

$$\psi = \psi(F, \vartheta). \tag{3.125}$$

Legendre Transform

We shall now show that when the function ψ is given, we can determine the values of internal energy and entropy by prescribing the deformation gradient and the temperature. We require two relationships. We obtain the first from simple geometry. From the $\mathcal{E}$–η diagram of the isochor, we see that

$$\psi = \mathcal{E} - \vartheta\eta. \tag{3.126}$$

We obtain the second relationship by considering two adjacent states on an isochor. We label these states (0) and (1). From Eq. (3.126), we obtain

$$\begin{aligned}\psi(F,\vartheta_1) - \psi(F,\vartheta_0) &= \mathcal{E}(F,\eta_1) - \mathcal{E}(F,\eta_0) - \vartheta_1\eta_1 + \vartheta_0\eta_0 \\ &= \mathcal{E}(F,\eta_1) - \mathcal{E}(F,\eta_0) - \vartheta_1(\eta_1 - \eta_0) - \eta_0(\vartheta_1 - \vartheta_0).\end{aligned}$$

Rearranging terms we get

$$\left[\frac{\psi(F,\vartheta_1) - \psi(F,\vartheta_0)}{\vartheta_1 - \vartheta_0} + \eta_0\right](\vartheta_1 - \vartheta_0) = \left[\frac{\mathcal{E}(F,\eta_1) - \mathcal{E}(F,\eta_0)}{\eta_1 - \eta_0} - \vartheta_1\right](\eta_1 - \eta_0).$$

In the limit as state (1) approaches state (0), we have

$$\left[\frac{\partial\psi}{\partial\vartheta}(F,\vartheta_0) + \eta_0\right](\vartheta_1 - \vartheta_0) = \left[\frac{\partial\mathcal{E}}{\partial\eta}(F,\eta_1) - \vartheta_1\right](\eta_1 - \eta_0).$$

The right-hand side of this result is zero; therefore,

$$\eta = -\frac{\partial\psi}{\partial\vartheta}. \tag{3.127}$$

This is our second relationship.

If we start with the Helmholtz energy function $\psi(F,\vartheta)$, we can use the values of F and ϑ to determine the internal energy and the entropy. The process is as follows:

- $\psi(F,\vartheta)$ is determined by Eq. (3.125).
- $\eta(F,\vartheta)$ is determined by Eq. (3.127).
- $\mathcal{E}(F,\vartheta)$ is determined by Eq. (3.126).

We can also reverse this process. If we start with the internal energy function $\mathcal{E}(F,\eta)$, we can use the values of F and η to determine the Helmholtz energy and the temperature as follows:

- $\mathcal{E}(F,\eta)$ is determined by Eq. (3.33).
- $\vartheta(F,\eta)$ is determined by Eq. (3.35).
- $\psi(F,\eta)$ is determined by Eq. (3.126).

Therefore, we can use the Helmholtz energy $\psi(F,\vartheta)$ to represent the same state surface as the internal energy $\mathcal{E}(F,\eta)$. This procedure, which only works in stable regimes of the state surface, is an example of a *Legendre transformation*. Specifically, we have used the Legendre transform, Eq. (3.126), of the internal energy $\mathcal{E}$ to define the Helmholtz energy ψ.

Both $\mathcal{E}(F,\eta)$ and $\psi(F,\vartheta)$ describe the same state surface. We can derive all of the quantities associated with an equilibrium state from either function. However, if we write $\mathcal{E}$ as a function of F and ϑ or if we write ψ as a function of F and η, some quantities cannot be evaluated. For example, we cannot evaluate η from $\mathcal{E}(F, \vartheta)$, and we cannot evaluate ϑ from $\psi(F, \eta)$. For this reason, the function $\mathcal{E}(F, \eta)$ is called the *fundamental form* of the internal energy, and $\psi(F, \vartheta)$ is called the fundamental form of the Helmholtz energy. *In previous sections, we only used the fundamental form of the internal energy. Throughout this section, we shall only use the fundamental form of the Helmholtz energy.*

3.5.2 *Thermoviscous Material*

A thermoviscous material is a viscous material that conducts heat. To incorporate heat conduction into the theory of dissipative materials, we must write constitutive equations that use temperature as an independent variable. As we have shown in the previous section, if we use the temperature and the deformation gradient as independent variables, we must represent the state surface by the Helmholtz energy

$$\psi = \mathcal{E} - \vartheta\eta.$$

This transformation only works over a stable regime of a state surface.

In Section 3.2 we initiated the development of the constitutive model for a viscous material by assuming a functional dependence for the constitutive equations. Then we identified all possible admissible thermodynamic processes and used them to determine a set of constraints for these constitutive equations. Our application of these constraints to the constitutive equations of the viscous material ensured that no admissible process could violate the second law of thermodynamics. Here we shall use the same procedure to develop a constitutive model for a thermoviscous material. We start with the following assumptions for the constitutive equations:

$$\psi = \psi(F, D, \vartheta, \partial\hat{\vartheta}/\partial X), \tag{3.128}$$
$$\eta = \eta(F, D, \vartheta, \partial\hat{\vartheta}/\partial X), \tag{3.129}$$
$$T = T(F, D, \vartheta, \partial\hat{\vartheta}/\partial X), \tag{3.130}$$
$$q = q(F, D, \vartheta, \partial\hat{\vartheta}/\partial X). \tag{3.131}$$

We include the dependence on D to account for viscosity, and we include the dependence on $\partial\hat{\vartheta}/\partial X$ to account for Fourier heat conduction. Notice that we have ignored Eq. (3.127), which shows that $\eta = -\partial\psi/\partial\vartheta$. In what follows, we shall demonstrate an alternative method for deriving this expression.

To examine the constraints imposed by the second law of thermodynamics, we must first define all possible admissible thermodynamic processes. Remember that in Section 3.2 the motion $\hat{x}(X, t)$ and the entropy $\hat{\eta}(X, t)$ defined an admissible process; however, for these particular constitutive equations, we leave it as an exercise for you to show that any choices of the motion $\hat{x}(X, t)$ and the temperature $\hat{\vartheta}(X, t)$ define an admissible process (see Exercise 3.36).

We substitute the equation for the rate of change of entropy, Eq. (3.6), into the second law of thermodynamics, Eq. (3.30), to obtain

$$\frac{\mathcal{D}}{\vartheta} = \frac{\partial\hat{\eta}}{\partial t} + \frac{1}{\rho_0}\frac{\partial}{\partial X}\left(\frac{\widehat{q}}{\hat{\vartheta}}\right) - \frac{r}{\vartheta} \geq 0. \tag{3.132}$$

This is called the *Clausius–Duhem inequality*. We also substitute the Legendre transform Eq. (3.126) into the balance of energy Eq. (3.10) to obtain

$$\frac{\partial \hat{\psi}}{\partial t} + \eta \frac{\partial \hat{\vartheta}}{\partial t} + \vartheta \frac{\partial \hat{\eta}}{\partial t} = \frac{TD}{\rho} - \frac{1}{\rho_0} \frac{\partial \hat{q}}{\partial X} + r. \tag{3.133}$$

Combining the previous two relationships to eliminate r, we see that

$$\mathcal{D} = -\frac{\partial \hat{\psi}}{\partial t} - \eta \frac{\partial \hat{\vartheta}}{\partial t} - \frac{q}{\rho_0 \vartheta} \frac{\partial \hat{\vartheta}}{\partial X} + \frac{TD}{\rho} \geq 0.$$

Now we employ the chain rule to expand $\partial \hat{\psi}/\partial t$, obtaining

$$\begin{aligned}\mathcal{D} = &-\frac{\partial \psi}{\partial D} \frac{\partial \hat{D}}{\partial t} - \frac{\partial \psi}{\partial(\partial \hat{\vartheta}/\partial X)} \frac{\partial^2 \hat{\vartheta}}{\partial t \partial X} - \left(\eta + \frac{\partial \psi}{\partial \vartheta}\right) \frac{\partial \hat{\vartheta}}{\partial t} \\ &+ \left(\frac{T}{\rho} - F \frac{\partial \psi}{\partial F}\right) D - \frac{q}{\rho_0 \vartheta} \frac{\partial \hat{\vartheta}}{\partial X} \geq 0.\end{aligned}$$

Remembering that $\hat{x}(X, t)$ and $\hat{\vartheta}(X, t)$ define an admissible process, we can apply the arguments that we used in Section 3.2. Because $\partial \hat{D}/\partial t$ appears explicitly in this inequality and none of the remaining terms depend on $\partial \hat{D}/\partial t$, the coefficient of $\partial \hat{D}/\partial t$ must be zero. Thus ψ cannot depend on D. Similarly, the coefficient of $\partial^2 \hat{\vartheta}/\partial t \partial X$ must be zero, and ψ cannot depend on $\partial \hat{\vartheta}/\partial X$. Next, we find that the coefficient of $\partial \hat{\vartheta}/\partial t$ must be zero. This yields the constraint that we previously presented in Eq. (3.127). Thus the constitutive assumptions are constrained to be

$$\psi = \psi(F, \vartheta), \tag{3.134}$$

$$T = T(F, D, \vartheta, \partial \hat{\vartheta}/\partial X), \tag{3.135}$$

$$\eta = -\partial \psi/\partial \vartheta, \tag{3.136}$$

$$q = q(F, D, \vartheta, \partial \hat{\vartheta}/\partial X), \tag{3.137}$$

and the reduced dissipation inequality for the thermoviscous material is

$$\mathcal{D} = \left(\frac{T}{\rho} - F \frac{\partial \psi}{\partial F}\right) D - \frac{q}{\rho_0 \vartheta} \frac{\partial \hat{\vartheta}}{\partial X} \geq 0. \tag{3.138}$$

Stress for Thermoelastic Material

By definition, in the thermoelastic material, we assume that the stress and the heat flux are *not* functions of D. When we introduce these conditions into the reduced dissipation inequality, we find that D appears explicitly in the inequality, while no other term depends on D. Because we can choose processes that change the sign of D without affecting other terms in the inequality, the dissipation inequality will be violated unless we require that

$$T = \rho_0 \frac{\partial \psi}{\partial F}. \tag{3.139}$$

For the thermoelastic material, the dissipation inequality becomes

$$\mathcal{D} = -\frac{q}{\rho_0 \vartheta} \frac{\partial \hat{\vartheta}}{\partial X} \geq 0.$$

This is called *Fourier's inequality*. It requires that we select a constitutive equation for q that ensures heat flows down a temperature gradient. It is important to remember that Fourier's inequality is a special case of the reduced dissipation inequality for a thermoviscous material, Eq. (3.138). We see that during dissipative processes, Fourier's inequality does not have to hold.

Definition of Equilibrium

In a thermoviscous material, we define an equilibrium state to be one in which both D and $\partial\hat{\vartheta}/\partial X$ are zero. Thus, from Eq. (3.138), at equilibrium $\mathcal{D} = 0$. Because the second law of thermodynamics requires that $\mathcal{D} \geq 0$, the dissipation $\mathcal{D}$ is a minimum at equilibrium. Hence the quantity $\frac{T}{\rho} - F\frac{\partial\psi}{\partial F}$ must change sign when D changes sign, and q must change sign when $\partial\hat{\vartheta}/\partial X$ changes sign. Consequently, at an equilibrium state

$$T(F, 0, \vartheta, 0) = T_e, \qquad q(F, 0, \vartheta, 0) = 0,$$

where the equilibrium stress is

$$T_e = \rho_0 \frac{\partial\psi}{\partial F}. \tag{3.140}$$

As with the viscous material, we find that the stress can be represented by the additive decomposition

$$T = T_e + \mathcal{T},$$

where the viscous stress is

$$\mathcal{T} = \mathcal{T}(F, D, \vartheta, \partial\hat{\vartheta}/\partial X),$$

and

$$\mathcal{T}(F, 0, \vartheta, 0) = 0.$$

In a reversible process, the quantities D and $\partial\hat{\vartheta}/\partial X$ are sufficiently small so that we can approximate the stress and heat flux as $\mathcal{T} = 0$, $T = T_e$, and $q = 0$.

Energy and Dissipation Relations

In terms of the equilibrium stress, the equation for the balance of energy, Eq. (3.133), becomes

$$\begin{aligned}\rho_0\vartheta\frac{\partial\hat{\eta}}{\partial t} &= (T - T_e)FD - \frac{\partial\hat{q}}{\partial X} + \rho_0 r \\ &= \mathcal{T}FD - \frac{\partial\hat{q}}{\partial X} + \rho_0 r,\end{aligned} \tag{3.141}$$

and the reduced dissipation inequality, Eq. (3.138), becomes

$$\begin{aligned}\rho_0\mathcal{D} &= (T - T_e)FD - \frac{q}{\vartheta}\frac{\partial\hat{\vartheta}}{\partial X} \geq 0 \\ &= \mathcal{T}FD - \frac{q}{\vartheta}\frac{\partial\hat{\vartheta}}{\partial X} \geq 0.\end{aligned} \tag{3.142}$$

3.5.3 *Equilibrium States and Processes*

Recall that when we represented the state surface with the internal energy $\mathcal{E}(F, \eta)$, we expressed the thermodynamic stiffnesses, specific heats, and expansion coefficients as derivatives of $\mathcal{E}$. We determined these relationships by examining the isentropic, isochoric, isothermal, and isobaric processes. Now that we have represented the state surface with the Helmholtz energy $\psi(F, \eta)$, we repeat this exercise to determine these quantities in terms of the derivatives of ψ. These four processes are reversible processes, and we recall that during a reversible process, $q = 0$ and $T = T_e$. Because the constitutive equation for the entropy is $\eta = \eta(F, \vartheta)$, the balance of energy reduces to

$$\vartheta\left(\frac{\partial \eta}{\partial F}\frac{\partial \hat{F}}{\partial t} + \frac{\partial \eta}{\partial \vartheta}\frac{\partial \hat{\vartheta}}{\partial t}\right) = r. \tag{3.143}$$

Isochoric Process

During an isochoric process $D = 0$, and the equation of the balance of energy, Eq. (3.143), becomes

$$\vartheta\frac{\partial \eta}{\partial \vartheta}\frac{\partial \hat{\vartheta}}{\partial t} = r.$$

When we compare this expression to Eq. (3.56),

$$c_v\frac{\partial \hat{\vartheta}}{\partial t} = r,$$

we see that the specific heat at constant volume is

$$c_v = \vartheta\frac{\partial \eta}{\partial \vartheta} = -\vartheta\frac{\partial^2 \psi}{\partial \vartheta^2}. \tag{3.144}$$

Isothermal Process

For a reversible process, the stress is given by Eq. (3.140). Because $\psi = \psi(F, \vartheta)$, the change in stress for a constant temperature process is

$$\frac{\partial \hat{T}_e}{\partial t} = \rho_0 F\frac{\partial^2 \psi}{\partial F^2}D.$$

When we compare this result to Eq. (3.58), we find that the longitudinal stiffness at constant temperature is

$$C^{\vartheta} = \rho_0 F\frac{\partial^2 \psi}{\partial F^2}. \tag{3.145}$$

Isobaric Process

The equilibrium stress T_e is constant during an isobaric process:

$$\frac{\partial}{\partial t}\left(\frac{\hat{T}_e}{\rho_0}\right) = \frac{\partial^2 \psi}{\partial F^2}\frac{\partial \hat{F}}{\partial t} + \frac{\partial^2 \psi}{\partial \vartheta \partial F}\frac{\partial \hat{\vartheta}}{\partial t} = 0.$$

Using Eq. (3.145) and solving for $D = \frac{1}{F}\frac{\partial \hat{F}}{\partial t}$, we obtain

$$D = \alpha \frac{\partial \hat{\vartheta}}{\partial t},$$

where the longitudinal coefficient of thermal expansion is

$$\alpha = -\frac{\rho_0}{C^{\vartheta}} \frac{\partial^2 \psi}{\partial \vartheta \partial F}. \tag{3.146}$$

During this process, we determine the change in entropy with Eq. (3.143):

$$-FD\vartheta \frac{\partial^2 \psi}{\partial \vartheta \partial F} + \vartheta \frac{\partial \eta}{\partial \vartheta} \frac{\partial \hat{\vartheta}}{\partial t} = r.$$

We combine Eq. (3.144) and the last three expressions to obtain

$$c_T \frac{\partial \hat{\vartheta}}{\partial t} = r, \tag{3.147}$$

where the specific heat at constant stress is

$$c_T = c_v + \frac{\vartheta C^{\vartheta} \alpha^2}{\rho}. \tag{3.148}$$

Isentropic Process

Along an isentrope, entropy is constant:

$$\frac{\partial \hat{\eta}}{\partial t} = \frac{\partial \eta}{\partial \vartheta} \frac{\partial \hat{\vartheta}}{\partial t} + \frac{\partial \eta}{\partial F} \frac{\partial \hat{F}}{\partial t} = 0.$$

We substitute Eqs. (3.144), (3.146), and (3.66) to obtain

$$\frac{1}{\vartheta} \frac{\partial \hat{\vartheta}}{\partial t} = -\Gamma D. \tag{3.149}$$

The change in the stress $T = T_e(F, \vartheta)$ during a reversible isentropic process is

$$\frac{\partial \hat{T}_e}{\partial t} = \rho_0 \frac{\partial^2 \psi}{\partial F^2} \frac{\partial \hat{F}}{\partial t} + \rho_0 \frac{\partial^2 \psi}{\partial F \partial \vartheta} \frac{\partial \hat{\vartheta}}{\partial t}.$$

Substitution of Eq. (3.145) gives

$$\frac{\partial \hat{T}_e}{\partial t} = C^{\vartheta} D - \rho_0 \frac{\partial \eta}{\partial F} \frac{\partial \hat{\vartheta}}{\partial t}.$$

Then applying Eqs. (3.146) and (3.149), we obtain a relationship that is identical to Eq. (3.49):

$$\frac{\partial \hat{T}_e}{\partial t} = C^{\eta} D, \tag{3.150}$$

where

$$C^{\eta} = \frac{C^{\vartheta}}{c_v} \left(c_v + \frac{\vartheta C^{\vartheta} \alpha^2}{\rho} \right).$$

We can demonstrate the equivalence of Eqs. (3.49) and (3.150) by substituting Eq. (3.148) to obtain Eq. (3.63),

$$\frac{c_T}{c_v} = \frac{C^\eta}{C^\vartheta},$$

and by dividing Eq. (3.148) by $c_T C^\vartheta$ to obtain Eq. (3.67),

$$\frac{1}{C^\vartheta} = \frac{1}{C^\eta} + \frac{\vartheta\alpha^2}{\rho c_T}.$$

3.5.4 *Rayleigh-Line Process*

We have found that dissipation produces structure in steady waves. In a viscous material, compression waves steepen and form steady waves rather than shock discontinuities. In this section, we show that heat conduction also contributes to structure in steady waves. Indeed, in the absence of viscosity, we find that heat conduction alone produces structure in steady waves.

Balance of Energy for Steady Waves

We have already obtained steady-wave solutions for the equations for the conservation of mass and balance of linear momentum. These results apply to both conducting and nonconducting materials. However, heat conduction does introduce a new term in the equation for the balance of energy. Hence we must derive new results for the balance of energy of a steady wave. When heat conduction is included, the equation for the balance of energy, Eq. (3.10), becomes

$$\rho_0 \frac{\partial \hat{\mathcal{E}}}{\partial t} = T\frac{\partial \hat{F}}{\partial t} - \frac{\partial \hat{q}}{\partial X},$$

where we assume that the specific external heat supply r is zero. Because we have expressed this equation in the material description, we assume a steady-wave solution exists in which the internal energy $\mathcal{E}$, stress T, deformation gradient F, and heat flux q are solely functions of the variable $\hat{\varphi}$,

$$\hat{\varphi} = X - C_S t.$$

Then we differentiate the equation of the balance of energy by the chain rule to obtain

$$\rho_0 C_S \frac{d\mathcal{E}}{d\hat{\varphi}} = C_S T \frac{dF}{d\hat{\varphi}} + \frac{dq}{d\hat{\varphi}}.$$

Next, remember that the steady-wave solution for the balance of linear momentum, Eq. (3.99), is

$$T - T_- = \rho_0 C_S^2 (F - F_-),$$

where T_- and F_- represent the initial state of the material ahead of the steady wave. We combine these two equations and integrate the result to obtain

$$q = C_S\left[\rho_0(\mathcal{E} - \mathcal{E}_-) - T_-(F - F_-) - \frac{1}{2}\rho_0 C_S^2 (F - F_-)^2\right], \tag{3.151}$$

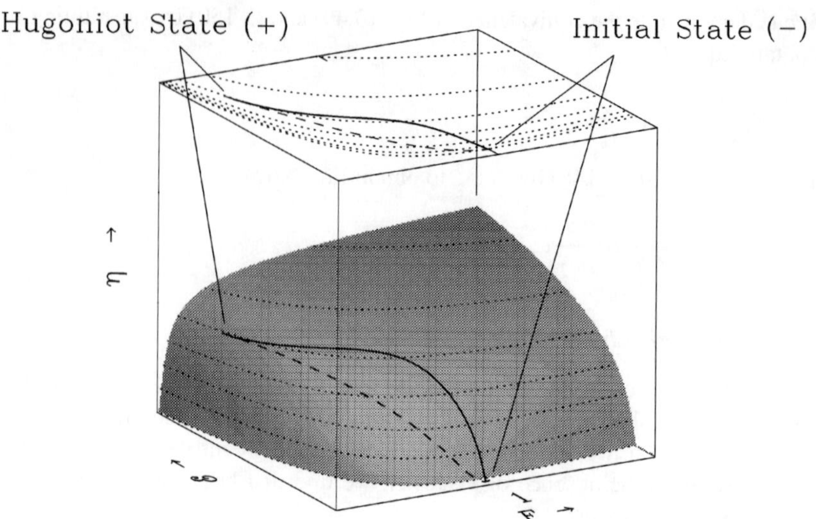

Figure 3.31 State surface with isentropes (· · ·) and a Rayleigh-line process for a thermoviscous material (—). For comparison, we show a Rayleigh-line process for a viscous material (– –).

where, because the initial state is also an equilibrium state, $q_- = 0$. Moreover, we can also write the balance of linear momentum in terms of the viscous stress,

$$\mathcal{T} = \rho_0 C_S^2 (F - F_-) + T_- - T_e. \tag{3.152}$$

The forms of the last two expressions are particularly appealing because the right sides of both equations are algebraic expressions that do not depend upon $dF/d\hat{\varphi}$ or $d\vartheta/d\hat{\varphi}$ while, through the constitutive equations for $\mathcal{T}$ and q, the left sides do depend upon these derivatives. Thus we have two ordinary differential equations in F and ϑ that we can integrate to evaluate the solution for the steady wave. We show the results of such a calculation in Figure 3.31. We use a solid line to represent the Rayleigh-line process for the thermoviscous material. We obtain this curve by specifying particular constitutive equations for the heat flux and the viscous stress such that $q \neq 0$ and $\mathcal{T} \neq 0$. However, if we require $q = 0$, we can also evaluate the Rayleigh-line process for the viscous material, which is represented by the dashed line. We see that the thermoviscous and the viscous processes have the same initial and Hugoniot states, but we also see that between these two states large differences exist between the two processes. We even show that it is possible for the entropy of the thermoviscous process to overshoot the value of entropy at the Hugoniot state.

Notice that we can write Eq. (3.151) as

$$q = C_S \left[\rho_0 (\mathcal{E} - \mathcal{E}_-) - \frac{1}{2}(T + T_-)(F - F_-) \right].$$

Because the Hugoniot state is also an equilibrium state, the heat flux at the Hugoniot state is $q_+ = 0$. When we use this expression to evaluate the Hugoniot state, we obtain the Rankine–Hugoniot equation, Eq. (3.112).

Linear Viscosity and Heat Conduction

Let us consider the following constitutive equations for the viscous stress and the heat flux:

$$\mathcal{T} = -BD = -\frac{B}{F}\frac{\partial \hat{F}}{\partial t},$$
$$q = -K\frac{\partial \vartheta}{\partial x} = -\frac{K}{F}\frac{\partial \hat{\vartheta}}{\partial X},$$

where the viscosity coefficient B and the heat conduction coefficient K are constants. This expression for the viscous stress is similar to Eq. (3.97) for the Newtonian fluid, but in the present context, we are also considering solids. The expression for heat flux is Fourier's law of heat conduction, Eq. (3.124). Notice that these constitutive relations satisfy the reduced dissipation inequality, Eq. (3.138), for arbitrary positive values of B and K. The steady-wave equations, Eqs. (3.151) and (3.152), become

$$\frac{dF}{d\hat{\varphi}} = -\frac{F}{BC_S}G, \qquad \frac{d\vartheta}{d\hat{\varphi}} = -\frac{FC_S}{K}N, \tag{3.153}$$

where

$$N = \rho_0(\mathcal{E} - \mathcal{E}_-) - T_-(F - F_-) - \frac{1}{2}\rho_0 C_S^2(F - F_-)^2,$$
$$G = \rho_0 C_S^2(F - F_-) + T_- - T_e.$$

We divide the second equation by the first to obtain

$$\frac{d\vartheta}{dF} = C_S^2\left(\frac{B}{K}\right)\left(\frac{N}{G}\right). \tag{3.154}$$

We can integrate this equation to obtain the steady wave profile, but not without some care. First notice that the equation $N = 0$ yields the Rayleigh-line process for material in which $\mathcal{T} \neq 0$ and $q = 0$. This is the steady-wave response of a viscous material. We illustrate it with the dashed curve in Figure 3.32. Then notice that the equation $G = 0$ yields the Rayleigh-line process for material in which $\mathcal{T} = 0$ and $q \neq 0$. This is the steady-wave response of a thermoelastic material. We illustrate it with the dotted curve. These solutions cross at the initial ($-$) and Hugoniot states ($+$). With the exception of these two states, $d\vartheta/dF$ is zero everywhere on the dashed line and singular everywhere on the dotted line.

The solid curve is the steady-wave solution for the thermoviscous material. Obviously this curve must also connect the initial state to the Hugoniot state. From our previous discussion, it is also clear that this curve cannot cross either the thermoelastic curve ($\cdots$) or the viscous curve ($- - -$). In the illustration the thermoviscous curve is contained in the region between the thermoelastic and viscous curves, and, in general, we find that solutions that lie outside this region do not connect the initial state to the Hugoniot state.

Comparative Importance of Viscosity and Heat Conduction

Inspection of Eq. (3.154) reveals that the steady-wave solution for a thermoviscous material depends on the ratio B/K. In Figure 3.32, the relative position of this solution between

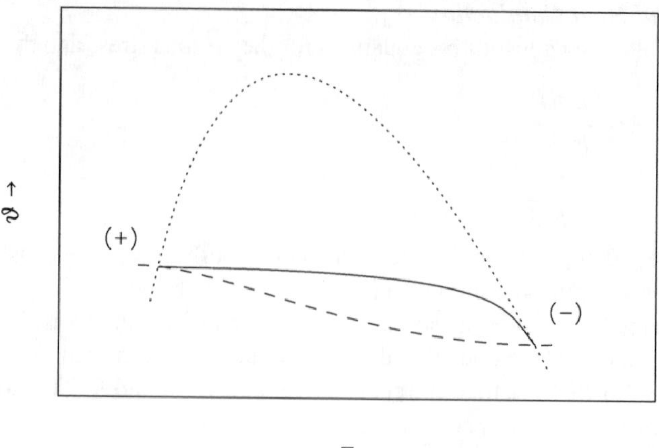

Figure 3.32 The ϑ–F diagram of the steady wave solution (—). We also show the solution for the viscous material (– –) and the thermoelastic material (· · ·).

the viscous and thermoelastic responses is determined solely by B/K. As $B/K \to \infty$, the relative contribution of heat conduction decreases, and the steady-wave solution approaches that of the purely viscous material. As $B/K \to 0$, the steady-wave solution approaches that of a purely conductive material. Depending on the value of B/K, the Rayleigh-line process for the thermoviscous material predicts very different temperature profiles. We even see that the temperature within the steady wave can exceed the temperature at the Hugoniot state.

When we estimate B/K for a variety of materials, we generally observe values that place the response of the thermoviscous material quite close to the purely viscous response. We might then conclude that heat conduction is relatively unimportant to the Rayleigh-line process. However, because our conclusion is based on simple linear models for viscosity and heat conduction, we should at least be skeptical of this result. The cause of structure in steady waves, be it viscosity, heat conduction, or some other phenomena, remains an open question. To understand why this is such a difficult issue to resolve, we need only notice that each of the waves we have considered is indistinguishable in the T–F diagram, which is given by the balance of linear momentum

$$T - T_- = \rho_0 C_S^2 (F - F_-). \tag{3.155}$$

Thus it is hard to distinguish between the effects of viscosity and heat conduction without direct measurements of temperatures within the steady wave profile. Such measurements are exceedingly difficult to obtain.

Integration of the Thermoviscous Equation

We can solve the thermoviscous equation for the steady wave, Eq. (3.154), by numerical integration. However, because $N = G = 0$ at both the initial state and the Hugoniot state, we must initiate integration at a small but finite distance from these states. To do this, we shall evaluate the solution in the neighborhood of these states with the following first-order

Taylor's series expansion of Eqs (3.153):

$$\frac{d}{d\hat{\varphi}}(F - F_\pm) = -\frac{F_\pm}{BC_S}\left[\left(\frac{\partial G}{\partial F}\right)_\pm (F - F_\pm) + \left(\frac{\partial G}{\partial \vartheta}\right)_\pm (\vartheta - \vartheta_\pm)\right],$$

$$\frac{d}{d\hat{\varphi}}(\vartheta - \vartheta_\pm) = -\frac{F_\pm C_S}{K}\left[\left(\frac{\partial N}{\partial F}\right)_\pm (F - F_\pm) + \left(\frac{\partial N}{\partial \vartheta}\right)_\pm (\vartheta - \vartheta_\pm)\right].$$

We use the "$\pm$" sign to indicate that the partial derivatives and functions are evaluated either at the initial state or the Hugoniot state. Next, we assume the following solutions to these linear equations:

$$F - F_\pm = F_0 e^{\lambda\varphi},$$

$$\vartheta - \vartheta_\pm = \vartheta_0 e^{\lambda\varphi},$$

where F_0, ϑ_0, and λ are arbitrary constants. We substitute these solutions into the Taylor's series to obtain

$$\begin{bmatrix} \frac{F_\pm}{BC_S}\left(\frac{\partial G}{\partial F}\right)_\pm + \lambda & \frac{F_\pm}{BC_S}\left(\frac{\partial G}{\partial \vartheta}\right)_\pm \\ \frac{F_\pm C_S}{K}\left(\frac{\partial N}{\partial F}\right)_\pm & \frac{F_\pm C_S}{K}\left(\frac{\partial N}{\partial \vartheta}\right)_\pm + \lambda \end{bmatrix} \begin{bmatrix} F_0 \\ \vartheta_0 \end{bmatrix} = 0.$$

A nontrivial solution exists if the determinant of the matrix is zero:

$$\lambda^2 + \left[\frac{F_\pm}{BC_S}\left(\frac{\partial G}{\partial F}\right)_\pm + \frac{F_\pm C_S}{K}\left(\frac{\partial N}{\partial \vartheta}\right)_\pm\right]\lambda$$

$$+ \frac{F_\pm^2}{BK}\left[\left(\frac{\partial G}{\partial F}\right)_\pm \left(\frac{\partial N}{\partial \vartheta}\right)_\pm - \left(\frac{\partial G}{\partial \vartheta}\right)_\pm \left(\frac{\partial N}{\partial F}\right)_\pm\right] = 0. \tag{3.156}$$

We obtain two roots for λ for both the initial state and the Hugoniot state. To integrate from the neighborhood of the initial state to the Hugoniot state, we must integrate in the negative-$\hat{\varphi}$ direction. If the values of λ are negative at the initial state, the solution will grow exponentially and not intersect the Hugoniot state. For most constitutive equations this is the situation at the initial state. If we start the integration in the neighborhood of the Hugoniot state, solutions for negative λ at the Hugoniot state will decay and connect with the initial state. We cannot initiate the integration exactly at the Hugoniot state (ϑ_+, F_+), but we can start at some arbitrary point $(\vartheta_+ + \Delta\vartheta, F_+ + \Delta F)$ where from the second Taylor's series

$$\Delta\vartheta = \left[\frac{\frac{F_\pm C_S}{K}\left(\frac{\partial N}{\partial F}\right)_\pm}{\frac{F_\pm C_S}{K}\left(\frac{\partial N}{\partial \vartheta}\right)_\pm - \lambda}\right]\Delta F. \tag{3.157}$$

Thus we select a value of ΔF, which is arbitrarily small with respect to $F_+ - F_-$, and then evaluate $\Delta\vartheta$. This determines the point at which we can initiate numerical integration of the steady wave solution of the thermoviscous material, Eq. (3.154).

Thermoelastic Steady Waves

A thermoelastic material conducts heat ($q \neq 0$), but the viscous stress is $\mathcal{T} = 0$. In a thermoelastic material the constitutive equations do not depend on D, and the stress T is

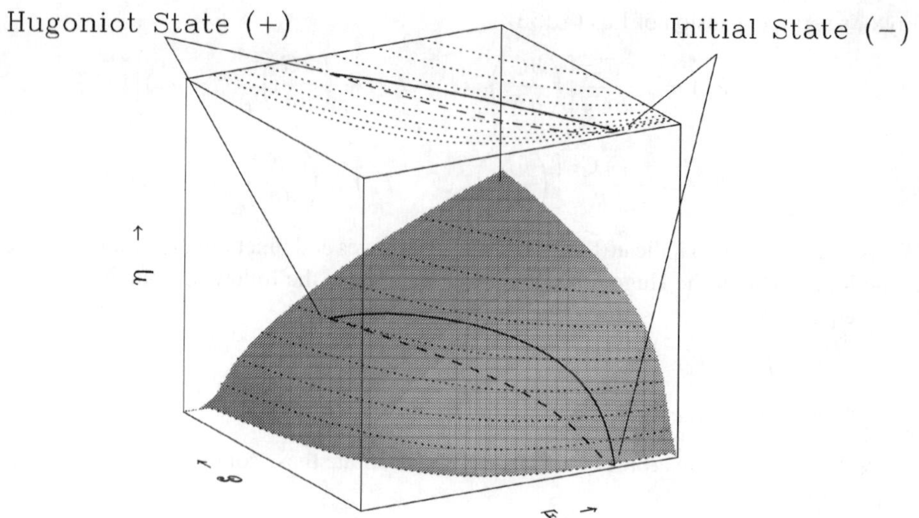

Figure 3.33 Comparison of the Rayleigh lines for a thermoelastic material (—) and a viscous material (– –) on a state surface with isentropes (· · ·).

always equal to the equilibrium stress T_e. We can determine the position of the Rayleigh line on the state surface using only the steady-wave relationship for the balance of linear momentum, Eq. (3.155). We illustrate a Rayleigh-line process for a thermoelastic material in Figure 3.33 with a solid line. We compare it to the Rayleigh-line process for the viscous material represented by the dashed line. Recall that the Rayleigh line for the viscous material is tangent to the isentropes at the initial and Hugoniot states, and the entropy on the Rayleigh line monotonically increases from the initial state to the Hugoniot state. In contrast, we see that the Rayleigh line for the thermoelastic material is not tangent to the isentropes at either the initial or the Hugoniot states, and it has a maximum point between these states. The entropy on the Rayleigh line *overshoots* the entropy at the Hugoniot state. Thus although both the viscous material and the thermoelastic material have the same initial and Hugoniot states, steady waves cause each material to follow significantly different Rayleigh-line processes between these states.

In the thermoelastic material, we require that $\mathcal{T} = 0$, and we obtain the value of the heat flux from the balance of energy, Eq. (3.151). Notice that we can determine the heat flux q *without* knowledge of the constitutive equation for q. A similar situation exists in the viscous material, where we can evaluate the viscous stress $\mathcal{T}$ without knowledge of the constitutive equation for $\mathcal{T}$. However, in the thermoelastic material, we have one additional complication. From the balance of energy, we know that the heat flux is zero only at the initial state and the Hugoniot state. It is not zero anywhere between these two states.

The problem becomes evident when we reexamine the dotted curve for the thermoelastic material in Figure 3.32. We see that the temperature peaks between the initial and Hugoniot states. At this point the temperature gradient $\partial\hat{\vartheta}/\partial X$ is zero. We determine this curve solely by the equation for the balance of linear momentum, Eq. (3.155). Once we have solved this equation, we use the balance of energy to evaluate q. This leads us to the conclusion that $q \neq 0$ when $\partial\hat{\vartheta}/\partial X = 0$. Unfortunately, this is in direct conflict with the requirement of the dissipation inequality that $q = 0$ when $\partial\hat{\vartheta}/\partial X = 0$. Thus we must conclude that this steady-wave solution is not admissible because it violates the dissipation inequality.

Recall that we use the dissipation inequality to constrain the constitutive equations. From it we know that $q = 0$ when $\partial\hat{\vartheta}/\partial X = 0$. Suppose we try to evaluate the Rayleigh-line process by employing a constitutive equation that obeys this requirement. This is easily illustrated by attempting to integrate the first expression in Eq. (3.153),

$$\frac{d\vartheta}{d\hat{\varphi}} = -\frac{FC_S}{K}N,$$

while simultaneously requiring $G = 0$. This will yield the same Rayleigh-line process as Eq. (3.155) unless the process contains a point where $d\vartheta/d\varphi = 0$. If such a point exists, the solution cannot approach this point, because the left side of the equation is required to be zero while the right side cannot be zero. This fact is reflected in Figure 3.33. Notice that the shape of the state surface in this figure is different from the state surface we have used in other figures in this chapter. To generate this figure we altered the state surface so that the temperature gradient $\partial\hat{\vartheta}/\partial X$ would not be zero on this Rayleigh-line process.

We now derive a condition for the velocity C_S of the steady wave, which guarantees that $\partial\hat{\vartheta}/\partial X \neq 0$ everywhere on the Rayleigh line between the initial state and the Hugoniot state. This is done by first determining the way in which temperature changes along the Rayleigh line. Recall that the equilibrium stress in terms of the Helmholtz energy is

$$T_e = \rho_0 \frac{\partial \psi}{\partial F}.$$

For a thermoelastic material, $T = T_e(F, \vartheta)$ on the Rayleigh line. Thus we can take the derivative of the equation of the balance of linear momentum, Eq. (3.155), with respect to the steady-wave variable $\hat{\varphi}$ to obtain

$$\frac{\partial^2 \psi}{\partial F^2}\frac{dF}{d\hat{\varphi}} + \frac{\partial^2 \psi}{\partial \vartheta \partial F}\frac{d\vartheta}{d\hat{\varphi}} = C_S^2 \frac{dF}{d\hat{\varphi}}.$$

Substituting Eqs. (3.145) and (3.146) into this result, we find that

$$\frac{d\vartheta}{d\hat{\varphi}} = \frac{\rho_0}{C^\vartheta \alpha}\left(C_I^2 - C_S^2\right)\frac{dF}{d\hat{\varphi}}, \tag{3.158}$$

where

$$C_I^2 = \frac{C^\vartheta}{\rho_0 F}.$$

The quantity C_I is called the *isothermal velocity*. We shall use it as a convenient notation, but you should understand that it does not represent the velocity of propagation of any wave that we have identified in this work. From Eqs. (3.60) and (3.88), if $c_v > 0$, we obtain

$$C^2 > C_I^2. \tag{3.159}$$

Now we can determine a constraint for the wave velocities that ensures $\partial\hat{\vartheta}/\partial X \neq 0$ everywhere between the initial state and the Hugoniot state. Because $dF/d\varphi \neq 0$ between the initial and Hugoniot states, Eq. (3.158) shows us that the temperature gradient is nonzero when

$$C_S^2 \neq C_I^2.$$

This means that along the Rayleigh line, we must require that either $C_S^2 > C_I^2$ or $C_S^2 < C_I^2$. To determine the correct choice, we shall use the supersonic and subsonic conditions for the velocity of a steady wave in a viscous material:

$$C_+^2 \geq C_S^2 \geq C_-^2. \tag{3.160}$$

Here $C_\pm$ represent the values of C at the initial and Hugoniot states. Although we derived these conditions for the viscous material, they also hold for the thermoelastic material. This is because both these types of material obey the same constitutive relation $\mathcal{E} = \mathcal{E}(F, \eta)$ and the same jump condition $[\![T]\!] = \rho_0 C_S^2 [\![F]\!]$. For prescribed values of $F_\pm$, we know that $C_\pm$ and $[\![T]\!]$ are determined by the constitutive relationship for $\mathcal{E}$, and C_S is determined by the jump condition.

The supersonic condition and Eq. (3.159) require that $C_S^2 \geq C^2 > C_I^2$ at the initial state. It follows that $C_S^2 > C_I^2$ must hold at the initial state and everywhere else along the Rayleigh line. Indeed, we find that

$$\begin{aligned} &C_S^2 \geq C_-^2 > C_I^2 \quad \text{(initial state)}, \\ &C_+^2 \geq C_S^2 > C_I^2 \quad \text{(Hugoniot state)}. \end{aligned} \tag{3.161}$$

The condition that $C_S^2 > C_I^2$ everywhere on the Rayleigh-line process is a rather strict requirement on the constitutive equations. For many materials this condition often fails. When it does, Eq. (3.158) shows that $dF/d\hat{\varphi}$ must become infinite in order to maintain a nonzero value of $d\vartheta/d\hat{\varphi}$ at the state where $C_S^2 = C_I^2$. We shall address this issue again in the next chapter.

The supersonic and subsonic conditions, Eq. (3.160), also imply that entropy of a thermoelastic material attains a maximum value between the initial and Hugoniot states. To demonstrate this, we write the equilibrium stress in terms of the internal energy:

$$T_e = T_e(F, \eta) = \rho_0 \frac{\partial \mathcal{E}}{\partial F}.$$

Because $T = T_e$ in a thermoelastic material, we take the derivative of the balance of linear momentum, Eq. (3.155), with respect to the steady-wave variable $\hat{\varphi}$ to obtain

$$\frac{\partial^2 \mathcal{E}}{\partial F^2}\frac{dF}{d\hat{\varphi}} + \frac{\partial^2 \mathcal{E}}{\partial \eta \partial F}\frac{d\eta}{d\hat{\varphi}} = C_S^2 \frac{dF}{d\hat{\varphi}}.$$

After substitution of Eqs. (3.65) and (3.87), we solve for $d\eta/d\hat{\varphi}$ to obtain

$$\frac{d\eta}{d\hat{\varphi}} = \frac{\rho_0 c_v}{\vartheta C^\vartheta \alpha}\left(C^2 - C_S^2\right)\frac{dF}{d\hat{\varphi}}.$$

Suppose we express entropy on the Rayleigh line as a function of F,

$$\eta = \eta_R(F).$$

From the last two expressions, we see that

$$\frac{d\eta_R}{dF} = \frac{\rho_0 c_v}{\vartheta C^\vartheta \alpha}\left(C^2 - C_S^2\right).$$

By substituting Eq. (3.161), we obtain

$$\frac{d\eta_R}{dF} \le 0 \quad \text{(initial state)},$$

$$\frac{d\eta_R}{dF} \ge 0 \quad \text{(Hugoniot state)}.$$

This demonstrates that $d\eta_R/dF = 0$ between the initial and Hugoniot states, and therefore the entropy overshoots its final value at the Hugoniot state.

A similar procedure can be used to demonstrate that the heat flux in a steady wave is in the direction of wave propagation:

$$q \ge 0 \quad (\text{positive } C_S),$$

$$q \le 0 \quad (\text{negative } C_S).$$

We leave the derivation of this result as an exercise (see Exercise 3.39).

3.5.5 *Exercises*

3.34 We know that because $\vartheta > 0$, the state surface can be represented by the constitutive equation for the entropy, $\eta = \eta(F, \mathcal{E})$. Consider a viscous material without heat conduction.

(a) Combine the second law of thermodynamics and the equation for the balance of energy to obtain a dissipation inequality for this material.

(b) Show that the temperature and the equilibrium stress are given by

$$\frac{1}{\vartheta} = \frac{\partial \eta}{\partial \mathcal{E}}, \qquad T_e = -\rho_0 \vartheta \frac{\partial \eta}{\partial F}.$$

(The inverse temperature is called *coldness*.)

(c) Determine the reduced dissipation inequality for this material.

(d) Determine C^η, C^ϑ, c_T, and c_v.

3.35 Consider the following constitutive equation for the internal energy:

$$\mathcal{E} = \mathcal{E}_0 \rho^{NR/c_v} e^{\eta/c_v}.$$

Because $\mathcal{E} = \mathcal{E}(\rho, \eta)$, this is the fundamental form for this internal energy.

(a) Determine the fundamental form for the Helmholtz free energy.

(b) Use these particular forms of $\mathcal{E}$ and ψ to evaluate

$$\vartheta = \frac{\partial \mathcal{E}(\rho, \eta)}{\partial \eta}, \qquad \eta = -\frac{\partial \psi(\rho, \vartheta)}{\partial \vartheta}.$$

(c) Use these particular forms of $\mathcal{E}$ and ψ to evaluate

$$p_e = \rho^2 \frac{\partial \mathcal{E}(\rho, \eta)}{\partial \rho}, \qquad p_e = \rho^2 \frac{\partial \psi(\rho, \vartheta)}{\partial \rho}.$$

(d) Determine C^ϑ, C^η, $c_p = c_T$, and c_v from the internal energy $\mathcal{E} = \mathcal{E}(\rho, \eta)$.

(e) Determine C^ϑ, C^η, $c_p = c_T$, and c_v from the Helmholtz energy $\psi = \psi(\rho, \eta)$.

3.36 Show that arbitrary choices of $\hat{x}(X, t)$ and $\hat{\vartheta}(X, t)$ define an admissible thermodynamic process for the thermoviscous material.

3.37 Verify that we can write the equation for the balance of total energy in the spatial description as

$$\frac{\partial}{\partial t}\left[\rho\left(\mathcal{E}+\frac{1}{2}v_m v_m\right)\right]+\frac{\partial}{\partial x_k}\left[\rho v_k\left(\mathcal{E}+\frac{1}{2}v_m v_m\right)\right]=\frac{\partial}{\partial x_k}(T_{km}v_m)-\frac{\partial q_k}{\partial x_k},$$

where we assume that b and r are equal to zero. Assume that a steady wave solution exists in which all of the dependent variables in this equation are solely functions of $\varphi = x_k N_k - c_s t$. Here N_k is the direction of propagation and c_S is the spatial wave velocity of the steady wave. Obtain the following relationship for this steady wave by integrating the equation for the balance of total energy:

$$\rho(N_k v_k - c_S)\left(\mathcal{E}+\frac{1}{2}v_m v_m\right)-(T_{km}v_m - q_k)N_k = \text{constant}. \tag{3.162}$$

This is the three-dimensional form of the balance of energy for a steady wave.

3.38 At the beginning of this chapter, we discussed the difficultly that Newton encountered while trying to determine the speed of sound in air. Show that Newton computed the isothermal velocity C_I instead of c_0.

3.39 Use Eq. (3.158),

$$\frac{d\vartheta}{d\hat{\varphi}}=\frac{\rho_0}{C^{\vartheta}\alpha}\left(C_I^2-C_S^2\right)\frac{dF}{d\hat{\varphi}},$$

to verify that in a thermoelastic material the heat flux in a steady wave is in the direction of wave propagation.

3.6 Enthalpy and Gibbs Energy

We describe two additional representations of the state surface in this section. They are the enthalpy and the Gibbs energy. Heat conduction is omitted.

3.6.1 *Enthalpy*

The next energy function we shall examine is obtained by interchanging the roles of the conjugate pair, T and F, with a Legendre transform. This transformation results in a new energy function called the *enthalpy* $\mathcal{H}$, where

$$\mathcal{H} \equiv \mathcal{E} - TF/\rho_0. \tag{3.163}$$

The constitutive assumptions for this formulation are

$$\begin{aligned}\mathcal{H} &= \mathcal{H}(T, \partial\hat{T}/\partial t, \eta),\\ F &= F(T, \partial\hat{T}/\partial t, \eta),\\ \vartheta &= \vartheta(T, \partial\hat{T}/\partial t, \eta).\end{aligned}$$

When we substitute the Legendre transform, Eq. (3.163), into the equation for the balance of energy, Eq. (3.10), we obtain

$$\frac{\partial\hat{\mathcal{H}}}{\partial t} = -\frac{F}{\rho_0}\frac{\partial\hat{T}}{\partial t}+r. \tag{3.164}$$

For this constitutive formulation, we can define an admissible thermodynamic process by prescribing $\hat{T}(X,t)$ and $\hat{\eta}(X,t)$. When the balance of energy, Eq. (3.164), is substituted into the second law of thermodynamics, Eq. (3.31), we obtain

$$\mathcal{D} = -\frac{\partial \mathcal{H}}{\partial(\partial \hat{T}/\partial t)}\frac{\partial^2 \hat{T}}{\partial t^2} + \left(\vartheta - \frac{\partial \mathcal{H}}{\partial \eta}\right)\frac{\partial \hat{\eta}}{\partial t} - \left(\frac{F}{\rho_0} + \frac{\partial \mathcal{H}}{\partial T}\right)\frac{\partial \hat{T}}{\partial t} \geq 0.$$

From this inequality, we deduce the following constraints on the constitutive equations:

$$\mathcal{H} = \mathcal{H}(T,\eta),$$
$$F = F(T, \partial\hat{T}/\partial t, \eta),$$
$$\vartheta = \partial\mathcal{H}/\partial\eta.$$

The reduced dissipation inequality is

$$\mathcal{D} = -\left(\frac{F}{\rho_0} + \frac{\partial \mathcal{H}}{\partial T}\right)\frac{\partial \hat{T}}{\partial t} \geq 0. \tag{3.165}$$

At an equilibrium state, during a reversible process, or when the deformation gradient is not a function of $\partial\hat{T}/\partial t$, the reduced dissipation inequality requires that the deformation gradient be

$$F = F_e = -\rho_0 \frac{\partial \mathcal{H}}{\partial T}.$$

We can derive expressions for the specific heats and stiffnesses by examining the four basic processes. We list the results in Section 3.7.

3.6.2 *Gibbs Energy*

We obtain the Gibbs energy $\mathcal{G}$ by exchanging the roles of T for F and ϑ for η with the Legendre transformation

$$\mathcal{G} = \mathcal{E} - \vartheta\eta - TF/\rho_0. \tag{3.166}$$

The constitutive assumptions for this formulation are

$$\mathcal{G} = \mathcal{G}(T, \partial\hat{T}/\partial t, \vartheta),$$
$$F = F(T, \partial\hat{T}/\partial t, \vartheta),$$
$$\eta = \eta(T, \partial\hat{T}/\partial t, \vartheta).$$

We substitute the Legendre transformation, Eq. (3.166), into the equation for the balance of energy, Eq. (3.10), to obtain

$$\frac{\partial \hat{\mathcal{G}}}{\partial t} + \eta\frac{\partial \hat{\vartheta}}{\partial t} + \vartheta\frac{\partial \hat{\eta}}{\partial t} = -\frac{F}{\rho_0}\frac{\partial \hat{T}}{\partial t} + r.$$

For this formulation, we can define an admissible thermodynamic process by prescribing $\hat{T}(X,t)$ and $\hat{\vartheta}(X,t)$. When we substitute this balance of energy into the second law of thermodynamics, Eq. (3.31), we obtain

$$\mathcal{D} = -\frac{\partial \mathcal{G}}{\partial(\partial \hat{T}/\partial t)}\frac{\partial^2 \hat{T}}{\partial t^2} - \left(\eta + \frac{\partial \mathcal{G}}{\partial \vartheta}\right)\frac{\partial \vartheta}{\partial t} - \left(\frac{F}{\rho_0} + \frac{\partial \mathcal{G}}{\partial T}\right)\frac{\partial \hat{T}}{\partial t} \geq 0.$$

From this inequality, we deduce the following constraints on the constitutive equations:

$$\mathcal{G} = \mathcal{G}(T, \vartheta),$$
$$F = F(T, \partial\hat{T}/\partial t, \vartheta),$$
$$\eta = -\partial\mathcal{G}/\partial\vartheta.$$

The reduced dissipation inequality is

$$\mathcal{D} = -\left(\frac{F}{\rho_0} + \frac{\partial\mathcal{G}}{\partial T}\right)\frac{\partial\hat{T}}{\partial t} \geq 0.$$

At an equilibrium state, during a reversible process, or when the deformation gradient is not a function of $\partial\hat{T}/\partial t$, the reduced dissipation inequality requires that the deformation gradient be

$$F = F_e = -\rho_0\frac{\partial\mathcal{G}}{\partial T}.$$

We list expressions for the specific heats and stiffnesses in Section 3.7.

3.7 Summary

The Internal Energy $\mathcal{E}$ has a fundamental form that is a function of the deformation gradient F and the entropy η,

$$\mathcal{E} = \mathcal{E}(F, \eta).$$

The constitutive equation for the stress is

$$T = T(F, D, \eta).$$

The first and second partial derivatives of the internal energy function are

$$\frac{\partial\mathcal{E}}{\partial\eta} = \vartheta, \qquad \frac{\partial\mathcal{E}}{\partial F} = \frac{T_e}{\rho_0}, \tag{3.167}$$

$$\frac{\partial^2\mathcal{E}}{\partial\eta^2} = \frac{\vartheta}{c_v}, \qquad \frac{\partial^2\mathcal{E}}{\partial F\partial\eta} = -\frac{\vartheta C^\vartheta\alpha}{\rho_0 c_v}, \qquad \frac{\partial^2\mathcal{E}}{\partial F^2} = \frac{C^\eta}{\rho_0 F}. \tag{3.168}$$

The Helmholtz Energy ψ has a fundamental form that is a function of the deformation gradient F and the temperature ϑ. It is defined by the Legendre transformation

$$\psi = \psi(F, \vartheta) = \mathcal{E} - \vartheta\eta.$$

The constitutive equation for the stress is

$$T = T(F, D, \vartheta).$$

The first and second partial derivatives of the Helmholtz energy function are

$$\frac{\partial\psi}{\partial\vartheta} = -\eta, \qquad \frac{\partial\psi}{\partial F} = \frac{T_e}{\rho_0},$$

$$\frac{\partial^2\psi}{\partial\vartheta^2} = -\frac{c_v}{\vartheta}, \qquad \frac{\partial^2\psi}{\partial F\partial\vartheta} = -\frac{C^\vartheta\alpha}{\rho_0}, \qquad \frac{\partial^2\psi}{\partial F^2} = \frac{C^\vartheta}{\rho_0 F}.$$

***The Enthalpy* $\mathcal{H}$** has a fundamental form that is a function of the stress T and the entropy η. It is defined by the Legendre transformation

$$\mathcal{H} = \mathcal{H}(T, \eta) = \mathcal{E} - TF/\rho_0.$$

The constitutive equation for the deformation gradient is

$$F = F(T, \partial \hat{T}/\partial t, \eta).$$

The first and second partial derivatives of the enthalpy function are

$$\frac{\partial \mathcal{H}}{\partial \eta} = \vartheta, \qquad \frac{\partial \mathcal{H}}{\partial T} = -\frac{F_e}{\rho_0},$$

$$\frac{\partial^2 \mathcal{H}}{\partial \eta^2} = \frac{\vartheta}{c_T}, \qquad \frac{\partial^2 \mathcal{H}}{\partial T \partial \eta} = -\frac{\vartheta \alpha}{\rho c_T}, \qquad \frac{\partial^2 \mathcal{H}}{\partial T^2} = -\frac{1}{\rho C^\eta}.$$

***The Gibbs Energy* $\mathcal{G}$** has a fundamental form that is a function of the stress T and the temperature ϑ. It is defined by the Legendre transformation

$$\mathcal{G} = \mathcal{G}(T, \vartheta) = \mathcal{E} - TF/\rho_0 - \vartheta \eta.$$

The constitutive equation for the deformation gradient is

$$F = F(T, \partial \hat{T}/\partial t, \vartheta).$$

The first and second partial derivatives of the Gibbs energy function are

$$\frac{\partial \mathcal{G}}{\partial \vartheta} = -\eta, \qquad \frac{\partial \mathcal{G}}{\partial T} = -\frac{F_e}{\rho_0},$$

$$\frac{\partial^2 \mathcal{G}}{\partial \vartheta^2} = -\frac{c_T}{\vartheta}, \qquad \frac{\partial^2 \mathcal{G}}{\partial T \partial \vartheta} = -\frac{\alpha}{\rho}, \qquad \frac{\partial^2 \mathcal{G}}{\partial T^2} = -\frac{1}{\rho C^\vartheta}.$$

3.7.1 *Exercises*

3.40 Using the graphical procedure of Section 3.5, derive the Legendre transformation for enthalpy.

3.41 Show that $\hat{T}(X, t)$ and $\hat{\eta}(X, t)$ define an admissible process for the alternative formulation that uses the enthalpy.

3.42 Show that $\hat{T}(X, t)$ and $\hat{\vartheta}(X, t)$ define an admissible process for the alternative formulation that uses the Gibbs energy.

CHAPTER 4

Constitutive Models

Throughout our development of mechanics and thermomechanics we have used constitutive models to describe waves in materials. In Chapter 2 we used the constitutive model for an elastic material to describe linear-elastic waves, structured waves, and shock discontinuities. However, we could not describe the steady wave with this constitutive model. In Chapter 3 we introduced more complex constitutive models for materials that dissipate work and conduct heat. We used the viscous material, the thermoviscous material, and the thermoelastic material to extend our description of nonlinear waves. These constitutive models are quite general descriptions that include gases, liquids, and solids. More specific constitutive models are discussed in this chapter. These models include the ideal gas and the Mie–Grüneisen solid. We also study plastic flow, porosity, and detonation.

We often calculate nonlinear wave phenomena with complex numerical computer codes. With such powerful capabilities you may be tempted to simply characterize the energy state of a material with large tables of data that yield values of equilibrium stress, temperature, stiffnesses, specifics heats, and thermal expansion coefficients for given values of F and η. Accurate computations of some very common materials rely heavily upon this method. The computation of nonlinear pressure waves in water is an important example. However, even with this capability, questions arise about the constitutive descriptions of these materials. For example, if water is heated by a shock wave, it often spontaneously "flashes" to steam when the pressure drops again in response to a release wave. We know that phase changes are associated with unstable regions of the state surface. Numerical solutions are unstable in these regions unless we employ appropriate approximations of these phase transitions. Indeed, you will often find it is much easier to maintain thermodynamic consistency with analytic constitutive models, rather than with tables. Moreover, certain simplified constitutive relationships have been eminently successful in predicting a wide range of phenomena. It is important for you to understand these models. We have already studied one of them, the nonlinear-elastic material. We now turn to another, the ideal gas. It is simple, elegant, and extremely useful.

4.1 Ideal Gas

In equilibrium, an ideal gas only supports positive pressure and does not support shear stresses. The most commonly recognized forms of the constitutive equations for an ideal gas are

$$p_e = NR\rho\vartheta, \qquad \mathcal{E} = c_v\vartheta, \tag{4.1}$$

where p_e is the equilibrium pressure and N, R, and c_v are constants. The new constants in these equations are the mole number N, which is the inverse of the molecular weight of the gas in moles/kg, and the gas constant R, which is equal to 8.3143 J/K/mole. Notice that these expressions only include a description of the equilibrium pressure p_e. This pressure does not depend upon the deformation rate D, and therefore dissipation is not represented by these relationships. Also, we have written the internal energy as a function of temperature alone. It is not written in the fundamental form as a function of F and η. We begin this section by examining the fundamental form of the internal energy for an ideal gas.

4.1.1 *Fundamental Statement*

The fundamental form of the constitutive equation for the internal energy of an ideal gas is

$$\int_{\mathcal{E}_0}^{\mathcal{E}} \frac{d\mathcal{E}'}{\theta(\mathcal{E}')} = NR\ln(\rho/\rho_0) + \eta - \eta_0, \tag{4.2}$$

where $\mathcal{E}'$ denotes an integration parameter. In this expression we have written the internal energy as a function of the variables ρ and η. The quantities $\mathcal{E}_0$ and η_0 are the internal energy and the entropy at the reference state where $\rho = \rho_0$. The function $\theta(\mathcal{E})$ is a prescribed property of the ideal gas. For example, if we prescribe

$$\theta(\mathcal{E}) = \mathcal{E}/c_v, \tag{4.3}$$

where c_v is constant, then

$$c_v \ln(\mathcal{E}/\mathcal{E}_0) = NR\ln(\rho/\rho_0) + \eta - \eta_0. \tag{4.4}$$

From our discussions of the nonlinear-elastic fluid (see Section 2.1.3) and the Newtonian fluid (see Section 3.4.3), we recall that an internal energy function that depends on the density and not directly on the tensor components of the deformation gradient will result in a spherical stress configuration at equilibrium.

The fundamental form of the internal energy, Eq. (4.2), completely defines the equilibrium behavior of an ideal gas. We can derive all other forms of the internal energy as well as the relationship for equilibrium pressure p_e from this equation. From the first derivatives of Eq. (4.2), we can determine the temperature and the equilibrium pressure of the gas. We start with the Leibniz theorem for differentiation of an integral and evaluate the partial derivative of Eq. (4.2) with respect to η:

$$\frac{1}{\theta(\mathcal{E})}\frac{\partial \mathcal{E}}{\partial \eta} = 1, \tag{4.5}$$

or

$$\vartheta = \theta(\mathcal{E}). \tag{4.6}$$

Thus the temperature of the gas is equal to our prescribed function $\theta(\mathcal{E})$. Similarly, after we substitute $F = \rho_0/\rho$, the partial derivative of Eq. (4.2) with respect to F yields

$$\frac{1}{\theta(\mathcal{E})}\frac{\partial \mathcal{E}}{\partial F} = -\frac{NR}{F}. \tag{4.7}$$

Then we substitute this result into Eq. (3.40) to obtain

$$p_e = -\rho_0 \frac{\partial \mathcal{E}}{\partial F} = NR\rho\vartheta. \tag{4.8}$$

Now, using the second derivatives of the internal energy, we evaluate the stiffnesses, specific heats, thermal expansion coefficient, and Grüneisen coefficient. First, we take the derivative of Eq. (4.5) with respect to η to obtain

$$\frac{\partial^2 \mathcal{E}}{\partial \eta^2} = \frac{d\theta(\mathcal{E})}{d\mathcal{E}} \frac{\partial \mathcal{E}}{\partial \eta}.$$

Then we solve for $d\theta(\mathcal{E})/d\mathcal{E}$ and substitute the definition of the isochoric specific heat, Eq. (3.55), to get

$$\frac{d\theta(\mathcal{E})}{d\mathcal{E}} = \frac{1}{\vartheta} \frac{\partial^2 \mathcal{E}}{\partial \eta^2} = \frac{1}{c_v}. \tag{4.9}$$

For the special case when c_v is a constant, we integrate this expression and find that

$$\mathcal{E} = c_v \vartheta, \qquad c_v = \text{constant}. \tag{4.10}$$

Next we evaluate the second derivative of $\mathcal{E}$ with respect to F by taking the derivative of Eq. (4.7) with respect to F to obtain

$$\frac{\partial^2 \mathcal{E}}{\partial F^2} = \frac{NR\vartheta}{F^2} - \frac{NR}{F} \frac{d\theta(\mathcal{E})}{d\mathcal{E}} \frac{\partial \mathcal{E}}{\partial F}.$$

After substitution of Eqs. (3.50), (4.8), and (4.9), we see that

$$C^\eta = \gamma p_e, \tag{4.11}$$

where we have defined the *adiabatic gas coefficient* γ,

$$\gamma \equiv \frac{NR}{c_v} + 1. \tag{4.12}$$

When c_v is a constant, γ is called the *adiabatic gas constant.* As we shall demonstrate, γ not only represents the ratio of the isentropic stiffness to the equilibrium pressure, it also represents the ratio of the specific heats and defines the shape of the isentrope on the state surface.

We evaluate the last of the second derivatives of the internal energy by taking the derivative of Eq. (4.5) with respect to F:

$$\frac{\partial^2 \mathcal{E}}{\partial \eta \partial F} = \frac{d\theta(\mathcal{E})}{d\mathcal{E}} \frac{\partial \mathcal{E}}{\partial F} = -\frac{NR\vartheta}{F c_v},$$

where we have used Eqs. (4.6), (4.8), and (4.9). Then we substitute the definition of the Grüneisen coefficient [Eq. (3.53)] to obtain

$$\Gamma = \frac{NR}{c_v} = \gamma - 1. \tag{4.13}$$

When c_v is constant, this result for Γ along with Eqs. (4.8) and (4.10) leads us to another useful relationship for p_e:

$$p_e = \Gamma \rho \mathcal{E} = \Gamma c_v \rho \vartheta, \qquad c_v = \text{constant}. \tag{4.14}$$

We emphasize that although Eq. (4.8) holds in general, Eq. (4.14) holds only when c_v is a constant.

The isothermal stiffness is given by Eqs. (3.60), (4.8), (4.11), and (4.13):

$$C^{\vartheta} = p_e. \tag{4.15}$$

Because the ideal gas is a fluid, we notice that $\mathcal{K}^{\vartheta} = C^{\vartheta}, \mathcal{K}^{\eta} = C^{\eta}$, and $c_T = c_p$. Therefore, we can determine the specific heat at constant pressure c_p by combining this result with Eqs. (3.63) and (4.11) to obtain

$$\frac{c_p}{c_v} = \frac{C^{\eta}}{C^{\vartheta}} = \gamma, \tag{4.16}$$

which also yields

$$c_p - c_v = NR. \tag{4.17}$$

We evaluate the thermal expansion coefficient α from Eqs. (3.66) and (4.13):

$$\alpha = 1/\vartheta. \tag{4.18}$$

This completes our derivation of the equilibrium pressure, the temperature, and the thermodynamic coefficients from the fundamental form of the internal energy, Eq. (4.2). Because R is a universal constant, you only need to prescribe the molar number N and the function $\theta(\mathcal{E})$ to completely define these quantities. For the special case when c_v is a constant, we can completely determine the equilibrium response of an ideal gas by prescribing values for N and c_v.

4.1.2 *An Atomic Model*

Let us pause in our discussion of this constitutive formulation to consider a simple physical picture of the ideal gas. Here we envision the ideal gas as a collection of atoms in a container (see Figure 4.1). These atoms are elastic spheres that have both mass and velocity. As these atoms travel through the container, they collide with each other and with the walls of the container. The average distance traveled by an atom between collisions with neighbors is called the *mean free path*. For example, the mean free path of air at standard atmospheric conditions is 65 nm.

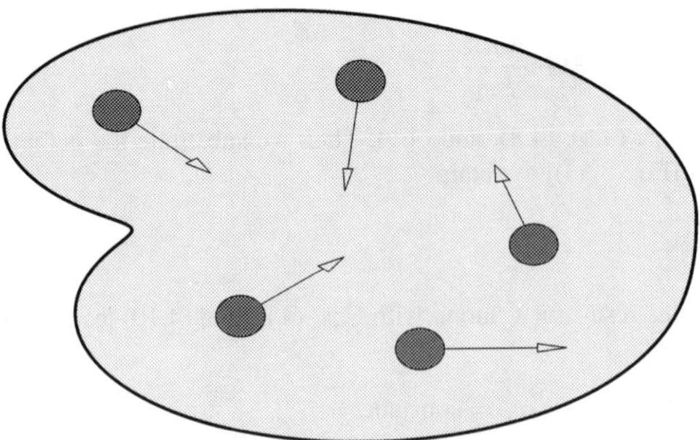

Figure 4.1 A collection of atoms in a container.

This picture of a gas was proposed by Daniel Bernoulli in the first half of the eighteenth century. It represents the genesis of the kinetic theory of gases and is the prime motivation behind the early development of statistical mechanics. In this book, it is not our intention to develop a constitutive theory for the ideal gas from the kinetic theory of gases. Rather, we pause briefly here to consider a simple analysis of atomic motion in order to obtain some insight and intuition. We assume that each atom in the container has the same mass m. Suppose a particular atom has the Cartesian velocity components v_i. Then the magnitude of the velocity V of the atom is given by

$$V^2 = v_1^2 + v_2^2 + v_3^2. \tag{4.19}$$

Although we recognize that the atoms travel in random directions at different velocities, we can greatly simplify our discussion by assuming that all have the same value of V. Therefore, the kinetic energy of each atom is $\frac{1}{2}mV^2$. The internal energy of the gas in the container is simply the sum of the kinetic energies of the individual atoms. If M represents the sum of the masses of the atoms, then the internal energy of the gas in the container is

$$M\mathcal{E} = \frac{1}{2}MV^2.$$

We can determine the average kinetic energy of the atoms with a macroscopic measurement of the temperature. Suppose these two quantities are proportional as in Eq. (4.10). Then

$$\mathcal{E} = \frac{1}{2}V^2 = c_v\vartheta. \tag{4.20}$$

Because we have just determined the temperature of the gas in terms of the velocity V of the atoms, let us see if we can obtain a similar relationship for the pressure. Pressure arises from collisions of the atoms with the walls of the container. Instead of separately analyzing the collisions of the individual atoms, suppose we simplify the model one step further by lumping all of the atoms together into one giant surrogate atom (see Figure 4.2). The mass of this atom is the summed mass M. The velocity of the surrogate atom is determined by first evaluating Eq. (4.19) for every individual atom. Next we take the average of these equations. Because of the large number of atoms, and the random distributions of the directions of

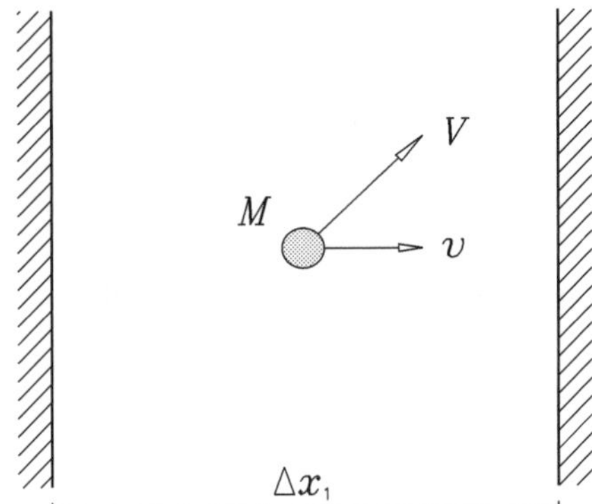

Figure 4.2 A surrogate atom with the average velocity and combined mass of the individual atoms.

their motions, we assume that

$$\langle v_1^2 \rangle = \langle v_2^2 \rangle = \langle v_3^2 \rangle,$$

where the symbol $\langle\,\rangle$ indicates the average values. Let the velocity components of the surrogate atom obey this result. The individual components can have either a positive or negative sign. Using the simplified notation $v^2 = \langle v_1^2 \rangle$, we obtain

$$v^2 = \frac{1}{3}V^2. \tag{4.21}$$

Figure 4.2 illustrates the surrogate atom as it approaches the right wall of the container ($v > 0$). Collision with the wall of the container changes the sign of v but not the magnitude. The velocity components parallel to the wall do not change during the collision. We can determine the force f on the wall resulting from the impact by Newton's second law of motion,

$$f = -M\frac{dv}{dt},$$

where we use the minus sign because we are evaluating the force on the wall and not the force on the atom. We determine the pressure in the gas by measuring f over a period of time T that includes many collisions of the atom on the wall of the container. Thus the quantity that we measure is the average force

$$\langle f \rangle \equiv \frac{1}{T}\int_0^T f\,dt = -\frac{1}{T}\int_0^T M\,dv.$$

Because each collision with the right wall causes a velocity change of $-2v$, we integrate this equation to obtain the following equilibrium pressure:

$$p_e = \frac{\langle f \rangle}{A} = \frac{n}{AT}(2Mv),$$

where n is the number of collisions with the right wall during the time interval T and A is the area of the wall. The distance between the two walls of the container is Δx_1. The period of time between collisions at the right wall is $2\Delta x_1/v$. Therefore

$$n = \frac{vT}{2\Delta x_1}.$$

We combine the two previous results to obtain

$$p_e = \rho v^2, \tag{4.22}$$

where the density of the gas is $\rho = M/A\Delta x_1$. Substituting Eqs. (4.20) and (4.21), we find that

$$p_e = \frac{2}{3}c_v\rho\vartheta.$$

This is the desired result for the pressure of the gas in the container. Notice that from Eqs. (4.13) and (4.14)

$$p_e = (\gamma - 1)c_v\rho\vartheta.$$

When you compare the previous two results you will see that the adiabatic gas constant for our simple model is $\gamma = 5/3$. This model is often called the *monoatomic gas*. Other values of γ arise from models of more complex gases. For example, the *diatomic gas* is composed of paired atoms. Each pair is called a molecule. The molecules of the diatomic gas move about the container in a manner that is similar to motion of the individual atoms in a monoatomic gas. However, in addition to the kinetic energy associated with this motion, the two atoms in each diatomic molecule can spin about each other. This gives rise to additional kinetic energy and a different value of γ. For a diatomic gas, $\gamma = 7/5$. Other values of γ follow from even more complex models.

It is interesting to notice that you can determine the velocity V of the individual atoms in a gas from macroscopic measurements of temperature. For example, recall that

$$p_e = \rho v^2 = \frac{1}{3}\rho V^2 = NR\rho\vartheta.$$

Although we derived this relationship for the monoatomic gas, inspection of our derivation will show that it also applies to gases containing molecules. Thus, the general expression for the velocity of the molecules in a gas is

$$V^2 = 3NR\vartheta. \tag{4.23}$$

To illustrate a particular case, suppose we calculate the velocity of atoms of hydrogen H_2 when $\vartheta = 273$ K. For hydrogen, $N = 500$ moles/kg. Recalling that $R = 8.31$ J/K/mole, we obtain

$$V^2 = 3 \times 500 \times 8.31 \times 273.$$

Thus

$$V = 1.84 \text{ km/s}.$$

This velocity is quite high. However, we should not be totally surprised. After all, the velocities of the atoms must necessarily be greater than the velocity of the waves in a gas.

4.1.3 *Equilibrium States and Processes*

When c_v is a constant, the fundamental relationship for an ideal gas is

$$\frac{\eta - \eta_0}{c_v} = \ln\frac{\mathcal{E}}{\mathcal{E}_0} + (\gamma - 1)\ln\frac{\rho_0}{\rho}. \tag{4.24}$$

We show the state surface for $\gamma = 1.4$ in Figure 4.3. This is the value of γ for air. We have clipped the image of the surface at a minimum entropy value of $(\eta - \eta_0)/c_v = -4$. However, in Eq. (4.24), we see that $\eta \to -\infty$ as either $\mathcal{E} \to 0$ or $F \to 0$. We also notice that it is not necessary to assign a specific value to c_v to construct this image. Recall that in Chapter 3 we defined the four reversible processes. They are the isentropic process, the isochoric process, the isothermal process, and the isobaric process. We now review these processes in the context of the ideal gas.

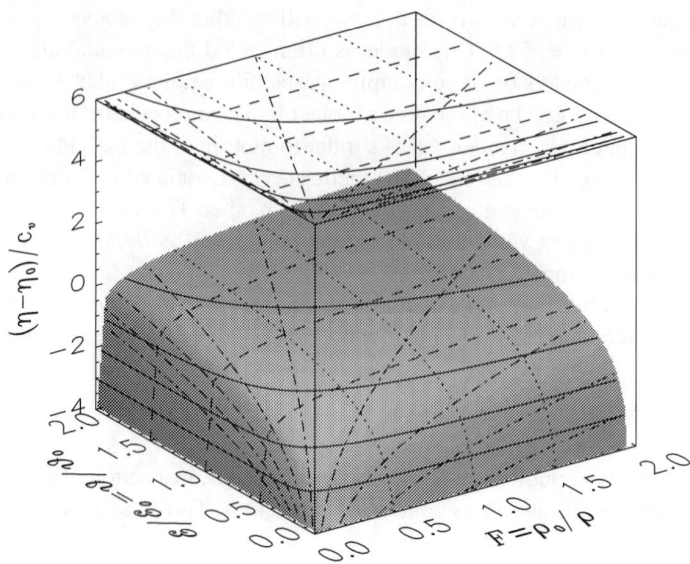

Figure 4.3 The state surface for $\gamma = 1.4$ with isentropes (—), isochors (· · ·), isotherms (– –), and isobars (– · –).

Isochoric Process

In an isochoric process, the density is constant. We illustrate the isochors with dotted lines that are spaced at uniform increments of F. From Eq. (4.8), the relationship between the pressure and the temperature along a process for constant ρ is

$$p_e/\vartheta = p_0/\vartheta_0,$$

where p_0 and ϑ_0 are reference values of the pressure and temperature. In the $\mathcal{E}$–F diagram the isochors appear as vertical dotted lines. This expression shows that heating an ideal gas in a confined volume will increase its pressure.

Isothermal Process

In an isothermal process the temperature is constant. From Eq. (4.10), we see that $\mathcal{E}$ is constant when ϑ is constant. We show the isotherms as dashed lines that are spaced at uniform increments of ϑ. From Eq. (4.8), the relationship between the pressure and the density along an isotherm is

$$p_e/\rho = p_0/\rho_0. \tag{4.25}$$

This relationship is called *Boyle's law*. In the $\mathcal{E}$–F diagram the isotherms appear as horizontal dashed lines.

Isentropic Process

In an isentropic process the entropy is constant. For constant η, Eq. (4.24) gives

$$\frac{\mathcal{E}}{\mathcal{E}_0}F^{\gamma-1} = \text{constant}.$$

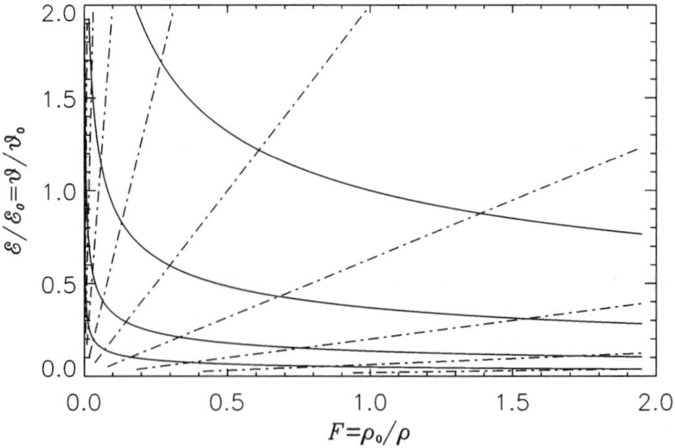

Figure 4.4 The $\mathcal{E}$–F diagram of the isentropes (—) and the isobars (– · –).

After we substitute Eqs. (4.8) and (4.10) into this expression, we obtain

$$p_e/\rho^\gamma = p_0/\rho_0^\gamma. \tag{4.26}$$

By comparing this expression to Boyle's law, Eq. (4.25), we see that the two expressions differ only by the exponential factor γ. This is also reflected in the difference between the isentropic stiffness $C^\eta = \gamma p_e$ and the isothermal stiffness $C^\vartheta = p_e$ (see Exercise 4.3). We use solid lines to plot the isentropes on the state surface and in the $\mathcal{E}$–F diagram (see Figure 4.4). They are spaced at uniform increments of η.

The expression for an isentropic process, Eq. (4.26), is the constitutive model we introduced in Chapter 2 to describe an adiabatic gas [see Eq. (2.27)]. Recall that linear-elastic waves and nonlinear structured waves are isentropic processes. Thus when these waves propagate within an ideal gas they obey the equation for the isentropic process, Eq. (4.26), which is also the constitutive model for an adiabatic gas.

Isobaric Process

In an isobaric process the pressure is constant. When the pressure is constant, Eq. (4.8) gives

$$\rho\vartheta = \rho_0\vartheta_0.$$

This is called *Charles' law*. We use dot-dashed lines to represent the isobars. They are determined from Eq. (4.14) for the pressures $p_e = \rho\Gamma\mathcal{E} = \rho_0\Gamma(0.02 \times 10^{i/2})$ where $i = 0, 1, \ldots, 8$. Notice that they are straight lines in the $\mathcal{E}$–F diagram.

4.1.4 *Elastic-Wave Process*

We know that the elastic wave process is an isentropic process. The material wave velocity C is related to the isentropic stiffnesses $C^\eta = \mathcal{K}^\eta$ by [see Eq. (3.88)]

$$C = \left(\frac{C^\eta}{\rho_0 F}\right)^{1/2}. \tag{4.27}$$

Substitution of Eq. (4.11) yields

$$C = \frac{\rho}{\rho_0}\left(\frac{\gamma p_e}{\rho}\right)^{1/2}. \tag{4.28}$$

The spatial wave velocity $c = FC$ is

$$c = \left(\frac{\gamma p_e}{\rho}\right)^{1/2}. \tag{4.29}$$

When we substitute Eq. (4.26) into the previous two results, we recover the expressions for the wave velocities in the adiabatic gas [see Eqs. (2.108) and (2.109)]. We obtain other useful expressions by substituting Eqs. (4.8), (4.12), and (4.16) to obtain

$$\begin{aligned} C &= \frac{\rho}{\rho_0}[(\gamma - 1)c_p\vartheta]^{1/2}, \\ c &= [(\gamma - 1)c_p\vartheta]^{1/2}. \end{aligned} \tag{4.30}$$

Recall that if c_v is a constant, then γ and c_p are also constant. Then we find that the spatial wave velocity depends only upon the temperature.

Monoatomic Gas

We can substitute Eqs. (4.13) and (4.16) into the previous result to obtain

$$c^2 = \gamma N R \vartheta.$$

From our results for the velocity V of atoms in a monoatomic gas, we know that [see Eq. (4.23)]

$$V^2 = 3NR\vartheta.$$

The adiabatic gas constant of a monoatomic gas is $\gamma = 5/3$; therefore,

$$\left(\frac{V}{c}\right)^2 = \frac{9}{5}.$$

Thus we find that the velocities of the atoms are slightly greater than the wave velocity c.

Acoustic Waves

A linear-elastic wave in an ideal gas is also called an *acoustic wave*. Like all elastic waves, acoustic waves cause the ideal gas to follow an isentrope. We can obtain the velocity of sound for an acoustic wave from the previous expressions for c and C. When we let $\rho = \rho_0$, then $p_e = p_0$. Thus $C = c = c_0$, where the velocity of sound c_0 is given by either of the two equivalent relationships:

$$c_0 = \left(\frac{\gamma p_0}{\rho_0}\right)^{1/2} \tag{4.31}$$

or

$$c_0 = [(\gamma - 1)c_p\vartheta_0]^{1/2}. \tag{4.32}$$

The acoustic impedance $z = \rho_0 c_0$ is

$$z_0 = (\gamma \rho_0 p_0)^{1/2}. \tag{4.33}$$

The velocity of sound depends only on the temperature ϑ_0 of the ideal gas. The change in temperature resulting from a propagating elastic wave is given by Eq. (3.94),

$$\frac{\Delta\vartheta}{\vartheta_0} = (1-\gamma)\frac{\partial \hat{u}}{\partial X}.$$

Newton's Speed of Sound In the beginning of Chapter 3 we refer to Newton's difficulty with the prediction of the speed of sound in air. Newton calculated this speed using data for the density and pressure of air. His data for both quantities were reasonably accurate; however, his analysis led him to believe that†

> The velocities of pulses propagated in an elastic fluid are in a ratio compounded of the square root of the ratio of the elastic force directly, and the square root of the ratio of the density inversely; supposing the elastic force of the fluid to be proportional to its condensation.

Consequently, he used $C^{\vartheta} = p_e$ in his calculation of the speed of sound instead of $C^{\eta} = \gamma p_e$ as in Eq. (4.27). This introduced an error of $\sqrt{\gamma}$ into his result.

His estimate of the speed of sound was 979 ft/s (298 m/s). For air, $\gamma = 1.4$, and using his data, the corrected estimate is $979 \times \sqrt{1.4} = 1158$ ft/s (353 m/s). This number compares well with the current estimate of 1130 ft/s (345 m/s) for the speed of sound at sea level and 290 K. Unfortunately neither Newton nor anyone else during this period of history understood the difference between the isentropic and isothermal process. Newton, like everyone else, had an occasional bad moment.

Properties of the Atmosphere People have used rockets and weather balloons to collect data for the pressure and density of the atmosphere at high altitudes.‡ From these data, we can calculate the temperature, the velocity of sound, and the acoustic impedance. We show data for pressure and density in Figure 4.5. The measured values of pressure, density, and temperature at sea level (0 km) are

$$p_e = 0.1037 \text{ MPa}, \qquad \rho = 1.22 \text{ kg/m}^3, \qquad \vartheta = 290 \text{ K}.$$

From these values, we can calculate NR:

$$NR = p_e/\rho\vartheta = 0.1037 \times 10^6/1.22/290 = 293 \text{ J/kg/K}.$$

We determine the temperature, velocity of sound, and acoustic impedance using

$$\vartheta = p_e/NR\rho, \qquad c = (\gamma p_e/\rho)^{1/2}, \qquad z = \rho c,$$

where $\gamma = 1.4$. We show the results of these calculations in Figures 4.5 and 4.6. Notice that at high altitudes, the temperature of the atmosphere is high. This results in a large velocity

† See p. 589 in *Newton's Principia for the Common Reader* by Chandrasekhar, S. (Oxford University Press, 1995).

‡ See "The pressure, density, and temperature of the Earth's atmosphere to 160 kilometers" by Havens, R. J., Koll, R. T., and LaGow H. E. in *Journal of Geophysical Research* **57**, pp. 59–72 (1952).

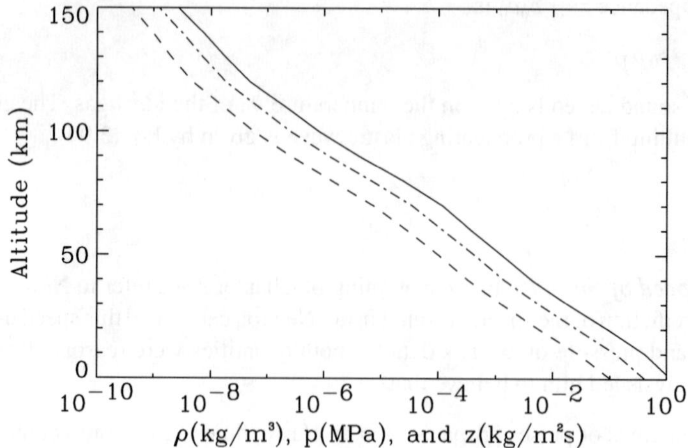

Figure 4.5 Density ρ (—), pressure p_e (– –), and acoustic impedance z (– · –) of the atmosphere.

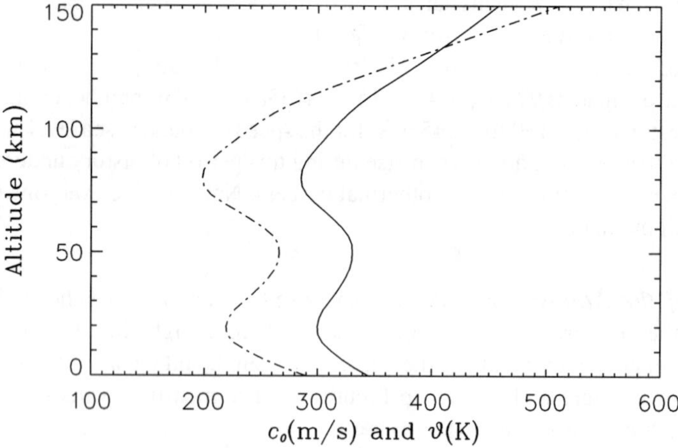

Figure 4.6 Velocity of sound c (—) and temperature ϑ (– · –) of the atmosphere.

of sound. However, the acoustic impedance z is extremely low, because the density is low. Recalling Eq. (3.95) in Section 3.4.2, we know that the energy that can be imparted to an elastic wave in a half space is proportional to z. Thus although sound propagates rapidly at these altitudes, because z is low, these waves have very little energy. Hence it is next to impossible to either produce or hear sound.

4.1.5 *Hugoniot Relationships for Constant Specific Heat*

Consider an ideal gas with a constant specific heat c_v at an initial state where the pressure p_0, the density ρ_0, and the temperature ϑ_0 are known. The Hugoniot relationship describes the locus of equilibrium states that can be attained by either a shock discontinuity or a steady wave. At the Hugoniot state the pressure is p_+, the density is ρ_+, and the temperature is ϑ_+. We can determine the Hugoniot state with the jump condition for the balance of linear momentum [see Eq. (2.147)],

$$-NR(\rho_+\vartheta_+ - \rho_0\vartheta_0) = \rho_0 C_S^2(F_+ - 1),$$

and the jump condition for the balance of energy [see Eq. (3.111)],

$$\rho_0 c_v(\vartheta_+ - \vartheta_0) = -NR\rho_0\vartheta_0(F_+ - 1) + \frac{1}{2}\rho_0 C_S^2(F_+ - 1)^2,$$

where $F_+ = \rho_0/\rho_+$. Using these two expressions, we can determine the Hugoniot curve for an ideal gas with a constant specific heat c_v.

Recalling that $NR = (\gamma - 1)c_v$ and $c_0^2 = \gamma(\gamma - 1)c_v\vartheta_0$, we rewrite these two jump conditions as

$$\begin{aligned} \frac{\vartheta_+}{\vartheta_0} &= F_+ + \gamma F_+(1 - F_+)M^2, \\ \frac{\vartheta_+}{\vartheta_0} &= 1 + (\gamma - 1)(1 - F_+) + \frac{1}{2}\gamma(\gamma - 1)(1 - F_+)^2 M^2, \end{aligned} \tag{4.34}$$

where

$$M = \frac{|C_S|}{c_0} = \frac{U_S}{c_0}.$$

The ratio M is called the *Mach number*. We eliminate ϑ_+/ϑ_0 from Eqs. (4.34) to obtain

$$(1 - F_+)\left\{\left[F_+ + \frac{1}{2}(\gamma - 1)(F_+ - 1)\right]M^2 - 1\right\} = 0. \tag{4.35}$$

This equation has two roots. One root is $F_+ = 1$, which represents the initial state. The other root is

$$1 - F_+ = \frac{2}{\gamma + 1}\left(1 - \frac{1}{M^2}\right), \tag{4.36}$$

which gives the density $\rho_+ = \rho_0/F_+$ at the Hugoniot state in terms of the wave velocity C_S. When we solve this result for M^2 and substitute the result into the first of Eqs. (4.34), we obtain the following expression for the temperature ϑ_+ at the Hugoniot state:

$$\frac{\vartheta_+}{\vartheta_0} = F_+\left(\frac{1 - F_l F_+}{F_+ - F_l}\right), \tag{4.37}$$

where

$$F_l = \frac{\gamma - 1}{\gamma + 1}.$$

These Hugoniot relationships for the ideal gas exhibit an interesting feature. For very strong shocks the Mach number M approaches infinity. As this limit is approached, Eqs. (4.36) and (4.37) become

$$F_+ \to F_l$$

and

$$\vartheta_+ \to \infty.$$

Thus

$$\frac{\rho_+}{\rho_0} \le \frac{\gamma+1}{\gamma-1}. \tag{4.38}$$

This is the upper limit for the density that can be attained by a either a shock wave or a steady wave. As the shock wave approaches this limit, the temperature in the gas approaches infinity. This is called the *shock-compression limit.* Thus, for air where $\gamma = 1.4$, it is impossible for a shock wave to increase the density by more than a factor of 6.

Weak Shock Waves

Consider an ideal gas at rest ($v = 0$) in an initial state where $F = 1$, $p = p_0$, and $\vartheta = \vartheta_0$. This gas encounters a shock wave that causes it to move to a state on the Hugoniot where $F = F_+$, $p = p_+$, and $\vartheta = \vartheta_+$. Now suppose an identical volume of gas at the same initial state encounters a simple compression wave that causes that gas to move along a isentrope to a final state where $F = F_+$. In general, we know that the pressure and temperature of the gas that encounters the compression wave will not be equal to the pressure and temperature of the gas that encounters the shock wave. Although both waves cause the same change in the deformation gradient, the shock wave causes the entropy of the gas to increase and results in a higher pressure and temperature. However, if the change in F caused by these waves is small, we know that the change in entropy resulting from the shock wave can be approximated by the Taylor's series [see Eq. (3.119)]

$$[\![\eta]\!] = \frac{1}{12\rho_0}\left(\frac{1}{\vartheta}\frac{\partial^2 T}{\partial F^2}\right)_- [\![F]\!]^3 + \cdots.$$

Because this jump in entropy depends upon the third power of $[\![F]\!]$, shock waves that produce small changes in F result in negligible changes in η. Such shock waves are called weak shocks. In Chapter 2 we derived the following expression for the velocity of a weak shock in an adiabatic gas [see Eq. (2.171)]:

$$U_S = c_0 + \frac{\gamma+1}{4} v_+. \tag{4.39}$$

From the preceding discussion, it is clear that this expression for a weak shock must be a special case of the general Hugoniot relationship, Eq. (4.36). We shall now show that this is so.

We substitute $v_+/U_S = 1 - F_+$ and $M = U_S/c_0$ into Eq. (4.36) to obtain

$$\frac{\gamma+1}{2} v_+ = U_S\left(1 - \frac{c_0^2}{U_S^2}\right) = \frac{[c_0 + (U_S - c_0)]^2 - c_0^2}{c_0 + (U_S - c_0)}$$

$$= \frac{2c_0(U_S - c_0) + (U_S - c_0)^2}{c_0\left(1 + \frac{U_S - c_0}{c_0}\right)}.$$

We assume that $U_S - c_0 < c_0$ so that the denominator can be represented by a binomial series to obtain

$$\frac{\gamma+1}{2} v_+ = \left[\frac{2c_0(U_S - c_0) + (U_S - c_0)^2}{c_0}\right]\left[1 - \frac{U_S - c_0}{c_0} + \cdots\right]$$

$$= 2(U_S - c_0)\left(1 - \frac{U_S - c_0}{2c_0} + \cdots\right),$$

where we have omitted terms containing third and higher powers of $(U_S - c_0)$. When $U_S - c_0 \ll 2c_0$, we obtain Eq. (4.39) by only retaining the first term on the right side.

Recall that, for a shock wave in a gas that is initially at rest, the jump in pressure across a shock wave is $[\![p]\!] = \rho_0 U_S v_+$. When we substitute the expression for the velocity of sound [Eq. (4.32)] into this jump in pressure, we obtain

$$\frac{[\![p]\!]}{p_0} = \frac{\gamma M v_+}{c_0}.$$

If the wave is a weak shock, then [see Eq. (4.39)]

$$\frac{v_+}{c_0} = \frac{4}{\gamma + 1}(M - 1), \tag{4.40}$$

and if the wave is a strong shock, then [see Eq. (4.36)]

$$\frac{v_+}{c_0} = \frac{2}{\gamma + 1}\left(M - \frac{1}{M}\right). \tag{4.41}$$

In Figure 4.7 we illustrate the region in which the weak shock [Eq. (4.40)] is a valid approximation to the actual Hugoniot solution [Eq. (4.41)]. Here we compare the weak- and strong-shock solutions for $\gamma = 1.4$. For example, when the shock wave doubles the pressure, the weak-shock solution overestimates the jump in pressure by about 10 percent.

Compression-Limited Shock Waves

In the previous section we studied the weak shock. It represents a limiting case in the analysis of the ideal gas where the change in η can be neglected. In this section we shall examine another limiting case. This is a strong shock wave that causes the deformation gradient to change from an initial state at $F_- = 1$ to a Hugoniot state at

$$F_+ = \frac{\rho_0}{\rho_+} = \frac{\gamma - 1}{\gamma + 1}. \tag{4.42}$$

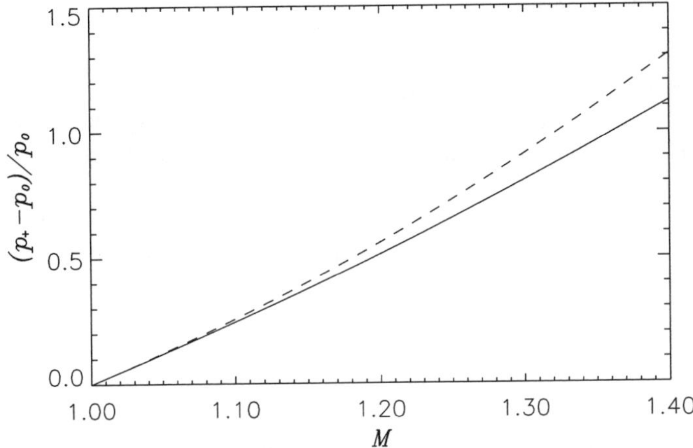

Figure 4.7 Comparison of the Hugoniot pressures for the strong shock (—) and weak shock (– –) when $\gamma = 1.4$.

Recall that this is the deformation gradient at the shock-compression limit [see Eq. (4.38)]. Because

$$\frac{p_+}{\rho_+\vartheta_+} = \frac{p_0}{\rho_0\vartheta_0},$$

we find that the temperature at the Hugoniot state is

$$\frac{\vartheta_+}{\vartheta_0} = \left(\frac{\gamma - 1}{\gamma + 1}\right)\frac{p_+}{p_0}. \tag{4.43}$$

Here we can let the pressure p_+ at the Hugoniot state be any value provided it is large enough to cause the deformation gradient F_+ to fall close to the compression limit given by Eq. (4.42).

Consider an ideal gas at rest. Suppose it encounters a shock wave that compresses it close to this limit. The kinematic compatibility condition for this shock wave is [see Eq. (2.139)]

$$(F_+ - 1)C_S = -v_+.$$

We substitute Eq. (4.42) to obtain

$$v_+ = \frac{2C_S}{\gamma + 1}. \tag{4.44}$$

The jump condition for the balance of linear momentum is [see Eq. (2.147)]

$$-p_+ + p_0 = \rho_0 C_S^2 (F_+ - 1).$$

We assume that $p_+ \gg p_0$ and ignore p_0 in this expression. Thus

$$p_+ = \rho_0 C_S^2 \left(\frac{2}{\gamma + 1}\right). \tag{4.45}$$

We shall find that Eqs. (4.42)–(4.45) are particularly useful for the analysis of very strong shock waves in an ideal gas. We shall demonstrate this in the next section.

4.1.6 *Shock-Tube Experiment*

Recall that a shock tube is a long hollow cylinder divided into two chambers by a diaphragm (see Figure 4.8). The chamber to the left is called the driver section. The one to the right is called the test section. Previously, we assumed both chambers contained identical gases. Here we allow different gases in each chamber. Initially, the temperatures in both chambers are equal to ϑ_0. However, the mole numbers, adiabatic gas constants, initial densities, sound speeds, and initial pressures are different. A prime (′) is affixed to these quantities to denote values in the driver section. Whereas the temperatures in each chamber are equal, the pressures are quite different and obey

$$p_0' \gg p_0.$$

We illustrate the x–t and the p–v diagrams for the shock tube in Figures 4.9 and 4.10. The diaphragm is released at $t = 0$. We illustrate its path with a dashed line in Figure 4.9. After the diaphragm is released, a shock wave propagates into the test section, and a centered rarefaction wave propagates into the driver section. These waves divide the x–t diagram

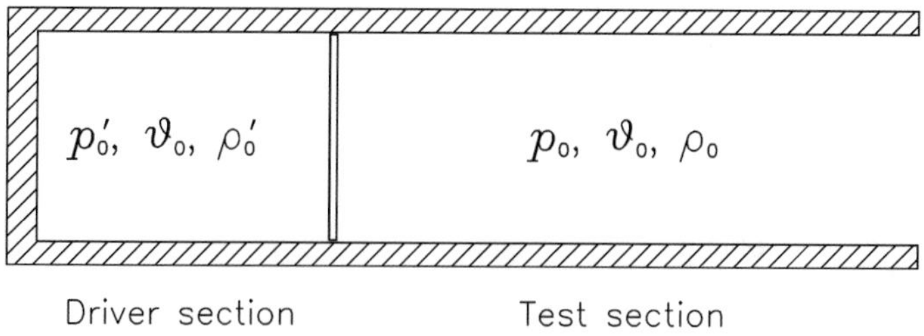

Figure 4.8 The shock-tube experiment.

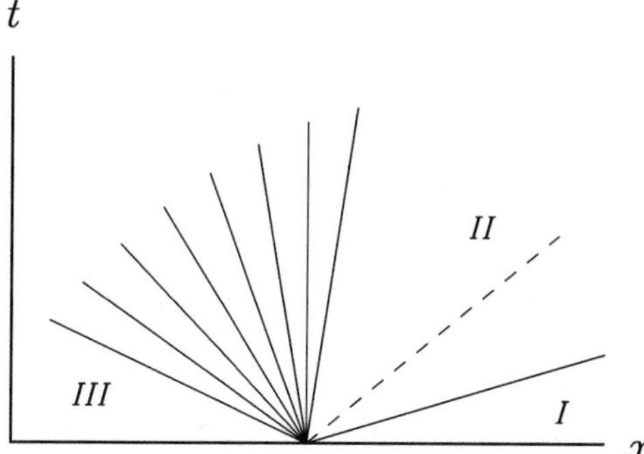

Figure 4.9 The x–t diagram for the shock-tube experiment.

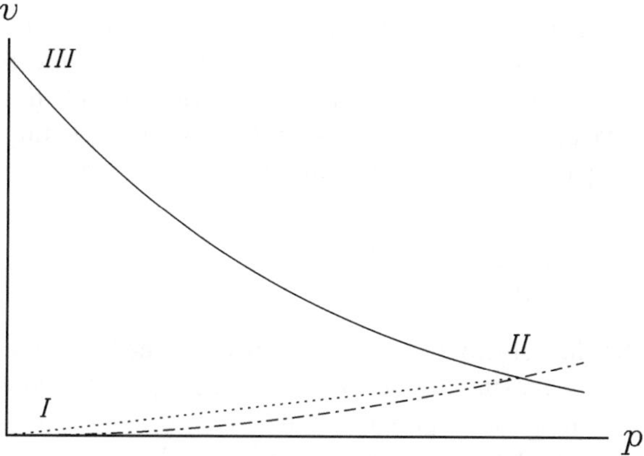

Figure 4.10 The p–v diagram for the shock-tube experiment.

into three regions. Initially, all of the test section is in region *I*, and all of the driver section is in region *III*. The initial conditions determine the solutions in regions *I* and *III*. Our objective is to determine the solution in region *II*. We do this by first determining how the rarefaction wave changes the solution from region *III* to region *II*, and then we determine how the shock wave changes the solution from region *I* to region *II*.

The centered rarefaction wave is a simple wave for which the Riemann integral $J_+ = 0$. From Eq. (2.112) in Section 2.5.4, we obtain

$$v = -\frac{2}{\gamma' - 1}(c - c_0').$$

From Eq. (2.109), we also obtain

$$\frac{c}{c_0'} = F^{-\frac{\gamma'-1}{2}} = \left(\frac{p}{p_0'}\right)^{\frac{\gamma'-1}{2\gamma'}}.$$

Combining these two expression, we find that

$$v = \frac{2c_0'}{\gamma' - 1}\left[1 - \left(\frac{p}{p_0'}\right)^{\frac{\gamma'-1}{2\gamma'}}\right]. \tag{4.46}$$

This is the expression for the response path of the rarefaction fan. It is represented by a solid line in the p–v diagram in Figure 4.10. This line has a negative slope because the rarefaction wave propagates in the negative-X direction.

If we apply Eq. (4.46) to region *II*, we obtain one equation for the unknown velocity and pressure in region *II*. We need another expression that contains these two unknowns. The jump conditions across the shock wave yield the required expression. Because $p_0' \gg p_0$, we can use the approximate expressions for a very strong shock. From Eqs. (4.44) and (4.45), we obtain

$$p_{II} = \frac{\gamma + 1}{2}\rho_0 v_{II}^2.$$

This is the Hugoniot of the shock wave. It is illustrated with a dot-dash line in Figure 4.10. It has a positive slope because the shock wave propagates in the positive-X direction.

The state at region *II* is determined by the intersection of the Hugoniot with the response path of the rarefaction wave. We obtain analytic expressions for this state by letting $v = v_{II}$ and $p = p_{II}$ and combining Eq. (4.46) with the previous expression to obtain

$$v_{II} = \frac{2c_0'}{\gamma' - 1}\left\{1 - \left[\frac{(\gamma + 1)\rho_0 v_{II}^2}{2p_0'}\right]^{\frac{\gamma'-1}{2\gamma'}}\right\}. \tag{4.47}$$

We can solve this equation for the velocity in region *II*. Notice that the term in the square brackets is a positive quantity. The maximum value of the velocity v_{II} is attained when this term approaches zero. This occurs when the pressure in the driver section is very large and the density in the test section is very low. The upper limit for the velocity is

$$v_{II} = \frac{2c_0'}{\gamma' - 1}. \tag{4.48}$$

In Chapter 2 we obtained the same expression for the velocity of a half space of adiabatic gas that was suddenly allowed to freely expand [see Eq. (2.125)]. It is called the escape velocity.

Now that we know the velocity in region *II*, let us determine the pressure and the temperature just behind the shock wave in the gas in the test section. *We will evaluate these quantities only for the limiting velocity, Eq. (4.48)*. First we must evaluate the velocity of the shock wave. From Eq. (4.44), it is

$$C_S = \frac{\gamma + 1}{\gamma' - 1} c_0'.$$

Then from Eq. (4.45), the pressure in region *II* is

$$p_{II} = 2\rho_0 \left(c_0'\right)^2 \frac{\gamma + 1}{(\gamma' - 1)^2},$$

and from Eq. (4.43), the temperature behind the shock wave in the test section is

$$\frac{\vartheta_+}{\vartheta_0} = \frac{2\rho_0 \left(c_0'\right)^2}{p_0} \frac{\gamma - 1}{(\gamma' - 1)^2}.$$

Then we substitute Eqs. (4.8) and (4.31) into this expression to obtain

$$\frac{\vartheta_+}{\vartheta_0} = \frac{2\gamma'(\gamma - 1)N'}{(\gamma' - 1)^2 N}. \tag{4.49}$$

This result represents the upper limit of the temperature that can be attained in the shock tube.

As an example, suppose hydrogen is used in the driver section. Then $\gamma' = 7/5$ and $N' = 500$ moles/kg. The test section contains air where $\gamma' = 7/5$ and $N' = 55.5$ moles/kg. When we substitute these values into Eq. (4.49), we obtain

$$\frac{\vartheta_+}{\vartheta_0} = 63.$$

Suppose $\vartheta_0 = 300$ K; then $\vartheta_+ = 18{,}900$ K. Gases will ionize at such high temperatures. Ionization absorbs energy and reduces the actual temperatures that are attainable in a shock tube.

4.1.7 *Viscous Ideal Gas*

So far, we have restricted our discussion of the ideal gas to equilibrium states. Even the Hugoniot relation only contains expressions for equilibrium quantities. We have determined these quantities solely with the fundamental equation for the ideal gas, Eq. (4.2). Now let us consider the constitutive equations for a viscous material [see Eqs. (3.33)–(3.35)]:

$$\begin{aligned} \mathcal{E} &= \mathcal{E}(\rho, \eta), \\ T &= T(F, D, \eta), \\ \vartheta &= \partial \mathcal{E}/\partial \eta, \end{aligned}$$

where we assume that the internal energy is restricted to be a function of ρ and not the deformation gradient tensor. We shall fit the equilibrium description for the ideal gas into this constitutive description for a viscous material.

We see that this is a relatively easy task, because the first and last of these equations are given by the fundamental equation for the internal energy, Eq. (4.4), and the prescribed function for the temperature, Eq. (4.6). Furthermore, we remember that at an equilibrium state or during a reversible process, the stress T is equal to the equilibrium pressure $-p_e$. Consequently, we can represent the stress T by the additive decomposition

$$T = -p_e + \mathcal{T}. \tag{4.50}$$

Because the equilibrium pressure p_e is also determined by the fundamental equation, we need only prescribe a constitutive equation for the dissipative stress $\mathcal{T}$ that obeys the reduced dissipation inequality, Eq. (3.44):

$$\mathcal{T} D \geq 0. \tag{4.51}$$

Recall that a requirement resulting from this inequality is

$$\mathcal{T}(F, 0, \eta) = 0. \tag{4.52}$$

Rayleigh-Line Process for Constant Specific Heat

Consider a steady wave in a viscous ideal gas with constant specific heat (see Figure 4.11). To determine the profile of this wave, we must know the constitutive equation for the viscous stress. However, for the present, let us see what we can learn about this wave without using this constitutive equation. Recall that the steady-wave solution for the equation of balance of linear momentum is [see Eq. (2.189)]

$$T + p_0 = \rho_0 C_S^2 (F - 1),$$

where $F = \rho_0/\rho$ and $F_- = 1$. Suppose we substitute Eq. (4.52) and the expression for the equilibrium pressure, Eq. (4.8), to obtain

$$\mathcal{T} = \rho_0 C_S^2 (F - 1) + (\gamma - 1) c_v (\rho\vartheta - \rho_0\vartheta_0),$$

where $NR = (\gamma - 1)c_v$. Recall that the velocity of sound is $c_0^2 = \gamma(\gamma - 1)c_v\vartheta_0$. Thus

$$\frac{\mathcal{T}}{(\gamma - 1)\rho c_v \vartheta_0} = \gamma F(F - 1)M^2 - F + \frac{\vartheta}{\vartheta_0}. \tag{4.53}$$

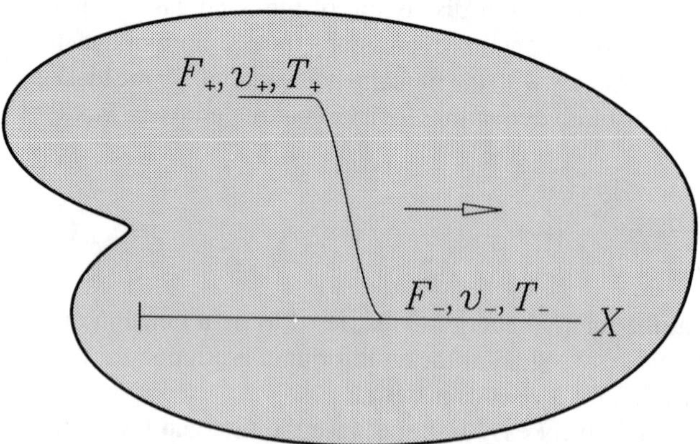

Figure 4.11 A viscous ideal gas with a steady compression wave.

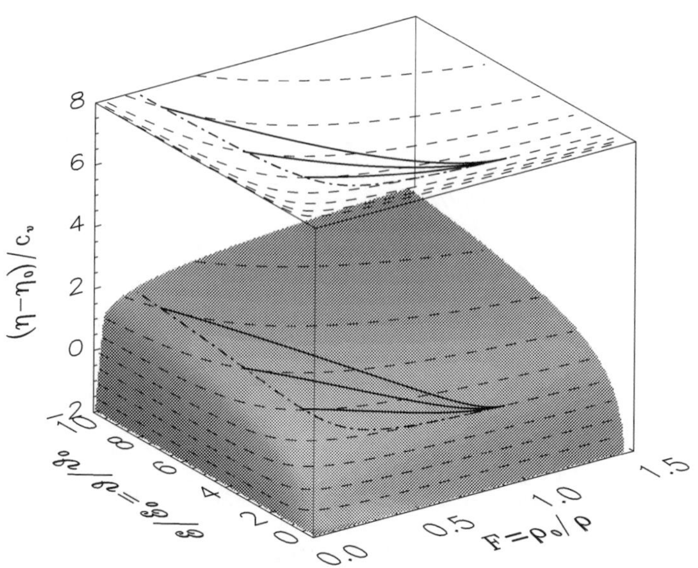

Figure 4.12 The state surface for $\gamma = 1.4$ with isentropes (— —), Hugoniot (— · —), and Rayleigh lines (—) for $M = 3.19$, 4.65, and 6.37.

By a similar procedure, we obtain the steady-wave solution for the balance of energy [see Eq. (4.34)]:

$$\frac{\vartheta}{\vartheta_0} = 1 + (\gamma - 1)(1 - F) + \frac{1}{2}\gamma(\gamma - 1)(1 - F)^2 M^2. \tag{4.54}$$

Because we now know the temperature in terms of F, we can determine the internal energy from $\mathcal{E} = c_v \vartheta$ and the entropy from the fundamental equation, Eq. (4.24). Thus we know the location of the Rayleigh-line process on the state surface. As an example, we show a state surface with constant c_v and $\gamma = 1.4$ in Figure 4.12. We also show the Hugoniot and the Rayleigh-line processes for three different steady compression waves. The Rayleigh-line processes are defined by values of the Mach number M. The initial state is located at $\rho = \rho_0$ and $\mathcal{E} = \mathcal{E}_0$.

To determine the value of the viscous stress along the Rayleigh-line process, let us combine the two previous equations to obtain

$$-\frac{F\mathcal{T}}{\rho_0 c_0^2} = (1 - F)\left\{\left[F + \frac{1}{2}(\gamma - 1)(F - 1)\right]M^2 - 1\right\}.$$

Recall that the Mach number M and the deformation gradient F_+ at the Hugoniot state are related by Eq. (4.36). We substitute this expression into the previous result to obtain

$$\frac{2F\mathcal{T}}{(\gamma + 1)\rho_0 U_S^2} = (F - 1)(F - F_+). \tag{4.55}$$

It is very important for you to remember that this result is not a constitutive relationship for $\mathcal{T}$. Rather, it is a condition that ensures the steady wave satisfies the balances of linear momentum and energy. For a Rayleigh-line process, this simple expression yields the viscous

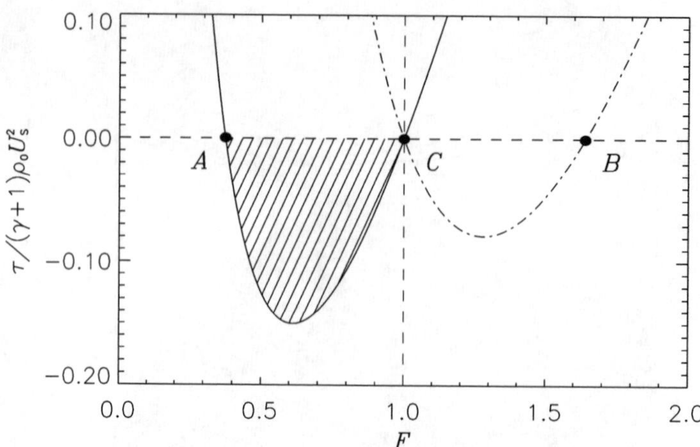

Figure 4.13 The viscous stress $\mathcal{T}$ for a steady compression wave, $M = 2$ (—), and for a steady rarefaction wave, $M = 0.75$ (− · −). For both processes, the viscous stresses are negative between the initial state C and the Hugoniot states at A and B.

stress $\mathcal{T}$ in terms of the deformation gradient F. As required by Eq. (4.52), the viscous stress is zero at both the initial state $F = 1$ and the Hugoniot state $F = F_+$. In Figure 4.13 we illustrate two plots of the viscous stress $\mathcal{T}$. The curve for $M = 2$ has a Hugoniot state A where $F_+ < 1$ and $\mathcal{T} = 0$. This is a steady compression wave. The curve for $M = 0.75$ has a Hugoniot state B where $F_+ > 1$ and $\mathcal{T} = 0$. This is a steady rarefaction wave. Notice that the steady compression wave satisfies the supersonic condition $U_S \geq c_0$, whereas the steady rarefaction wave violates it [see Eq. (3.108)].

Existence of Steady Compression Wave When a steady compression wave propagates through an ideal gas

$$F_+ \leq F \leq 1.$$

Because $\partial \hat{F}/\partial t \leq 0$, we find that the reduced dissipation inequality, Eq. (4.51), requires

$$\mathcal{T} \leq 0.$$

From Eq. (4.55) and the example for $M = 2$ shown in Figure 4.13, we see that this condition is satisfied. This means that steady compression waves can exist in a viscous ideal gas.

Nonexistence of the Steady Rarefaction Wave When a steady rarefaction wave propagates through an ideal gas

$$F_+ \geq F \geq 1.$$

Because $\partial \hat{F}/\partial t \geq 0$, we find that the reduced dissipation inequality, Eq. (4.51), requires

$$\mathcal{T} \geq 0.$$

From Eq. (4.55) and the example for $M = 0.75$ shown in Figure 4.13, we see that this condition is not satisfied. This means that steady rarefaction waves cannot exist in a viscous ideal gas.

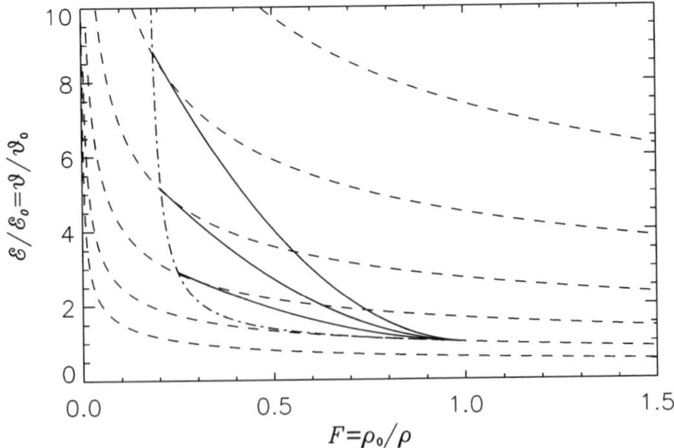

Figure 4.14 The $\mathcal{E}$–F diagram of the Rayleigh-line processes (—), the Hugoniot (– · –), and the isentropes (– –).

Shock-Compression Limit We have already shown that during one-dimensional motion, shock waves and steady waves cannot compress an ideal gas beyond the shock-compression limit [Eq. (4.38)]

$$\frac{\rho_+}{\rho_0} \leq \frac{\gamma + 1}{\gamma - 1}.$$

This is evident from the $\mathcal{E}$–F diagram of the Rayleigh-line processes (see Figure 4.14). The Hugoniot approaches an asymptote where the temperature becomes infinite. This illustrates the effect of *shock heating*. Although we can compress the ideal gas either isothermally or isentropically beyond this limit, the steady wave acts through $\mathcal{T}$ to produce heat, to increase temperature, and to limit the compression.

Let us use a plot of the pressures of the Hugoniot and the Rayleigh-line processes to illustrate this point. In Figure 4.15 we compare the Rayleigh-line processes to the isentrope and the isotherm that intersect the initial state. We plot the stress T and the equilibrium pressure p_e for three different Rayleigh-line processes. The differences between the curves for T and the curves for p_e represent the viscous stresses $\mathcal{T}$. Notice that $\mathcal{T}$ is often larger than p_e. In contrast, when we move the gas along either the isotherm or the isentrope, then $\mathcal{T} = 0$. Thus we can compress the gas along these reversible processes with significantly smaller pressures and without encountering a limiting value of density. The temperatures produced by shock heating result in a remarkable effect. Indeed, densities that we can produce by isentropic compression of air with an ordinary bicycle tire pump are not achievable by shock compression.

Steady-Wave Profiles

To evaluate the profile of a steady wave in a viscous ideal gas, we must know the constitutive equation for the viscous stress. Suppose we assume the following relationship:

$$\mathcal{T} = \frac{\rho_0 b_n}{F}\frac{\partial \hat{F}}{\partial t}\left|\frac{\partial \hat{F}}{\partial t}\right|^{n-1}, \tag{4.56}$$

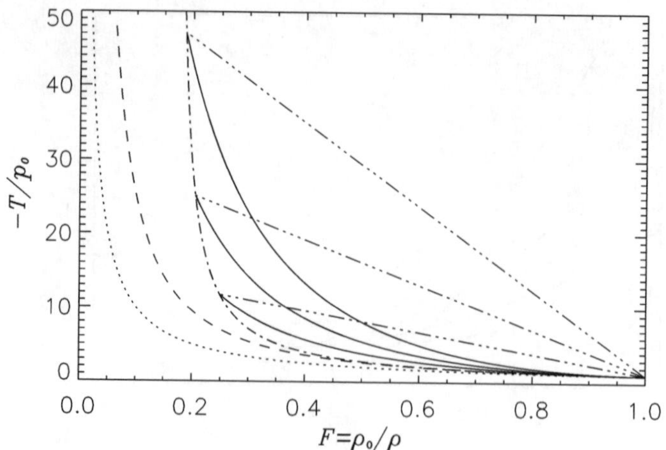

Figure 4.15 Hugoniot pressure (– · –) compared to the isentropic (– –) and isothermal (· · ·) pressures. We show the total stress T (– · · · –) and the equilibrium pressure p_e (—) for three different Rayleigh-line processes.

where n and b_n are constants. Provided that b_n is a positive constant, this constitutive equation satisfies the dissipation inequality, Eq. (4.51). It also satisfies the equilibrium condition, Eq. (4.52). This is a rather restrictive assumption because $\mathcal{T}$ is not a function of η.

In this section we arbitrarily limit the values of n to either 1 or 2. When $n = 1$, the viscous stress has a linear dependence upon $\partial\hat{F}/\partial t$. This relationship is equivalent to the viscous stress for the Newtonian fluid where [see Eq. (3.97)]

$$b_1 = \lambda' + \frac{4}{3}\mu'.$$

The coefficient b_1 is called the *linear viscosity coefficient*. When $n = 2$, the viscous stress has a quadratic dependence upon $\partial\hat{F}/\partial t$, and b_2 is called the *quadratic viscosity coefficient*. These names are derived from numerical analysis, where Eq. (4.56) has been used to implement certain finite-difference solutions. While we see that b_1 can be directly related to the kinematic viscosities of the ideal gas, the choice of values of these coefficients are often based solely upon considerations associated with numerical solution methods (see Appendix A). However, that it not our intent here, and we assume that values of b_n exist that represent physical qualities of the ideal gas.

If the steady-wave variable $\hat{\varphi} = X - C_S t$ is substituted into Eq. (4.56), and the result is combined with Eq. (4.55), we obtain

$$\frac{1}{(1-F)(F-F_+)}\frac{dF}{d\hat{\varphi}}\left|\frac{dF}{d\hat{\varphi}}\right|^{n-1} = \frac{\gamma+1}{2b_n}U_S^{2-n}, \tag{4.57}$$

where we have assumed that the steady wave propagates in the positive-X direction so that $C_S > 0$ and $U_S = C_S$. Having established this model, we can calculate profiles of the steady waves and compare the relative effects of linear and quadratic viscosity.

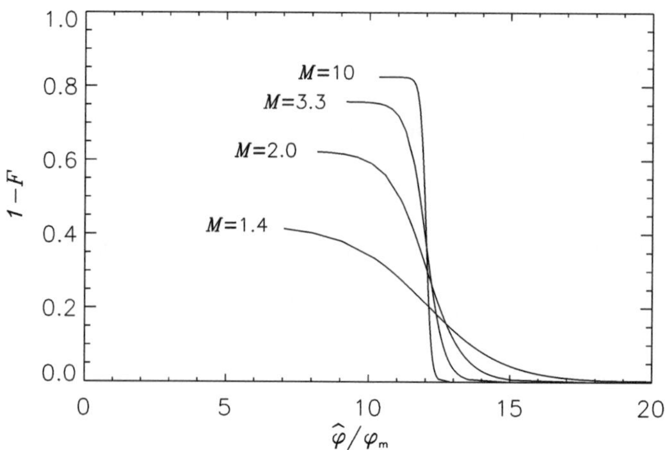

Figure 4.16 Steady wave profiles in air with a linear viscosity.

Linear Viscosity When $n = 1$, we can write Eq. (4.57) as

$$\frac{dF}{(1 - F)(F - F_+)} = \frac{\gamma + 1}{2b_1} U_S \, d\hat{\varphi}.$$

The left side of this equation is a function of F only, and the right side is a function of $\hat{\varphi}$ only. We integrate this expression to find that

$$\frac{1}{F_+ - 1} \ln\left(\frac{1 - F}{F - F_+}\right) = \frac{\gamma + 1}{2b_1} U_S \hat{\varphi},$$

where we have set the constant of integration to zero. We illustrate the steady wave profiles in air for four values of M (see Figure 4.16). Here we have used the measured values of the velocity of sound c_0, the kinematic viscosities λ' and μ', the ratio of specific heats γ, and the mean free path φ_m in air at standard atmospheric conditions:

$$c_0 = 345\,\text{m/s}, \qquad \gamma = 1.4,$$
$$b_1 = \lambda' + \frac{4}{3}\mu' = 22.5\,\text{mm}^2/\text{s}, \qquad \varphi_m = 65\,\text{nm}.$$

The mean free path φ_m provides a convenient length scale for displaying the results. We obtain real values of $\hat{\varphi}$ only in the region where the Rayleigh-line process can exist, $1 \geq F \geq F_+$ and $\mathcal{T} < 0$. Notice that the values of $\hat{\varphi}$ are *not* bounded. Strictly interpreted, these waves have infinite thicknesses; however, as shown in the illustration, from a practical point of view, they have finite thicknesses. We notice that the calculated thicknesses are strongly coupled to the amplitude $1 - F_+$. The weakest steady wave has a thickness of about $\hat{\varphi} = X = 7\varphi_m$. The strongest wave has a thickness that is less then ϕ_m. However, the actual mean free path at the Hugoniot state is much shorter than ϕ_m, and therefore, the thickness of this steady wave is still greater than the mean free path at the Hugoniot state.

Recall that we defined the Lagrangian strain as [see Eq. (1.98) in Section 1.5.1]

$$E = \frac{1}{2}(F^2 - 1).$$

The strain at the initial state is $E_- = 0$, and the strain at the Hugoniot state is

$$E_+ = \frac{1}{2}\left(F_+^2 - 1\right).$$

If we consider the weakest steady wave in Figure 4.16, the strain at the Hugoniot state is

$$E_+ = \frac{1}{2}(0.6^2 - 1) = -0.42.$$

The time required for this wave to pass a fixed point in the gas is approximately

$$\Delta t = \frac{7\phi_m}{U_S} = \frac{7 \times 65 \times 10^{-9}}{345 \times 1.4} = 9.4 \times 10^{-10}\ \text{s}.$$

An often quoted number for steady waves is the *strain rate* of the wave,

$$\left|\frac{E_+ - E_-}{\Delta t}\right| = 3.4 \times 10^8\ \text{s}^{-1}.$$

Experimental measurements yield data that are in reasonably good agreement with this estimate of strain rate. These data also confirm the strong dependence of shock thickness upon shock strength that we illustrated in Figure 4.16.

Quadratic Viscosity When $n = 2$, we write Eq. (4.57) as

$$\frac{(dF)^2}{(1 - F)(F - F_+)} = \frac{\gamma + 1}{2b_2}(d\hat{\varphi})^2.$$

For a steady compression wave propagating in the positive-X direction, we find that $dF/d\hat{\varphi} \geq 0$. Thus we use the positive root of this expression,

$$\frac{dF}{[(1 - F)(F - F_+)]^{1/2}} = \left(\frac{\gamma + 1}{2b_2}\right)^{1/2} d\hat{\varphi}.$$

The left side of this equation is a function of F only, and the right side is a function of $\hat{\varphi}$ only. Integration yields

$$2\tan^{-1}\sqrt{\frac{F - F_+}{1 - F}} = \left(\frac{\gamma + 1}{2b_2}\right)^{1/2} \hat{\varphi},$$

where we have set the constant of integration to zero. This solution only yields real values for $\hat{\varphi}$ in the region where the Rayleigh-line process exists, $1 \geq F \geq F_+$. Because the values of $\hat{\varphi}$ are bounded, the thicknesses of these waves are finite. We show steady-wave profiles for the same four values of M that we used in the example for linear viscosity (see Figure 4.17). These wave thicknesses vary by less than a factor of 2. This differs significantly from the results we obtained for the linear-viscosity profiles.

Finally, we must emphasize again that the location of the Rayleigh-line process on the state surface is determined by the value of M and not the constitutive function for $\mathcal{T}$. Both linear and the quadratic viscosity produce the same Rayleigh-line process in the viscous ideal gas even though the corresponding steady wave profiles differ significantly. By using different values of the viscosity coefficients b_n or even different expressions for $\mathcal{T}$, we cannot change the location of the Rayleigh-line process, but we can change the rate at which the steady wave causes the gas to follow this process.

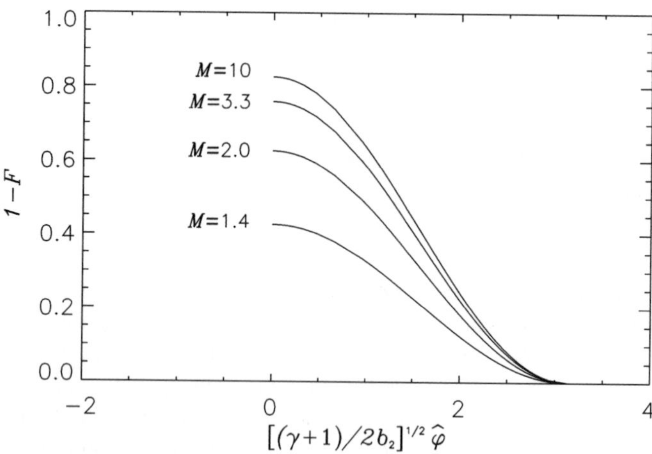

Figure 4.17 Steady wave profiles in air with a quadratic viscosity.

4.1.8 *Thermoelastic Ideal Gas*

In the previous section, we modeled the ideal gas as a viscous material, and we ignored heat conduction. We also found that the position of the Rayleigh-line process on the state surface was unaffected by the constitutive equation for $\mathcal{T}$. However, the position of the Rayleigh-line process will change if we treat the ideal gas as a thermoelastic material. In a thermoelastic gas, we allow heat conduction but ignore viscosity. According to Eqs. (3.134)–(3.137) and (3.139), the constitutive equations of a thermoelastic ideal gas are

$$\begin{aligned} \psi &= \psi(\rho, \vartheta), \qquad -T = p_e = -\rho_0 \partial\psi/\partial F, \\ \eta &= -\partial\psi/\partial\vartheta, \qquad q = q(F, \vartheta, \partial\hat{\vartheta}/\partial X). \end{aligned} \tag{4.58}$$

To fit the equilibrium description of the ideal gas into this constitutive description, we must first derive an expression for the Helmholtz energy $\psi = \mathcal{E} - \vartheta\eta$. For an ideal gas with a constant specific heat, the internal energy is $\mathcal{E} = c_v\vartheta$, and the entropy is given by the fundamental equation, Eq. (4.4):

$$\eta = c_v \ln\frac{\vartheta}{\vartheta_0} + (\gamma - 1)c_v \ln\frac{\rho_0}{\rho} + \eta_0. \tag{4.59}$$

Thus the Helmholtz energy for an ideal gas is

$$\psi = c_v\vartheta\left[1 - \ln\frac{\vartheta}{\vartheta_0} - (\gamma - 1)\ln\frac{\rho_0}{\rho} - \frac{\eta_0}{c_v}\right]. \tag{4.60}$$

Because $\psi = \psi(\rho, \vartheta)$, this is the fundamental form of the Helmholtz energy. It is the first constitutive equation in Eqs. (4.58). The second constitutive equation then yields the familiar relationship for the equilibrium pressure:

$$p_e = (\gamma - 1)c_v\rho\vartheta = NR\rho\vartheta,$$

and the third constitutive equation yields the expression for the entropy, Eq. (4.59). To complete the constitutive description of a thermoelastic ideal gas, we must specify a constitutive relationship for the heat flux q that satisfies the dissipation inequality

$$-q\frac{\partial\hat{\vartheta}}{dX} \geq 0,$$

which, as we recall, also implies that

$$q(F, \vartheta, 0) = 0.$$

Rayleigh's Existence Condition for Steady Waves

Consider a steady wave propagating in a thermoelastic ideal gas. We require the steady-wave solutions to be continuous and differentiable functions of the steady-wave variable $\hat{\varphi} = X - C_S t$. In Section 3.5 we learned that heat flux q in a steady wave can only be zero at the initial and Hugoniot states. We also found that at states where this condition is violated, the steady-wave solution becomes unbounded. This behavior of the solution for a steady wave in a thermoelastic gas with a constant specific heat was discovered by Rayleigh. His classical argument is often used to discount the relevance of heat conduction to the description of steady waves in all types of materials. We present Rayleigh's argument below; however, as we proceed, it should be remembered that, like Rayleigh, we are restricting our discussion to ideal gases with constant specific heat and no viscosity.

We obtain the steady-wave solution of the equation for the balance of linear momentum for a thermoelastic ideal gas from Eq. (4.53) by requiring the viscous stress to be zero:

$$\frac{\vartheta}{\vartheta_0} = F - \gamma F(F-1)M^2, \tag{4.61}$$

where $1 \geq F \geq F_+$. We obtain the Mach number $M = U_S/c_0$ from Eq. (4.36),

$$1 - F_+ = \frac{2}{\gamma+1}\left[1 - \frac{1}{M^2}\right]. \tag{4.62}$$

Thus the Hugoniot condition and the balance of linear momentum determine the temperature profile of the steady wave. In Figure 4.18 we show a graph of this solution for $\gamma = 1.4$ and several values of F_+. Notice that the temperature in a particular steady wave is strictly a function of F. Thus the temperature gradient in the steady wave is

$$\frac{\partial\hat{\vartheta}}{\partial X} = \frac{d\vartheta(F)}{dF}\frac{\partial\hat{F}}{\partial X}.$$

Because $\partial\hat{F}/\partial X = 0$ at the initial and Hugoniot states, we find that $\partial\hat{\vartheta}/\partial X = 0$ at these states. However, $\partial\hat{F}/\partial X \neq 0$ between the initial and Hugoniot states, and therefore $\partial\hat{\vartheta}/\partial X \neq 0$ unless $d\vartheta/dF = 0$. Several profiles in Figure 4.18 do exhibit maximum values of temperature where $d\vartheta/dF = 0$. In fact, we have graphed one of the profiles as a dashed line, because its maximum occurs at the Hugoniot state. The profiles above this one exhibit maximum values between the initial and Hugoniot states, and those below exhibit maximum values only at the Hugoniot state.

Each profile in Figure 4.18 is a solution that must also satisfy the balance of energy Eq. (3.151),

$$q = C_S\left[\rho_0 c_v(\vartheta - \vartheta_0) + p_0(F-1) - \frac{1}{2}\rho_0 C_S^2(F-1)^2\right].$$

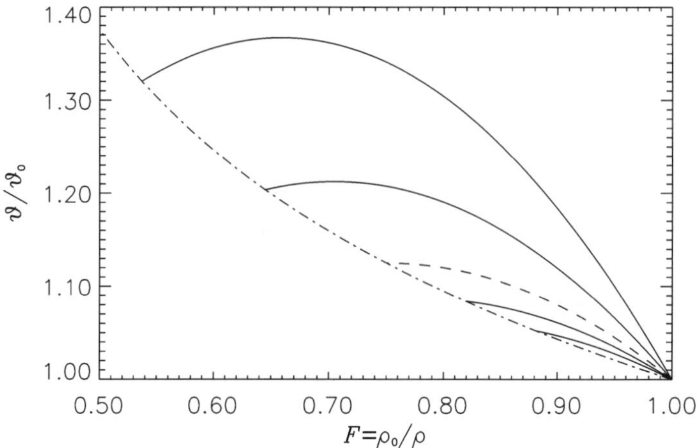

Figure 4.18 Temperature profiles of steady waves in a thermoelastic ideal gas. Each profile connects the same initial state to different Hugoniot states on the Hugoniot curve (— · —). Profiles above the dashed profile exhibit a maximum temperature between the initial and Hugoniot states, and those below do not.

We notice that $q = 0$ at the initial state, and by definition $q = 0$ at the Hugoniot state. Between the initial and Hugoniot states, $q \neq 0$. For the profiles in Figure 4.18 that exhibit a maximum value between the initial and Hugoniot state, the balance of energy then implies that a point exists where $q(F, \vartheta, 0) \neq 0$. This violates the dissipation inequality. We must conclude that although the temperature profiles that exhibit a point where $d\vartheta/dF = 0$ do satisfy the balance of linear momentum, the waves they represent cannot exist. Indeed, if we insist upon applying Fourier heat conduction to describe q, we will obtain the following ordinary differential equation for the balance of energy:

$$-\frac{K}{F}\frac{d\vartheta}{d\hat{\varphi}} = C_S\left[\rho_0 c_v(\vartheta - \vartheta_0) + p_0(F-1) - \frac{1}{2}\rho_0 C_S^2(F-1)^2\right],$$

where K is the heat conduction coefficient. As we have discussed in Section (3.5), if we attempt to integrate this equation past the state where $d\vartheta/dF = 0$, the solution will become unbounded.

The Rayleigh-line process that we have graphed as a dashed line in Figure 4.18 is the limiting solution that satisfies the dissipation inequality. For this profile, we see that $d\vartheta/dF = 0$ at the Hugoniot state. At this particular Hugoniot state, we see that the derivative of Eq. (4.61) with respect to F gives

$$\frac{1}{\vartheta_0}\left(\frac{d\vartheta}{dF}\right)_+ = 1 + \gamma(1 - 2F_+)M^2 = 0.$$

When we use Eq. (4.62) to eliminate F_+, we find that

$$\gamma(3-\gamma)M^2 = 3\gamma - 1.$$

This relationship yields the Mach number for the Rayleigh-line process represented by the dashed line. When $\gamma \geq 3$ or $\gamma < \frac{1}{3}$, no real values of M satisfy this equation. Consequently the dashed profile does not exist for these values of γ, and the thermoelastic material never violates the dissipation inequality. However, when $\frac{1}{3} < \gamma < 3$, positive values of M^2, which satisfy this equation, do exist, and thermoelastic material only satisfies the dissipation inequality over a limited region. In this case we can use Eq. (4.62) to evaluate the deformation gradient at the limiting Hugoniot state,

$$1 - F_+ = \frac{2(\gamma - 1)}{3\gamma - 1}.$$

Then we employ the jump condition for the balance of linear momentum,

$$p_+ - p_0 = \rho_0 C_S^2 (1 - F_+),$$

to obtain

$$\frac{p_+}{p_0} = \frac{1 + \gamma}{3 - \gamma}.$$

When $\frac{1}{3} < \gamma < 3$ and the ratio of the Hugoniot pressure to the initial pressure is less than this value, the temperature profile of the Rayleigh-line process does not have a maximum between the initial and the Hugoniot states. Thus the dissipation inequality is satisfied and the steady wave exists. In contrast, when $\frac{1}{3} < \gamma < 3$ and the pressure ratio is greater than this value, the temperature profile has a maximum. This contradicts the dissipation inequality. Thus the steady wave cannot exist. Indeed, if we consider the example of air where $\gamma = 1.4$, we see just how strong a constraint this represents. For air, the limiting pressure ratio is only $p_+/p_0 = 1.5$. Because this is a relatively modest strength for a steady wave, we conclude that the thermoelastic ideal gas cannot explain the structure of most steady waves in air. People often use these conclusions to justify the omission of heat conduction as a structuring mechanism for steady waves *in all types of materials*. However, you should notice that these arguments are special in that they only apply to ideal gases in which $\frac{1}{3} < \gamma < 3$.†

4.1.9 *Thermoviscous Ideal Gas*

We have shown that heat conduction *alone* cannot explain the structure of steady waves in many typical ideal gases. However, this does not mean that heat conduction cannot influence the structure of a steady wave in these gases. In this section we analyze the thermoviscous ideal gas. In this gas both viscous stress and heat conduction are present.

According to Eqs. (3.134)–(3.137) in Section 3.5.2, the constitutive equations of a thermoviscous ideal gas have the following forms:

$$\begin{aligned} \psi &= \psi(\rho, \vartheta), \qquad & T &= T(F, D, \vartheta, \partial\hat{\vartheta}/\partial X), \\ \eta &= -\partial\psi/\partial\vartheta, \qquad & q &= q(F, D, \vartheta, \partial\hat{\vartheta}/\partial X). \end{aligned} \tag{4.63}$$

The Helmholtz energy for an ideal gas is [see Eq. (4.60)]

$$\psi = c_v \vartheta \left[1 - \ln \frac{\vartheta}{\vartheta_0} - (\gamma - 1) \ln \frac{\rho_0}{\rho} - \frac{\eta_0}{c_v} \right].$$

† You can find an excellent discussion of this topic in "Steady, structured shock waves. Part 1: Thermoelastic materials" by Dunn, J. E., and Fosdick R. L. in *Archive for Rational Mechanics and Analysis* **104**, pp. 295–365 (1988).

This fundamental equation yields the equilibrium pressure

$$p_e = (\gamma - 1)c_v\rho\vartheta = NR\rho\vartheta.$$

As with the viscous ideal gas, we represent the stress by the additive decomposition

$$T = -p_e + \mathcal{T}.$$

We assume that the viscous stress is given by the expression for the Newtonian fluid, Eq. (3.97), and the heat flux is given by Fourier's law of heat conduction, Eq. (3.124). The steady-wave equations for the balance of energy and linear momentum are [see Eq. (3.153)]

$$\frac{d\vartheta}{d\hat{\varphi}} = -\frac{FC_S}{K}N, \qquad \frac{dF}{d\hat{\varphi}} = -\frac{F}{BC_S}G, \tag{4.64}$$

where

$$N = \rho_0 c_v(\vartheta - \vartheta_0) + p_0(F - 1) - \frac{1}{2}\rho_0 C_S^2(F - 1)^2,$$

$$G = \rho_0 C_S^2(F - 1) + (\gamma - 1)c_v(\rho\vartheta - \rho_0\vartheta_0),$$

$$B = \rho_0\left(\lambda' + \frac{4}{3}\mu'\right).$$

Following the method of analysis outlined in Chapter 3, we evaluate the following derivatives at the initial and Hugoniot states (see Section 3.5.4):

$$\left(\frac{\partial N}{\partial F}\right)_{\pm} = p_{\pm}, \qquad \left(\frac{\partial N}{\partial \vartheta}\right)_{\pm} = \rho_0 c_v,$$

$$\left(\frac{\partial G}{\partial F}\right)_{\pm} = \rho_0 C_S^2, \qquad \left(\frac{\partial G}{\partial \vartheta}\right)_{\pm} = (\gamma - 1)\rho c_v.$$

We use these factors to evaluate Eq. (3.156):

$$\lambda^2 + \rho_0 F_{\pm} C_S\left[\frac{1}{B} + \frac{c_v}{K}\right]\lambda + \frac{\rho_0 c_v}{BK}\left[\rho_0 F_{\pm}^2 C_S^2 - (\gamma - 1)p_{\pm}F_{\pm}\right] = 0.$$

This is a quadratic equation in λ. To determine the coefficients in this equation, we prescribe an initial state for the gas and the velocity C_S of the steady wave. Then we solve the quadratic equation and obtain two values of λ at both the initial and Hugoniot states. For example, suppose we select $C_S > 0$. We find that, at the Hugoniot state, the values of λ are of opposite sign. To integrate Eqs. (4.64) from the Hugoniot state back to the initial state, we must integrate in the positive-$\hat{\varphi}$ direction. As explained in Chapter 3, this requires that we select the negative root of λ. We can then initiate the numerical integration at the point

$$F = F_+ + \Delta F, \qquad \vartheta = \vartheta_+ + \Delta\vartheta,$$

where $\Delta F > 0$ is an arbitrarily small quantity with respect to $1 - F_+$ and [see Eq. (3.157)]

$$\Delta\vartheta = \frac{p_+\Delta F}{\rho_0 c_v - \frac{\lambda K}{F_+ C_S}}.$$

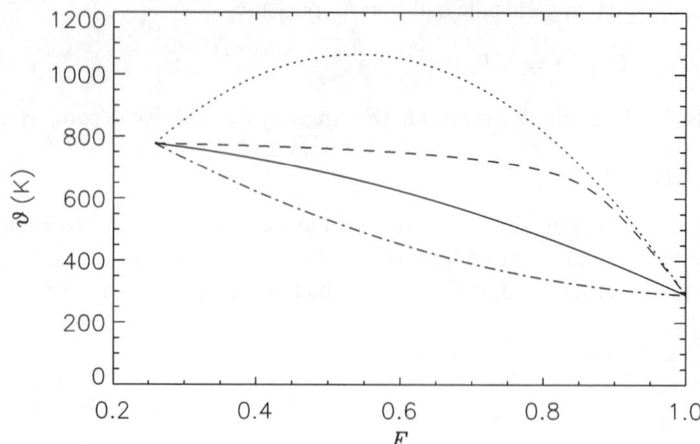

Figure 4.19 A Rayleigh-line process in a thermoviscous ideal gas (—) compared to a viscous ideal gas (— · —) and a thermoelastic ideal gas (· · ·). We also show a thermoviscous process for $10 \times K$ (— —).

To obtain the profile of the steady wave, either we can integrate the differential relations [Eqs. (4.64)] as shown or we can integrate their ratio

$$\frac{d\vartheta}{dF} = C_S^2 \frac{B}{K} \frac{N}{G}. \tag{4.65}$$

As an example of this latter procedure, we integrate Eq. (4.65) using the following parameters for air (see Figure 4.19):

$$K = 25\,\text{mW/m-K}, \qquad p_0 = 0.1037\ \text{MPa},$$
$$\rho_0 = 1.22\,\text{kg/m}^3, \qquad \gamma = 1.4,$$
$$\vartheta_0 = 290\,\text{K}, \qquad M = 3,$$
$$c_v = 733\,\text{J/kg-K}, \qquad \varphi_m = 65\,\text{nm},$$
$$\lambda' + \frac{4}{3}\mu' = 22.5\,\text{mm}^2/\text{s}.$$

We use a solid line to illustrate the ϑ–F diagram of the Rayleigh-line process for a Mach number of $M = 3$. The initial state is at the lower right and the Hugoniot state is at the upper left. We use the dashed line to illustrate another calculation in which we have increased the thermal conductivity K by a factor of 10. In this calculation, the temperature rises more quickly from the initial state and then levels off as the Hugoniot state is approached. Both these solutions fall between two bounding curves. The lower dot-dashed curve is the Rayleigh-line process followed by a viscous ideal gas. We obtain it by first setting the thermal conductivity to zero ($K = 0$) and then evaluating the algebraic equation $N = 0$. The upper dotted curve is the process followed by a thermoelastic material. We obtain this curve by setting $\lambda' + 4\mu'/3 = 0$ and evaluating the algebraic equation $G = 0$. The thermoelastic curve exhibits a maximum temperature between the initial and Hugoniot states. As we discussed in the previous section, at this state, heat must flow in the absence of a temperature gradient. This violates the dissipation inequality, and thus the dotted line represents a process that cannot be followed by a steady wave.

Because the dot-dashed line represents the Rayleigh-line process for a viscous ideal gas, the shape of this curve is not affected by the constitutive equation for the viscous stress. The viscous ideal gas must always follow this process. As is evident by comparing the solid and dashed curves, the thermoviscous ideal gas can follow different Rayleigh-line processes. Inspection of Eq. (4.65) reveals that the shape of the process is determined by the ratio of the thermal conductivity K and the viscosity coefficient B.

Suppose we now integrate Eqs. (4.64), using the same physical parameters. We show the resulting steady-wave profiles for ϑ and T in Figures 4.20 and 4.21. Once again, we use the mean free path to scale the length dimension $\hat{\varphi}$. We do not illustrate the thermoelastic curves because these differential equations cannot be integrated past the state where the

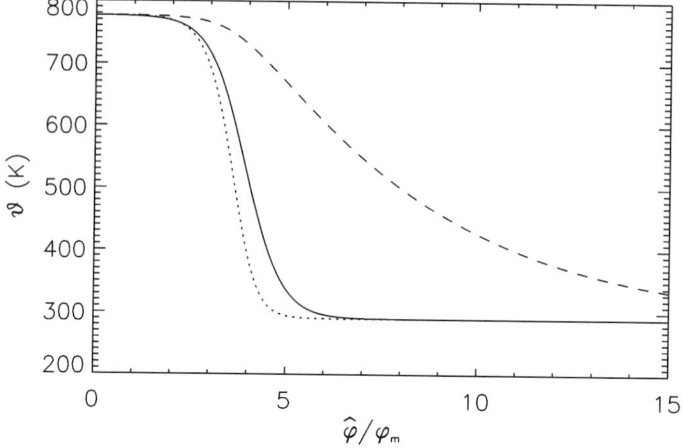

Figure 4.20 Temperature profiles of steady waves in a thermoviscous gas for $K = 25$ mW/m-K (—) and $K = 250$ mW/m-K (– –) compared to a steady wave in a viscous gas (· · ·).

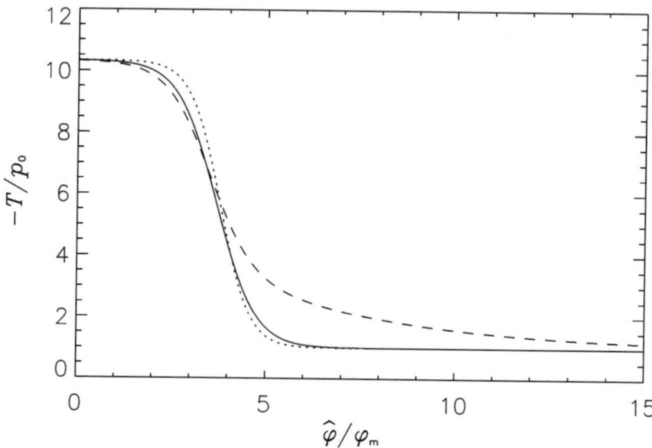

Figure 4.21 Stress profiles of steady waves in a thermoviscous gas for $K = 25$ mW/m-K (—) and $K = 250$ mW/m-K (– –) compared to a steady wave in a viscous gas (· · ·).

temperature reaches a maximum value. When the thermoviscous profiles are compared to the viscous profiles, we find that heat conduction tends to increase the thickness of the steady wave. We also see that heat is pushed ahead of the pressure front of the wave. Thus the thickness of the temperature profile can differ significantly from the thickness of the stress profile. From the ϑ–F diagram and the steady-wave profiles, we conclude that heat conduction can significantly influence the Rayleigh-line process in an ideal gas. However, the smallest influences are observed in the stress profiles. This is unfortunate, because at present, our experimental capabilities are limited to measurements of pressure and temperature at Hugoniot states and to measurements of the stress profiles of waves. Further research in this area awaits the development of techniques to measure the temperature profile in a wave†.

4.1.10 *Supersonic Space Vehicle*

To close our discussion of the constitutive model for an ideal gas, let us consider a simple analysis of a supersonic space vehicle reentering the earth's atmosphere. We shall use this example to review what we have learned about the ideal gas. Suppose this vehicle is 40 km above sea level traveling at a velocity of 6.5 km/s. We start by determining the initial state of the atmosphere before the arrival of the vehicle. The properties of the atmosphere at 40 km (131,000 ft) above sea level are (see Figures 4.5 and 4.6)

$$\rho_0 = 4.3\ \text{g/m}^3, \qquad \vartheta_0 = 260\,\text{K},$$
$$p_0 = 328\,\text{Pa}\ (0.046\,\text{psi}), \qquad c_0 = 325\,\text{m/s},$$
$$\gamma = 1.4.$$

The Mach number of the vehicle is

$$M = 6{,}500/325 = 20.$$

Because the vehicle is traveling much faster than the velocity of sound at this elevation, a steady wave or *bow shock* forms in front of the vehicle (see Figure 4.22). Determining the shape and location of this bow shock with respect to the vehicle requires a solution of our three-dimensional equations. We shall not attempt such a complex solution here. Rather, we shall estimate the temperature and pressure of the air between the leading portion of the shock wave and the bow of the vehicle with a one-dimensional approximation.

We observe that the bow shock is traveling at the same velocity as the vehicle and

$$U_S = 6.5\,\text{km/s}.$$

As a volume of air crosses the bow shock it undergoes a Rayleigh-line process that causes it to move from the initial state described above to a Hugoniot state that we shall now determine. We determine the deformation gradient at the Hugoniot state from Eq. (4.36):

$$1 - F_+ = \frac{2}{1.4+1}\left[1 - \left(\frac{1}{20}\right)^2\right] = 0.8313.$$

† Indeed, for some people this raises an even more fundamental concern about the proper definition of temperature for highly irreversible processes. Following a strict interpretation of the axioms of thermodynamics, we could argue that temperature and internal energy are concepts that only have meaning at equilibrium states, and therefore, the measurement of the temperature profile of a shock wave is meaningless. However, for many people who actually solve problems, that point of view is dogma.

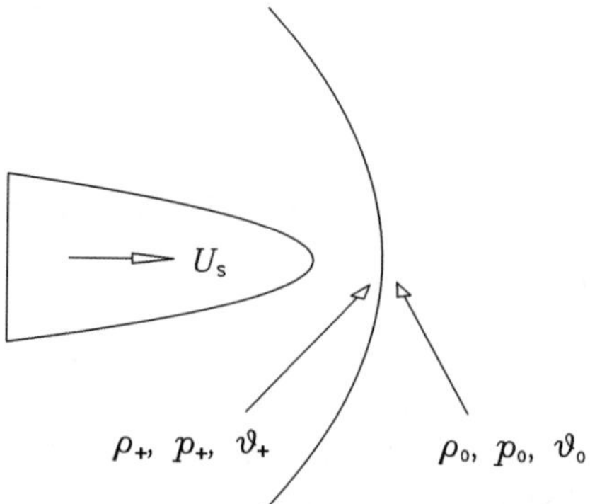

Figure 4.22 A supersonic vehicle driving an atmospheric bow shock.

The ratio of the Hugoniot density to the initial density is

$$\frac{\rho_0}{\rho_+} = F_+ = 0.1688.$$

This density ratio is quite close to the value for the shock-compression limit 1/6 [see Eq. (4.38)]. We obtain the ratio of the Hugoniot temperature to the initial temperature from Eq. (4.34):

$$\frac{\vartheta_+}{\vartheta_0} = 1 + (1.4 - 1)(1 - F_+) + \frac{1.4(1.4 - 1)}{2}(20)^2(1 - F_+)^2 = 78.7.$$

From Eq. (4.8), we determine the ratio of the Hugoniot pressure to the initial pressure:

$$\frac{p_+}{p_0} = \frac{\rho_+}{\rho_0}\frac{\vartheta_+}{\vartheta_0} = (1/0.1688)(78.7) = 466.$$

Thus the properties at the Hugoniot state are

$$\rho_+ = 25.5\,\text{g/m}^3, \qquad \vartheta_+ = 20.5\,\text{kK},$$
$$p_+ = 0.153\,\text{MPa (21 psi)} \qquad .$$

Notice that ϑ_+ is enormous. Indeed, anyone whose seen a meteor burn up in the atmosphere should not be totally surprised by this result. It is clear that this temperature poses a difficult problem for designers of these types of vehicles. Our results also suggest that the air will ionize and release electrons. The process of ionization requires energy that could significantly reduce the Hugoniot temperature.

From Eq. (4.28), the material wave velocity at the Hugoniot state is

$$C_+ = (1.4 \times 0.153 \times 10^6/0.0255)^{1/2}/.1688 = 17.2 \text{ km/s},$$

and from Eq. (4.28), the spatial wave velocity at the Hugoniot state is

$$c_+ = (1.4 \times 0.153 \times 10^6/0.0255)^{1/2} = 2.9 \text{ km/s}.$$

Notice that

$$C_+ > U_S,$$

which satisfies the subsonic condition for a steady wave, Eq. (3.109).

The velocity of the air between the bow shock and the vehicle is given by the jump condition

$$v_+ = U_S(1 - F_+) = 5.4\,\text{km/s}.$$

The velocity of the vehicle relative to the velocity of the air behind the bow shock is

$$U_S - v_+ = 6.5 - 5.4 = 1.1\,\text{km/s}.$$

Between the bow shock and the vehicle, the velocity of sound c_0 is determined by substituting the pressure and density of the Hugoniot state into Eq. (4.31). We find that the velocity of sound at this state is $c_0 = c_+ = 2.9$ km/s. With respect to the Hugoniot state, the vehicle is in subsonic flight traveling at a Mach number of $M = 1.1/2.9 = 0.38$.

So far, we have only used the Hugoniot jump conditions and the equilibrium constitutive relations for air. Further analysis requires more constitutive information. For example, suppose we wish to determine the location of the Rayleigh-line process on the state surface. To do this, we must decide whether or not heat conduction is important. Suppose we decide it is not important. We could then use Eq. (4.54) to obtain the following relationship for the Rayleigh-line process:

$$\begin{aligned}\frac{\vartheta}{\vartheta_0} &= 1 + (1.4 - 1)(1 - F) + \frac{1.4(1.4 - 1)}{2}(20)^2(1 - F)^2 \\ &= 1 + 0.4 \times (1 - F) + 112 \times (1 - F)^2.\end{aligned}$$

We illustrate this result in Figure 4.23. Recall that we do not need the constitutive equation for the dissipative stress in order to determine the location of the Rayleigh line for a viscous ideal gas. However, this is not true if heat conduction is important. Furthermore, we need a complete constitutive description to determine the bow-shock profiles for pressure and temperature. These factors may become very important if the air surrounding the vehicle exhibits a constitutive response that is more complicated than that of an ideal gas. For

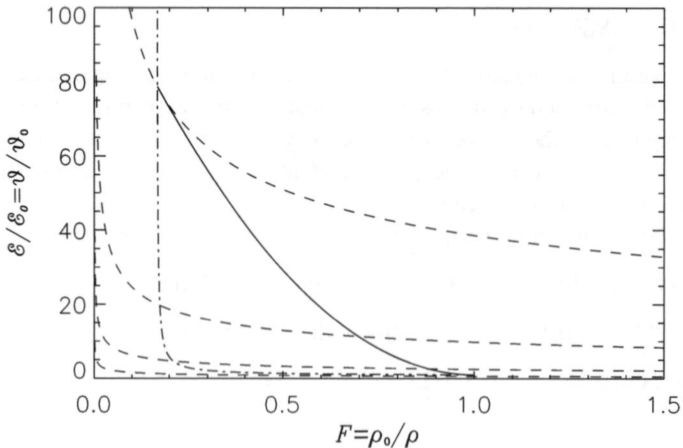

Figure 4.23 Hugoniot (– · –) and Rayleigh line (—) of the bow shock, and the $\mathcal{E}$–F diagram of the isentropes (– –).

example, we have suggested that ionization may occur. If it does, the exact location of the Rayleigh-line process on the state surface could become quite important.

4.1.11 *Exercises*

4.1 The adiabatic gas constant of air is 1.4, and the specific heat c_v is 733 J/kg/K. Determine the molecular weight of air.

4.2 The mole number of oxygen O_2 is 31.3 moles/kg. The specific heat is $c_p = 918$ J/K/kg. Suppose the temperature is $\vartheta = 295$ K and the pressure is $p = 0.1037$ MPa.

(a) Determine the density, adiabatic gas constant, thermal expansion coefficient, and Grüneisen constant.

(b) Determine the isentropic and isothermal stiffnesses.

(c) Determine the speed of sound.

4.3 Using the definitions Eqs. (3.49) and (3.58), determine C^η and C^ϑ from the pressure–density relations Eqs. (4.25) and (4.26).

4.4 In its fundamental form, the enthalpy $\mathcal{H}$ of an ideal gas is a function of pressure and entropy. Assume that c_v is constant.

(a) Verify that the enthalpy of an ideal gas is

$$\mathcal{H} = \gamma \mathcal{E}_0 e^{\frac{\eta-\eta_0}{\gamma c_v}} \left(\frac{p}{p_0} \right)^{\frac{\gamma-1}{\gamma}}.$$

(b) Verify that at an equilibrium state

$$\vartheta = \frac{\partial \mathcal{H}}{\partial \eta}, \qquad \frac{1}{\rho} = \frac{\partial \mathcal{H}}{\partial p}.$$

4.5 In its fundamental form, the Gibb's energy $\mathcal{G}$ of an ideal gas is a function of pressure and temperature. Assume that c_v is constant.

(a) Verify that the Gibb's energy of an ideal gas is

$$\mathcal{G} = \vartheta \left\{ c_v \left[\gamma - \gamma \ln \frac{\vartheta}{\vartheta_0} + (\gamma - 1) \ln \frac{p}{p_0} \right] - \eta_0 \right\}.$$

(b) Verify that at an equilibrium state

$$\eta = -\frac{\partial \mathcal{G}}{\partial \vartheta}, \qquad \frac{1}{\rho} = \frac{\partial \mathcal{G}}{\partial p}.$$

4.6 For the monoatomic gas, the specific internal energy is [see Eqs. (4.20)–(4.21)]

$$\mathcal{E} = \frac{3}{2} v^2.$$

An atom of a monoatomic gas has *three degrees of freedom* – one for each spatial dimension. The motion in each degree of freedom contributes $\frac{1}{2}v^2$ to the specific internal energy. The internal energy of a diatomic gas is

$$\mathcal{E} = \frac{5}{2} v^2.$$

Each molecule of a diatomic gas is composed of a pair of atoms connected together by a molecular bond. Not only can these pairs of atoms translate in each of three directions, they can also twirl about each other on two possible axes. Here there are five degrees of freedom that contribute equal amounts of energy. More complex molecules have more degrees of freedom. Letting α represent the number of degrees of freedom, then

$$\mathcal{E} = \frac{\alpha}{2}v^2.$$

We also found that the equilibrium pressure is [see Eq. (4.22)]

$$p_e = \rho v^2.$$

This expression holds for a gas with an arbitrary number of degrees of freedom. Use Eqs. (4.14) and (4.29) to show that the magnitude of the velocity of the molecule V and the speed of sound are related by

$$\frac{1}{3}\left(\frac{V}{c_0}\right)^2 = \left(\frac{v}{c_0}\right)^2 = \frac{1}{\gamma} = \frac{\alpha}{\alpha + 2}.$$

4.7 Verify the relationships for the speed of sound in Eq. (4.30).

4.8 Verify the Hugoniot expressions in Eqs. (4.35) and (4.37).

4.9 We give the expressions for the velocities v_+ behind a weak shock and a strong shock in Eqs. (4.40) and (4.41).

(a) Derive an expression for the ratio of the pressure behind a weak shock divided by the pressure behind the strong shock.

(b) Plot this ratio for $1 \leq M \leq 2$.

4.10 Consider three separate containers of oxygen initially at rest. We describe the initial state of each container in Exercise 4.2. Suppose the pressure in each container is increased by a factor of ten. The first container is compressed isothermally. The second container is compressed isentropically. The third container is compressed by a shock wave. Determine the final densities and temperatures in each container.

4.11 A steady wave propagates through a volume of air that is at standard atmospheric conditions:

$$\gamma = 1.4, \qquad p_e = 0.1037 \text{ MPa}, \qquad \rho = 1.22 \text{ kg/m}^3, \qquad \vartheta = 290 \text{ K}.$$

The Mach number of the wave is $M = 5$.

(a) Neglect heat conduction, and assume that air is a viscous gas. Plot the ϑ–F diagram of the Rayleigh-line process for this steady wave.

(b) In the same diagram, plot the isentropes of the initial and Hugoniot states. Show that they are tangent to the Rayleigh line.

(c) Compare the curvatures of the processes in this diagram to verify that the supersonic and subsonic conditions are satisfied.

(d) Neglect viscous stresses, and assume that air is a thermoelastic gas. Plot the Rayleigh-line process for this gas on the same ϑ–F diagram.

(e) Can this Rayleigh-line process for the thermoelastic gas exist?

4.12 The driver section of a shock tube contains hydrogen H_2 at a pressure of 1 MPa. For hydrogen $N = 500$ moles/kg and $c_p = 14.3$ kJ/kg/K. The test section contains air. The Grüneisen constant, density, and pressure of air are 0.4, 1.22 kg/m^3, and 0.1037 MPa, respectively. The temperature in both sections is 290 K. Use Eq. (4.47) to determine the velocity of gas behind the shock wave after the diaphragm is released.

4.13 Using the expressions for the compression-limited shock wave, repeat the calculations for the supersonic space vehicle in Section 4.1.10 and determine F_+, p_+, and ϑ_+.

4.14 In Figure 4.19 we illustrate two solutions for a steady wave in a thermoviscous ideal gas. The physical parameters we used to obtain these profiles are listed in the discussion that accompanies this figure. Let $\Delta F = 10^{-9}$ and use a numerical computer routine to integrate Eq. (4.65). Plot your results and compare them to Figure 4.19.

4.2 Mie–Grüneisen Solid

Liquids and solids are called *condensed materials*. When compared to a gas, the atoms and molecules in a condensed material are much closer together. For example, the interatomic spacing of atoms in a solid is approximately 10^{-10} m, and the mean free path of air at standard atmospheric conditions is about 600 times greater than this distance. At such small distances, interatomic forces dominate and bind each atom in the solid to its nearest neighbors. Because of these forces, the solid can resist both compression and expansion, whereas the random atomic collisions in gases only resist compression. Thus one manifestation of the dominance of interatomic forces is that a condensed material can exist at an equilibrium state of zero pressure, whereas a gas cannot. The interatomic forces also result in material stiffnesses that are vastly greater than that of the gas. Indeed, although is it easy to increase the density of a gas by a factor of two, it is nearly impossible to double the density of a solid. In this section we discuss one model for a condensed material: the Mie–Grüneisen solid. Historically, this model has been at the center of efforts to connect the theoretical predictions of microstructural models of atomic lattices to experimental observations of the macroscopic behavior of metals. Today, we consider the Mie–Grüneisen model to be too specialized to accurately model atomic structure; however, it is still a good intuitive model of a solid and often serves as a useful empirical model for the shock-wave behavior of many materials.

The predominant application of the Mie–Grüneisen model has been to solids under large compressive loads. This allows us to greatly simplify the analysis by approximating the equilibrium stress in the solid with the equilibrium bulk response [see Eq. (2.25)]:

$$T_e \approx \sigma_e.$$

Although the interatomic forces in a solid result in shear stresses and produce a configuration of triaxial stress during uniaxial deformation, we shall ignore the deviatoric components of stress at equilibrium. Hence we assume the equilibrium stress tensor is spherical and represented by the pressure

$$p_e = -T_e \approx -\sigma_e. \tag{4.66}$$

Later, in Section 4.3, we shall consider states where we cannot ignore the deviatoric stresses at equilibrium.

As we discovered in our discussion of the Newtonian fluid (see Section 3.4.3), a configuration of spherical stress implies that the internal energy is a function of the density ρ alone and does not depend upon the separate components of the deformation gradient F_{ij}. Superficially, this means the solid behaves like a liquid, and because of this similarity to a liquid, you may be tempted to apply this theory to liquids. However, the theory is generally ill suited to such an application and usually offers results that are comparable to those obtained from a nonlinear-elastic constitutive model.

The constitutive model for the Mie–Grüneisen solid contains features of the two materials that we have already studied, the nonlinear-elastic material and the ideal gas. For this material, we assume that the internal energy is composed of two parts:

$$\mathcal{E} = \mathcal{E}_c(\rho) + \mathcal{E}_\vartheta(\rho, \eta), \tag{4.67}$$

where $\mathcal{E}_c$ is the *cold energy*, and $\mathcal{E}_\vartheta$ is the *thermal energy*. The cold energy $\mathcal{E}_c$ represents interatomic forces. We assume this energy to be independent of the temperature, and thus, we assume it exists undiminished even at temperatures near to and including absolute zero. The thermal energy $\mathcal{E}_\vartheta$ represents forces resulting from the thermal vibrations of atoms due to heating. As an essential feature of the Mie–Grüneisen model, we also assume that the thermal energy has the following form:

$$\mathcal{E}_\vartheta = A(\rho)B(\eta),$$

where $A(\rho)$ and $B(\eta)$ are arbitrary functions. Although it is not immediately obvious, we shall show that the separation of $\mathcal{E}_\vartheta$ into the product of $A(\rho)$ and $B(\eta)$ results in a thermal energy that is similar to the internal energy of an ideal gas. To demonstrate this, we evaluate the logarithm of the thermal energy $\mathcal{E}_\vartheta$. Using the last two expressions, we find that

$$\ln(\mathcal{E} - \mathcal{E}_c) = \ln A(\rho) + \ln B(\eta).$$

Next, suppose we introduce two arbitrary functions $\Gamma(\rho)$ and $c_\eta(\eta)$ such that

$$\ln\left(\frac{\mathcal{E} - \mathcal{E}_c}{\mathcal{E}_0}\right) = \int_{\rho_0}^{\rho} \frac{\Gamma(\rho')}{\rho'}\, d\rho' + \int_{\eta_0}^{\eta} \frac{d\eta'}{c_\eta(\eta')}, \tag{4.68}$$

where $\mathcal{E}_0$ is an arbitrary constant. Because $\Gamma(\rho)$ and $c_\eta(\eta)$ are arbitrary functions, the right side of this expression is equivalent to the logarithm of the original product $A(\rho)B(\eta)$. If we select $\Gamma(\rho)$ and $c_\eta(\eta)$ to be the constants Γ_0 and c_v, then we can integrate the right side to obtain

$$c_v \ln\left(\frac{\mathcal{E} - \mathcal{E}_c}{\mathcal{E}_0}\right) = c_v\Gamma_0 \ln(\rho/\rho_0) + \eta - \eta_0. \tag{4.69}$$

When $\mathcal{E}_c = 0$ and $c_v\Gamma_0 = NR$, we obtain the fundamental equation for an ideal gas with constant specific heat, Eq. (4.4). In contrast, if we let $\Gamma = 0$ and $c_\eta \to \infty$, then Eq. (4.68) becomes

$$\mathcal{E} = \mathcal{E}_c(\rho).$$

This is the internal energy expression for a nonlinear-elastic fluid. *Thus the fundamental form of the internal energy of the Mie–Grüneisen solid is a combination of two familiar terms. The cold energy is similar to a nonlinear-elastic fluid, and the thermal energy is similar to an ideal gas.*

The fundamental form of the internal energy, Eq. (4.68), completely defines the equilibrium behavior of the Mie–Grüneisen solid. From the first derivatives of Eq. (4.68), we can determine the temperature and the equilibrium pressure of the solid. We apply the Leibniz theorem for differentiation of an integral [Eq. (2.143)] to obtain the partial derivative of Eq. (4.68) with respect to η:

$$\left(\frac{1}{\mathcal{E} - \mathcal{E}_c}\right)\frac{\partial \mathcal{E}}{\partial \eta} = \frac{1}{c_\eta}.$$

Using the definition of temperature, $\vartheta = \partial\mathcal{E}/\partial\eta$, we obtain

$$\mathcal{E} - \mathcal{E}_c = \mathcal{E}_\vartheta = c_\eta \vartheta. \tag{4.70}$$

Notice that $\mathcal{E} = \mathcal{E}_c$ when $\vartheta = 0$, which reinforces our use of the terminology "cold energy" for $\mathcal{E}_c$. Similarly, the partial derivative of Eq. (4.68) with respect to F yields

$$\left(\frac{1}{\mathcal{E} - \mathcal{E}_c}\right)\left(\frac{\partial\mathcal{E}}{\partial F} - \frac{d\mathcal{E}_c}{dF}\right) = \frac{\Gamma}{\rho}\frac{\partial\rho}{\partial F} = -\frac{\rho}{\rho_0}\Gamma$$

and

$$\frac{\partial\mathcal{E}}{\partial F} = \frac{d\mathcal{E}_c}{dF} - \frac{\Gamma c_\eta \vartheta}{F}. \tag{4.71}$$

After multiplication by ρ_0, we use the definition of the equilibrium pressure $p_e = -\rho_0 \partial\mathcal{E}/\partial F$ to obtain

$$p_e = p_c + \rho\Gamma(\mathcal{E} - \mathcal{E}_c), \tag{4.72}$$

where we have defined the *cold pressure* p_c to be

$$p_c \equiv -\rho_0 \frac{d\mathcal{E}_c}{dF}. \tag{4.73}$$

We can also write the equilibrium pressure as

$$p_e = p_c + p_\vartheta, \tag{4.74}$$

where we have defined the *thermal pressure* p_ϑ to be

$$p_\vartheta \equiv \rho\Gamma c_\eta \vartheta = \rho\Gamma\mathcal{E}_\vartheta.$$

Using the second derivatives of the internal energy, we can evaluate the stiffnesses, specific heats, and Grüneisen coefficient. We take the partial derivative of Eq. (4.71) with respect to η to obtain

$$\frac{\partial\vartheta}{\partial F} = -\frac{\Gamma}{F}\frac{\partial}{\partial\eta}(c_\eta \vartheta),$$

and by substitution of Eq. (4.70), we see that

$$\frac{\partial\vartheta}{\partial F} = -\Gamma\frac{\vartheta}{F}. \tag{4.75}$$

This result is identical to Eq. (3.53) in Section 3.3.1, and it demonstrates that the function $\Gamma(\rho)$ is indeed the Grüneisen parameter. Thus, in the Mie–Grüneisen solid, the Grüneisen parameter is not a function of the entropy.

We obtain the second derivative of the internal energy with respect to η by differentiating Eq. (4.70) with respect to η:

$$\frac{\partial\vartheta}{\partial\eta} = \frac{\vartheta}{c_\eta}\left(1 - \frac{dc_\eta}{d\eta}\right).$$

From the definition of the isochoric specific heat c_v, Eq. (3.55) in Section 3.3.2, we obtain

$$c_v = c_\eta\left(1 - \frac{dc_\eta}{d\eta}\right)^{-1}.$$

Thus the isochoric specific heat is strictly a function of entropy. For the special case when c_η is a constant, $c_v = c_\eta$.

We obtain the second derivative of the internal energy with respect to F by differentiating Eq. (4.71) with respect to F:

$$\frac{\partial^2 \mathcal{E}}{\partial F^2} = \frac{d^2 \mathcal{E}_c}{dF^2} - \frac{\partial}{\partial F}\left(\frac{\Gamma c_\eta \vartheta}{F}\right).$$

We expand the partial derivative on the right side and substitute Eq. (4.75) to obtain

$$\frac{\partial^2 \mathcal{E}}{\partial F^2} = \frac{d^2 \mathcal{E}_c}{dF^2} + \left[\Gamma^2 + \frac{d}{d\rho}(\rho\Gamma)\right]\frac{c_\eta \vartheta}{F^2}.$$

We now define the *cold stiffness* to be

$$C^C \equiv \rho_0 F \frac{d^2 \mathcal{E}_C}{dF^2} = -F\frac{dp_c}{dF}. \tag{4.76}$$

Then we use the definition of the isentropic stiffness, Eq. (3.50), in Section 3.3.1, to obtain

$$C^\eta = C^C + \Phi p_\vartheta, \tag{4.77}$$

where

$$\Phi \equiv \Gamma + \frac{1}{\Gamma}\frac{d}{d\rho}(\rho\Gamma). \tag{4.78}$$

From Eq. (3.60) in Section 3.3.3, we see that the isothermal stiffness is

$$C^\vartheta = C^C + \left(\Phi - \Gamma\frac{c_v}{c_\eta}\right)p_\vartheta.$$

In typical applications of the Mie–Grüneisen model both c_η and Γ are viewed as empirical quantities. It is commonly assumed that the specific heat is a constant so that $c_v = c_\eta$. With respect to Γ, one of two assumptions is often used: Either $\Gamma = \Gamma_0$ is constant or $\rho\Gamma = \rho_0\Gamma_0$ is constant. Under these conditions, we find that

$$\left.\begin{aligned} C^\eta &= C^C + (\Gamma_0 + 1)p_\vartheta \\ C^\vartheta &= C^C + p_\vartheta \end{aligned}\right\} \text{ when } c_v \text{ and } \Gamma \text{ are constant}$$

and

$$\left.\begin{aligned} C^\eta &= C^C + \Gamma p_\vartheta \\ C^\vartheta &= C^C \end{aligned}\right\} \text{ when } c_v \text{ and } \rho\Gamma \text{ are constant.}$$

As we shall see, although either of these assumptions may produce useful results, they can also lead to results that are anomalous and in striking contradiction to physical observation. In the next section we present a simple analysis that illustrates how Γ can be fundamentally linked to the nonlinear response of the material.

4.2.1 *A Lattice Model*

During our development of the constitutive relation for an ideal gas, we paused to examine a simple physical picture of the gas as a collection of molecules. Using this model we found that the temperature of the gas was equal to the sum of the kinetic energies of molecules. Recall that, to simplify the analysis, the individual molecules were combined

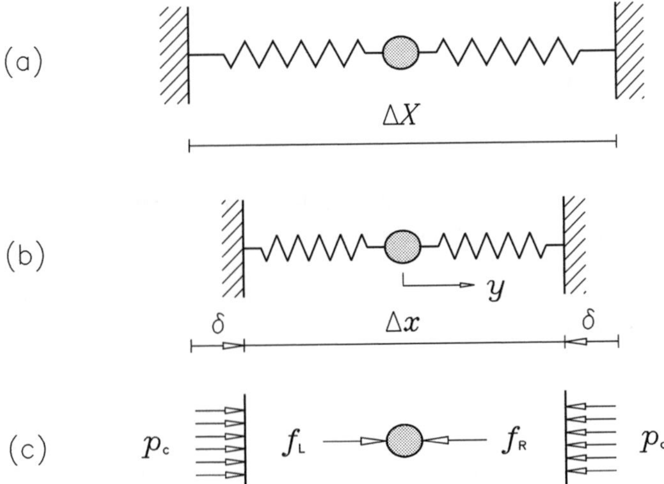

Figure 4.24 A simple harmonic oscillator contained between two movable walls. Diagram (a) shows the walls in the reference configuration. Diagram (b) shows each wall compressed a distance δ. Diagram (c) illustrates the tractions acting on the walls and the spring forces acting on the atom.

into a single giant surrogate molecule (see Figure 4.2). The Mie–Grüneisen model is based upon a similar picture of atomic motion. The most general analysis is based upon the quantum statistical mechanics of a lattice of oscillating atoms. A detailed treatment of such a system for even the simpliest of solids is beyond the scope of this book. However, we can analyze a very elementary model of a solid to illustrate the motivation behind Eq. (4.67) in which we have partitioned the internal energy into cold energy and thermal energy. Using this model, which was originally proposed by Johnson,† we can also illustrate how the Grüneisen parameter Γ is linked to cold response of the solid.

Suppose the basic building block of a solid lattice is the simple spring–mass oscillator shown in Figure 4.24. The oscillator is composed of an atom of mass m connected by two identical springs to two movable walls. It is one element of a periodic array of oscillators, which is virtually infinite in extent. The walls define the boundaries between this oscillator and identical oscillators to the left and right that are not shown in the figure. The springs represent the interatomic forces that bind this atom to its nearest neighbors. Two types of energy are present. If the atom is stationary, we can store energy by moving the walls to compress or expand the springs. If instead the walls are stationary, we can store energy in the oscillating motion of the atom. Like the ideal gas, the motion of the atom represents thermal energy. The remaining energy due to compression of the springs represents the cold energy of the interatomic forces.

To simplify the discussion, we assume that the motions of the walls are symmetric about the center of the oscillator. These motions are described by the quantity δ. The walls are initially separated by the infinitesimal distance ΔX, and during deformation of the body, the separation of the walls changes to Δx. Therefore, the deformation gradient of the body

† See "Single-particle model of a solid: The Mie–Grüneisen equation" by Johnson, J. N. in *American Journal of Physics* **36**, pp. 917–919 (1968).

is

$$F = \frac{\Delta x}{\Delta X} = \frac{\Delta X - 2\delta}{\Delta X} = 1 - \frac{2\delta}{\Delta X}. \tag{4.79}$$

Cold Response

We use the parameter y to describe the motion of the atom m. Energy associated with y represents the thermal energy of the body. When we hold m fixed at $y = 0$, the thermal energy is zero. We can then measure the cold response of the solid by applying a uniform pressure p_c on each wall. We observe the change in the displacement δ while we hold $y = 0$. Let us represent this response with the smooth function

$$p_c = p_c(F).$$

The Taylor's series expansion of p_c about $F = 1$ is

$$p_c = \frac{dp_c}{dF}(F-1) + \frac{1}{2}\frac{d^2 p_c}{dF^2}(F-1)^2 + \cdots,$$

where $p_c(1) = 0$. You should understand that these derivatives are evaluated at $F = 1$. We can also write this expansion in terms of δ by substituting Eq. (4.79) to obtain

$$p_c = -\frac{dp_c}{dF}\frac{2\delta}{\Delta X} + \frac{d^2 p_c}{dF^2}\frac{2\delta^2}{\Delta X^2} + \cdots.$$

Because the two springs are identical, we know that the forces f_L and f_R (defined as positive in compression as shown in Figure 4.24) in each spring are equal to the product of p_c and the cross-sectional area A of each wall:

$$p_c A = f_L = f_R = (k_0\delta + k_1\delta^2)A,$$

where we have ignored the higher-order terms in δ and defined

$$k_0 = -\frac{2}{\Delta X}\frac{dp_c}{dF}, \qquad k_1 = \frac{2}{\Delta X^2}\frac{d^2 p_c}{dF^2}. \tag{4.80}$$

By applying the constraints in Eqs. (2.22) and (2.23) for a nonlinear-elastic material, we require both k_0 and k_1 to be positive. This is the required form of the cold response of the material.

Thermal Response

To evaluate the thermal response, we hold δ fixed, let the mass move, and determine the thermal energy of the body. The motion y produces a different force in each spring. The forces in the left and right springs are

$$f_L = \left[k_0(\delta - y) + k_1(\delta - y)^2\right]A,$$
$$f_R = \left[k_0(\delta + y) + k_1(\delta + y)^2\right]A.$$

Newton's law of motion for the mass m then becomes

$$m\frac{d^2 y}{dt^2} = f_L - f_R$$
$$= -2A(k_0 + 2k_1\delta)y.$$

Now let us assume the material is at an equilibrium state where F and δ are constant while y is free to change. Without loss of generality, we can also assume that $y = 0$ at $t = 0$. Under these conditions, the solution to this ordinary differential equation is

$$y = y_0 \sin(\omega t),$$

where y_0 is an arbitrary constant and

$$\omega = [2A(k_0 + 2k_1\delta)/m]^{1/2}. \tag{4.81}$$

The mass m of an atom is very small, and, consequently, the frequency ω is very high. This suggests that we can only measure the average force in each of the springs. The average force in each spring is

$$\langle f_i \rangle \equiv \frac{1}{T}\int_0^T f_i \, dt,$$

where $T = 2\pi/\omega$ is the period of oscillation of the atom. Noticing that

$$\frac{1}{T}\int_0^T \sin^2(\omega t)\, dt = \frac{1}{2},$$

we can easily show that

$$\langle f_L \rangle = \langle f_R \rangle = \left(k_0\delta + k_1\delta^2 + \frac{1}{2}k_1 y_0^2\right)A. \tag{4.82}$$

The average pressure resulting from this average force is

$$p_e = \frac{\langle f_R \rangle}{A} = p_c + \frac{1}{2}k_1 y_0^2. \tag{4.83}$$

Because we are holding δ fixed, the total energy E of the atom and the springs is equal to the following expression for the kinetic energy of the atom plus the work the atom does against the springs:

$$\begin{aligned} E &= \frac{m}{2}\left(\frac{dy}{dt}\right)^2 + \int_0^y (f_R - f_L)\, dy \\ &= \frac{m}{2}\left(\frac{dy}{dt}\right)^2 + \int_0^y 2A(k_0 + 2k_1\delta) y\, dy \\ &= \frac{m}{2}\left[\left(\frac{dy}{dt}\right)^2 + \omega^2 y^2\right]. \end{aligned}$$

We substitute the solution for y to obtain

$$E = \frac{1}{2}m(\omega y_0)^2.$$

The thermal energy $\mathcal{E}_\vartheta$ per unit mass of the body is

$$\mathcal{E}_\vartheta \equiv \frac{E}{m} = \frac{1}{2}(\omega y_0)^2.$$

Notice that Eq. (4.83) yields a result that is identical to Eq. (4.72):

$$p_e = p_c + \rho\Gamma\mathcal{E}_\vartheta,$$

where

$$\rho = \frac{m}{A\Delta x}, \qquad \Gamma = \frac{Ak_1\Delta x}{m\omega^2}.$$

Substitution of Eqs. (4.80) and (4.81) gives

$$\Gamma = -\frac{F}{2}\frac{d^2p_c/dF^2}{dp_c/dF}, \tag{4.84}$$

where we have approximated ω with its value at $\delta = 0$. This is the key result of the lattice model. It establishes the link between the Grüneisen ratio and the cold response of the lattice. Notice that this result fits within the framework of the Mie–Grüneisen material because Γ is strictly a function of ρ.

Now suppose for a moment we recall our discussion of the model for a monoatomic gas. We found that it led to a value for the adiabatic gas constant $\gamma = 5/3$ (see Section 4.1.2). We noticed that other values of γ are derived from more complex models such as the diatomic gas. A similar conclusion follows here. The expression for Γ in Eq. (4.84) is special to the particular atomic lattice we have analyzed in this section. Other lattice models lead to different values of Γ. In particular, several other lattice models have been proposed to describe metals. Most notable among these is the model proposed by Slater[†], where

$$\Gamma = -\frac{F}{2}\frac{d^2p_c/dF^2}{dp_c/dF} - \frac{2}{3}, \tag{4.85}$$

and the models by proposed by Dugdale and Mac Donald[‡] and Rice, McQueen, and Walsh[#] where

$$\Gamma = -\frac{F}{2}\frac{d^2p_c/dF^2}{dp_c/dF} - 1. \tag{4.86}$$

In our analysis of the simple lattice model, we found that the Grüneisen parameter Γ can also be expressed in terms of the harmonic frequency ω. Indeed, from Eq. (4.81), we find that

$$\frac{d\omega}{d\delta} = \frac{2k_1A}{m\omega}.$$

Then

$$\frac{F}{\omega}\frac{d\omega}{d\delta}\frac{d\delta}{dF} = -\frac{k_1A\Delta x}{m\omega^2},$$

and consequently

$$\Gamma = -\frac{d(\ln\omega)}{d(\ln F)}. \tag{4.87}$$

The result in Eq. (4.84) is special and only yields the Grüneisen parameter for the simple lattice model analyzed in this section. Although we shall not show it here, it is possible to

[†] See pp. 451–456 in *Introduction to Chemical Physics* by Slater, J. C. (McGraw-Hill, 1939).

[‡] See "The thermal expansion of solids" by Dugdale, J. S. and Mac Donald D. K. C. in *Physical Review* **59**, p. 832–834 (1953).

[#] See "Compression of solids by strong shock waves" by Rice, M. H., McQueen, R. G., and Walsh, J. M. in *Solid State Physics* 6, ed. Seitz, F. and Turnbull, D., pp. 1–63 (Academic Press, 1958).

demonstrate that Eq. (4.87) is more general. Each of the different lattice models we have mentioned in the previous paragraph obey Eq. (4.87).

4.2.2 *Equilibrium States and Processes*

To illustrate a state surface for a Mie–Grüneisen solid, we must prescribe the cold energy $\mathcal{E}_c$. Solely for the purpose of illustration, let us assume that

$$\mathcal{E}_c = \left(\frac{c_0}{s}\right)^2 \left\{ \ln[1 - s(1 - F)] + \frac{1}{1 - s(1 - F)} \right\}, \tag{4.88}$$

where c_0 and s are arbitrary constants. Our reason for selecting this function becomes apparent when we evaluate the cold pressure p_c by substituting Eq. (4.88) into Eq. (4.73):

$$p_c = \frac{\rho_0 c_0^2 (1 - F)}{[1 - s(1 - F)]^2}. \tag{4.89}$$

Notice the similarity between this result and the stress expression for the linear-Hugoniot model, Eq. (2.179) in Section 2.7.7. Later in this section, we shall find that this cold pressure only yields the linear-Hugoniot model when we select $\Gamma = 0$. Now we differentiate Eq. (4.89) to obtain

$$\frac{dp_c}{dF} = -\rho_0 c_0^2 \frac{1 + s(1 - F)}{[1 - s(1 - F)]^3},$$

$$\frac{d^2 p_c}{dF^2} = 4\rho_0 c_0^2 s \frac{1 + \frac{1}{2}s(1 - F)}{[1 - s(1 - F)]^4}.$$

Therefore, the cold stiffness C^C is [see Eq. (4.76)]

$$C^C = \rho_0 c_0^2 F \frac{1 + s(1 - F)}{[1 - s(1 - F)]^3}. \tag{4.90}$$

From the analysis of the simple lattice model, we can also evaluate the Grüneisen parameter. Using Eq. (4.84), we obtain

$$\Gamma = 2Fs \frac{1 + \frac{1}{2}s(1 - F)}{1 - s^2(1 - F)^2}.$$

However, instead of using this expression for Γ, let us assume that both Γ and the specific heat $c_\eta = c_v$ are constants that we can arbitrarily prescribe. Most often this is typical of the way in which the Mie–Grüneisen model is used. The Grüneisen parameter is treated simply as an empirical parameter, rather than a function derived from an atomic lattice model.

We show the resulting state surface for this Mie–Grüneisen solid in Figure 4.25. We have clipped the bottom of the surface at the isentrope that passes through the point $F = 1$ and $\vartheta = \vartheta_0$. Thus the response of the material at $\vartheta = 0$ is not illustrated. In contrast, we include the response of the cold material in the $\mathcal{E}$–F diagram (see Figure 4.26). Here we use the bold line to illustrate the coincident isentrope and isotherm at $\vartheta = 0$ K. The parameters we use to generate this surface do not match the properties of any particular material, but they are typical of many metals such as steel. On both the state surface and the $\mathcal{E}$–F diagram, we have also drawn the four reversible processes representing the isentropes (—), isochors (···),

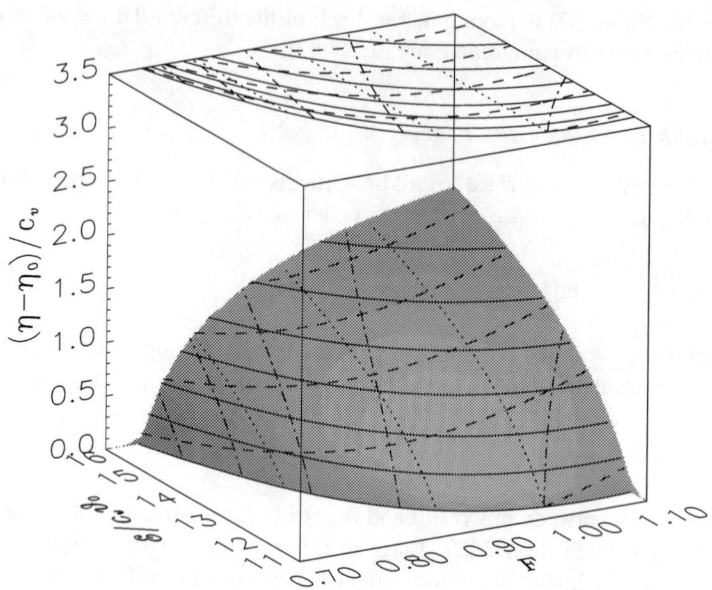

Figure 4.25 The state surface with isentropes (—), isochors (· · ·), isotherms (— —), and isobars (— · —). Here $\Gamma = 2$, $c_v = 2$ kJ/kg/K, $s = 2$, $c_0 = 5$ km/s, $\rho_0 = 8$ Mg/m^3, $\vartheta_0 = 300$ K, and $\eta_0 = c_v \ln(c_v \vartheta_0)$.

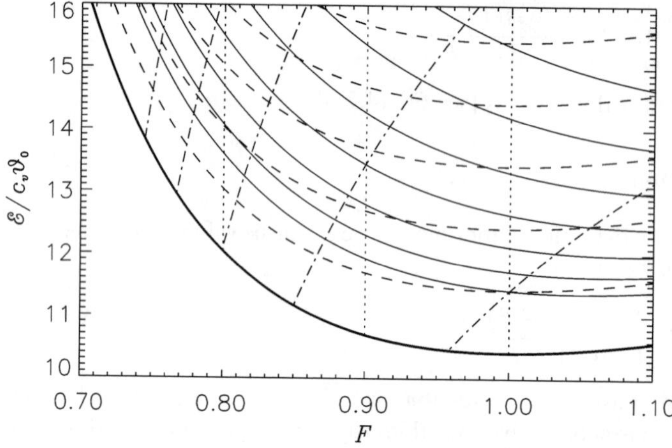

Figure 4.26 The $\mathcal{E}$–F diagram with isentropes (—), isochors (· · ·), isotherms (— —), and isobars (— · —). The bold curve is the isentrope and isotherm at $\vartheta = 0$ K.

isotherms (- - -), and isobars (— · —). Notice that we have included states where $F = 1.1$. Typically, the stretching of metals at room temperature will cause them to fracture before this value of F is reached. The isotherms in the figures are evenly spaced at an increment of 300 K. We show all of the isotherms from 0 K to 1,500 K in Figure 4.26, but we show only the isotherms from 300 K to 1,500 K in Figure 4.25. At the higher temperatures, melting is possible, but we do not include this in the state surface. The isobars begin at zero pressure on the right and increase to the left by evenly spaced increments of 50 GPa. The pressure of the atmosphere is approximately 0.1 MPa, and thus the left-most isobar represents a pressure that is a factor of 2×10^6 times atmospheric pressure. In contrast to the ideal gas, enormous changes in pressure only cause modest changes in temperature. For example, consider a material that initially resides at zero pressure on the lowest isentrope in Figure 4.25 (the lowest thin-lined isentrope in Figure 4.26). When we compress this material along this isentrope to the left-most isobar, the temperature will increase by only 200 K. Similarly, we also see that the thermal expansion is relatively small. The zero-pressure isobar shows us that if the temperature of the material is increased from 300 K to 600 K, the deformation gradient F will only increase from 1.0 to 1.06.

Isentropic Process

During an isentropic process, entropy is constant. For an isentropic process, we can derive a relationship between the temperature and the deformation gradient without requiring any knowledge of the cold response. Let us assume the specific heat is constant, so that $c_v = c_\eta$. We consider two cases. First, we suppose that $\Gamma = \Gamma_0$, and then we suppose that $\rho\Gamma = \rho_0\Gamma_0$.

***Constant* $\boldsymbol{\Gamma = \Gamma_0}$** For this case, we use the fundamental equation, Eq. (4.68), to obtain

$$\ln\vartheta = \Gamma_0 \ln\left(\frac{\rho}{\rho_0}\right) + a,$$

where a is a constant. Here we have used $\mathcal{E} - \mathcal{E}_c = c_v\vartheta$ and $\eta =$ constant. By requiring that $\vartheta = \vartheta_0$ at $\rho = \rho_0$, we obtain

$$\ln\vartheta = \Gamma_0 \ln\left(\frac{\rho}{\rho_0}\right) + \ln\vartheta_0.$$

Therefore,

$$\frac{\vartheta}{\vartheta_0} = F^{-\Gamma_0}. \tag{4.91}$$

***Constant* $\boldsymbol{\rho\Gamma = \rho_0\Gamma_0}$** For this case, we use the fundamental equation, Eq. (4.68), to obtain

$$\ln\vartheta = \rho_0\Gamma_0 \int_{\rho_0}^{\rho} \frac{d\rho'}{(\rho')^2} + a,$$

where a is a constant. We integrate this equation to obtain

$$\ln\vartheta = \Gamma_0(1 - F) + \ln\vartheta_0,$$

where we have required that $\vartheta = \vartheta_0$ at $F = 1$. This yields the following result:

$$\frac{\vartheta}{\vartheta_0} = e^{\Gamma_0(1-F)}. \tag{4.92}$$

In Figure 4.27 we compare the two sets of isentropes, Eqs. (4.91) and (4.92), for three values of Γ_0. Notice that as we compress the material from the initial state where $F = 1$, higher temperatures are attained for larger values of Γ_0. We also find that the condition of constant Γ results in larger temperature increases than the condition of constant $\rho\Gamma$.

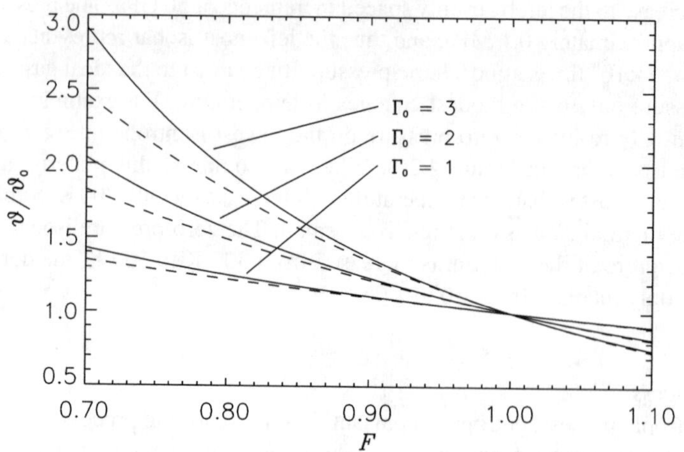

Figure 4.27 Comparison of isentropes where $\Gamma = \Gamma_0$ (—) and $\rho\Gamma = \rho_0\Gamma_0$ (− −).

Isothermal Process

During an isothermal process, temperature is constant. When the specific heat is also constant, then $c_v = c_\eta$, and the equilibrium pressure is [see Eq. (4.74)]

$$p_e = p_c + \rho\Gamma c_v \vartheta.$$

When both c_v and $\rho\Gamma$ are constant, the equilibrium pressure along an isotherm differs from the cold pressure by a constant. Thus the equilibrium pressure is solely a function of F, and the curve $p_e(\rho)$ is parallel to the curve $p_c(\rho)$. If instead we require Γ to be constant, then $p_e(\rho)$ is not parallel to $p_c(\rho)$. Notice that the isothems in Figure 4.26 are not parallel, because we chose Γ, and not $\rho\Gamma$, to be constant.

Isochoric Process

During an isochoric process, the density and therefore the Grüneisen parameter are both constant. When the specific heat is also constant, then $c_v = c_\eta$, and [see Eq. (4.74)]

$$p_e = p_c + \rho_0\Gamma_0 c_v \vartheta.$$

Consider an initial state where $\vartheta = \vartheta_0$. The pressure at this state is

$$p_0 = p_c + \rho_0\Gamma_0 c_v \vartheta_0.$$

Thus the pressure at any other point on the isochor is

$$p_e = p_0 + \rho_0\Gamma_0 c_v (\vartheta - \vartheta_0).$$

Isobaric Process

During an isobaric process, pressure is constant. When the specific heat is constant, then $c_v = c_\eta$, and the equilibrium pressure is [see Eq. (4.74)]

$$p_e = p_c + \rho\Gamma c_v \vartheta.$$

Consider an initial state where $F = 1$ and $\vartheta = \vartheta_0$. The following relationship holds on the isobar that passes through this initial state:

$$p_c(\rho) + \rho\Gamma c_v\vartheta = p_c(\rho_0) + \rho_0\Gamma_0 c_v\vartheta_0.$$

For the special case when $\rho\Gamma = \rho_0\Gamma_0$, we see that

$$p_c(\rho) - p_c(\rho_0) = \rho_0\Gamma_0 c_v(\vartheta_0 - \vartheta). \tag{4.93}$$

4.2.3 *Elastic-Wave Process*

The elastic-wave process is an isentropic process. The material wave velocity C is related to the isentropic stiffness C^η by Eq. (3.88),

$$C = \left(\frac{C^\eta}{\rho_0 F}\right)^{1/2}.$$

Substituting Eq. (4.77), we obtain

$$C = \left(\frac{C^C + \Phi p_\vartheta}{\rho_0 F}\right)^{1/2}, \tag{4.94}$$

where

$$\Phi \equiv \Gamma + \frac{1}{\Gamma}\frac{d}{d\rho}(\rho\Gamma).$$

The spatial wave velocity $c = FC$ is

$$c = \left(\frac{C^C + \Phi p_\vartheta}{\rho}\right)^{1/2}. \tag{4.95}$$

Recall that we have approximated the stress in the Mie–Grüneisen solid with the mean normal stress σ. Thus the equilibrium pressure is

$$p_e \approx -\sigma_e. \tag{4.96}$$

Because we ignore the deviatoric stresses, we represent the isentropic response of the material with the bulk response, and as $F \to 1$ the pressure is given by [see Eqs. (2.25) and (2.26)]

$$p_e = \mathcal{K}^\eta(1 - F),$$

where [see Eq. (3.72]

$$\mathcal{K}^\eta \equiv \lambda + \frac{2}{3}\mu. \tag{4.97}$$

We use $\mathcal{K}^\eta$ to define the *bulk velocity*

$$c_b \equiv \left(\frac{\lambda + \frac{2}{3}\mu}{\rho_0}\right)^{1/2}.$$

At first, you might think that the bulk velocity represents the velocity of linear-elastic waves in the Mie–Grüneisen solid. However, we already know that linear-elastic waves in a solid travel at the velocity of sound

$$c_0 = \left(\frac{\lambda + 2\mu}{\rho_0}\right)^{1/2}.$$

These two velocities are only equal for fluids where the deviatoric stresses are identically zero and $\mu = 0$. Thus, for a solid where $\mu \neq 0$, the bulk velocity does not represent the velocity of a physical wave. Rather, it is an artifact of the approximation, Eq. (4.96).

Example 4.1 Consider a block of AISI-304 stainless steel with the following properties:

$$\lambda = 109\,\text{GPa}, \qquad \beta = 14.7 \times 10^{-6}\,\text{K}^{-1},$$
$$\mu = 76.8\,\text{GPa}, \qquad \rho_0 = 7.89\,\text{Mg/m}^3,$$
$$c_T \approx c_p = 460\,\text{J/kg/K}.$$

Let us determine the isothermal stiffness, the bulk velocity, the velocity of sound, and the Grüneisen ratio.

There are several subtle issues we wish to illustrate with this example. We have assumed that the deviatoric stresses can be ignored. This means that we must consistently ignore the contributions of the deviatoric stresses at all states and not just at states of large compression. Therefore, we approximate the isentropic stiffness of the block with the isentropic bulk stiffness [see Eq. (4.97)]

$$C^\eta \approx \mathcal{K}^\eta = 109 + \frac{2}{3} \times 76.8 = 160 \text{ GPa}.$$

Then from Eq. (3.67),

$$\frac{1}{C^\vartheta} = \frac{1}{C^\eta} + \frac{\vartheta\alpha^2}{\rho c_T}.$$

Therefore, when $\vartheta = 295$ K

$$C^\vartheta = \left[\frac{1}{160 \times 10^9} + \frac{295 \times (3 \times 14.7 \times 10^{-6})^2}{7890 \times 460}\right]^{-1} = 156 \text{ GPa}.$$

The stiffnesses and the specific heats are related by Eq. (3.63),

$$\frac{c_T}{c_v} = \frac{C^\eta}{C^\vartheta} = \frac{160}{156} = 1.03.$$

Notice that the ratio of the specific heats is approximately equal to one. This is a typical value for most solids. Thus it makes little difference if we compute the bulk velocity correctly with the isentropic stiffness as

$$c_b = \left(\frac{C^\eta}{\rho_0}\right)^{1/2} = \left(\frac{160 \times 10^9}{7890}\right)^{1/2} = 4.50\,\text{km/s}$$

or *incorrectly* with the isothermal stiffness as $c_b = \sqrt{156 \times 10^9/7890} = 4.45$ km/s. This is in direct contrast to the ideal gas, where a similar mistake leads us to a 20% error in the computation of the speed of sound for air. Notice, however, that bulk speed is substantially different from the speed of sound c_0 for this stainless steel:

$$c_0 = \left(\frac{109 \times 10^9 + 2 \times 76.8 \times 10^9}{7890}\right)^{1/2} = 5.77\,\text{km/s}.$$

The Grüneisen parameter Γ_0 is determined from Eq. (3.66):

$$\Gamma_0 = \frac{\alpha C^\eta}{\rho c_p} = \frac{3 \times 14.7 \times 10^{-6} \times 160 \times 10^9}{7890 \times 460} = 1.94. \tag{4.98}$$

4.2.4 *Hugoniot Relations*

The equilibrium pressure, given by Eq. (4.72), is

$$p_e = p_c + \rho\Gamma(\mathcal{E} - \mathcal{E}_c). \tag{4.99}$$

Here we say the pressure p_e and energy $\mathcal{E}$ are *referenced to the cold curve* because the pressure difference $p_e - p_c$ is expressed in terms of the energy difference $\mathcal{E} - \mathcal{E}_c$. The product $\rho\Gamma$ controls the process by which the solid *departs* from the cold curve. Suppose that we subject this Mie–Grüneisen solid to a shock wave. This shock wave causes the state of the material to jump from the initial state, where $F = 1$ and $\mathcal{E} = \mathcal{E}_-$, to the Hugoniot state, where $F = F_+$ and $\mathcal{E} = \mathcal{E}_+$. Because the Hugoniot curve determines the locations of the initial and Hugoniot states of the material, let us see if we can write a relationship where the pressure p_e and energy $\mathcal{E}$ are *referenced to the Hugoniot curve* instead of the cold curve.

Suppose we define two functions:

$$p_H(F) \equiv \rho_0 U_S^2(1 - F), \qquad \mathcal{E}_H(F) \equiv \frac{p_H}{2\rho_0}(1 - F) + \mathcal{E}_-. \tag{4.100}$$

At the initial state

$$p_- = p_H(1) = 0, \qquad \mathcal{E}_- = \mathcal{E}_H(1),$$

and at the Hugoniot state

$$p_+ = p_H(F_+), \qquad \mathcal{E}_+ = \mathcal{E}_H(F_+).$$

We can also evaluate the pressure at the Hugoniot state with Eq. (4.99):

$$p_+ = p_c + \rho\Gamma(\mathcal{E}_+ - \mathcal{E}_c) = p_c + \rho\Gamma[(\mathcal{E}_+ - \mathcal{E}_-) - (\mathcal{E}_c - \mathcal{E}_-)],$$

which we can rewrite as

$$p_c - \rho\Gamma(\mathcal{E}_c - \mathcal{E}_-) = p_H - \rho\Gamma(\mathcal{E}_H - \mathcal{E}_-) = p_H\left[1 - \frac{\Gamma}{2}\left(\frac{1 - F}{F}\right)\right]. \tag{4.101}$$

Next, combining this result with Eq. (4.99), we find that

$$\begin{aligned} p_e &= p_c + \rho\Gamma[(\mathcal{E} - \mathcal{E}_-) - (\mathcal{E}_c - \mathcal{E}_-)] \\ &= p_H\left[1 - \frac{\Gamma}{2}\left(\frac{1 - F}{F}\right)\right] + \rho\Gamma(\mathcal{E} - \mathcal{E}_-). \end{aligned} \tag{4.102}$$

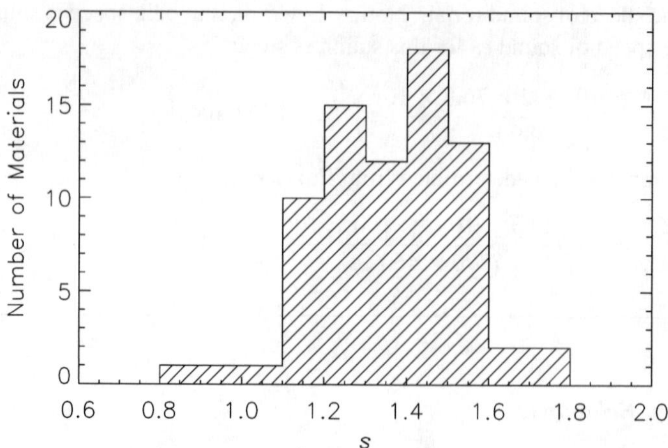

Figure 4.28 Histogram of values of s for solids that do not exhibit a phase transformation during shock loading. From Jeanloz, R., 1989, Shock wave equation of state and finite strain Theory, *Journal of Geophysical Research*, **94**, p. 5,873.

Then we substitute Eq. (4.100) to obtain

$$p_e = p_H + \rho\Gamma(\mathcal{E} - \mathcal{E}_H). \tag{4.103}$$

This is the desired expression for the response of the solid referenced to the Hugoniot curve.

The Linear-Hugoniot Model

Recall that we introduced the linear-Hugoniot model in Section 2.7.7. We found in a wide variety of materials that the experimental observation of strong shock waves reveals shock velocities U_S that obey the simple empirical relationship

$$U_S = c_b + sv_+, \tag{4.104}$$

where $v_- = 0$. Because we are ignoring the deviatoric components of stress, we have replaced the velocity of sound c_0 by the bulk velocity c_b. This straight-line relationship is an especially good representation of the shock velocities of metals. Values of c_b and s for a large variety of materials are reported by Marsh†. Interestingly, the values of s for many materials fall into a narrow range. In Figure 4.28 we show the distribution of values of s reported by Marsh. In Figure 4.29 we show a typical example of a material that exhibits this straight-line relationship. Also included in the diagram is the linear expression

$$U_S = 4.58 + 1.49v_+,$$

which illustrates just how closely we can represent the data with a straight line. The intercept of the straight line with the vertical axis is located at $U_S = 4.58$ km/s. This value compares reasonably well to the bulk velocity for this material, $c_b = 4.50$ km/s, which we determined in Example 4.1.

† See *LASL Shock Hugoniot Data*, ed. Marsh, S. P. (University of California Press, 1980).

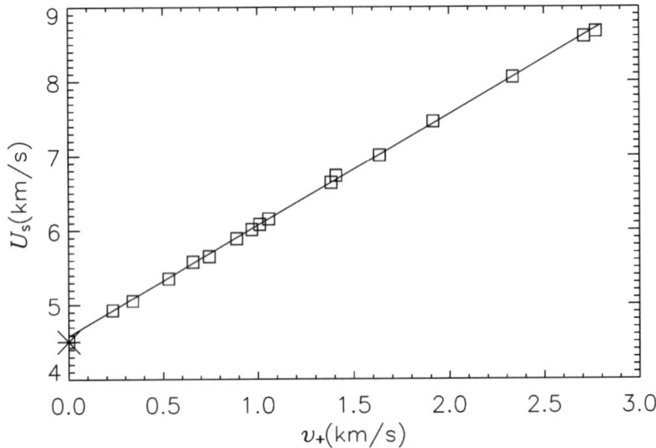

Figure 4.29 The U_S–v_+ diagram for AISI-304 stainless steel. The squares denote individual experimental observations. The solid line represents the empirical fit $U_S = 4.58 + 1.49v_+$. The asterisk is the calculated bulk velocity.

Because the comparison of the data to the linear Hugoniot is quite good when viewed in the U_S–v_+ diagram, let us determine the Hugoniot pressure from the data and compare it to the linear-Hugoniot model again. We use the kinematic compatibility condition, Eq. (2.139), and the momentum jump condition, Eq. (2.146), to determine the deformation gradient and the pressure directly from the data as follows:

$$F_+ = 1 - \frac{v_+}{U_S}, \qquad p_+ = \rho_0 U_S v_+. \tag{4.105}$$

When the data are processed through these relationships, we obtain values of p_+ and F_+ that do not depend upon any constitutive assumption. To determine the pressure from the linear-Hugoniot model, we use Eq. (2.177):

$$p_+ = \frac{\rho_0 c_b^2 (1 - F_+)}{[1 - s(1 - F_+)]^2}, \tag{4.106}$$

where $c_b = 4.58$ km/s and $s = 1.49$. We compare the data to the linear-Hugoniot model again in Figure 4.30.

Because we have now characterized the Hugoniot pressure with the linear-Hugoniot model, let us determine the equilibrium pressure p_e at states that are not on the Hugoniot curve. To achieve this result, we return to Eq. (4.103), which expresses the pressure in the solid using the Hugoniot curve as a reference,

$$p_e = p_H + \rho\Gamma(\mathcal{E} - \mathcal{E}_H). \tag{4.107}$$

From Eqs. (4.100), we find that

$$\begin{aligned} p_H(F) &\equiv \frac{\rho_0 c_b^2 (1 - F)}{[1 - s(1 - F)]^2}, \\ \mathcal{E}_H(F) &\equiv \frac{p_H}{2\rho_0}(1 - F) + \mathcal{E}_-. \end{aligned} \tag{4.108}$$

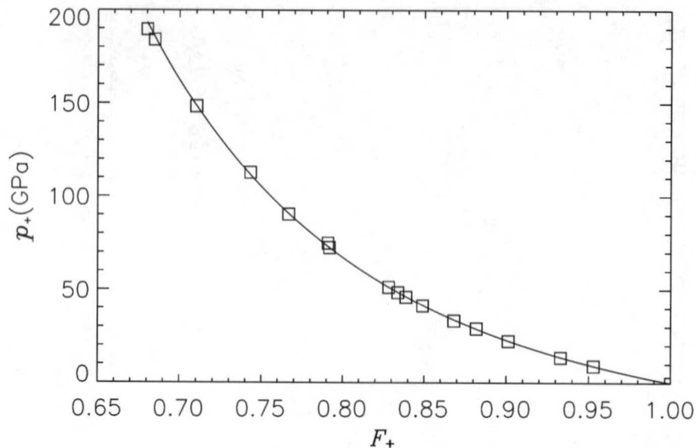

Figure 4.30 The p_+–F_+ diagram for AISI-304 stainless steel. The rectangles denote individual experimental observations. The solid line is the linear-Hugoniot model.

Grüneisen Parameter The only unknown function in the expression for the equilibrium pressure is the Grüneisen parameter Γ. To determine Γ, let us recall our discussion of the atomic lattice model. There we found that Γ could be expressed in terms of derivatives of the cold compression curve. In particular, suppose we use the following expression obtained from Eq. (4.86):

$$\Gamma = -\frac{F}{2}\frac{d^2 p_H/dF^2}{dp_H/dF} - 1,$$

where we have replaced the derivatives of p_c with derivatives of p_H. This estimation of Γ was proposed by Rice, McQueen, and Walsh. We substitute Eq. (4.108) to obtain

$$\Gamma = 2Fs\frac{1+\frac{1}{2}s(1-F)}{1-s^2(1-F)^2} - 1. \tag{4.109}$$

By evaluating this result at the initial state $F = 1$, we obtain

$$\Gamma_0 = 2s - 1.$$

For our example of AISI-304 stainless steel, in which $s = 1.49$, we obtain $\Gamma_0 = 1.98$. This value is in close agreement with Eq. (4.98). This level of accuracy has been demonstrated for a large number of metallic elements and alloys. However, other types of materials do not obey this relationship for Γ even if they do exhibit a linear-Hugoniot relationship.

Compression Limit Consider a shock wave where

$$F_+ = 1 - \frac{1}{s}.$$

When we substitute this value of the deformation gradient into Eq. (4.106), we find that $p_+ \to \infty$. This means that the material exhibits a compression limit when subjected to a shock wave. We also encountered this phenomenon in our study of the ideal gas.

The compression limit is a result of heating due to the shock wave. However, as a typical material is compressed isentropically to $F = 1 - \frac{1}{s}$ by a structured wave, it does not exhibit a compression limit. Thus p_e must remain finite during an isentropic process. Inspection of Eq. (4.102) reveals that p_e can only remain finite if

$$\frac{\Gamma}{2}\left(\frac{1-F}{F}\right) = 1, \quad \text{when } F = 1 - \frac{1}{s}.$$

Thus

$$\Gamma = 2s - 2, \quad \text{when } F = 1 - \frac{1}{s}.$$

Notice that this result contradicts the result of Eq. (4.109), which yields $\Gamma \to \infty$ at the compression limit. This demonstrates a fundamental flaw with all of the estimations of the Grüneisen parameter that are based on atomic lattice models. It is for this reason that we usually treat Γ as an empirical parameter.

Cold Energy We can obtain a differential equation for the cold energy $\mathcal{E}_c$ of the linear-Hugoniot model by combining the definition of the cold pressure, Eq. (4.73), with Eq. (4.101) to obtain

$$\frac{d\mathcal{E}_c}{dF} + \left(\frac{\Gamma}{F}\right)(\mathcal{E}_c - \mathcal{E}_-) = -\frac{p_H}{\rho_0}\left[1 - \frac{1}{2}\left(\frac{\Gamma}{F}\right)(1-F)\right]. \tag{4.110}$$

This is a linear, first-order, ordinary differential equation. If we prescribe the function $\Gamma(F)$, we can integrate this differential equation to obtain the cold energy $\mathcal{E}_c(F)$ of the linear-Hugoniot model.

4.2.5 *Shock Invariant*

The first comprehensive study of the thicknesses of shock waves in a solid was reported by Barker[†] in 1968. He prepared several aluminum target plates by vapor depositing thin mirrors on their back surfaces. He then used a flyer-plate to generate a shock wave in each aluminum target (see Figure 4.31). In each experiment, he assumed the thickness of the target plate was sufficient to allow the wave generated by the impact to evolve into a steady wave before it reached the back free boundary of the target. As the steady wave interacted with the back surface of the target, it caused the mirror to move. This motion was measured using a special type of optical interferometer called a *velocity interferometer*. This interferometer gave a direct measurement of the velocity of the mirror. Recall that the reflection of a steady wave from a free boundary results in a release wave (see Section 2.8.1). Therefore, the motion of the mirror was simultaneously influenced by both the compression and expansion characteristics of the target material.

Through detailed analysis of the interaction of the incident steady wave with the reflected rarefaction wave, we can determine the profile of the incident steady wave from the velocity measurements of the mirror. This is usually accomplished by numerical analysis of the wave interaction problem using a digital computer (see Appendix A). From this type of

[†] See "Fine structure of compression and release wave shapes in aluminum measured by the velocity interferometer technique" by Barker, L. M. in *Behavior of Dense Media Under High Dynamic Pressure*, pp. 483–504 (Gordon and Breach, 1968).

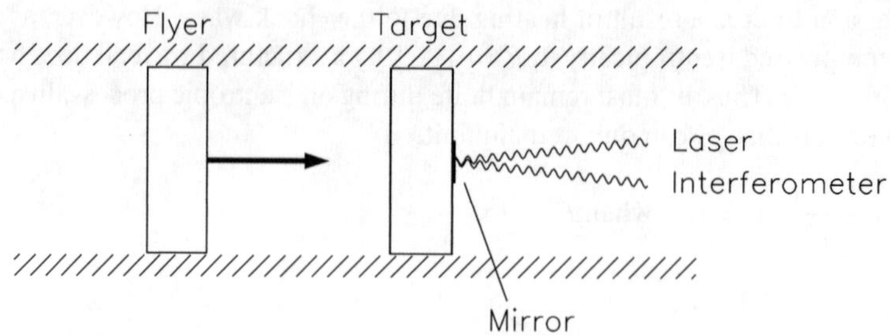

Figure 4.31 The flyer-plate experiment. The velocity of the back surface of the target plate is measured with a velocity interferometer.

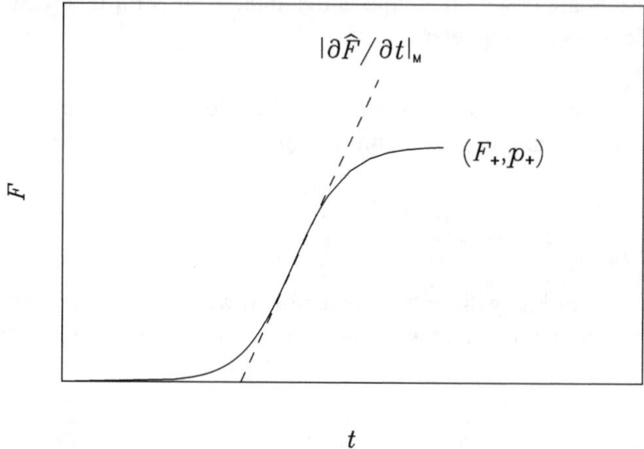

Figure 4.32 The maximum value of rate of change of the deformation gradient of a steady wave.

analysis, we are able to determine two quantities for each of the measured wave profiles: the maximum absolute value of the rate of change of the deformation gradient $|\partial \hat{F}/\partial t|_M$, and the pressure at the Hugoniot state p_+ (Figure 4.32). This experiment has been repeated by others using a variety of materials. Some of these data are summarized by Swegle and Grady (see Figure 4.33).

Grady† has noted that these data are reasonably approximated by the empirical relationship

$$\left| \frac{\partial \hat{F}}{\partial t} \right|_M \propto p_+^4. \tag{4.111}$$

Along with the data for each material, we show a solid line that represents a fit to this empirical relationship. We might argue that some of the data are too sparse to offer any conclusions about Grady's observation. We might also argue that any plot that uses logarithmic scales on both the horizontal and vertical axis illustrates only crude trends in data. However,

† See "Strain-rate dependence of the effective viscosity under steady-wave shock compression" by Grady, D. E. in *Applied Physics Letters* **38**, pp. 825–826 (1981).

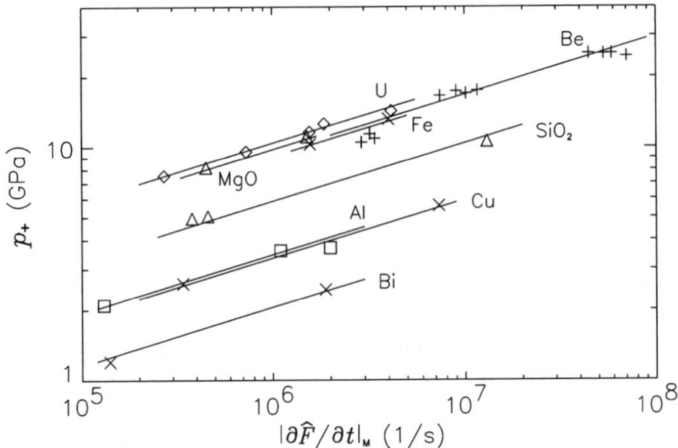

Figure 4.33 Hugoniot pressure versus the maximum rate of change of deformation gradient for various solids. Solid lines represent Eq. (4.113). Reprinted from Swegle, J. W. and Grady, D. E., 1985, Shock viscosity and the prediction of shock wave rise times, *Journal of Applied Physics* **58**, p. 693, with kind permission of the American Institute of Physics.

the combined weight of these observations from a wide variety of solids is compelling and may offer us some insight into the fundamental constitutive behavior of solids. At present, we can only offer Eq. (4.111) as an experimental observation and place it alongside the linear-Hugoniot relationship $U_S = c_0 + sv_+$ to form a set of curious empirical rules that seem to apply to a surprisingly wide variety of condensed materials.

Grady used Eq. (4.111) as motivation to define a quantity he called the *shock invariant* A:

$$A = \rho_0 \delta\mathcal{E}\,\delta t. \tag{4.112}$$

Here $\delta\mathcal{E}$ is the energy dissipated by the passage of the shock wave, δt is the time over which the energy is dissipated at a material point, and A is a constant. We shall now demonstrate that this statement is equivalent to Eq. (4.111). Consider a material whose pressure under isentropic compression is equal to the Hugoniot pressure [see Eq. (4.107)]

$$p_e = p_H. \tag{4.113}$$

This means we have ignored the thermal pressure $\rho\Gamma(\mathcal{E} - \mathcal{E}_H)$ and used the weak-shock assumption. Next we substitute Eq. (4.108) and then take a Taylor's series expansion of the denominator of the result about $F = 1$ to obtain

$$p_e = \rho_0 c_b^2(1 - F) + 2s\rho_0 c_b^2(1 - F)^2 + \cdots. \tag{4.114}$$

Now let us compare the response of this material to two types of waves. First, we consider a structured wave. Then we consider a steady wave.

When a structured compression wave propagates through this material, the deformation gradient changes from $F = 1$ to $F = F_+$. This causes the internal energy to change from its initial value $\mathcal{E}_-$ to a final value $\mathcal{E}_S$, which we can obtain from the balance of energy [see

Eq. (3.10)]

$$\rho_0 \frac{\partial \hat{\mathcal{E}}}{\partial t} = -p_e \frac{\partial \hat{F}}{\partial t}.$$

By integrating this relationship, we obtain

$$\rho_0(\mathcal{E}_S - \mathcal{E}_-) = \frac{1}{2}\rho_0 c_b^2 (1 - F_+)^2 + \frac{2}{3} s \rho_0 c_b^2 (1 - F_+)^3 + \cdots.$$

None of this change of internal energy is dissipated in the material.

Now we set aside this structured-wave solution for a moment and consider a steady wave propagating through this material. This wave also changes the deformation gradient from $F = 1$ to $F = F_+$. However, in this case the internal energy changes from its initial value $\mathcal{E}_-$ to a different final value $\mathcal{E}_+$. The final value of the internal energy is [see Eq. (3.112)]

$$\rho_0(\mathcal{E}_+ - \mathcal{E}_-) = \frac{1}{2} p_e (1 - F_+) = \frac{1}{2}\rho_0 c_b^2 (1 - F_+)^2 + s \rho_0 c_b^2 (1 - F_+)^3 + \cdots.$$

The final internal energy of the steady wave $\mathcal{E}_+$ differs from the final internal energy of the structured wave $\mathcal{E}_S$ because the steady wave dissipates energy whereas the structured wave does not. The difference

$$\rho_0 \delta\mathcal{E} \equiv \rho_0(\mathcal{E}_+ - \mathcal{E}_S) = \frac{1}{3} s \rho_0 c_b^2 (1 - F_+)^3 + \cdots$$

represents the energy dissipated by the steady wave. Moreover, the time required for the steady wave to pass a material point is approximately

$$\delta t \approx \frac{1 - F_+}{|\partial \hat{F}/\partial t|_M}.$$

The product of these two quantities yields the shock invariant

$$A = \frac{1}{3} s \rho_0 c_b^2 \frac{(1 - F_+)^4}{|\partial \hat{F}/\partial t|_M} + \cdots.$$

If we ignore the higher-order terms, we find that

$$\left|\frac{\partial \hat{F}}{\partial t}\right|_M = \frac{s \rho_0 c_b^2 (1 - F_+)^4}{3A}.$$

Substituting Eq. (4.114), we obtain

$$\left|\frac{\partial \hat{F}}{\partial t}\right|_M = \frac{s p_+^4}{3A\left(\rho_0 c_b^2\right)^3},$$

where we have again ignored high-order terms. Because the shock invariant A is a constant, we see that

$$A' \equiv \frac{s}{3A\left(\rho_0 c_b^2\right)^3}$$

is also a constant and

$$\left|\frac{\partial \hat{F}}{\partial t}\right|_M = A' p_+^4. \tag{4.115}$$

This is the desired result. It is equivalent to Eq. (4.111).

Viscous Solid

As noted in Eq. (4.66), we assume that the equilibrium stress in the Mie–Grüneisen solid is spherical and equal to the pressure p_e. Suppose we wish to represent this solid as a viscous material. Recalling that the stress in a viscous material is represented by an additive decomposition, we express the stress in the Mie–Grüneisen solid as

$$T = -p_e + \mathcal{T}.$$

If we decide to represent the viscous stress $\mathcal{T}$ by the expression for a Newtonian fluid, then [see Eq. (3.97)]

$$\mathcal{T} = \rho\left(\lambda' + \frac{4}{3}\mu'\right)D. \tag{4.116}$$

For materials where limited experimental data are available, this type of fluid representation is often employed. This is especially true of situations where we assume that the profile of the steady wave is unimportant, but where we require some representation of $\mathcal{T}$ to implement a numerical solution of the equations.

Although Eq. (4.116) is often used as an expedient representation of the viscous stress, the empirical observation Eq. (4.111) does offer us the motivation to derive an alternative expression for $\mathcal{T}$ that yields a more accurate representation of the profiles of steady waves. Of course, a representation of $\mathcal{T}$ based on physical arguments would be preferred. But often such a model is not available, and we must take a more pragmatic approach. We discuss a simple example here.

To determine a representation of viscous stress that yields profiles that obey Eq. (4.111), we notice that when the linear-Hugoniot relationship is combined with the relationship for the stress on the Rayleigh line, we obtain the following expression for the stress in a steady wave [see Eq. (2.188)]:

$$T = -\rho_0(c_b + sv_+)v.$$

Suppose we approximate the equilibrium pressure p_e by the Hugoniot pressure. Then we have

$$p_e = \rho_0(c_b + sv)v. \tag{4.117}$$

Thus the viscous stress is

$$\mathcal{T} = T + p_e = s\rho_0(v^2 - v_+v).$$

We see that $\mathcal{T} = 0$ at both the initial state $v = 0$ and the Hugoniot state $v = v_+$. A maximum absolute value $|\mathcal{T}|_M$ occurs between these two states. To determine this value, we take the derivative

$$\frac{d\mathcal{T}}{dv} = s\rho_0(2v - v_+).$$

We see that the maximum value occurs at $2v = v_+$. Thus,

$$|\mathcal{T}|_M = \frac{1}{4}s\rho_0v_+^2.$$

From Eq. (4.117), we approximate the Hugoniot pressure as $p_+ \approx \rho_0c_bv_+$ to obtain

$$|\mathcal{T}|_M \approx \frac{sp_+^2}{4\rho_0c_b^2}.$$

We substitute Eq. (4.115) to obtain

$$\left|\frac{\partial \hat{F}}{\partial t}\right|_M = \frac{16\rho_0^2 c_b^4 A'}{s^2} |\mathcal{T}|_M^2.$$

Suppose we assume this relationship to hold not only at the state where the maximum values occur but at every state. This suggests the following constitutive relationship:

$$\frac{\partial \hat{F}}{\partial t} = \frac{16\rho_0^2 c_b^4 A'}{s^2} \operatorname{sgn}(\mathcal{T}) |\mathcal{T}|^2, \tag{4.118}$$

which we can rewrite as

$$\mathcal{T} = \left(\frac{s^2}{16\rho_0^2 c_b^4 A'}\right)^{1/2} \operatorname{sgn}\left(\frac{\partial \hat{F}}{\partial t}\right) \left(\frac{\partial \hat{F}}{\partial t}\right)^{1/2},$$

where we use the sign function to ensure positive dissipation and to satisfy the dissipation inequality, Eq. (3.44). Either form of this constitutive function gives the desired fourth-power relationship between the maximum of the rate of change of the deformation gradient and the Hugoniot pressure. In Section 4.3 we shall demonstrate the usefulness of this result in calculating steady-wave profiles.

Dunn[†] has shown how fourth-power relationships can also be established for an infinite variety of viscous constitutive models of the type

$$\mathcal{T} = h(F)\left(\frac{\partial \hat{F}}{\partial t}\right)^m,$$

where m is any positive constant and $h(F)$ is an arbitrary function of the deformation gradient. Moreover, he has shown that thermoelastic materials obey Eq. (4.111). Thus it is important to remember that although the empirical observation Eq. (4.111) may provide insight into the nature of wave phenomena in solids, it alone cannot be used to uniquely determine the dissipative portion of the constitutive model for a solid.

4.2.6 *Square Wave in Stainless Steel*

We shall end this section with an example of a square wave propagating through a block of stainless steel (see Figure 4.34). This example is used to review what we have learned about the Mie–Grüneisen model. The square wave is composed of a leading shock wave that is followed by a rarefaction wave. We assume the rarefaction wave is a centered wave, and at the time $t = 0$, the rarefaction wave results in a sudden drop in pressure. At later times, the leading wave will still be a shock wave, but the rarefaction wave will disperse causing the pressure to drop over a finite distance. Our objective is to determine the distorted profile of the square wave at a later time.

Shock Wave

Suppose the material ahead of the square wave is at rest, where $F_- = 1$, and the shock wave compresses the material to $F_+ = 0.75$. The stainless-steel body obeys the linear-Hugoniot

[†] See "Implications and origins of shock structure in solids" by Dunn, J. E. in *Shock Compression of Condensed Matter*, ed. Schmidt, S. C., Johnson, J. N., and Davison, L. W., pp. 21–32 (Elsevier, 1990).

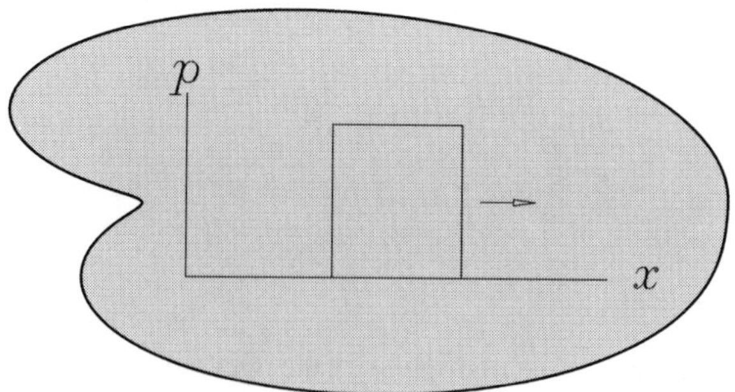

Figure 4.34 A square wave propagating in a block of stainless steel.

model

$$U_S = c_0 + s v_+,$$

where $c_0 = 4.58$ km/s and $s = 1.49$. The jump condition for the conservation of mass is

$$v_+ = U_S(1 - F_+).$$

These two relationships yield

$$U_S = \frac{c_0}{1 - s(1 - F_+)}.$$

Thus the velocity of the shock wave is

$$U_S = \frac{4580}{1 - 1.49 \times (1 - 0.75)} = 7.30 \text{ km/s},$$

and the velocity of the material behind the shock wave is

$$v_+ = 7300 \times (1 - 0.75) = 1.82 \text{ km/s}.$$

The jump condition for the balance of linear momentum across the shock wave is

$$p_+ = \rho_0 U_S v_+,$$

where $\rho_0 = 7.89$ Mg/m^3. Thus the pressure behind the shock wave is

$$p_+ = 7890 \times 7300 \times 1820 = 105 \text{ GPa}. \tag{4.119}$$

Notice that these quantities have been determined solely from the conservation of mass, balance of linear momentum, and the linear-Hugoniot relationship. We have not used the Mie–Grüneisen model. However, to determine the temperature behind the shock wave, we must employ the Mie–Grüneisen model to determine the cold response of the material.

Cold Response

We assume that the Grüneisen parameter obeys

$$\rho\Gamma = \rho_0\Gamma_0,$$

where $\Gamma_0 = 1.94$. We determine the cold energy by solving the differential equation Eq. (4.110),

$$\frac{d\mathcal{E}_c}{dF} + \Gamma_0(\mathcal{E}_c - \mathcal{E}_-) = -\frac{p_H}{\rho_0}\left[1 - \frac{1}{2}\Gamma_0(1-F)\right], \tag{4.120}$$

where

$$p_H(F) = \rho_0 U_S^2(1-F).$$

To solve this differential equation, we first multiply it by the integrating factor $e^{-\Gamma_0(1-F)}$ to obtain

$$\frac{d}{dF}[(\mathcal{E}_c - \mathcal{E}_-)e^{-\Gamma_0(1-F)}] = -\frac{p_H}{\rho_0}\left[1 - \frac{1}{2}\Gamma_0(1-F)\right]e^{-\Gamma_0(1-F)}.$$

Then we integrate this equation and find that

$$[(\mathcal{E}_c - \mathcal{E}_-)e^{-\Gamma_0(1-F)}]_1^F = c_v\vartheta_H,$$

where

$$c_v\vartheta_H \equiv -\int_1^F \frac{p_H}{\rho_0}\left[1 - \frac{1}{2}\Gamma_0(1-F)\right]e^{-\Gamma_0(1-F)}\,dF. \tag{4.121}$$

We denote the temperature ahead of the shock wave by ϑ_-, and from Eq. (4.70)

$$\mathcal{E}_- - \mathcal{E}_c(1) = c_v\vartheta_-,$$

where we assume that the specific heat is constant so that $c_\eta = c_v$. We combine the two previous relationships to obtain an expression for the cold energy of the body:

$$\mathcal{E}_c - \mathcal{E}_- = c_v(\vartheta_H - \vartheta_-)e^{\Gamma_0(1-F)}. \tag{4.122}$$

Now that we have determined the cold energy of the material, we can also determine the cold pressure p_c and the cold stiffness C^C. The cold pressure is obtained from Eq. (4.73):

$$p_c = -\rho_0\frac{d\mathcal{E}_c}{dF}.$$

Noticing Eq. (4.120), we obtain

$$p_c = \rho_0\Gamma_0(\mathcal{E}_c - \mathcal{E}_-) + p_H\left[1 - \frac{1}{2}\Gamma_0(1-F)\right]. \tag{4.123}$$

We use this expression for the cold pressure to determine the cold stiffness [Eq. (4.76)]

$$C^C = -F\frac{dp_c}{dF}.$$

Following a series of routine manipulations, we obtain the result

$$C^C = -F\left[1 - \frac{1}{2}\Gamma_0(1-F)\right]\frac{dp_H}{dF} + F\Gamma_0\left(p_c - \frac{1}{2}p_H\right). \tag{4.124}$$

Hugoniot Temperature

The Hugoniot pressure and temperature are related through Eq. (4.74),

$$p_+ = p_c(F_+) + \rho_0\Gamma_0 c_v \vartheta_+. \tag{4.125}$$

We can use this equation to determine the temperature ϑ_+ behind the shock wave. We begin by numerically integrating Eq. (4.121) to obtain

$$\vartheta_H(F_+) = 1{,}360\,\text{K}.$$

Noticing that $c_v = 447$ J/kg/K and $\vartheta_- = 295$ K, we evaluate Eq. (4.122) to obtain

$$\mathcal{E}_c(F_+) - \mathcal{E}_- = c_v[\vartheta_H(F_+) - \vartheta_-]e^{\Gamma_0(1-F_+)}$$

or

$$\mathcal{E}_c - \mathcal{E}_- = 447 \times (1{,}360 - 295)e^{1.94\times(1-0.75)} = 773\,\text{kJ/kg}.$$

Using Eqs. (4.119) and (4.123), we then evaluate the cold pressure at the Hugoniot state:

$$p_c(F_+) = 7{,}890 \times 1.94 \times (773 \times 10^3) + (105 \times 10^9) \times \left[1 - \frac{1}{2} \times 1.94 \times (1 - 0.75)\right]$$
$$= 91.3\text{ GPa}.$$

Finally, we solve Eq. (4.125) for the Hugoniot temperature:

$$\vartheta_+ = \frac{(105 - 91.3) \times 10^9}{7{,}890 \times 1.94 \times 447} = 2{,}000\,\text{K}.$$

Rarefaction Wave

Having determined the pressure, velocity, and temperature at the Hugoniot state behind the shock wave, we now evaluate the response of the material to the rarefaction wave. This wave results in an isentropic process. Because we assume that $\rho\Gamma$ is constant, the temperature during this isentropic process is given by Eq. (4.92):

$$\frac{\vartheta}{\vartheta_0}\frac{\vartheta_0}{\vartheta_+} = e^{\Gamma_0(1-F)}e^{-\Gamma_0(1-F_+)},$$

$$\frac{\vartheta}{\vartheta_+} = e^{\Gamma_0(F_+-F)}.$$

Because the rarefaction wave follows a reversible process, the pressure is

$$p = p_c + \rho_0\Gamma_0 c_v \vartheta.$$

In Figure 4.35 we show a graph of these results for the rarefaction wave. Consider a material point in the body when $t = 0$. This point is in front of the shock wave. Initially the pressure is zero and the temperature is 295 K. Let us assume that the shock reaches this point before the rarefaction wave catches the shock wave. When this material point encounters the shock wave, its pressure and temperature jump to $p = 105$ GPa and $\vartheta = 2{,}000$ K. These conditions are illustrated by the right terminus of the graph. At a later time, when the material point encounters the rarefaction wave, the resulting isentropic process causes the temperature and pressure to move along the curve to the left until the pressure returns to

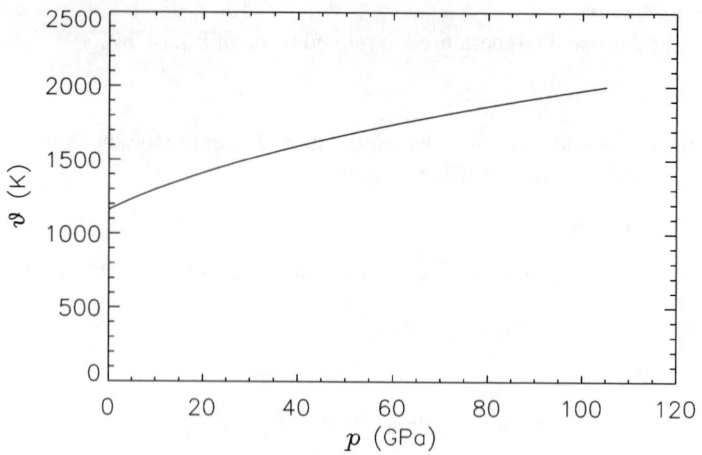

Figure 4.35 Temperature ϑ versus pressure p for the rarefaction wave.

zero and the temperature becomes 1,164 K. For comparison, we notice that at zero pressure, the melting point of stainless steel is about 1,690 K.

We can determine the material velocity of the rarefaction wave with Eq. (4.94),

$$C = \left(\frac{C^C + \rho_0 \Gamma^2 c_v \vartheta}{\rho_0 F} \right)^{1/2},$$

where C^C is given by Eq. (4.124). We determine the velocity of the material during the isentropic process from the Riemann integral for a simple wave [see Eq. (2.113) in Section 2.6.1]:

$$\begin{aligned} v - v_0 - (v^+ - v_0) &= -\int_1^F C(\bar{F})\, d\bar{F} + \int_1^{F_+} C(\bar{F})\, d\bar{F}, \\ v - v^+ &= -\int_{F_+}^F C(\bar{F})\, d\bar{F}. \end{aligned} \tag{4.126}$$

We numerically integrate this expression. In Figure 4.36 we show graphs of C and v versus the pressure p for the rarefaction wave. Before the pressure in the rarefaction wave reaches zero, the material velocity v changes sign. Recall that when the pressure returns to zero, the temperature is 1,164 K. This elevated temperature causes thermal expansion of the body that results in the reversal of the velocity of the material. We also show the shock wave velocity in this graph as a dotted line. Notice that part of the rarefaction wave propagates faster than the shock wave. Indeed, the velocity of the rarefaction wave at the Hugoniot state is $C_+ = 10.1$ km/s. Recall that the subsonic condition for a shock wave is [see Eq. (3.109)]

$$C_+ > U_S.$$

We also remember that the expression for the velocity of a weak shock is [see Eq. (2.166)]

$$U_S \approx \frac{1}{2}(C_+ + c_0) = \frac{10.1 + 4.58}{2} = 7.35 \text{ km/s},$$

which is in close agreement with the actual value of $U_S = 7.30$ km/s. This suggests that we

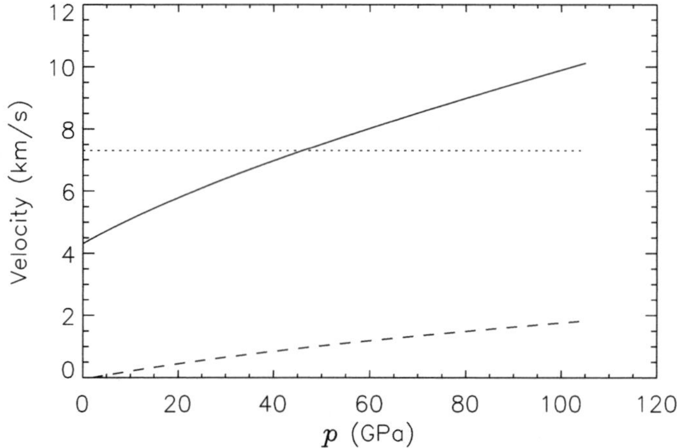

Figure 4.36 Comparison of the material wave velocity C (—), the material velocity v (– –), and the shock wave velocity (· · ·) for the square wave.

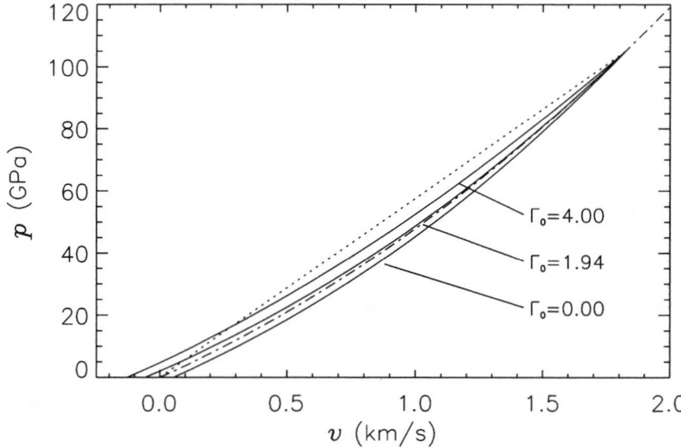

Figure 4.37 The p–v diagrams for the square wave. The isentrope for the rarefaction wave is illustrated for three values of Γ_0 (—). The Hugoniot curve (– · –) and the Rayleigh line (· · ·) are also shown.

can approximate this shock wave as a weak shock. You may find this somewhat surprising, considering the large value of the Hugoniot pressure. Recall that for a weak shock, the release path followed by a rarefaction wave can be approximated by the Hugoniot. We illustrate the Hugoniot curve and the release path resulting from this square wave in the p–v diagram in Figure 4.37. To illustrate the influence of the Grüneisen parameter, the comparsion is illustrated for three values of Γ_0. When $\Gamma_0 = 0$, the Mie–Grüneisen model reduces to a nonlinear-elastic material, and we notice that the isentrope for the rarefaction wave is slightly steeper than the Hugoniot curve. When $\Gamma_0 = 1.94$, the Hugoniot is steeper than the isentrope, but indeed, we find that the Hugoniot is a reasonable approximation of the isentrope.

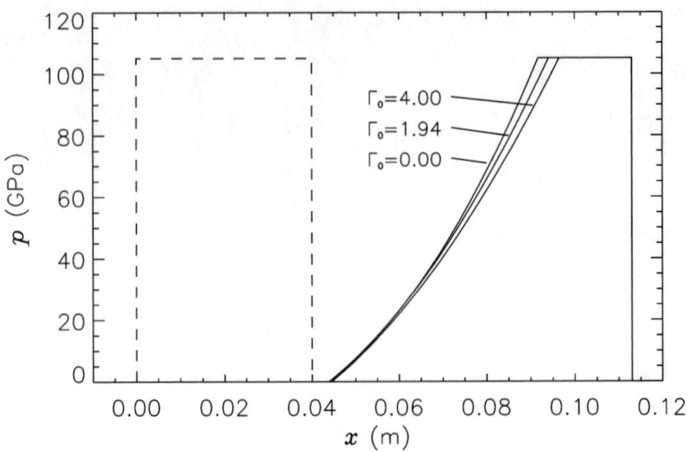

Figure 4.38 Profiles of the square wave at $t = 0$ (– –) and $t = 10\ \mu$s (—). Three values of the Grüneisen parameter are shown.

Wave Profiles

At $t = 0$, let us assume the rarefaction wave is located at $x = 0$ and the shock wave is located at $x = 0.04$ m. Both waves propagate to the right. At time t, the location of the shock wave is

$$x = U_S t + 0.04,$$

and the location of the rarefaction wave is

$$x(F) = [FC(F) + v(F)]t, \quad 0.75 \leq F \leq 1,$$

where we recall that the spatial wave velocity is $c = FC$. We show the resulting profiles for $t = 0$ and $t = 10\ \mu$s in Figure 4.38. We use three values of Γ_0 to illustrate the influence of Γ_0 on the profile of the wave. At later times, the rarefaction wave will overtake and attenuate the shock wave. Larger values of Γ_0 cause more rapid attenuation of the shock wave. Because our solution for the rarefaction wave depends upon the simple-wave solution of Eq. (4.126), it is not valid beyond the time when the rarefaction wave contacts the shock wave. However, we have found that the weak-shock assumption is valid. Hence interaction of the rarefaction wave with the shock wave does not result in a reflected wave of significant amplitude. Thus it is reasonable to continue to use the simple-wave solution for the rarefaction wave past the time of contact.

In these profiles, we have illustrated the shock wave as a discontinuous jump in pressure. We know that it is actually a steady wave of finite thickness. We can estimate the thickness of the steady wave using the data of Wise and Mikkola†. Using five measurements over pressures between 9.6 and 19.6 GPa, they estimate that

$$A' = 71\ (\text{GPa}^4\text{-s})^{-1}. \tag{4.127}$$

† See "Hugoniot and wave-profile measurements on shock-loaded stainless steel (21Cr-6Ni-9Mn)" by Wise, J. L. and Mikkola, D. E. in *Shock Waves in Condensed Matter 1987*, ed. Schmidt, S. C. and Holmes, N. C., pp. 261–264 (Elsevier Science, 1988).

Thus

$$\left|\frac{\partial \hat{F}}{\partial t}\right|_M = 71 p_+^4,$$

where p_+ is expressed in GPa. Using our value of $p_+ = 105$ GPa, we obtain

$$\left|\frac{\partial \hat{F}}{\partial t}\right|_M = 8.63 \times 10^9.$$

The time Δt required for F to change by 0.25 is

$$\Delta t = 0.25/8.63 \times 10^9 = 28.9 \times 10^{-12} \text{ s}.$$

The thickness Δx of the steady wave is

$$\Delta x = U_S \Delta t = 7{,}300 \times 28.9 \times 10^{-12} = 0.2\ \mu\text{m}.$$

This number is approximately 1,000 times greater than the interatomic distance between the atoms in stainless steel.

4.2.7 *Exercises*

4.15 The cold energy of a hypothetical Mie–Grüneisen solid is

$$\mathcal{E}_C = \frac{E_0}{\rho_0}(F - \ln F).$$

The specific heat is constant so that $c_v = c_\eta$.

(a) Determine the equilibrium expression for the cold pressure.

(b) Determine Γ using Eq. (4.84).

(c) Determine the cold stiffness, isothermal stiffness, and isentropic stiffness.

(d) Determine the coefficient of thermal expansion α.

4.16 The average interatomic forces in our simple lattice model are given by Eq. (4.82). Verify this result.

4.17 We estimate the oscillation frequency of an atom in Eq. (4.81). For a typical metal we find that $dp_c/dF \approx -4 \times 10^{11}$ N/m^2, $\rho_0 \approx 4 \times 10^3$ kg/m^3, and $\Delta X \approx 10^{-10}$ m. Estimate the frequency of oscillation of the atoms in such a material.

4.18 Consider a Mie–Grüneisen material in which $\rho\Gamma$ and c_v are constants.

(a) Demonstrate that

$$\alpha C^C = \alpha_0 C_0^C.$$

(b) Along an isobaric process, use Eq. (4.93) to demonstrate that

$$p_c(\rho) - p_c(\rho_0) = \alpha_0 C_0^C(\vartheta_0 - \vartheta).$$

(c) Use a Taylor's series expansion about the point $F = 1$ to verify that

$$F - 1 = \alpha_0(\vartheta - \vartheta_0)$$

on the isobar in the neighborhood of $F = 1$.

4.19 Verify Eq. (4.114).

4.20 Verify Eq. (4.124).

4.21 For the square wave in stainless steel discussed in Section 4.2.6 we determined the temperature ϑ_+ at the Hugoniot state by assuming that $\rho\Gamma$ was constant.

(a) Determine this temperature when Γ is constant.

(b) Suppose the stainless steel is compressed along an isentrope rather than along a Rayleigh line. Determine the temperatures at $F = 0.75$, assuming first that Γ is constant and then that $\rho\Gamma$ is constant.

4.22 Consider a block of aluminum with the following properties:

$$\lambda = 57.4\,\text{GPa}, \qquad \beta = 25 \times 10^{-6}\,\text{K}^{-1},$$
$$\mu = 27.8\,\text{GPa}, \qquad \rho_0 = 2.78\,\text{Mg/m}^3,$$
$$c_p = 900\,\text{J/kg/K}.$$

This material is at rest. It has a temperature of 295 K.

(a) Verify that the isentropic and isothermal stiffnesses are 75.9 GPa and 72.3 GPa, respectively.

(b) Verify that the bulk velocity and the speed of sound are 5.23 km/s and 6.37 km/s, respectively.

(c) Verify that the Grüneisen parameter Γ_0 as determined by Eq. (3.66) is 2.28.

4.23 In Section 4.2.6 we calculated the response of stainless steel to a square wave. Repeat this calculation, but this time use the material properties for aluminum that are given in the previous exercise. Use the linear-Hugoniot model with $s = 1.29$. Assume that $\rho\Gamma$ and c_v are constant. Let $\Gamma_0 = 2.28$. At $t = 0$, the thickness of the square wave is 0.04 m, and the deformation gradient behind the shock wave is $F_+ = 0.75$.

(a) Determine the pressure and temperature at the Hugoniot state.

(b) Determine the temperature behind the rarefaction wave where the pressure is zero.

(c) Will this material melt during this process?

4.24 We show the spatial representation of the pressure profiles of a square wave in stainless steel in Figure 4.38. Construct these profiles in the material representation. Remember that initially the shock wave is *not* located at $X = 0.04$ m.

4.3 Elastic-Plastic Solid

In the previous section we found that interatomic forces bind each atom of a solid to its nearest neighbors. Thus a solid can resist loads that produce deviatoric stresses. This ability to resist shear stress is a feature that distinguishes a solid from a liquid. However, this ability is limited. All solids have an upper bound to the magnitude of the deviatoric stress they can support. A typical limit for a structural steel is approximately 0.4 GPa. If we try to exceed this limit, the solid will either fail or flow like a liquid. We made use of this fact in the previous section. There we limited our discussion to waves that resulted in very large compressive stresses. Under these conditions, we could ignore the deviatoric components of stress in comparison to the mean normal stress $\sigma = \frac{1}{3}T_{kk}$. This greatly simplified the analysis. However, when the magnitude of σ is comparable to the deviatoric stress, we must include the effects of the deviatoric stresses in our analysis. One way of achieving this goal is to employ the *theory of metal plasticity*. This is the topic of this section.

As its name implies, metal plasticity is a theory that is most appropriately applied to the description of metals. Much work has been directed towards linking macroscopic observations of forces and deflections to microscopic models of the atomic structure of metals. The most notable efforts involve the *theory of dislocations*. We shall discuss some of this work in this section. With varying degrees of success, theories based on metal plasticity have also been developed for other types of materials such as polymers and geological materials.

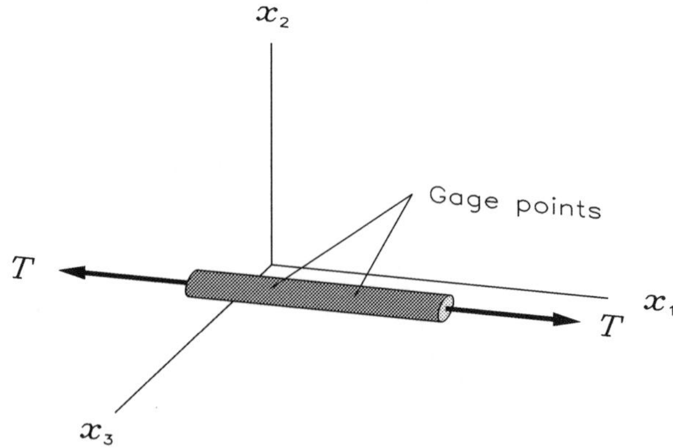

Figure 4.39 A tensile-bar test. The axial stress T is typically induced by a large hydraulic apparatus. The lateral stress $\tilde{T}$ is assumed to be zero.

4.3.1 *Uniaxial Stress*

Metal plasticity is most commonly characterized with a tensile-bar test. In this test, we employ an apparatus for applying a large axial load to a long slender rod (Figure 4.39). We determine the axial stress $T = T_{11}$, which is applied to the rod, by measuring the force produced by the apparatus and the cross-sectional area of the rod. We assume that all other components of stress are zero, including the lateral components $\tilde{T} = T_{22} = T_{33} = 0$. This is a uniaxial configuration of stress, which we have already studied in Chapter 1. We recall that the deviatoric components of such a stress configuration are [see Eq. (1.124) in Section 1.7.2]

$$\left[T'_{km}\right] = \begin{bmatrix} \frac{2}{3}T & 0 & 0 \\ 0 & -\frac{1}{3}T & 0 \\ 0 & 0 & -\frac{1}{3}T \end{bmatrix} \tag{4.128}$$

and the invariants of the deviatoric stress tensor are [see Eq. (1.125)]

$$I_{T'} = 0, \qquad II_{T'} = \frac{1}{3}T^2, \qquad III_{T'} = \frac{2}{27}T^3.$$

The maximum value of the shear stress is [see Eq. 1.129)]

$$\tau_{\max} = \frac{1}{2}|T|. \tag{4.129}$$

We can estimate the deformation gradient $F = F_{11}$ by measuring the distance between two marked *gage points* on the rod. As the rod is stretched by the apparatus, we can divide the current distance between the two gage points Δx by the original distance between the two gage points ΔX to obtain

$$F = \frac{\Delta x}{\Delta X}.$$

We show a hypothetical graph of the measurements of T versus $F - 1$ in Figure 4.40. Tests on several identical rods of the same material are illustrated. In the first test, we load a rod from point A to point B. When we remove the load, the rod returns to point A. This

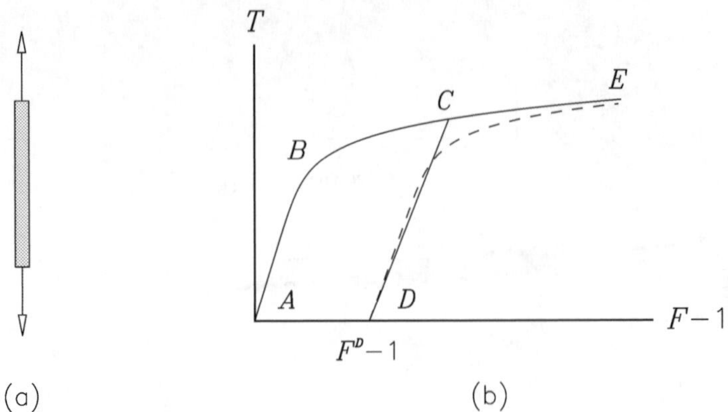

Figure 4.40 A tensile bar (a) in which $T > 0$ and $\tilde{T} = 0$. The stress-deformation gradient curve (b) illustrates the initial elastic deformation followed by plastic deformation.

represents the response of an elastic material. In the next test, we load the rod with a larger force from point A past point B to point C. When we remove this load, the rod does not return to point A. Instead, it moves to point D where $F = F^D$. Thus, by pulling harder, we have permanently stretched the rod. The original distance between the gage points has increased from ΔX to $\Delta x = F^D \Delta X$. This permanent stretch is called *plastic deformation*. Now suppose we reload the same rod. It will follow the dashed line from point D to point E. We can also reach point E by starting with yet another rod at point A and continuously stretching it along the path through points B and C to point E. Of course, if we continue to stretch the rod past point E, then ultimately, the material in the rod will break into two or more pieces.

The material behavior we illustrate in Figure 4.40 is very complicated. Let us approximate this behavior with a simple model, which we call the *elastic-plastic material*. The elastic-plastic material obeys a simple set of rules. We assume that the rod exhibits a property called the *yield stress* Y, and then we require that the magnitude of T cannot exceed Y:

$$T^2 \le Y^2.$$

This is called a *yield condition*. The rod is elastic when the magnitude of T is smaller than Y. When $T^2 = Y^2$, we assume that the rod deforms without any additional resistance to the applied stress. This ensures that

$$II_{T'} \le \frac{1}{3} Y^2, \qquad \tau_{\max} \le \frac{1}{2} Y, \tag{4.130}$$

where our reasons for emphasizing the second invariant $II_{T'}$ will become apparent later. We show a graph of T versus $F - 1$ in Figure 4.41a. Let us see how a rod of the elastic-plastic material behaves as we load it in the apparatus. For example, suppose we load it so that $T^2 < Y^2$. In this case the rod moves between points A and B. The deformation of the rod is elastic, and when we release the load, the rod returns to point A. Here we recover the deformation experienced by the rod when we unload the rod. This type of deformation is called *elastic deformation*. Now suppose we reload the rod and deform it to point C. This time, the elastic–plastic rod resists the deviatoric stresses until we pass point B; then the rod deforms without additional resistance. The point B is called the *yield point* of the

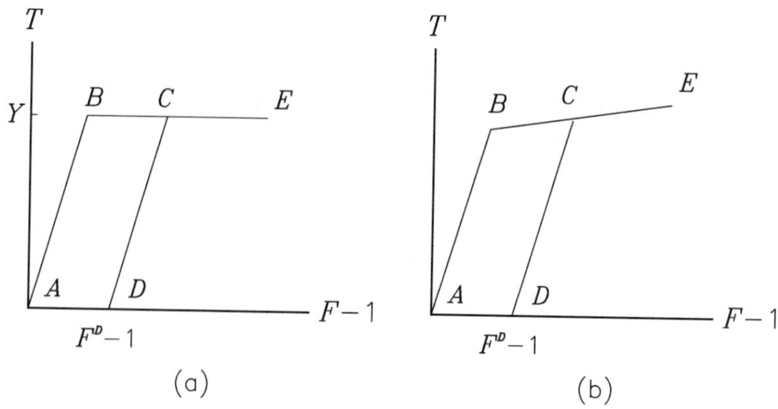

Figure 4.41 The response of an elastic-plastic rod during loading with a uniaxial stress (a) without work hardening and (b) with work hardening.

material. The additional deformation, which the rod experiences while moving from point B to point C, is called *plastic deformation*. The deformation gradient at point C is composed of both an elastic portion and a plastic portion. We can determine the plastic portion of the deformation gradient at point C by releasing the applied load and allowing the rod to move to point D. Because the rod is elastic along the path from point C to point D, the plastic portion of the deformation gradient does not change as we release the load, and the value $F = F^D$ that we measure at point D represents the plastic deformation at every point on the path connecting points C and D. When we reload the rod from point D, it returns along the same path from D to C and behaves like an elastic material until it passes point C. As we stretch it from point C to point E, we produce additional plastic deformation.

In Figure 4.41a the yield stress is a constant and independent of the amount of plastic deformation. Relatively few materials exhibit this type of behavior. In contrast, in Figure 4.41b, we illustrate a different situation. Here the yield stress Y is not constant. As we load the rod past point B and produce plastic deformation, the yield stress Y increases. Thus the value of Y depends on the amount of plastic deformation. A material with a higher value of Y is said to be harder. Because work is required to plastically deform the rod and increase Y, this effect is called *work hardening*.

4.3.2 *Uniaxial Deformation*

Using the tensile-bar test, we have illustrated plastic deformation under conditions of uniaxial stress, but one-dimensional structured waves and shock waves result in uniaxial deformation and not uniaxial stress. We shall now examine how plastic deformation arises during uniaxial deformation. Suppose we compress an elastic-plastic body while constraining the lateral motion (see Figure 4.42). The deviatoric components of the applied stress are [see Eq. 1.124)]

$$\left[T'_{km}\right] = \begin{bmatrix} \frac{2}{3}(T - \tilde{T}) & 0 & 0 \\ 0 & -\frac{1}{3}(T - \tilde{T}) & 0 \\ 0 & 0 & -\frac{1}{3}(T - \tilde{T}) \end{bmatrix}.$$

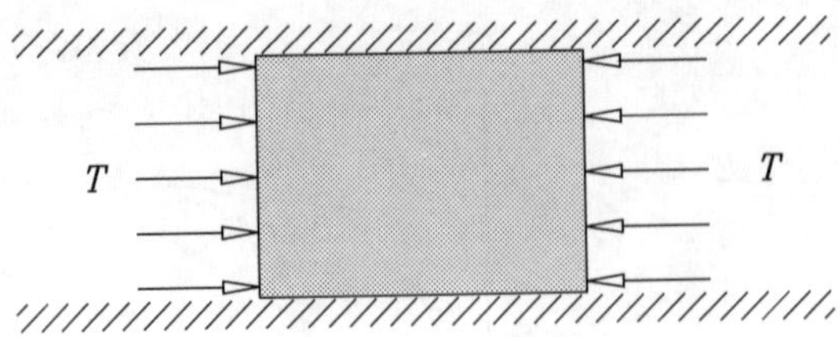

Figure 4.42 An elastic-plastic body during uniaxial deformation.

The invariants of this deviatoric stress tensor are [see Eq. (1.125)]

$$I_{T'} = 0, \qquad II_{T'} = \frac{1}{3}(T - \tilde{T})^2, \qquad III_{T'} = \frac{2}{27}(T - \tilde{T})^3,$$

and the maximum value of the shear stress is [see Eq. (1.129)]

$$\tau_{\max} = \frac{1}{2}|T - \tilde{T}|.$$

The mean normal stress is

$$\sigma = \frac{1}{3}(T + 2\tilde{T}).$$

In Figure 4.42 we have illustrated a compressive load ($T < 0$). During this type of loading, the magnitude of the applied compressive stress T can be increased without limit. Under these conditions, the magnitude of the mean normal stress σ will continue to increase, but the deviatoric stresses will reach limiting values. For uniaxial deformation, we modify the yield condition for the elastic-plastic material. Here we require that

$$(T - \tilde{T})^2 \leq Y^2. \tag{4.131}$$

Notice that the tensile-bar yield condition $T^2 \leq Y^2$ also obeys this rule, because $\tilde{T} = 0$ for uniaxial stress. As a result of Eq. (4.131), during uniaxial deformation, the second invariant $II_{T'}$ and the maximum shear stress $\tau_{\max}$ obey

$$II_{T'} \leq \frac{1}{3}Y^2, \qquad \tau_{\max} \leq \frac{1}{2}Y,$$

which are identical to the tensile-bar conditions Eq. (4.130)]. We show a graph of T and σ versus $1 - F$ in Figure 4.43. Here we illustrate an elastic-plastic material that is compressed from point A to point C. The solid line represents the axial stress T; the dashed line represents the mean normal stress σ. Notice the slope of the solid curve is discontinuous at point B. This is the yield point of the material. Between points A and B the material experiences elastic deformation, and between point B and C, the material experiences both elastic and plastic deformation. The axial component of the deviatoric stress T'_{11} is given by

$$T'_{11} = T - \sigma = \frac{2}{3}(T - \tilde{T}),$$

which is the vertical distance between the solid line and the dashed line. As we load the material from point B to point C, this distance is constant and equal to $\frac{2}{3}Y$. This indicates that the yield stress of the material is constant, and work hardening is not present. After

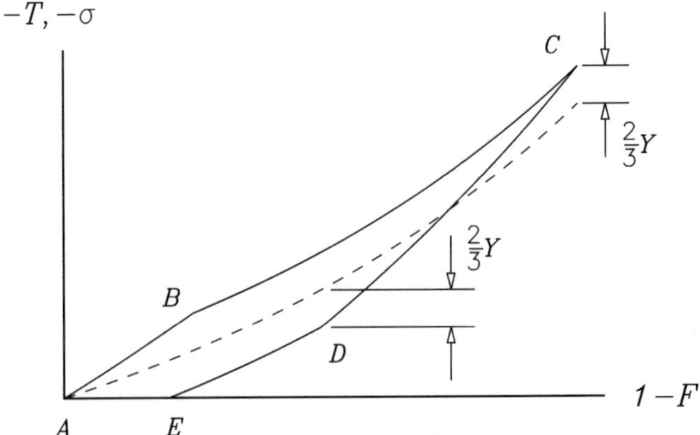

Figure 4.43 The uniaxial compression of an elastic-plastic body. The mean normal stress (– –) and the axial stress T (—) are shown.

we reach point C, we then slowly remove the load. Then the material follows the solid line from point C past point D to point E. Point D is another yield point. Between points C and D, the deformation is elastic. Along this portion of the unloading path, the deviatoric stress changes from $\frac{2}{3}Y$ to $-\frac{2}{3}Y$. After we pass point D, the material again experiences plastic deformation until we completely remove the applied stress T and reach point E.

Curvature of the Uniaxial Response

Recall that in Chapter 2, for a nonlinear-elastic material, we impose two conditions upon the shape of the curve relating the stress T to the deformation gradient F [see Eqs. (2.22) and (2.23)]:

$$\frac{dT}{dF} > 0, \qquad \frac{d^2T}{dF^2} \leq 0.$$

Let us compare these conditions directly to the curves in Figure 4.43. It is easily demonstrated that these conditions yield

$$\frac{d(-T)}{d(1-F)} > 0, \qquad \frac{d^2(-T)}{d(1-F)^2} \geq 0,$$

which means that the slope of the curve must be positive and the concave side of the curve must face up. Now consider point B in Figure 4.43. The loading curve, which extends from point A to point C, is discontinuous and its slope and curvature are not defined at point B. This discontinuity is an artifact of the elastic-plastic model. We need only compare Figure 4.40 to Figure 4.41 to understand this fact. Indeed, the actual response of the material is represented by a smooth curve at point B, which violates the inequality on the second derivative, so that

$$\frac{d^2T}{dF^2} = -\frac{d^2(-T)}{d(1-F)^2} > 0. \tag{4.132}$$

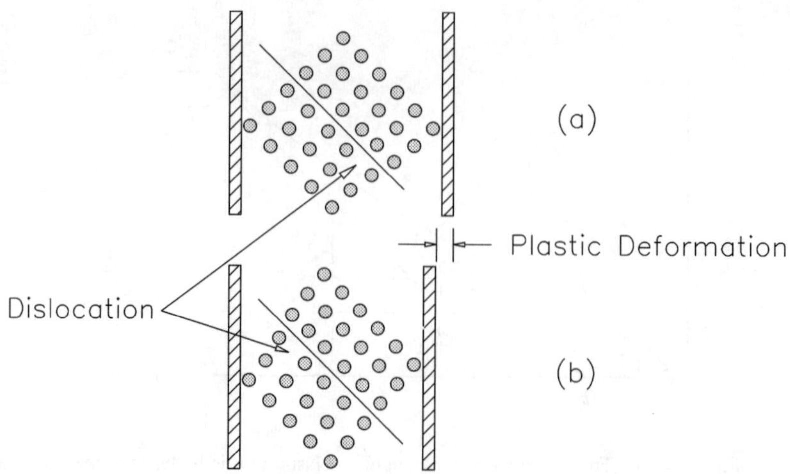

Figure 4.44 A lattice of atoms (a) before and (b) after plastic deformation. An imaginary slip plane (—) separates the lattice rows containing six atoms from the rows containing five atoms. Plastic deformation results when the dislocation moves along the imaginary slip plane.

The sign of $\partial^2 T/\partial F^2$ has been a central issue in determining the existence of shock waves and steady waves [see Eq. (3.120) in Section 3.4.5]. Thus Eq. (4.132) suggests that these waves cannot exist in the region of point B. We shall return to this issue later in this section.

4.3.3 *An Imperfect Lattice*

There is a simple and elegant explanation of metal plasticity that was first presented by Taylor in 1934[†]. Metals are a polycrystalline material. You might first suppose that each crystal is a lattice of atoms in perfect alignment. However, Taylor noticed that imperfections, which lower the value of the yield stress, are always present in crystals. He called these imperfections *dislocations*. To illustrate, let us consider a single crystal with a single dislocation (see Figure 4.44). In the top image, the crystal is held between two rigid plates. An imaginary slip plane passes through the crystal at the position of the dislocation. Above the slip plane, we have drawn six atoms in each row of the crystal. Below the slip plane, we have drawn five atoms in each row. The dislocation is evident as a misalignment of atoms, which otherwise would form a perfect cubic array. Taylor demonstrated that this misalignment causes a weakening of the interatomic forces in the crystal.

Now suppose we compress the crystal between the rigid plates. At first we will deform the lattice without causing any realignment of the relative positions of the atoms. This represents elastic deformation of the crystal. However, if we continue to increase the compression, eventually the bonds between some of the atoms will fail. Because the interatomic forces are weakest near the dislocation, these bonds will break first. The bottom image shows what happens. The atoms rearrange themselves so that the dislocation moves up and to the left along the imaginary plane. If we remove the compressive force, the crystal will remain in a permanently altered configuration. This represents plastic deformation of the crystal.

[†] See "The mechanism of plastic deformation of crystals" by Taylor, G. I. in *Proceedings of the Royal Society* **145**, pp. 362–387 (1934).

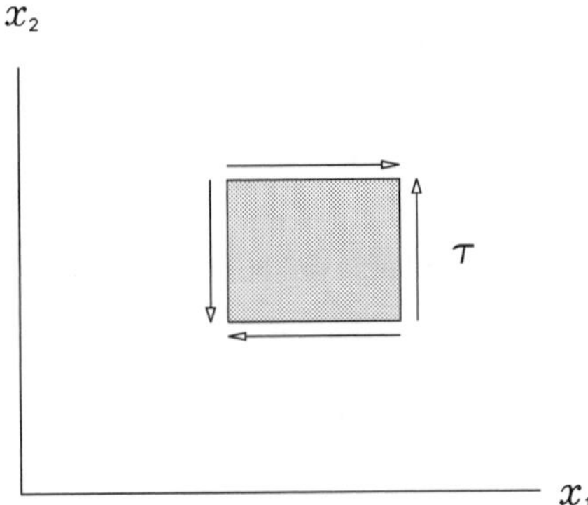

Figure 4.45 A block of crystalline material subjected to pure shear where $T_{12} = T_{21} = \tau$.

4.3.4 *Pure Shear*

With our discussion of the crystal lattice, we have demonstrated that deviatoric stresses cause dislocations to propagate through a crystal. In fact, Taylor introduced these concepts by analyzing a crystal under the influence of shear stress alone. Such a configuration of stress is called *pure shear*. We have demonstrated that a single yield condition, Eq. (4.131), can be used to describe the response of an elastic-plastic solid to both the tensile-bar test and uniaxial deformation. We shall now examine the yield condition for pure shear.

Suppose we subject a metal composed of many crystals to the following configuration of pure shear [see Figure 4.45]:

$$[T_{km}] = \left[T'_{km}\right] = \begin{bmatrix} 0 & T_{12} & 0 \\ T_{12} & 0 & 0 \\ 0 & 0 & 0 \end{bmatrix}.$$

The second invariant $II_{T'}$ and the maximum shear stress $\tau_{\max}$ are

$$II_{T'} = T_{12}^2, \qquad \tau_{\max} = T_{12}. \tag{4.133}$$

The yield condition for pure shear is

$$T_{12}^2 \le k^2,$$

where k is a constitutive property of the material. Thus

$$II_{T'} \le k^2, \qquad \tau_{\max} \le k. \tag{4.134}$$

Notice that $I_{T'} = III_{T'} = 0$. It is inconsistent to have different yield conditions for different configurations of stress. Thus we must establish a connection between the yield condition Eq. (4.131), in which we use the yield stress Y, and the yield condition Eq. (4.134), in which we use the yield stress k. We shall examine two possible choices.

Tresca Yield Condition
Suppose we replace the two yield conditions, Eqs. (4.131) and (4.134), by a yield condition that allows plastic deformation when $\tau_{\max}$ reaches a limiting value. Then

$$\tau_{\max} \leq \frac{1}{2}Y$$

is the yield condition for both the tensile-bar test and uniaxial deformation, and

$$\tau_{\max} \leq k$$

is the yield condition for pure shear. These two equations represent the same yield condition provided that

$$k = \frac{1}{2}Y.$$

This is called the *Tresca yield condition.*

Mises Yield Condition
Suppose we replace the two yield conditions, Eqs. (4.131) and (4.134), by another yield condition that allows plastic deformation when $II_{T'}$ reaches a limiting value. Then

$$II_{T'} \leq \frac{1}{3}Y^2$$

is the yield condition for both the tensile-bar test and uniaxial deformation, and

$$II_{T'} \leq k^2 \tag{4.135}$$

is the yield condition for pure shear. These two equations represent the same yield condition provided that

$$k = \frac{1}{\sqrt{3}}Y.$$

This is called the *Mises yield condition.* With respect to metals, when we compare the Tresca yield condition and the Mises yield condition, we usually find that the Mises yield condition is in better agreement with experimental data.

4.3.5 *Decomposition of Motion*

In the previous sections we demonstrated how the movements of dislocations cause crystals to plastically deform. In this section we illustrate how this phenomenon can be modeled by decomposition of the motion into elastic and plastic parts. Recall that the motion connects the current position x_k of material point $\mathbf{X}$ to the reference position X_k [see Eq. (1.37) in Section 1.2.1]:

$$x_k = \hat{x}_k(X_m, t).$$

The displacement of material point $\mathbf{X}$ is

$$u_k = x_k - X_k.$$

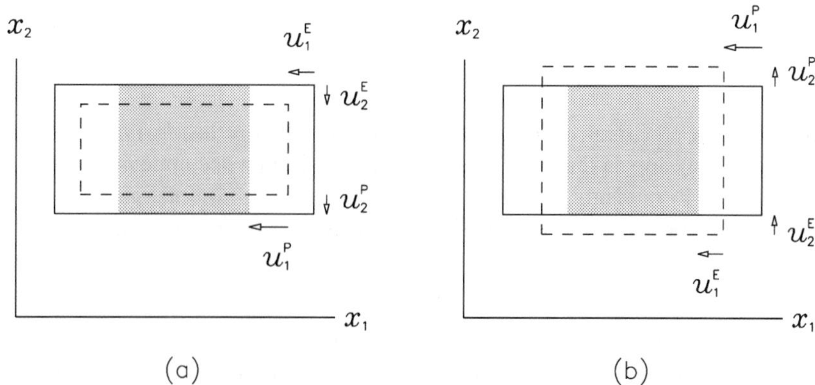

Figure 4.46 The decomposition of motion into an elastic part and a plastic part: (a) the elastic-first decomposition; (b) the plastic-first decomposition.

To represent an elastic-plastic material, we decompose the motion into an elastic motion and a plastic motion. Two possible ways of doing this are illustrated in Figure 4.46. Initially, let us concentrate on the decomposition in Figure 4.46a. Here we illustrate a block of material that we have constrained to undergo uniaxial compression. Thus the only nonzero component of displacement is u_1. We illustrate the undeformed reference configuration of the material with a solid line. We illustrate the current configuration with a shaded area. The block is transformed from the reference configuration to the current configuration in two steps. In the first step, an elastic deformation transforms the block from the reference configuration to the *intermediate configuration*. We use a dashed line to represent the shape of the block in the intermediate configuration. Here u_1^E and u_2^E denote the components of the displacements between the reference configuration and the intermediate configuration. We do not require the intermediate configuration to be one of uniaxial deformation. Thus both axial and lateral components of the elastic motion are present. In the second step of the decomposition of motion, a plastic deformation transforms the block from the intermediate configuration to the current configuration. The lateral component u_2^P of the plastic displacement cancels the lateral component u_2^E of the elastic displacement. The combination of these two motions results in uniaxial compression. Mathematically, we represent this decomposition of the motion as follows:

$$x_k^E = \hat{x}_k^E(X_m, t),$$

$$x_j = \check{x}_j^P\left(x_k^E, t\right).$$

Here x_k^E defines the intermediate configuration. We call this the *elastic-first decomposition*. In general, x_k^E does not represent a physical configuration. Indeed, we see that the intermediate configuration in the illustration is not physically achievable within the constraints of uniaxial deformation. Also, the ordering of the transformations does not imply that elastic deformation precedes plastic deformation. Both plastic and elastic deformations can occur simultaneously.

Figure 4.46b illustrates an alternate method of decomposing the same motion. In this case the plastic transformation precedes the elastic transformation. We represent this decomposition of the motion by

$$x_k^P = \hat{x}_k^P(X_m, t),$$
$$x_j = \check{x}_j^E\left(x_k^P, t\right).$$

Here x_k^P defines the intermediate configuration. We call this the *plastic-first decomposition*. Although it may appear that the elastic-first and plastic-first decompositions lead to different mathematical formalisms, we shall see that as long as we confine our analysis to uniaxial deformation, both decompositions of motion lead to identical formulations.

4.3.6 *Decomposition of Deformation Gradient*

To understand why the ordering in the decomposition of motion is not important, let us evaluate the deformation gradients for each decomposition. We begin with the elastic-first decomposition. By applying the chain rule to the definition of the deformation gradient, we obtain

$$F_{mn} = \frac{\partial \hat{x}_m}{\partial X_n} = \frac{\partial \check{x}_m^P}{\partial x_k^E}\frac{\partial \hat{x}_k^E}{\partial X_n}.$$

Next we define

$$\check{F}_{mk}^P \equiv \frac{\partial \check{x}_m^P}{\partial x_k^E}, \qquad F_{kn}^E \equiv \frac{\partial \hat{x}_k^E}{\partial X_n},$$

so that

$$F_{mn} = \check{F}_{mk}^P F_{kn}^E. \tag{4.136}$$

We can also begin with the plastic-first decomposition, and use an identical procedure to obtain

$$F_{mn} = \check{F}_{mk}^E F_{kn}^P, \tag{4.137}$$

where

$$\check{F}_{mk}^E \equiv \frac{\partial \check{x}_m^E}{\partial x_k^P}, \qquad F_{kn}^P \equiv \frac{\partial \hat{x}_k^P}{\partial X_m}.$$

Recall that for uniaxial deformation

$$[F_{mn}] = \begin{bmatrix} F & 0 & 0 \\ 0 & 1 & 0 \\ 0 & 0 & 1 \end{bmatrix}.$$

Suppose we also require the elastic and plastic deformation gradients to have the following forms:

$$\left[F_{mk}^P\right] = \left[\check{F}_{mk}^P\right] = \begin{bmatrix} F^P & 0 & 0 \\ 0 & \tilde{F}^P & 0 \\ 0 & 0 & \tilde{F}^P \end{bmatrix},$$

$$\left[F_{mk}^E\right] = \left[\check{F}_{mk}^E\right] = \begin{bmatrix} F^E & 0 & 0 \\ 0 & \tilde{F}^E & 0 \\ 0 & 0 & \tilde{F}^E \end{bmatrix}.$$

Because these forms of the deformation gradients have nonzero components only on the major diagonal, the order of multiplication on the right sides of Eqs. (4.136) and (4.137) does not affect the result. Indeed, both the elastic-first and plastic-first compositions satisfy the following relationships:

$$F = F^E F^P, \qquad 1 = \tilde{F}^E \tilde{F}^P. \tag{4.138}$$

Thus we can proceed with our study of the elastic-plastic model without concern for which of the two decompositions we are using. However, we should notice that in more general types of deformation, $F^E_{mk} \neq \breve{F}^E_{mk}$ and $F^P_{mk} \neq \breve{F}^P_{mk}$. In these cases, the method that we use to decompose the motion is important.

Bulk Response

In metal plasticity, we assume that a state of spherical deformation cannot result in plastic deformation. We shall now demonstrate how this assumption constrains the values of the components of the plastic deformation gradient. The volume strain is [see Eq. (1.80) in Section 1.4.5]

$$e = \frac{dV - dV_0}{dV_0} = III_F - 1,$$

which yields

$$e = F^E (\tilde{F}^E)^2 F^P (\tilde{F}^P)^2 - 1.$$

When the material is in the reference configuration, we know that $F^P = \tilde{F}^P = 1$. If we assume that e cannot be a function of the plastic deformation, we must conclude that the product $F^P(\tilde{F}^P)^2$ cannot change. Therefore, we find that

$$F^P (\tilde{F}^P)^2 = 1 \tag{4.139}$$

holds not only for the reference configuration, but for all configurations. This is not always an appropriate assumption. For example, if the material has a significant amount of porosity, the application of a spherical stress can crush the pores and produce a permanent change in volume. We discuss this possibility in Section 4.4.

Decomposition of Deformation in Linear Elasticity

Recall that when we consider a body in uniaxial deformation, the elastic and plastic components of the deformation gradient assume the form of triaxial deformation. Thus the components of the elastic Lagrangian strain are [see Eqs. (1.98) in Section 1.5.1]

$$E^E = \frac{1}{2}[(F^E)^2 - 1], \qquad \tilde{E}^E = \frac{1}{2}[(\tilde{F}^E)^2 - 1].$$

In linear elasticity $F^E = 1 + \Delta F^E$, where ΔF^E is infinitesimal. It is easily demonstrated that $(F^E)^2 \approx 2\Delta F^E + 1$. Using this approximation, we obtain

$$E^E = F^E - 1, \qquad \tilde{E}^E = \tilde{F}^E - 1.$$

Similarly,

$$E^P = F^P - 1, \qquad \tilde{E}^P = \tilde{F}^P - 1,$$

and

$$E = F - 1, \qquad \tilde{E} = 0.$$

Therefore, Eqs. (4.138) become

$$E = E^E + E^P, \qquad 1 = \tilde{E}^E + \tilde{E}^P,$$

where we have ignored the terms containing products of strain. When a similar procedure is applied to Eq. (4.139), we obtain

$$E^P + 2\tilde{E}^P = 0.$$

Notice that whereas the decomposition of the deformation gradients involves the *product* $F = F^E F^P$, the decomposition of the linear-elastic strains involves the *sum* $E = E^E + E^P$. This latter relationship is often of considerable advantage in the analysis of waves in elastic-plastic materials.

4.3.7 *Constitutive Relations*

We arbitrarily assume the following set of constitutive relationships for an elastic plastic material that undergoes one-dimensional motion:

$$\begin{aligned} \mathcal{E} &= \mathcal{E}\left(F_{11}^E, F_{22}^E, F_{33}^E, \eta\right), \\ \mathcal{T} &= \mathcal{T}\left(F_{11}^E, F_{22}^E, F_{33}^E, D_{11}^E, D_{22}^E, D_{33}^E, \eta\right), \\ D^P &= D^P\left(F_{11}^E, F_{22}^E, F_{33}^E, \eta\right), \end{aligned} \tag{4.140}$$

where the elastic and plastic deformation rates are defined to be [see Eq. (1.87)]

$$D_{ij}^E = \frac{1}{F_{ij}^E}\frac{\partial \hat{F}_{ij}^E}{\partial t} \quad \text{(no summation)},$$

$$D^P = \frac{1}{F_{11}^P}\frac{\partial \hat{F}_{11}^P}{\partial t}.$$

There are several important features of these constitutive equations that you should notice. First, the internal energy $\mathcal{E}$ is only a function of the elastic portion of the deformation gradient and the entropy. From our previous discussions in Chapter 3, we know that $\mathcal{E}$ cannot be a function of the deformation rate. We also know that we can obtain the temperature and the equilibrium stresses from the internal energy. Thus, for example, the temperature is

$$\vartheta = \frac{\partial \mathcal{E}}{\partial \eta}. \tag{4.141}$$

We also intend that the internal energy represent the strain-energy potential of a nonlinear-elastic material. Thus, motivated by Eq. (2.14) in Section 2.1.3 and (3.82) in Section 3.4.2, we require

$$(T_{km})_e = \rho F_{mj}^E \frac{\partial \mathcal{E}}{\partial F_{kj}^E}, \tag{4.142}$$

where $(T_{km})_e$ denotes the equilibrium values of the components of stress. From the symmetry of the stress tensor, we know that the only nonzero components are

$$\begin{aligned} T_e &= (T_{11})_e = \rho F^E \frac{\partial \mathcal{E}}{\partial F^E}, \\ \tilde{T}_e &= (T_{22})_e = (T_{33})_e = \rho \tilde{F}^E \frac{\partial \mathcal{E}}{\partial \tilde{F}^E}. \end{aligned} \tag{4.143}$$

In addition, we have prescribed a constitutive equation for the axial component of the viscous stress $\mathcal{T}$. The additive decomposition of this stress requires that

$$T_{11} = T_e + \mathcal{T}. \tag{4.144}$$

We shall find that a constitutive equation for the lateral component of dissipative stress is not required to solve the system of equations for one-dimensional motion. The last of Eqs. (4.140) is a constitutive relationship for the axial component of the plastic deformation rate. An expression of this type is often called a *plastic flow rule*. We discuss the plastic flow rule in more detail later.

It is helpful to observe that we can express the various deformation gradients in this formulation in terms of just two quantities, F and F^P. Using the expressions for the decomposition of motion, Eqs.(4.138), and the relationship for the bulk response of the plastic deformation, Eq. (4.139), we obtain the following expressions for the nonzero components of the deformation gradients:

$$\begin{aligned} F_{11}^E &= F^E = F/F^P, \\ F_{22}^E &= F_{33}^E = \tilde{F}^E = \sqrt{F^P}, \\ F_{22}^P &= F_{33}^P = \tilde{F}^P = 1/\sqrt{F^P}. \end{aligned} \tag{4.145}$$

Because the deformation gradients are always positive, we must always use the positive square root. We obtain similar relationships for the deformation rates by evaluating the derivatives of the logarithms of these equations:

$$\begin{aligned} D_{11}^E &= D^E = D - D^P, \\ D_{22}^E &= D_{33}^E = \tilde{D}^E = D^P/2, \\ D_{22}^P &= D_{33}^P = \tilde{D}^P = -D^P/2. \end{aligned} \tag{4.146}$$

When we introduce these expressions into Eq. (4.140), the constitutive relationships assume the following forms:

$$\begin{aligned} \mathcal{E} &= \bar{\mathcal{E}}(F, F^P, \eta), \\ \mathcal{T} &= \bar{\mathcal{T}}(F, F^P, D, D^P, \eta), \\ D^P &= \bar{D}^P(F, F^P, \eta). \end{aligned} \tag{4.147}$$

Mean Normal Stress

The equilibrium value of the mean normal stress is

$$\sigma_e = \frac{1}{3}(T_e + 2\tilde{T}_e).$$

The constitutive assumptions, Eqs. (4.140), require that σ_e is strictly a function of the elastic deformation gradient and the entropy. Recall that for a Mie–Grüneisen material, we assumed that the mean normal stress was a function of density and entropy:

$$\sigma_e = \sigma_e(\rho, \eta). \tag{4.148}$$

We shall find it useful to also apply this assumption to the elastic-plastic material. However, this is possible only if we can express the density strictly as a function of the elastic deformation gradient. Indeed, Eq. (4.139) does require that

$$\frac{\rho_0}{\rho} = F^E(\tilde{F}^E)^2,$$

and, therefore, we can write the equilibrium value of the mean normal stress of an elastic-plastic material solely as a function of density and entropy. We shall assume that Eq. (4.148) holds throughout the remainder of our discussion of elastic-plastic materials.

4.3.8 *Transformability of Energy*

Recall from Chapter 3 that we require two steps to establish the transformability of energy in a material. First, we must determine all possible processes that satisfy the governing equations. We can chose these processes without regard for the initial and boundary conditions. Second, we must constrain the constitutive equations so that these processes cannot violate the second law of thermodynamics.

Defining a Process

We now show that any definition of the motion $\hat{x}(X, t)$ and the entropy $\hat{\eta}(X, t)$ throughout the volume of the elastic-plastic material results in an admissible thermodynamic process. After we have chosen the functions $\hat{x}(X, t)$ and $\hat{\eta}(X, t)$, the governing equations are satisfied as follows:

- The velocity v and deformation gradient F are determined from their definitions.
- The acceleration $\partial\hat{v}/\partial t$, deformation rate D, and rate of entropy $\partial\hat{\eta}/\partial t$ are then evaluated.
- The flow rule for D^P, Eq. (4.147), is integrated to obtain D^P and F^P.
- The two remaining constitutive assumptions are used to evaluate $\mathcal{E}$ and $\mathcal{T}$.
- The temperature ϑ and equilibrium stresses T_e and $\tilde{T}_e$ are evaluated from Eq. (4.141) and Eq. (4.142).
- The body force b is selected so that the balance of linear momentum is satisfied.
- The external heat supply r is selected so that the balance of energy is satisfied.

Thus the motion $\hat{x}(X, t)$ and the entropy $\hat{\eta}(X, t)$ can be chosen independently, and for every such choice there is a corresponding admissible thermodynamic process.

Dissipation Inequality

We constrain the constitutive equations with the dissipation inequality. We determine this inequality by combining the equation for the balance of energy with the second law of thermodynamics. In the absence of heat conduction, the balance of energy is

$$\rho \frac{\partial \hat{\mathcal{E}}}{\partial t} = T_{11} D_{11} + \rho r.$$

Recalling that the internal energy is given by Eq. (4.140), we use the chain rule to expand the right side of the balance of energy:

$$\rho F_{11}^E \frac{\partial \hat{\mathcal{E}}}{\partial F_{11}^E} D_{11}^E + \rho F_{22}^E \frac{\partial \hat{\mathcal{E}}}{\partial F_{22}^E} D_{22}^E + \rho F_{33}^E \frac{\partial \hat{\mathcal{E}}}{\partial F_{33}^E} D_{33}^E + \rho \vartheta \frac{\partial \hat{\eta}}{\partial t} = T_{11} D_{11} + \rho r.$$

Then we substitute Eqs. (4.144)–(4.146) to obtain

$$\rho \vartheta \frac{\partial \hat{\eta}}{\partial t} - (T_e - \tilde{T}_e) D^P - \mathcal{T} D = \rho r. \tag{4.149}$$

This is the required form of the balance of energy. We combine this result with the second law of thermodynamics [see Eq. (3.31)],

$$\vartheta \frac{\partial \hat{\eta}}{\partial t} \geq r,$$

to obtain the dissipation inequality

$$(T_e - \tilde{T}_e) D^P + \mathcal{T} D \geq 0. \tag{4.150}$$

4.3.9 *Plastic Flow Rule*

Both D^P and $\mathcal{T}$ are prescribed by constitutive relations. *Let us assume that each term in Eq. (4.150) must separately satisfy the dissipation inequality. Therefore*

$$\mathcal{T} D \geq 0, \qquad (T_e - \tilde{T}_e) D^P \geq 0. \tag{4.151}$$

The first of these inequalities constrains the constitutive relationship for the dissipative stress. It is typical of constraints that we have obtained for other material models in this book. The second inequality is unique to the elastic-plastic material. It constrains the plastic flow rule. Our objective here is to write a flow rule that satisfies this inequality.

Notice that the equilibrium stresses appear in the second inequality. This suggests that the yield condition should be stated in terms of the equilibrium stress. We assume that the stress state satisfies the Mises yield condition, Eq. (4.135):

$$II_{T_e'} \leq k^2,$$

where the second invariant of the equilibrium stress is

$$II_{T_e'} = \frac{1}{2} T'_{(ij)e} T'_{(ij)e}.$$

Thus the components of the equilibrium stress satisfy

$$\left(T'_{11}\right)_e^2 + \left(T'_{22}\right)_e^2 + \left(T'_{33}\right)_e^2 = 2k^2, \quad \text{at a yield point.}$$

Now imagine a coordinate system with axes T'_{11}, T'_{22}, and T'_{33}. In this coordinate system, the previous expression states that the locus of all possible yield points is a sphere. We illustrate

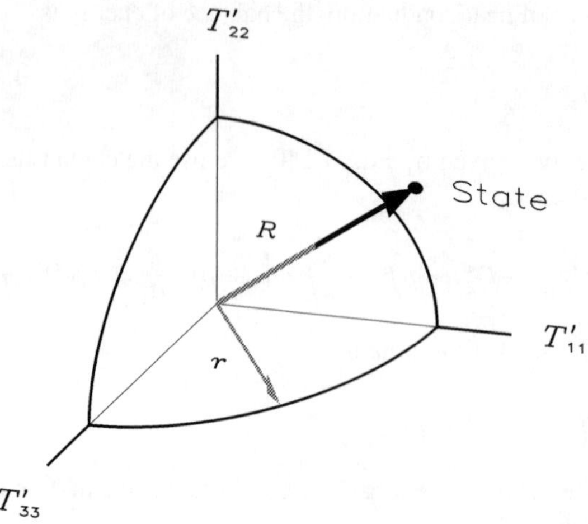

Figure 4.47 The positive quadrant of the Mises yield surface.

a portion of this sphere in Figure 4.47. The sphere is called a *yield surface*. Its radius r is

$$r = \sqrt{2}k = \sqrt{\frac{2}{3}}\,Y.$$

We represent the current state of stress in the material with a point in this coordinate system. Obviously, if the material is not at equilibrium, viscous stresses are present that we cannot illustrate in this coordinate system. The state of the equilibrium stress can exist either inside, outside, or on the yield surface. The distance R between this state of stress and the origin is

$$R^2 = \left(T'_{11}\right)_e^2 + \left(T'_{22}\right)_e^2 + \left(T'_{33}\right)_e^2.$$

Using Figure 4.47, we can express the yield condition in simple geometrical terms: *If the state of stress falls inside the yield surface where $R < r$, plastic deformation does not occur ($D_P = 0$). If the state of stress falls outside the yield surface where $R \geq r$, plastic deformation occurs ($D_P \neq 0$).* Thus the material is elastic inside the yield surface and plastic outside the yield surface.

An example of a flow rule that only produces plastic deformation outside the yield surface is

$$D_{ij}^P = \begin{cases} b\left(\frac{R-r}{R}\right)\left(T'_{ij}\right)_e, & R \geq r, \\ 0, & R < r, \end{cases}$$

where b is an arbitrary constant. For uniaxial deformation, we notice that

$$R = \sqrt{\frac{2}{3}}\,|T_e - \tilde{T}_e|,$$

$$r = \sqrt{\frac{2}{3}}\,Y,$$

$$T'_{11} = \frac{2}{3}(T_e - \tilde{T}_e).$$

Thus the flow rule becomes

$$D^P = \begin{cases} \frac{2}{3}b(|T_e - \tilde{T}_e| - Y)\,\mathrm{sgn}(T_e - \tilde{T}_e), & |T_e - \tilde{T}_e| \geq Y, \\ 0, & |T_e - \tilde{T}_e| < Y. \end{cases} \tag{4.152}$$

To determine the conditions that guarantee that this plastic flow rule satisfies the dissipation inequality, we substitute the previous expression for D_P into the inequality Eq. (4.151) to obtain

$$b(|T_e - \tilde{T}_e| - Y)\,|T_e - \tilde{T}_e| \geq 0 \quad \text{for } |T_e - \tilde{T}_e| \geq Y.$$

Therefore, the flow rule satisfies the dissipation inequality provided

$$b \geq 0.$$

4.3.10 *Elastic-Wave Process*

In many elastic-plastic materials, we find that plastic deformation does not alter the spatial wave velocity of elastic waves. For example, suppose we first measure the time for a linear-elastic wave to traverse the thickness of a plate (see Figure 4.48a). We calculate the spatial wave velocity of this wave, which for the purpose of discussion we label c_0. After evaluating this velocity, we plastically deform the plate to a fraction of its original thickness while maintaining the original density. Now we send another linear-elastic wave through the deformed plate (see Figure 4.48b). We again determine the spatial wave velocity, which we now label c. After comparing the two measured values of velocity, we find that $c = c_0$. Let us examine our theory for the elastic-plastic material, to see if it predicts this important observation.

The material wave velocity is [see Eq. (3.86) in Section 3.4.2]

$$C = \left(\frac{1}{\rho_0}\frac{\partial T}{\partial F}\right)^{1/2}. \tag{4.153}$$

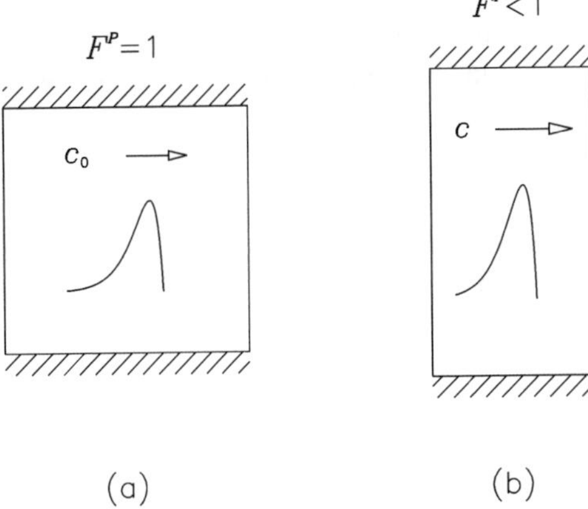

Figure 4.48 Two linear-elastic waves traversing the thickness of a plate (a) before plastic deformation and (b) after plastic deformation.

For an isentropic process, the stress T is [see Eq. (4.143)]

$$T = T_e = \rho F^E \frac{\partial \mathcal{E}}{\partial F^E} = \frac{\rho_0}{F^P}\frac{\partial \mathcal{E}}{\partial F^E}.$$

Let us first determine the velocity of an elastic wave in a material in which $F^P = 1$. For this elastic-wave process, Eq. (4.145) reveals that $F = F^E$ and $\tilde{F}^E = 1$. Therefore,

$$\frac{\partial T}{\partial F} = \frac{\partial T}{\partial F^E}$$

and

$$C_0 \equiv C = \left[\frac{\partial^2 \mathcal{E}}{(\partial F^E)^2}\right]^{1/2},$$

where we use the subscript "0" to denote the velocity when $F^P = 1$. The spatial wave velocity is

$$c_0 = FC_0 = F^E C_0.$$

Next consider the same material after it is plastically deformed. For the elastic-wave process in the deformed material, $F^P \neq 1$ is again constant. Consequently, $\tilde{F}^E$ is also constant, and the derivative of the stress is

$$\frac{\partial T}{\partial F} = \left[\frac{\partial T}{\partial F^E}\right]\frac{\partial F^E}{\partial F} = \left[\frac{\rho_0}{F^P}\frac{\partial^2 \mathcal{E}}{\partial (F^E)^2}\right]\frac{1}{F^P}.$$

Therefore, the material wave velocity C in the deformed material is related to the velocity C_0 in the undeformed material by

$$C = \frac{C_0}{F^P}.$$

The spatial wave velocities before and after plastic deformation are related by

$$c = FC = F^E C_0 = c_0.$$

Not only is this result in agreement with our original observation of linear-elastic waves in Figure 4.48, indeed, it is applicable to all elastic waves and not just linear-elastic waves.

4.3.11 *Plastic-Wave Process*

Imagine a process that results in plastic deformation. The rate of deformation during this process is sufficiently slow so that the state of stress is negligibly close to the yield surface and

$$|T - \tilde{T}| = Y + \epsilon. \tag{4.154}$$

The function ϵ is a positive small quantity in comparison to Y:

$$Y \gg \epsilon > 0.$$

Moreover, let us assume we can ignore the difference between the stress and the equilibrium stress. For such a process it is easily demonstrated that

$$T = \sigma_e + \frac{2}{3}(Y + \epsilon)\ \mathrm{sgn}(T - \tilde{T}),$$

where the equilibrium value of the mean normal stress is a function only of density and entropy [see Eq. (4.148)]. During this process, we require D^P to be monotonic and nonzero. Thus the flow rule requires that the sign of $T - \tilde{T}$ does not change during this process. Such a process is called a *plastic wave.*

The plastic wave is not an isentropic process. To demonstrate this fact, let us examine the equation for the balance of energy, Eq. (4.149):

$$\rho\vartheta\frac{\partial\hat{\eta}}{\partial t} = (T_e - \tilde{T}_e)D^P,$$

where we have assumed that $\mathcal{T} = 0$ and $r = 0$. For this process, the flow rule becomes

$$D^P = -\frac{2}{3}b\epsilon,$$

where, without loss of generality, we limit the discussion to the case when $\mathrm{sgn}(T - \tilde{T})$ is negative. We substitute this expression into the equation for the balance of energy to obtain

$$\frac{\partial\hat{\eta}}{\partial t} = \frac{2b\epsilon(Y + \epsilon)}{3\rho\vartheta} \approx \frac{2b\epsilon Y}{3\rho\vartheta},$$

where we notice that $Y + \epsilon \approx Y$. By taking the ratio of this result over D^P, we find that

$$\frac{\partial\hat{\eta}/\partial t}{D^P} \approx -\frac{Y}{\rho\vartheta}. \tag{4.155}$$

This demonstrates that the plastic wave is not an isentropic process. This fact poses a problem. Recall that the material wave velocity C and the spatial wave velocity c are defined for a structured wave that follows an isentropic process. During an isentropic process, the change in stress is given by the chain-rule expansion

$$\frac{\partial\hat{T}}{\partial t} = \frac{\partial T}{\partial F}\frac{\partial\hat{F}}{\partial t},$$

where the term containing $\partial T/\partial\eta$ does not appear because $\partial\hat{\eta}/\partial t = 0$. However, for a plastic wave, we have just shown that $\partial\hat{\eta}/\partial t \neq 0$. Consequently, the change in stress is

$$\frac{\partial\hat{T}}{\partial t} = \frac{\partial\hat{\sigma}_e}{\partial t} = \frac{\partial\sigma_e}{\partial\rho}\frac{\partial\hat{\rho}}{\partial t} + \frac{\partial\sigma_e}{\partial\eta}\frac{\partial\hat{\eta}}{\partial t},$$

where we have assumed that $Y + \epsilon$ is constant and noticed that the mean normal stress is a function of ρ and η. To evaluate the wave velocities C and c, we must either deal with the consequences of the second term on the right-hand side of this expression or show that it is negligible with respect to the first term. Suppose we examine the relative magnitudes of these two terms.

From Eqs. (3.69) and (3.74) we know that

$$\frac{\partial\hat{\sigma}_e}{\partial t} = -\frac{\mathcal{K}^\eta}{\rho}\frac{\partial\hat{\rho}}{\partial t} - \frac{\vartheta\beta\mathcal{K}^\vartheta}{c_v}\frac{\partial\hat{\eta}}{\partial t}. \tag{4.156}$$

A comparison of the previous two relationships shows that

$$\frac{\partial \sigma_e}{\partial \rho} = -\frac{\mathcal{K}^\eta}{\rho}, \qquad \frac{\partial \sigma_e}{\partial \eta} = -\frac{\vartheta \beta \mathcal{K}^\vartheta}{c_v}. \tag{4.157}$$

Now we substitute Eq. (4.155) into Eq. (4.156) to obtain

$$\frac{\partial \hat{T}}{\partial t} = \mathcal{K}^\eta D + \frac{\beta \mathcal{K}^\vartheta Y}{\rho c_v} D^P.$$

Noticing that $\mathcal{K}^\eta / c_p = \mathcal{K}^\vartheta / c_v$, we obtain

$$\frac{\partial \hat{T}}{\partial t} = \mathcal{K}^\eta \left(1 + \frac{Y\beta}{\rho_0 c_p}\right) D,$$

where for the plastic wave we have assumed that $D^P = D$. Thus we find that the contribution due to the change in entropy is negligible provided

$$\left|\frac{Y\beta}{\rho c_p}\right| \ll 1. \tag{4.158}$$

Provided this inequality holds, we can determine the material wave velocity of the plastic wave as follows:

$$C = \left(\frac{1}{\rho_0}\frac{\partial T}{\partial F}\right)^{1/2} = \frac{1}{F}\left(-\frac{\partial \sigma_e}{\partial \rho}\right)^{1/2} = \frac{1}{F}\left(\frac{\mathcal{K}^\eta}{\rho}\right)^{1/2}, \tag{4.159}$$

where we have used Eq. (4.157). The spatial wave velocity of the plastic wave is

$$c = \left(\frac{\mathcal{K}^\eta}{\rho}\right)^{1/2}. \tag{4.160}$$

Example 4.2 Consider a block of stainless steel with the following properties:

$$Y = 750 \text{ MPa}, \qquad c_p = 460 \text{ J/kg/K},$$
$$\beta = 4.9 \times 10^{-6}\ \text{K}^{-1}, \qquad \rho_0 = 7.89 \text{ Mg/m}^3.$$

When a plastic wave propagates through this material, we can determine the velocity of the wave with Eqs. (4.159) and (4.160) provided Eq. (4.158) is satisfied. By substitution into Eq. (4.160), we obtain

$$\frac{Y\beta}{\rho_0 c_p} = \frac{(750 \times 10^6)(4.9 \times 10^{-6})}{7{,}890 \times 460} = 1.0 \times 10^{-3}.$$

When we assume that $F \approx 1$, this result satisfies the inequality. Moreover, we have already calculated the velocity of a plastic wave for this material. Recall that in Example 4.1 we determined the bulk velocity c_b of stainless steel. Remember that c_b is the velocity of a linear-elastic wave when the contribution of the deviatoric stress is ignored. For the calculation of the bulk velocity, we used the values of the Lamé constants at the reference configuration. If we use the values of the Lamé constants at the yield surface, this calculation will yield the velocity of the plastic wave.

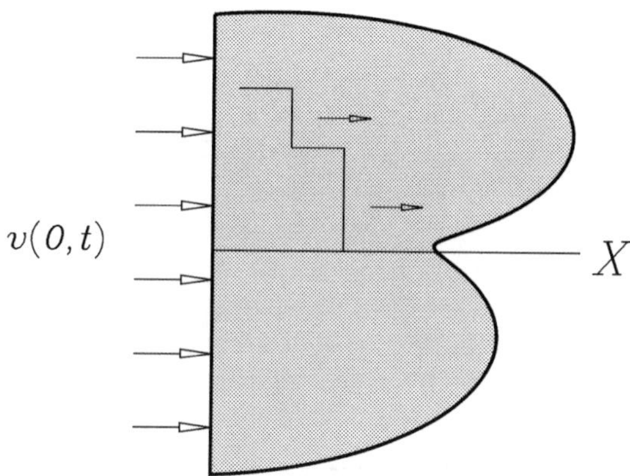

Figure 4.49 A half space of an elastic-plastic material that is subjected to sudden compression.

4.3.12 *Hugoniot Relations*

Consider a half space of an elastic-plastic material initially at rest (see Figure 4.49). At $t = 0$, we subject the boundary of this half space to a sudden jump in velocity. For the other materials we have studied thus far, this boundary condition produces a shock wave. However, we know that when the deviatoric stress in an elastic-plastic material approaches its limiting value, the shape of the curve $T(F)$ has the following curvature:

$$\frac{d^2T}{dF^2} = -\frac{d^2(-T)}{d(1-F)^2} > 0.$$

This curvature contradicts one of the requirements for the existence of a shock wave [see Eq. (4.132)]. Because of the curvature of $T(F)$ in the neighborhood of the yield point, the existence of a shock wave in this region is not necessarily guaranteed. Moreover, we have simplified the actual response of the material by introducing a yield point where the slope of the curve $T(F)$ is discontinuous and the second derivative does not even exist. As a consequence, the elastic-plastic material exhibits a response to this boundary condition that is significantly different from any material we have studied up to this point.

We illustrate the T–v diagram of the response of the elastic-plastic material in Figure 4.50. We represent the Hugoniot curve with a dot-dashed line. Notice the break in the slope of this curve at the yield point. As we illustrate in Figure 4.50, the yield point is reached when the boundary condition is $v(0, t) = v_B$. For boundary conditions that satisfy

$$v(0, t) \leq v_B,$$

the Hugoniot curve for the elastic-plastic material is identical to the Hugoniot curve for a nonlinear-elastic material. Under these conditions, the boundary condition produces a single shock wave in the material. When $v(0, t) = v_B$, the magnitude of the Hugoniot stress behind the shock is denoted by T^{HEL}. This stress is called the *Hugoniot elastic limit.* By traditional definition, it is a positive quantity. Thus the stress T^+ at the Hugoniot state is

$$T^+ = -T^{HEL} \quad \text{when } v(0, t) = v_B.$$

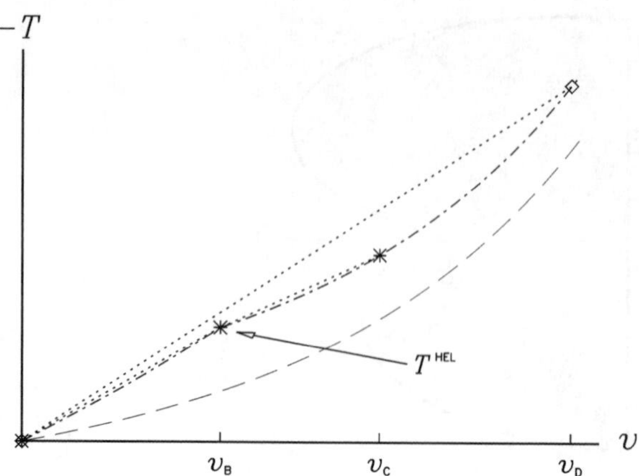

Figure 4.50 The T–v diagram for the Hugoniot (— · —) and the mean normal stress (— —) of an elastic-plastic material. Rayleigh lines for splitting ($* \cdots *$) and stable ($\diamond \cdots \diamond$) shock waves are shown.

Recall that the jump in entropy across a shock wave in a nonlinear-elastic material is third order in the jump of F [see Eq. (3.119)]. From experiments on elastic-plastic materials, we know that when $v(0, t) \leq v_B$, the jump in F is small enough that the jump in entropy can be neglected. This means that the isentrope and the Hugoniot curve are virtually coincident in the elastic region. Thus the boundary condition produces a weak shock wave. By comparing the isentrope for uniaxial-compression shown in Figure 4.43 to the Hugoniot curve, we obtain the following approximate relationship between the Hugoniot elastic limit T^{HEL}, the yield stress Y, and the mean normal stress σ_e at point B:

$$T^{HEL} \approx -\sigma_e + \frac{2}{3}Y,$$

where $\sigma_e < 0$, $T^{HEL} > 0$, and $Y > 0$.

Now consider the case when

$$v(0, t) > v_B.$$

Here we find that two shock waves are produced. Let us assume $v(0, t) = v_C$, where $v_C > v_B$. We show the Rayleigh line ($* \cdots *$) for this case in Figure 4.50. It is composed of two straight segments that are joined at the Hugoniot elastic limit. Each segment of the Rayleigh line has a unique slope. The portion of the Rayleigh line to the left of v_B is steeper than the portion to the right. We illustrate the resulting wave profile in Figure 4.51a. This is called a *splitting shock wave* because the jump in stress produced by the boundary condition splits into two shock waves. The leading shock wave is called the *elastic precursor*. It results from the portion of the Rayleigh line where $v < v_B$. The stress just behind this wave is equal to T^{HEL}. The trailing shock wave is called the *plastic shock wave*. It results from the portion of the Rayleigh line where $v > v_B$. The relative slopes of the Rayleigh line segments show that the elastic precursor propagates faster than the plastic shock wave, and with advancing time, the separation between the two waves increases.

In Figure 4.51b, we illustrate another case where the boundary condition is $v(0, t) = v_D$ (see Figure 4.50). Notice that we can connect the initial state and the Hugoniot state with a

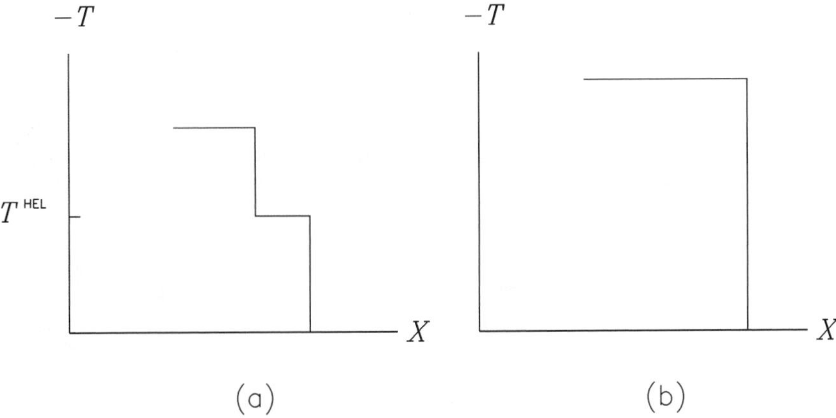

Figure 4.51 (a) A splitting shock wave, $v(0, t) = v_C$. (b) A stable shock wave, $v(0, t) = v_D$.

single straight Rayleigh line ($\diamond \cdots \diamond$). An elastic precursor does not form because the slope of this Rayleigh line, and consequently the velocity of this wave, is greater than the elastic precursor. Often, we say the elastic precursor is *overdriven* by the plastic shock wave.

Example 4.3 When we studied the Mie–Grüneisen material, we investigated the behavior of a square wave in a block of stainless steel. In that example, we assumed that the pressure resulting from the wave was large and that the deviatoric stresses could be ignored. Let us return to that example, but this time we shall consider a square wave with smaller values of stress that are comparable in magnitude to the yield stress Y. At time $t = 0$, we assume a square wave exists in a block of stainless steel (see Figure 4.52). Both the compression and rarefaction waves propagate in the positive-X direction. We assume the Hugoniot relation for the mean normal stress σ_e to be

$$\sigma_e = \frac{\rho_0 c_b^2 (F - 1)}{[1 - s(1 - F)]^2} \tag{4.161}$$

and the deviatoric stress to be

$$\frac{2}{3}(T - \tilde{T}) = \frac{4}{3}\mu(F^E - 1). \tag{4.162}$$

The values of the constants in these constitutive relationships are

$$c_b = 4.40\,\text{km/s}, \qquad Y = 900\,\text{MPa},$$

$$\mu = 78.6\,\text{GPa}, \qquad \rho_0 = 7.82\,\text{Mg/m}^3,$$

$$s = 1.44.$$

These values are slightly different from those that we used in the previous examples for stainless steel. For reasons that shall become apparent, we adopt values for the samples of stainless steel that were used by Wise and Mikkola (see the end of Section 4.2.6). Here we exclude a dependence upon entropy in the descriptions of σ_e and $\frac{2}{3}(T - \tilde{T})$ in order to simplify the discussion and concentrate on the essential features of the elastic-plastic model.

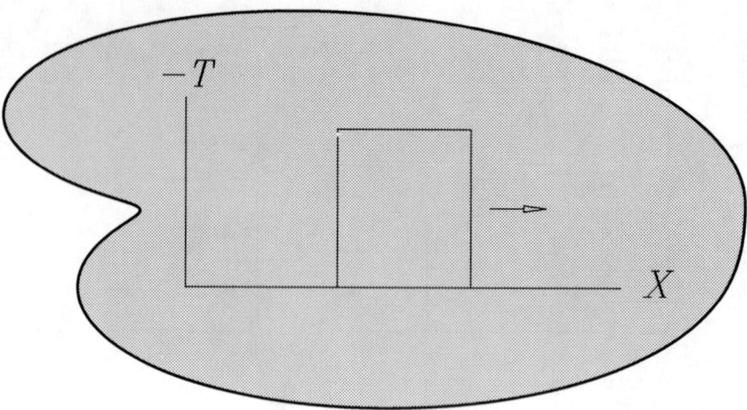

Figure 4.52 A square wave propagating in a block of stainless steel.

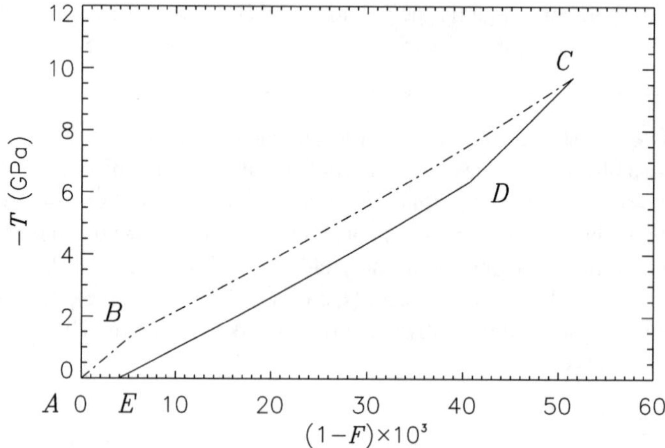

Figure 4.53 Hugoniot curve (– · –) and release path (—).

This is justified so long as we restrict our discussion to waves in which the magnitude of the stress is comparable to Y.

Notice that we have not prescribed a constitutive relationship for the plastic flow rule for D^P. Instead, we shall assume that the state of stress remains on the yield surface during plastic deformation. For this example, we shall require the stress behind the compression wave to be $T_+ = -9.7$ GPa. We show the response paths produced by the shock wave and the rarefaction wave in Figure 4.53. Compression from A to B is elastic. Compression from B to C is plastic. The response caused by the rarefaction wave is elastic from C to D and plastic from D to E. The values of the elastic and plastic components of the deformation

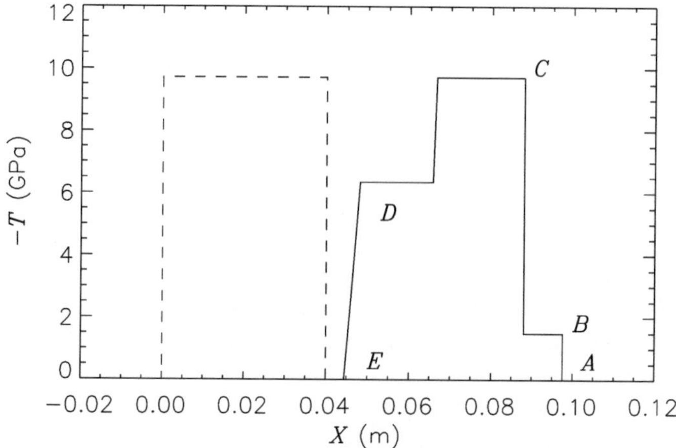

Figure 4.54 Profiles of the square wave at $t = 0$ (– –) and $t = 10$ μs (—). Labeled points correspond to those in Figure 4.53.

gradient are:

$$
\begin{aligned}
A \to B & \begin{cases} F^P = 1 \\ 1 \geq F^E \geq 1 - \kappa \end{cases} \\
B \to C & \begin{cases} 1 \geq F^P \geq 0.9485/(1-\kappa) \\ F^E = 1 - \kappa \end{cases} \\
C \to D & \begin{cases} F^P = 0.9485/(1-\kappa) \\ 1 - \kappa \leq F^E \leq 1 + \kappa \end{cases} \\
D \to E & \begin{cases} 0.9485/(1-\kappa) \leq F^P \leq 1 \\ F^E = 1 + \kappa \end{cases}
\end{aligned}
\tag{4.163}
$$

where

$$\kappa = \frac{Y}{2\mu}.$$

Recall that F^P is constant during the elastic processes and F^E is constant during the plastic processes. We illustrate the Hugoniot with a dot-dashed line. It represents a splitting shock wave. The velocity of the elastic precursor is 5.75 km/s; the velocity of the plastic shock wave is 4.8 km/s. We have determined these wave velocities from the slopes of the Rayleigh lines connecting A to B and B to C, respectively. We show the profile of the splitting shock wave at $t = 10$ μs in Figure 4.54. In this figure we also see that the rarefaction wave splits into an elastic wave and a plastic wave. Both are structured waves. We have determined the material velocity of the elastic wave from Eq. (4.153), and we have determined the velocity of the plastic wave from Eq. (4.159). Because the release paths are essentially straight, neither of these waves has much structure.

Steady Waves

We know that the yield point in an elastic-plastic material can cause a shock discontinuity to split into two waves. We can represent each of these waves as a steady wave. The dissipation inequality Eq. (4.150),

$$(T_e - \tilde{T}_e)D^P + \mathcal{T}D \geq 0,$$

tells us that the structures of these steady waves are determined by the viscous stress $\mathcal{T}$ and the plastic flow rule for D^P. Clearly, the structure of the *steady elastic precursor* is not influenced by the plastic flow rule, and it depends solely on the form of the constitutive equation for the viscous stress $\mathcal{T}$. In contrast, the structure of the *steady plastic wave* is determined by both the viscous stress and the plastic flow rule. *In the theory of metal plasticity, it is usually assumed that the viscous stress has a negligible influence upon the structure of a plastic wave.* In this section we shall examine how the plastic flow rule alone can influence the profile of the steady plastic wave.

Instead of using the plastic flow rule in Eq. (4.152), suppose we let the flow rule be

$$D^P = \begin{cases} b\left(\frac{2}{3}|T_e - \tilde{T}_e| - \frac{2}{3}Y\right)^2 \operatorname{sgn}(T_e - \tilde{T}_e), & |T_e - \tilde{T}_e| \geq Y, \\ 0, & |T_e - \tilde{T}_e| < Y. \end{cases} \tag{4.164}$$

This assumption is motivated by our earlier discussion of the shock invariant in which we saw that a fourth-power relationship exists between the maximum rate of change of the deformation gradient and the Hugoniot stress of steady waves in certain solids. Recall that we used this empirical observation as motivation for specifying the viscous stress with Eq. (4.118) in Section 4.2.5. Notice that the term $(\frac{2}{3}|T_e - \tilde{T}_e| - \frac{2}{3}Y)$ in Eq. (4.164) assumes the role of the viscous stress $\mathcal{T}$ in Eq. (4.118). In both cases, we have related the deformation gradient to the square of the stress. Moreover, this comparison reveals that for processes where $\partial\hat{F}/\partial t \approx D^P$, we can estimate b as

$$b = \frac{16\rho_0^2 c_b^4 A'}{s^2}. \tag{4.165}$$

Let us now analyze a steady plastic wave in an elastic-plastic solid. Consider the stainless-steel block that we discussed in the previous section. We shall determine the profile of the plastic compression wave in this material by deriving two equations for the elastic and plastic deformation gradients. This plastic compression wave results from process B–C in Figure 4.53. We determine the initial state for this wave with the values of deformation gradient F_- and stress T_- at point B. From Eq. (4.163), we find that

$$F_- = 1 - \frac{Y}{2\mu}, \qquad F^P = 1,$$

and from Eqs. (4.161) and (4.162),

$$T_- = \frac{\rho_0 c_b^2(F_- - 1)}{[1 - s(1 - F_-)]^2} + \frac{4}{3}\mu(F_- - 1).$$

After the plastic wave evolves into a steady wave, it propagates at the velocity C_S, and it obeys the steady wave solution for the balance of linear momentum [see Eq. (2.189) in

Section 2.9.1],

$$\frac{\rho_0 c_b^2(F^E F^P - 1)}{[1 - s(1 - F^E F^P)]^2} + \frac{4}{3}\mu(F^E - 1) = \rho_0 C_S^2(F^E F^P - F_-) + T_-. \tag{4.166}$$

This is the first of our required relationships. It is a nonlinear equation in two unknowns, F^E and F^P.

We obtain the second relationship from the plastic flow rule. Because we are interested in a steady wave propagating in the positive-X direction, the variables F^P and F^E must be functions of the steady-wave variable $\hat{\phi} = X - C_S t$. Thus the plastic flow rule yields

$$\frac{dF^P}{d\hat{\phi}} = -\frac{F^P}{C_S}\begin{cases}\frac{4}{9}b(|2\mu(F^E - 1)| - Y)^2, & |2\mu(F^E - 1)| \geq Y,\\ 0, & |2\mu(F^E - 1)| < Y,\end{cases} \tag{4.167}$$

where because the plastic wave results in compression, we know that $F^E < 1$. If we initiate integration slightly off the yield surface, we can numerically integrate Eqs. (4.166) and (4.167) to determine either the history or the profile of the steady wave. From the previous section, we know that $C_S = 4.74$ km/s. With the exception of b, we also know all of the material constants. We estimate the value of b from Eqs. (4.127) and (4.165):

$$b = \frac{16 \times (7{,}820)^2 \times (4{,}400)^4 \times 71 \times 10^{-36}}{(1.44)^2} = 1.26 \times 10^{-11}\ (\text{Pa}^2\text{-s})^{-1}.$$

We compare the integrated solution for stress to the experimental measurements of a steady stress wave in Figure 4.55. The data are from Experiment SSWP1 in Wise and Mikkola (see Section 4.2.6). The data show the elastic precursor, which of course is not present in the calculation of the steady plastic wave. The elastic precursor has a rise time of approximately 50 ns. This small but finite rise time is attributed to the crystalline microstructure of the solid. On this small time scale the material is heterogeneous, and each crystal causes a wave reflection. The combined effect of all of the reflections off the innumerable crystals is to

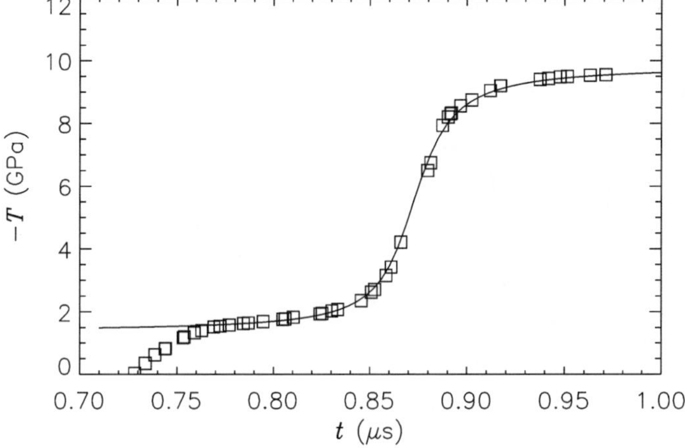

Figure 4.55 Comparison of the calculated history of the steady plastic wave (—) and the data of Wise and Mikkola (boxes). Original digital data files are courtesy of J. L. Wise, Sandia National Laboratories.

spread the wave over many crystal dimensions. This effect is called *wave dispersion*. Indeed, we see that the elastic precursor is particularly susceptible to wave dispersion because it is essentially a linear-elastic wave. In contrast, because the steady plastic wave is nonlinear, the effect of spreading caused by wave dispersion is overbalanced by nonlinear effects that cause the characteristic contours to converge and form a steady wave. The structure of the steady plastic wave is dominanted by plastic flow. We see that agreement between our calculation of the steady wave and the data is excellent. However, because we have determined the thickness of the calculated steady wave with b, we should not be totally surprised by this agreement between theory and experiment. After all, we evaluated b using the empirical fit of Wise and Mikkola, Eq. (4.127). However, the agreement is still excellent and captures more than just the thickness of the steady wave. In fact, the quality of this comparison can be reproduced for steady waves of greatly differing amplitudes without readjusting the constitutive parameters in the model. This is best illustrated in the work of Swegle and Grady (see Section 4.2.5). They have determined values of b for other materials by analyzing the measured profiles of plastic waves (see Appendix B). Their theory is essentially the same as ours with one exception: They incorporate the Mie–Grüneisen theory into their work, by including the thermal-pressure term in the mean normal stress. Moreover, their analysis is not restricted to steady waves but involves the numerical analysis of the partial differential equations describing the elastic-plastic material. This approach allows them to include the wave reflections at the various interfaces of the flyer-plate experiments used to generate and measure the wave profiles. We show their calculations and the experimental data for several wave histories in aluminum in Figure 4.56. These results illustrate the high degree of accuracy that can be attained in modeling the compression response of metals with an elastic-plastic theory.

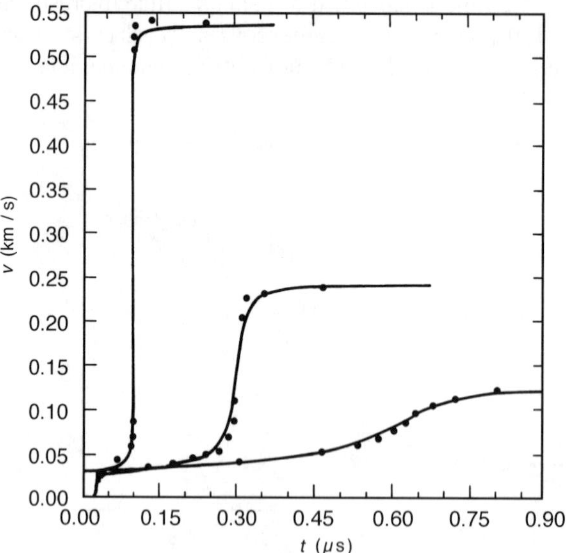

Figure 4.56 Comparison of experiment (solid circles) and calculation (—) for wave profiles in aluminum. Calculated stress levels are 9, 3.7, and 2.1 GPa. Reprinted from Swegle, J. W. and Grady, D. E., 1985, Shock viscosity and the prediction of shock wave rise times, *Journal of Applied Physics* **58**, p. 699, with kind permission of the American Institute of Physics.

In closing, you should remember that plastic flow rules such as Eqs. (4.152) and (4.164) are empirical. However, recent work has shown how such relationships can be motivated by the theory of dislocations. We hope the discussion in this section will provide a sufficient background for those who wish to investigate the research literature for greater detail.[†]

4.3.13 *Exercises*

4.25 Verify Eq. (4.133).

4.26 Consider an elastic-plastic material that does not conduct heat and does not have a viscous stress. Discuss how the decomposition of the deformation gradient into elastic and plastic components,

$$F = F^E F^P,$$

allows us to satisfy simultaneously the equation for the equilibrium stress and the equation for the stress along a Rayleigh-line process.

4.27 In a particular tensile-bar test, we observe the yield stress and density of a sample of material to be

$$Y = 0.71 \text{ GPa}, \qquad \rho_0 = 7{,}903 \text{ kg/m}^3.$$

We also observe that the velocities of longitudinal and transverse waves are

$$c_0 = 5.79 \text{ km/s}, \qquad c_{0T} = 3.16 \text{ km/s}.$$

Suppose we subject this material to a pure shear stress.

(a) If the material obeys the Tresca yield condition, determine the value of pure shear stress that will cause it to yield.

(b) If the material obeys the Mises yield condition, determine the value of pure shear stress that will cause it to yield.

(c) Assuming the material is linear-elastic below the yield stress, determine the Hugoniot elastic limit for each yield condition.

4.28 A certain type of brass has a density, yield stress, and transverse wave velocity of

$$\rho_0 = 8{,}450 \text{ kg/m}^3, \qquad Y = 0.48 \text{ GPa}, \qquad c_{0T} = 2.13 \text{ km/s}.$$

The bulk response of this material is

$$\sigma_e = \frac{118 \times (1 - F)}{[1 - 1.43 \times (1 - F)]^2} \text{ GPa}.$$

(a) Determine T^{HEL}.

(b) Determine the plastic wave velocity at the yield stress.

(c) Determine the speed of sound and the bulk velocity at the stress-free state.

4.29 A half space of a certain type of aluminum obeys

$$\sigma_e = \frac{75 \times (1 - F)}{[1 - 1.37 \times (1 - F)]^2} \text{ GPa},$$

$$T - \tilde{T} = 27.4 \times (1 - F) \text{ GPa},$$

$$Y = 0.27 \text{ GPa}.$$

[†] You can find a good treatment of this topic in the chapter by J. N. Johnson entitled "Micromechanical considerations in shock compression of solids" in *High-Pressure Shock Compression of Solids* (Springer-Verlag, 1992), Asay, J. R. and Shahinpoor, M. (editors), pp. 217–264.

Initially the half space is free of stress. Suppose a sudden jump in stress T_+ is applied to the boundary.

(a) Determine the range of stresses T_+ that result in a splitting shock wave.

(b) Determine the velocities of the two components of the splitting shock wave when $T_+ =$ 3 GPa.

(c) If a rarefaction wave follows at some finite distance behind the splitting shock wave of part (b), determine the material wave velocity of the leading portion of this wave.

4.30 In Figure 4.54, we show the wave profile at $t = 10\ \mu s$ for our example of stainless steel.

(a) Construct the profiles for $t = 5$ and $10\ \mu s$.

(b) Determine when the leading portion of the rarefaction wave will contact the trailing portion of the splitting shock wave.

4.31 Consider the following constitutive equations:

$$\mathcal{E} = \mathcal{E}\left(F_{ij}^E, \eta\right), \qquad T = T\left(F_{ij}^E, \eta\right),$$
$$\vartheta = \vartheta\left(F_{ij}^E, \eta\right), \qquad D_{km}^P = D_{km}^P\left(F_{ij}, F_{ij}^P, \eta\right),$$

where we have ignored heat conduction and viscosity. Assume the stress and the deformation gradient have the following forms:

$$T_{ij} = \begin{bmatrix} T & 0 & 0 \\ 0 & \tilde{T} & 0 \\ 0 & 0 & \tilde{T} \end{bmatrix}, \qquad F_{ij} = \begin{bmatrix} F & 0 & 0 \\ 0 & \tilde{F} & 0 \\ 0 & 0 & \tilde{F} \end{bmatrix}.$$

Use the three-dimensional equation for the balance of energy and the second law of thermodynamics to show that the constitutive equations must satisfy the following relationships:

$$\vartheta = \frac{\partial \mathcal{E}}{\partial \eta}, \qquad T = \rho F^E \frac{\partial \mathcal{E}}{\partial F^E}, \qquad \tilde{T} = \rho \tilde{F}^E \frac{\partial \mathcal{E}}{\partial \tilde{F}^E},$$

where the reduced dissipation inequality is

$$T D^P + 2\tilde{T}\tilde{D}^P \geq 0.$$

4.4 Saturated Porous Solid

Many materials are mixtures of other materials. For example, the alloy brass is a mixture of copper and zinc, and air is a mixture of nitrogen and oxygen with small amounts of argon, carbon dioxide, and hydrogen. These types of mixtures are called *miscible mixtures* because the physical components of the constituents are separable only on an atomic scale. Other types of mixtures have coarser internal structures. Blood is composed of cells floating in plasma, and concrete is produced by using cement to bind sand and gravel together. These are called *immiscible mixtures.*

Just as we can combine physical components to produce a mixture, we can combine the constitutive models of the components to describe the constitutive properties of the mixture. The result is called a *mixture theory*. In fact, we could argue that most constitutive theories are really mixture theories. For example, we have already studied waves propagating through stainless steel without any consideration of the separate responses of the iron, chromium, nickel, and manganese constituents in this material. The material properties we used in these calculations represent the *mixture values*, which are properties averaged over the separate constituents of this alloy. Indeed, given an ample supply of time, intelligence, and computing power, we could have avoided using this mixture model and analyzed this material down to the detail of the molecular scale of mixing, including the individual atomic dislocations and the separate crystalline boundaries. Usually such a detailed computation is not only impossible to achieve but yields microscopic detail that is not important to the primary purpose of the calculation. However, in some types of materials, a little additional information about microscale effects proves to be quite useful. For example, suppose a shock wave propagates through a water-saturated porous ceramic. The water contained in the pores of the ceramic is heated. A release wave follows behind the shock wave. As the compressive stress in the ceramic drops, the pore water, which has a larger coefficient of thermal expansion, expands faster than the ceramic matrix. Eventually, the expansion of the water against the walls of the pores causes the ceramic matrix to fracture. In contrast, if the pores of the ceramic are dry, the shock and release waves might not produce any damage to the matrix. Here the additional information that would help us analyze this mixture are the temperatures and thermal expansions of the components of the mixture.

The art of constructing a useful mixture theory is to find a way to model a little additional information about the mixture, such as the separate temperatures of each constituent, without the necessity of retaining an enormous amount of microscopic detail. Mixture theories that have been notably successful in this respect are immiscible mixture theories used to describe saturated porous solids. In this section we shall restrict our attention to this type of immiscible mixture. While at first this may seem to be a very special type of mixture, we notice that most constitutive models for geological media are based on this type of theory. This class of mixture also includes solid foams that are widely used to absorb impact.

Recall that the state of a material is defined by two conjugate pairs of variables. They are (1) the stress and deformation gradient (T, F) and (2) the temperature and entropy (ϑ, η). In an immiscible mixture we shall define these conjugate pairs for each constituent. Thus, as an example, imagine a mixture of porous rock and air. Because we retain these two conjugate pairs for both the rock and the air, when this mixture is compressed, we can account for the fact that the temperature could increase more in the air while the pressure increases more in the rock. Although we shall use the conjugate pairs (T, F) and (ϑ, η) to describe the state of each constituent, these variables are still not sufficient to describe the state of the mixture. Let us see why.

4.4.1 *Porous Locking Solid*

Consider the special case where the "fluid" in our mixture is a vacuum, so that only the porous solid remains (see Figure 4.57). We shall use this dry porous solid to illustrate just why the conventional conjugate pairs (T, F) and (ϑ, η) are insufficient to describe the state of this material. To accentuate the need for an additional conjugate pair of variables, we will assign some unusual properties to this material.

Suppose we ignore thermal effects and assume that the state of the solid does not depend upon (ϑ, η). Therefore, the state is defined solely by the conjugate pair (T, F). We might now characterize this porous solid as a nonlinear-elastic material. However, let us place two additional constraints on this material. First, we constrain the solid to uniaxial motion, and then we require the density of the solid to be constant; that is, the solid is incompressible. Indeed, we might now expect that $F = \rho_0/\rho$ is a constant, and as a result T is a constant. This means that we cannot use the remaining conjugate pair (T, F) to describe the state of the solid. However, because the solid has pores, we see that it is possible to change F and deform the material by collapsing the pores without changing the density of the solid. Thus

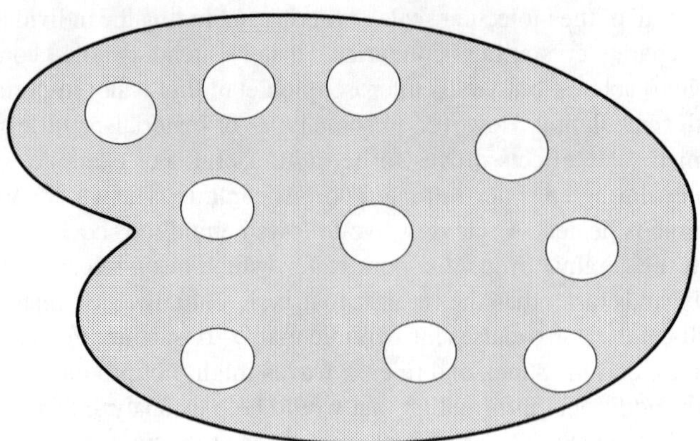

Figure 4.57 An element of volume of a porous solid.

we recognize that, because of the pores, the deformation gradient F is *no longer* tied directly to the density of the solid. Clearly, the conjugate pairs (ϑ, η) and (T, F) taken alone are not adequate to describe the behavior of this material. Another conjugate pair is required. We choose a new pair (Λ, φ) that contains a flux term called the *volume fraction* φ and a force term called the *interface pressure* Λ. Because we introduce these variables to model a special internal feature of the material, they are often called *internal state variables*. We shall introduce the volume fraction here and leave our discussion of the interface pressure until later.

Consider a small element of the porous solid that contains many pores (see Figure 4.57). The shaded region represents the volume occupied by the solid. Let the volume of the shaded region be V_S. The total volume of the element is V. We define the volume fraction φ of the porous solid to be the ratio

$$\varphi \equiv \frac{V_S}{V}. \tag{4.168}$$

Let the mass of the solid in this element be M. Then the *true density* of the solid is

$$\bar{\rho} \equiv \frac{M}{V_S},$$

and the *partial density* is

$$\rho \equiv \frac{M}{V}.$$

We combine these three relationships to obtain

$$\rho = \varphi\bar{\rho}. \tag{4.169}$$

You should recognize that our definition of the partial density ρ is in agreement with our conventional definition of density. Therefore, the conservation of mass becomes

$$F = \frac{\rho_0}{\rho} = \frac{\varphi_0\bar{\rho}_0}{\varphi\bar{\rho}}.$$

Hence, by introducing the volume fraction into our theory, we can require the true density of the solid to be constant, $\bar{\rho} = \bar{\rho}_0$, and still allow F to change:

$$F = \frac{\varphi_0}{\varphi}.$$

Obviously, the volume fraction φ must obey

$$0 < \varphi \leq 1,$$

and consequently, in the special case when the true density is constant, the deformation gradient obeys

$$F \geq \varphi_0.$$

We now define a *porous locking solid*. In this solid, the deviatoric stresses are negligible, and the stress is equal to the mean normal stress plus the viscous stress

$$T = \sigma_e(\bar{\rho}) + \mathcal{T}.$$

Here σ_e is only a function of the true density $\bar{\rho}$. For this special porous solid, we shall also require that the solid is incompressible so that $\bar{\rho} = \bar{\rho}_0$ regardless of the value of T. This means that the equilibrium stress is zero as long as $F > \varphi_0$, and the material is rigid at $F = \varphi_0$ (see Figure 4.58). Because the slope of the isentrope is infinite when $F = \varphi_0$, structured waves propagate at infinite velocity. However, as we demonstrate with the Rayleigh line, shock waves propagate with finite velocity.

Shock Wave Response

We can model the porous locking solid by introducing the volume fraction as an internal state variable. To illustrate the role of this variable, we analyze the response of this material to a shock wave. We show that even through the true density of the solid is constant, we can still determine the propagation velocity and amplitude of a shock wave.

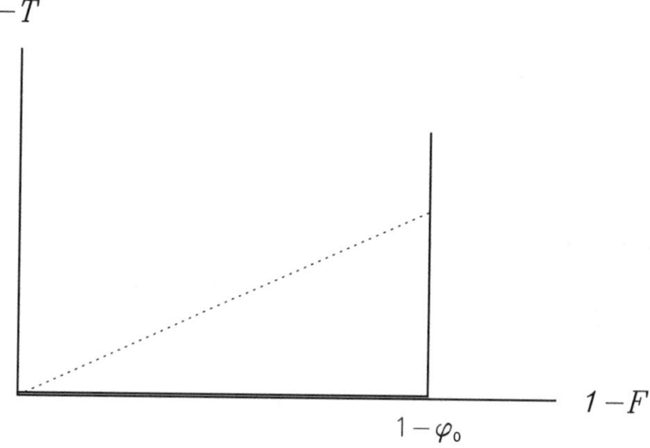

Figure 4.58 The T–F diagram of a porous locking solid. We show an isentrope (—) and a Rayleigh line (· · ·).

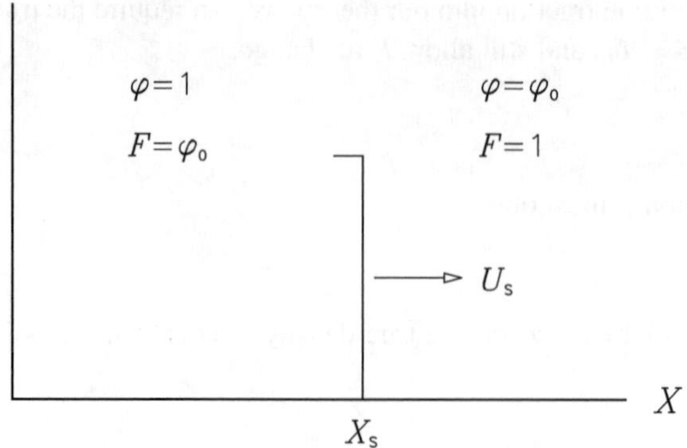

Figure 4.59 A shock wave in a porous locking solid.

Suppose that the porous solid is initially at rest and at a state of zero stress. A shock wave propagates in the positive-x direction. From Figure 4.58, we see that the stress in this wave causes the pores in the solid to completely collapse. Thus $\varphi = \varphi_0$ ahead of the wave, and $\varphi = 1$ behind the wave (see Figure 4.59). Let the material position of the shock wave be given by the function

$$X_S = X_S(t).$$

The propagation velocity of the shock wave is

$$U_S = \frac{dX_S}{dt}. \tag{4.170}$$

Ahead of the shock wave, the initial conditions yield

$$\left.\begin{aligned} \varphi_- &= \varphi_0 \\ v_- &= 0 \\ T_- &= 0 \\ F_- &= 1 \end{aligned}\right\} X > X_S.$$

At the shock wave, the jump conditions yield

$$\left.\begin{aligned} v_+ &= -U_S(\varphi_0 - 1) \\ T_+ &= -\rho_0 U_S v_+ \end{aligned}\right\} X = X_S. \tag{4.171}$$

Behind the shock wave, we have

$$\left.\begin{aligned} \varphi_+ &= 1 \\ F_+ &= \varphi_0 \end{aligned}\right\} X < X_S.$$

From the definitions of the velocity and the deformation gradient, we know that [see Eq. (1.132) in Section 1.8.1]

$$\frac{\partial \hat{v}}{\partial X} = \frac{\partial \hat{F}}{\partial t}.$$

Because $F = \varphi_0$ behind the shock wave, the right-hand side of this expression is zero, and we find that v does not depend on X:

$$v = v_+(t), \quad \text{when } X < X_S.$$

The balance of linear momentum is

$$\rho_0 \frac{\partial \hat{v}}{\partial t} = \frac{\partial \hat{T}}{\partial X}.$$

Because v is not a function of X behind the shock wave, we can integrate the momentum equation to obtain

$$T = \rho_0 \frac{dv_+}{dt} X + d(t), \quad \text{when } X < X_S,$$

where $d(t)$ is an arbitrary function of t. Suppose we require that $T = 0$ at $X = 0$. Then $d(t) = 0$, and

$$T = \rho_0 \frac{dv_+}{dt} X, \quad \text{when } X < X_S.$$

We now know the velocity and stress everywhere behind the shock wave in terms of the unknown function $v_+(t)$ and its derivative. To determine $v_+(t)$, we must analyze the jump conditions at $X = X_S$.

At the shock front, we can combine the previous expression with the second jump condition in Eq. (4.171) to obtain

$$T_+ = \rho_0 \frac{dv_+}{dt} X_S = -\rho_0 U_S v_+.$$

We substitute the first jump condition in Eq. (4.171) to obtain

$$-\frac{dU_S}{dt} X_S = U_S^2.$$

Then we substitute Eq. (4.170) to find that

$$X_S \frac{d^2 X_S}{dt^2} + \left(\frac{dX_S}{dt}\right)^2 = 0,$$

which reduces to

$$\frac{d}{dt}\left(X_S \frac{dX_S}{dt}\right) = 0.$$

The solution to this differential equation is

$$X_S = 2U_0\sqrt{tt_0 + B},$$

where U_0, t_0, and B are arbitrary constants. If we require that $X_S = 0$ at $t = 0$, then $B = 0$ and

$$X_S = 2U_0 t_0 \sqrt{\frac{t}{t_0}}.$$

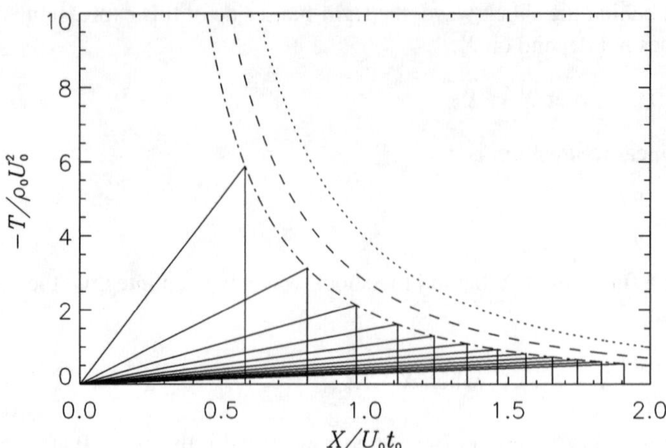

Figure 4.60 A shock wave propagating in a porous locking solid. We illustrate the stress T_+ at the Hugoniot state just behind the shock wave for $\varphi_0 = 0.01$ ($\cdots$), 0.30 ($- \, -$), and 0.50 ($- \cdot -$). Profiles for $\varphi_0 = 0.50$ are shown at increments of $t/t_0 = 0.075$.

Thus the velocity of propagation of the shock wave is [see Eq. (4.170)]

$$U_S = U_0 \sqrt{\frac{t_0}{t}};$$

the velocity of the material behind the shock wave is [see Eq. (4.171)]

$$v_+ = U_0(1 - \varphi_0)\sqrt{\frac{t_0}{t}};$$

and the stress behind the shock wave is [see Eq. (4.171)]

$$T_+ = -\rho_0 U_0^2 (1 - \varphi_0)\frac{t_0}{t}.$$

These are the desired results for the history of the propagating shock wave. We illustrate $T_+(t)$ versus $X(t)$ in Figure 4.60 for three different initial values of porosity. Notice that the shock wave attenuates as it propagates. Smaller values of φ_0, which indicate greater porosity, result in greater attenuation. For the largest value of porosity, $\varphi_0 = 0.50$, we show profiles of the shock wave and the trailing release wave. It is left as an exercise to show that the momentum of the material in the wave is a constant with respect to time, while the total energy of the wave decreases. Indeed, the dissipation of energy in porous solids can be very large, which makes them quite useful as tools for impact absorption.

4.4.2 *Immiscible Mixture Theory*

In the previous section we demonstrated the importance of the volume fraction φ by means of the analysis of a shock wave in a porous locking solid. In this section we shall use the concept of volume fraction to develop a theory for an immiscible mixture of a fluid in a porous solid. *We assume the solid is saturated by the fluid.*

Volume Fraction and Deformation

We defined the volume fraction of the porous solid in Eq. (4.168). To distinguish the solid s from the fluid f, let us denote the volume fraction of the solid by φ_s and the volume fraction of the fluid by φ_f. Because the solid is saturated with fluid,

$$\varphi_s + \varphi_f = 1.$$

As a matter of convenience, we shall often refer to the ξ-th constituent of the mixture, where $\xi = s$ or f. Thus, we can rewrite the previous expression as

$$\sum_{\xi} \varphi_\xi = 1. \tag{4.172}$$

By employing the arguments leading to Eq. (4.169), we define the true density $\bar{\rho}_\xi$ and partial density ρ_ξ of the ξ-th constituent, so that

$$\rho_\xi = \varphi_\xi \bar{\rho}_\xi. \tag{4.173}$$

Notice that throughout this section, repetition of the index ξ does not imply summation over the constituents.

Now we make an important assumption. We assume that *each constituent of the mixture has an identical motion*, $x_\xi = \hat{x}(X, t)$. As a result, the deformation gradient F is the deformation gradient of each constituent, and therefore,

$$F = \frac{\rho_{\xi 0}}{\rho_\xi} = \frac{\varphi_{\xi 0}\bar{\rho}_{\xi 0}}{\varphi_\xi \bar{\rho}_\xi} \tag{4.174}$$

holds for both the solid and the fluid constituent. We also define the *true deformation gradient* F^ξ to be

$$F^\xi \equiv F \frac{\varphi_\xi}{\varphi_{\xi 0}}. \tag{4.175}$$

These two results yield

$$F^\xi = \frac{\bar{\rho}_{\xi 0}}{\bar{\rho}_\xi}.$$

The deformation rate of the ξ-th constituent is defined to be

$$D^\xi \equiv \frac{1}{F^\xi}\frac{\partial \hat{F}^\xi}{\partial t}.$$

Taking the time derivative of the logarithm of Eq. (4.175) gives the following useful relationship:

$$D^\xi = D + \frac{1}{\varphi_\xi}\frac{\partial \hat{\varphi}_\xi}{\partial t}.$$

As a matter of convenience, we define the rate of change of the volume fraction H_ξ to be

$$H_\xi \equiv \frac{\partial \hat{\varphi}_\xi}{\partial t}.$$

Then

$$D^{\xi} = D + \frac{1}{\varphi_{\xi}} H_{\xi}. \tag{4.176}$$

Elastic and Plastic Volume Fractions

Porous solids are often crushed by shock waves and steady waves. This was illustrated in our example of the porous locking solid where $\varphi = 1$ everywhere behind the shock wave. As stress returned to zero, none of the original pore volume in this solid was recovered. However, in most porous materials some of the pore volume is recovered when the load is removed. Following the spirit of our treatment of the elastic-plastic material in Section 4.3, let us call the recoverable portion of φ_s the *elastic volume fraction* and denote it by the symbol φ_s^E. The part of φ_s that is not recovered we shall call the *plastic volume fraction* φ_s^P. Then we require that the deformation gradients of both the mixture and the porous solid must decompose into elastic and plastic components,

$$F \equiv F^E F^P, \qquad F^s \equiv F^{sE} F^{sP},$$

where we define

$$F^{sE} \equiv F^E \frac{\varphi_s^E}{\varphi_{s0}^E}, \qquad F^{sP} \equiv F^P \frac{\varphi_s^P}{\varphi_{s0}^P}.$$

For these definitions to be compatible with Eq. (4.175), we find that the volume fraction of the solid must obey the multiplicative decomposition

$$\varphi_s \equiv \varphi_s^E \varphi_s^P. \tag{4.177}$$

By taking the time derivative of the natural logarithm of Eq. (4.177), we obtain the useful relationship

$$\frac{1}{\varphi_s} H_s = \frac{1}{\varphi_s^E} H_s^E + \frac{1}{\varphi_s^P} H_s^P, \tag{4.178}$$

where

$$H_s^E \equiv \frac{\partial \hat{\varphi}_s^E}{\partial t}, \qquad H_s^P \equiv \frac{\partial \hat{\varphi}_s^P}{\partial t}.$$

Mass, Linear Momentum, and Energy

The mixture as a whole must obey the conservation of mass, the balance of linear momentum, and the balance of energy. In the absence of chemical reactions, the individual constituents of the mixture must also obey these principles. Let us next consider these relationships.

Each component of the mixture conserves mass. This means, for example, that we are excluding situations in which the solid melts and adds to the mass of the fluid. The conservation of mass for each component of the mixture is [see Eq. (4.174)]

$$\rho_{\xi} F = \rho_{\xi 0}. \tag{4.179}$$

The density of the mixture is equal to the sum of the partial densities of the constituents:

$$\rho \equiv \sum_{\xi} \rho_{\xi}. \tag{4.180}$$

When we sum Eq. (4.179) over the constituents, we obtain

$$\rho F = \rho_0,$$

which is the conservation of mass that we have used for all materials in this work.

Next let us apply a similar approach to the balance of linear momentum. We assume each constituent has a *true stress* denoted by $\bar{T}_{\xi}$. The constituent stress, which is sometimes called the *partial stress*, is

$$T^{\xi} \equiv \varphi_{\xi} \bar{T}_{\xi}.$$

In the fluid, where we can represent the true stress by a true pressure $\bar{p}_f = -\bar{T}_f$, the *partial pressure* is

$$p_f = \varphi_f \bar{p}_f.$$

The balance of linear momentum for each constituent of the mixture is

$$\rho_{\xi 0} \frac{\partial \hat{v}}{\partial t} = \frac{\partial \hat{T}^{\xi}}{\partial X} + \rho_{\xi 0} b_{\xi} + f_{\xi}. \tag{4.181}$$

Recall that we assume that each constituent of the mixture must have the same motion and consequently the same velocity v. This means that the solid and the fluid exchange momentum in such a way as to keep their respective motions equal even though their densities ρ_{ξ}, stresses T^{ξ}, and body forces b_{ξ} are different. We represent this exchange of momentum with the force f_{ξ}. Because each constituent must apply an equal and opposite force on the other constituent, we find that

$$\sum_{\xi} f_{\xi} = 0.$$

We define the stress T in the mixture to be equal to the sum of the partial stresses T^{ξ} of the constituents:

$$T \equiv \sum_{\xi} T^{\xi}.$$

The body force acting on the mixture is

$$\rho_0 b \equiv \sum_{\xi} \rho_{\xi 0} b_{\xi}.$$

Thus, when we sum Eq. (4.181) over the constituents, we obtain

$$\rho_0 \frac{\partial \hat{v}}{\partial t} = \frac{\partial \hat{T}}{\partial X} + \rho_0 b, \tag{4.182}$$

which we recognize as the conventional equation for the balance of linear momentum.

Now we assume that the internal energy $\mathcal{E}_{\xi}$ in each component of the mixture satisfies

$$\rho_{\xi 0} \frac{\partial \hat{\mathcal{E}}_{\xi}}{\partial t} = T^{\xi} FD + \rho_{\xi 0} r_{\xi} + e_{\xi} - \Lambda_{\xi} H_{\xi}, \tag{4.183}$$

where Λ_ξ is the interface pressure of the ξ-th constituent. This equation contains both familiar terms and terms we have not seen before. Let us examine each of the four separate terms on the right-hand side of this equation. The first term is a familiar expression. It is the stress working in the ξ-th constituent due to the conjugate pair (T^ξ, F). The second term is also familiar. It is the specific external heat supply to the ξ-th constituent. It is related to the external heat supply of the mixture by

$$\rho_0 r \equiv \sum_\xi \rho_{\xi 0} r_\xi.$$

The third term is new. It represents the exchange of heat between the constituents of the mixture. The heat lost by one constituent is gained by the other. Thus

$$\sum_\xi e_\xi = 0.$$

The fourth term is also new. It represents the work done by the conjugate pair $(\Lambda_\xi, \varphi_\xi)$. This means that a change in volume fraction cannot occur unless the constituent does work. Like the exchange of heat, the work done by one constituent must be received by the other:

$$\sum_\xi \Lambda_\xi H_\xi = 0.$$

Because the volume fractions of the constituents must add to one, we find that $\sum_\xi H_\xi = 0$ and

$$\Lambda_s = \Lambda_f. \tag{4.184}$$

The internal energy of the mixture is

$$\rho_0 \mathcal{E} \equiv \sum_\xi \rho_{\xi 0} \mathcal{E}_\xi. \tag{4.185}$$

When Eq. (4.183) is summed over the constituents, we obtain

$$\rho_0 \frac{\partial \hat{\mathcal{E}}}{\partial t} = TFD + \rho_0 r,$$

which is the conventional equation for the balance of energy.

You will find it instructive to pause for a moment and compare the two working terms $T^\xi D$ and $\Lambda_\xi H_\xi$ in Eq. (4.183). The first term arises from the conjugate pair (T^ξ, F). Although we have discussed stress in great detail in Chapter 1, stress is actually defined by this working term in the balance of energy. Indeed, we know that substitution of the equation for the balance of energy into the second law of thermodynamics yields a dissipation inequality that tells us the stress can be partitioned into equilibrium and viscous components. We even know how to obtain the equilibrium stress from the internal energy. As we proceed, we shall also find that the interface pressure Λ_ξ is defined by its working term in the balance of energy. It too can be partitioned into equilibrium and viscous components, and we shall determine the equilibrium component from the internal energy.

Constitutive Assumptions

We assume three separate sets of constitutive relationships for the saturated porous solid. The first set contains the three constitutive relationships for the solid:

$$\mathcal{E}_s = \mathcal{E}_s\left(F^s, \varphi_s^E, \varphi_s^P, \eta_s\right), \tag{4.186}$$
$$\bar{T}^s = \bar{T}^s\left(F^s, \varphi_s^E, \varphi_s^P, \eta_s\right), \tag{4.187}$$
$$\Lambda_s = \Lambda_s\left(F^s, \varphi_s^E, \varphi_s^P, \eta_s\right). \tag{4.188}$$

Notice that we have made the internal energy, *true* stress, and interface pressure functions of both the elastic portion and the plastic portion of the volume fraction. The second set of constitutive relationships contains three expressions for the fluid:

$$\mathcal{E}_f = \mathcal{E}(\bar{\rho}_f, \eta_f), \tag{4.189}$$
$$\bar{p}_f = \bar{p}_f(\bar{\rho}_f, \eta_f), \tag{4.190}$$
$$\Lambda_f = \Lambda_f(\bar{\rho}_f, \eta_f). \tag{4.191}$$

We assume that emplacement of the fluid into the pores of the solid does not alter the functional relationships for the internal energy $\mathcal{E}_f$ and the true pressure $\bar{p}_f$. We also assume that the fluid does not dissipate energy. Notice that only variables associated with the solid appear in the first set of constitutive equations, and only variables associated with the fluid appear in the second set. This natural grouping of variables and constitutive equations for an immiscible mixture is called the *immiscibility postulate*. In general, it does not apply to miscible mixtures.

Our last set of constitutive relationships contains two expressions:

$$e_s = e_s\left(F^s, \varphi_s^E, \varphi_s^P, \eta_s, \bar{\rho}_f, \eta_f\right), \tag{4.192}$$
$$H_s^P = H_s^P\left(F^s, \varphi_s^E, \varphi_s^P, \eta_s, \bar{\rho}_f, \eta_f\right). \tag{4.193}$$

The function e_s represents the flow of heat between the solid and the fluid. It must necessarily contain variables from both the solid and the fluid. The second expression represents the plastic collapse of pores in the solid. It is similar to the flow rule for an elastic-plastic material (see Section 4.3). Because the pore fluid resists the pore collapse, this expression must also depend upon the properties of the pore fluid. Notice that the pore-collapse rule for H_s^P and the heat flux e_s are the only sources of dissipation that we allow in the mixture. Indeed, we do not intend these three sets of constitutive functions to be the most general possible statements within the context of a viscous immiscible mixture. Rather, we select these functions because they are sufficient to illustrate the essential features of a saturated porous solid.[†]

Admissible Thermodynamic Processes

Any specification of the motion $\hat{x}(X, t)$ and entropies $\hat{\eta}_s(X, t)$ and $\hat{\eta}_f(X, t)$ throughout the volume of this material results in an admissible thermodynamic process. To illustrate

[†] A general theory for immiscible mixtures can be found in "A thermomechanical theory for reacting immiscible mixtures" by Drumheller, D. S. and Bedford, A. in *Archive for Rational Mechanics and Analysis* **73**, pp. 257–284 (1980).

this, let us first consider the relationship [see Eq. (4.184)]

$$\Lambda_s\left(F^s, \varphi_s^E, \varphi_s^P, \eta_s\right) = \Lambda_f(\bar{\rho}_f, \eta_f).$$

Substituting

$$F^s = F\frac{\varphi_s^E \varphi_s^P}{\varphi_{s0}}$$

and

$$\bar{\rho}_f = \frac{\rho_{f0}}{F\left(1 - \varphi_s^E \varphi_s^P\right)},$$

we obtain

$$\breve{\Lambda}_s\left(\varphi_s^E, \varphi_s^P, F, \eta_s\right) = \breve{\Lambda}_f\left(\varphi_s^E, \varphi_s^P, F, \eta_f\right).$$

We assume that we can solve this expression to obtain

$$\varphi_s^E = \mathcal{F}\left(\varphi_s^P, F, \eta_s, \eta_f\right). \tag{4.194}$$

Similar substitutions into the constitutive relationship Eq. (4.193) result in

$$\partial\hat{\varphi}_s^P/\partial t = \breve{H}_s^P\left(\varphi_s^E, \varphi_s^P, F, \eta_s, \eta_f\right). \tag{4.195}$$

Now we show how to solve the governing equations for an arbitrary selection of solutions for $\hat{x}(X,t)$, $\hat{\eta}_s(X,t)$, and $\hat{\eta}_f(X,t)$. They are solved as follows:

- The velocity v and deformation gradient F are determined from their definitions.
- Recognizing that F and η_ξ are known functions, we combine Eqs. (4.194) and (4.195) to yield an ordinary differential equation in the unknown φ_s^P. We integrate this equation to determine φ_s^P.
- We determine φ_s^E from Eq. (4.194).
- The volume fractions are $\varphi_s = 1 - \varphi_f = \varphi_s^P \varphi_s^E$.
- The partial densities are $\rho_\xi = \rho_{\xi 0}/F$.
- The true deformation gradient of the solid is $F^s = F\varphi_s/\varphi_{s0}$.
- The true density of the fluid is $\bar{\rho}_f = \rho_f/\varphi_f$.
- The internal energies $\mathcal{E}_\xi$, true stress of the solid $\bar{T}^s$, true pressure of the fluid $\bar{p}_f$, interface pressures Λ_ξ, and heat exchange e_s are determined from the constitutive relations.
- The mixture stress is $T = \varphi_s \bar{T}^s - \varphi_f \bar{p}_f$.
- The body force b is selected so that the equation for the balance of linear momentum, Eq. (4.182), is satisfied.
- The external heat supplies r_ξ are selected so that the equations for the balances of energy, Eq. (4.183), are satisfied.

Notice that these admissible processes satisfy the conventional forms of the equations for the conservation of mass and balance of momentum that apply to all materials and not just mixtures. However, each component of the mixture must satisfy a separate balance of energy. The sum of these balances of energy is identical to the conventional equation for the balance of energy.

Transformability of Energy

Recall that all admissible thermodynamic processes must satisfy the second law of thermodynamics. We achieve this by constraining the constitutive functions of the mixture to ensure positive dissipation. To determine these constraints, we shall substitute the balance of energy for each constituent into the dissipation inequality. However, prior to doing this, some preliminary manipulations of the energy relations are in order. Let us begin with the balance of energy for the solid, Eq. (4.183). We use the chain-rule expansion of the time derivative of $\mathcal{E}_s$, substitute Eq. (4.176), and rearrange terms to obtain

$$\rho_{s0}\vartheta_s \frac{\partial \hat{\eta}_s}{\partial t} - \rho_{s0} r_s = T^s F\left(D^s - \frac{H_s}{\varphi_s}\right) - \rho_{s0}\frac{\partial \mathcal{E}_s}{\partial F^s} F^s D^s$$
$$- \rho_{s0}\frac{\partial \mathcal{E}_s}{\partial \varphi_s^E} H_s^E - \rho_{s0}\frac{\partial \mathcal{E}_s}{\partial \varphi_s^P} H_s^P - \Lambda_s H_s + e_s.$$

We notice that $\rho_s F^s = \varphi_s \bar{\rho}_{s0}$ and regroup terms on the right-hand side to obtain

$$\rho_{s0}\vartheta_s \frac{\partial \hat{\eta}_s}{\partial t} - \rho_{s0} r_s = \varphi_s\left(\bar{T}^s - \bar{\rho}_{s0}\frac{\partial \mathcal{E}_s}{\partial F^s}\right) F D^s - (\bar{T}^s F + \Lambda_s) H_s$$
$$- \rho_{s0}\frac{\partial \mathcal{E}_s}{\partial \varphi_s^E} H_s^E - \rho_{s0}\frac{\partial \mathcal{E}_s}{\partial \varphi_s^P} H_s^P + e_s.$$

We obtain the desired form of the balance of energy of the solid by substituting Eq. (4.178):

$$\rho_{s0}\vartheta_s \frac{\partial \hat{\eta}_s}{\partial t} - \rho_{s0} r_s = \varphi_s\left(\bar{T}^s - \bar{\rho}_{s0}\frac{\partial \mathcal{E}_s}{\partial F^s}\right) F D^s$$
$$- \varphi_s^E\left(\bar{T}^s + \frac{\Lambda_s}{F} + \varphi_s^P \bar{\rho}_s \frac{\partial \mathcal{E}_s}{\partial \varphi_s^P}\right) F H_s^P \tag{4.196}$$
$$- \varphi_s^P\left(\bar{T}^s + \frac{\Lambda_s}{F} + \varphi_s^E \bar{\rho}_s \frac{\partial \mathcal{E}_s}{\partial \varphi_s^E}\right) F H_s^E + e_s.$$

We obtain the desired form of the balance of energy for the fluid by an equivalent process. The result is

$$\rho_{f0}\vartheta_f \frac{\partial \hat{\eta}_f}{\partial t} - \rho_{f0} r_f = \frac{\varphi_f F}{\bar{\rho}_f}\left(\bar{p}_f - \bar{\rho}_f^2 \frac{\partial \mathcal{E}_f}{\partial \bar{\rho}_f}\right)\frac{\partial \hat{\bar{\rho}}_f}{\partial t} + (\bar{p}_f F - \Lambda_f) H_f - e_s. \tag{4.197}$$

As we now have the required forms of the balances of energy, let us determine the dissipation inequality for the mixture. The second law of thermodynamics stipulates that the rate of increase of entropy in an element of material must be greater than or equal to the flux of entropy delivered from external sources [see Eq. (3.31) in Section 3.2.3]:

$$\rho_0 \frac{\partial \hat{\eta}}{\partial t} \geq \frac{\rho_0 r}{\vartheta}.$$

Because the entropy of the immiscible mixture is†

$$\rho_0 \eta = \sum_\xi \rho_{\xi 0} \eta_\xi, \tag{4.198}$$

and the flux of entropy delivered to the ξ-th constituent from its external heat supply is $\rho_{\xi 0} r_\xi / \vartheta_\xi$, the second law of thermodynamics for the immiscible mixture is

$$\sum_\xi \rho_{\xi 0} \frac{\partial \hat{\eta}_\xi}{\partial t} \geq \sum_\xi \frac{\rho_{\xi 0} r_\xi}{\vartheta \xi}.$$

Substituting the balances of energy, Eqs. (4.196) and (4.197), into this inequality, we obtain

$$\begin{aligned} &\frac{\varphi_s}{\vartheta_s}\left(\bar{T}^s - \bar{\rho}_{s0}\frac{\partial \mathcal{E}_s}{\partial F^s}\right) F D^s + \frac{\varphi_f F}{\vartheta_f \bar{\rho}_f}\left(\bar{p}_f - \bar{\rho}_f^2 \frac{\partial \mathcal{E}_f}{\partial \bar{\rho}_f}\right)\frac{\partial \hat{\bar{\rho}}_f}{\partial t} \\ &\quad - \frac{\varphi_s^E}{\vartheta_s}\left(\bar{T}^s + \frac{\Lambda_s}{F} + \varphi_s^P \bar{\rho}_s \frac{\partial \mathcal{E}_s}{\partial \varphi_s^P}\right) F H_s^P - \frac{\varphi_s^P}{\vartheta_s}\left(\bar{T}^s + \frac{\Lambda_s}{F} + \varphi_s^E \bar{\rho}_s \frac{\partial \mathcal{E}_s}{\partial \varphi_s^E}\right) F H_s^E \\ &\quad + \left(\frac{\bar{p}_f F - \Lambda_f}{\vartheta_f}\right) H_f + \left(\frac{\vartheta_f - \vartheta_s}{\vartheta_f \vartheta_s}\right) e_s \geq 0. \end{aligned} \tag{4.199}$$

This is the dissipation inequality for the mixture. We must construct the constitutive relations to ensure that it is never violated for any admissible thermodynamic process. *We achieve this by requiring every term in this expression to separately satisfy the inequality.* For example, the second term yields

$$\left(\bar{p}_f - \bar{\rho}_f^2 \frac{\partial \mathcal{E}_f}{\partial \bar{\rho}_f}\right)\frac{\partial \hat{\bar{\rho}}_f}{\partial t} \geq 0.$$

To satisfy this inequality consider a set of processes that contain the same values of $\bar{p}_f$, $\bar{\rho}_f$, and $\partial \mathcal{E}_f / \partial \bar{\rho}_f$. Different processes in this set also contain values of $\partial \hat{\bar{\rho}}_f / \partial t$ that have arbitrarily large positive and negative values. All of these processes cannot satisfy this constraint unless we require

$$\bar{p}_f = \bar{\rho}_f^2 \frac{\partial \mathcal{E}_f}{\partial \bar{\rho}_f}. \tag{4.200}$$

In addition, the first, fourth, and fifth terms in Eq. (4.199) yield

$$\bar{T}^s = \bar{\rho}_{s0} \frac{\partial \mathcal{E}_s}{\partial F^s}, \tag{4.201}$$

$$\bar{T}^s = -\frac{\Lambda_s}{F} - \varphi_s^E \bar{\rho}_s \frac{\partial \mathcal{E}_s}{\partial \varphi_s^E}, \tag{4.202}$$

$$\bar{p}_f = \frac{\Lambda_f}{F}. \tag{4.203}$$

† The entropy of a mixture does not always obey Eq. (4.198). In particular, we can show that the entropy of a mixture of two ideal gases that diffuse into each other on a molecular level is greater than the sum of the constituent entropies. The additional term that appears in the mixing rule is called the *entropy of mixing*. Here we ignore the entropy of mixing because we are concerned with immiscible mixtures. You can find a good discussion of this point in *Thermodynamics and Statistical Mechanics* by Sommerfeld, A. (Academic Press, 1964).

The remaining inequalities are

$$-\left(\bar{T}^s + \frac{\Lambda_s}{F} + \varphi_s^P \bar{\rho}_s \frac{\partial \mathcal{E}_s}{\partial \varphi_s^P}\right) H_s^P \geq 0, \tag{4.204}$$

$$(\vartheta_f - \vartheta_s) e_s \geq 0, \tag{4.205}$$

which we can satisfy by requiring the constitutive relations for H_s^P and e_s to have the same signs and zeros as their respective coefficients in these inequalities. These inequalities represent dissipation of energy in the mixture. Recall that $e_s = -e_f$ represents the flow of heat between the solid and the pore fluid, and both the energy equations for the solid, Eq. (4.196), and the fluid, Eq. (4.197), are directly affected by e_s. The inequality Eq. (4.205) requires that heat can only flow from the hotter to the colder constituent. In contrast, the inequality in Eq. (4.204) represents dissipation resulting from pore collapse, and only the energy equation of the solid is directly affected.

4.4.3 *A Mixture of Two Fluids*

We shall see that shock wave propagation in a porous solid can result in very high temperatures. Consequently, it is quite possible the solid will melt to produce a mixture of two fluids. Under these conditions we can replace the constitutive relations for the solid with the following relationships:

$$\mathcal{E}_s = \mathcal{E}_s(\bar{\rho}_s, \eta_s),$$
$$\bar{p}_s = \bar{p}_s(\bar{\rho}_s, \eta_s),$$
$$\Lambda_s = \Lambda_s(\bar{\rho}_s, \eta_s).$$

Then we can easily demonstrate that

$$\bar{p}_s = \Lambda_s/F = \Lambda_f/F = \bar{p}_f.$$

The mixture pressure is

$$p = \varphi_s \bar{p}_s + \varphi_f \bar{p}_f = \bar{p}_s = \bar{p}_f,$$

where we recall that $\sum_\xi \varphi_\xi = 1$. Thus we obtain

$$p_\xi = \varphi_\xi p.$$

It is possible to generalize this result to immiscible fluid mixtures containing an arbitrary number of constituents. This rule also applies to miscible mixtures of ideal gases, where it is called *Dalton's Law*.

Wave Velocities

If both liquid and gas are present, the mixture can exhibit exceedingly slow wave velocities. Propagation velocities as low as 33 m/s have been reported for steam–water mixtures. This happens because the mixture exhibits the stiffness of a gas combined with the density of a liquid. To calculate this velocity, we must first derive a relationship for the stiffness of each constituent of the mixture. Then we must derive another relationship that ensures that each constituent has the same true pressure. Recall that the isentropic stiffness of a fluid is

$$C^\eta = -F \frac{\partial p}{\partial F}.$$

If this fluid is a constituent of a mixture, we find that

$$C_\xi^\eta = \bar{\rho}_\xi \frac{\partial \bar{p}_\xi}{\partial \bar{\rho}_\xi}. \tag{4.206}$$

This is the first desired relationship. For the second relationship, we start by noticing that

$$\frac{\partial \bar{p}_1}{\partial F} = \frac{\partial \bar{p}_2}{\partial F}.$$

A structured wave is an isentropic process. During an isentropic process, we can use the chain rule to obtain

$$\frac{\partial \bar{p}_1}{\partial \bar{\rho}_1}\frac{\partial \bar{\rho}_1}{\partial F} = \frac{\partial \bar{p}_2}{\partial \bar{\rho}_2}\frac{\partial \bar{\rho}_2}{\partial F}.$$

We cast this expression into a more convenient form by noticing that the partial derivative of the logarithm of the conservation of mass Eq. (4.179) is

$$\frac{1}{\bar{\rho}_\xi}\frac{\partial \bar{\rho}_\xi}{\partial F} = -\frac{1}{\varphi_\xi}\frac{\partial \varphi_\xi}{\partial F} - \frac{1}{F}. \tag{4.207}$$

We combine these expressions to obtain the second desired relationship,

$$\bar{\rho}_1 \frac{\partial \bar{p}_1}{\partial \bar{\rho}_1}\left(\frac{1}{\varphi_1}\frac{\partial \varphi_1}{\partial F} + \frac{1}{F}\right) = \bar{\rho}_2 \frac{\partial \bar{p}_2}{\partial \bar{\rho}_2}\left(\frac{1}{\varphi_2}\frac{\partial \varphi_2}{\partial F} + \frac{1}{F}\right).$$

We substitute Eq. (4.206) to obtain

$$\left(\frac{C_1^\eta}{\varphi_1} + \frac{C_2^\eta}{\varphi_2}\right)\frac{\partial \varphi_1}{\partial F} = \frac{C_2^\eta - C_1^\eta}{F}, \tag{4.208}$$

where we notice that $\partial\varphi_1/\partial F = -\partial\varphi_2/\partial F$.

Now we derive the expression for the velocity of a structured wave. The material velocity C of this wave is given by

$$C^2 = -\frac{1}{\rho_0}\frac{\partial p}{\partial F},$$

where

$$\rho_0 = \sum_\xi \varphi_{\xi 0}\bar{\rho}_{\xi 0}.$$

Because the true pressures of each constituent are equal to the mixture pressure p, we obtain

$$C^2 = -\frac{1}{\rho_0}\frac{\partial \bar{p}_1}{\partial \bar{\rho}_1}\frac{\partial \bar{\rho}_1}{\partial F}.$$

We substitute Eqs. (4.206) and (4.207) to find that

$$C^2 = \frac{C_1^\eta}{\rho_0}\left(\frac{1}{\varphi_1}\frac{\partial \varphi_1}{\partial F} + \frac{1}{F}\right).$$

When we use Eq. (4.208) to eliminate $\partial\varphi_1/\partial F$, we obtain

$$C = \left[\rho_0 F\left(\frac{\varphi_1}{C_1^\eta} + \frac{\varphi_2}{C_2^\eta}\right)\right]^{-1/2}. \tag{4.209}$$

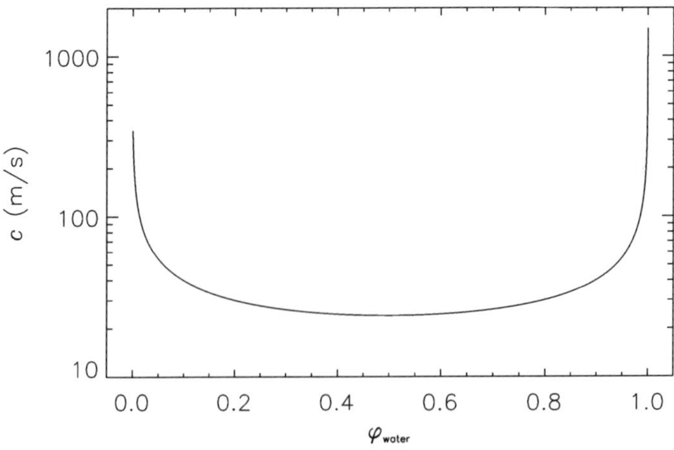

Figure 4.61 The speed of sound in an air–water mixture.

Speed of Sound in an Air–Water Mixture

Consider a mixture of air and water at standard atmospheric conditions. The isentropic stiffnesses and the densities of the constituents of this mixture are

$$C_1^\eta = 2.2 \text{ GPa}, \qquad \bar{\rho}_1 = 1.0 \text{ Mg/m}^3,$$
$$C_2^\eta = 0.15 \text{ MPa}, \qquad \bar{\rho}_2 = 1.2 \text{ kg/m}^3,$$

where constituent 1 is water and constituent 2 is air. By substituting these parameters into our expression for C, Eq. (4.209), we obtain the graph in Figure 4.61. Here we compute the speed of sound by letting $F = 1$ and varying the volume fraction of water $\varphi_1 \equiv \varphi_{\text{water}}$ from zero to one. When $\varphi_{\text{water}} = 0$, we obtain the speed of sound in air, $C = 345$ m/s, and when $\varphi_{\text{water}} = 1$, we obtain the speed of sound in water, $C = 1.48$ km/s. Notice that over most of the mixture compositions, the speed of sound of the mixture is below that of both constituents. The minimum value is $C = 24$ m/s. When $\varphi_{\text{water}} \approx 0$, the mixture consists of small water droplets in air, and when $\varphi_{\text{water}} \approx 1$, the mixture is a bubbly liquid. Typically, it is difficult to produce homogeneous mixtures between these limits. However, these calculations show that if only one percent of the volume of the mixture is occupied by air, the speed of sound drops dramatically from 1,480 m/s to 121 m/s. This result has been confirmed by experimental measurements on bubbly liquids.

4.4.4 *Dry Plastic Porous Solid*

Unlike the two-fluid mixture just discussed, determination of the eight constitutive relationships, Eqs. (4.186)–(4.193), for a fluid–solid mixture can be a very imposing task. Fortunately, even very simple constitutive models can yield quite useful results. Indeed, we have already obtained useful insight from the porous locking solid, which has constant true density. In this section we shall discuss another simple constitutive model for a dry porous solid. In this case, we allow the true density of the solid to change, but we eliminate any dependence on the elastic volume fraction φ_s^E. Hence any change in the volume fraction is plastic. We call this material the *plastic porous solid*. Throughout this section we assume that the plastic porous solid is dry.

For a dry porous solid, we find that $p_f = \Lambda_f/F = \Lambda_s/F = 0$. As a result, we combine Eqs. (4.201) and (4.202) to obtain

$$F^s \frac{\partial \mathcal{E}_s}{\partial F^s} = -\varphi_s^E \frac{\partial \mathcal{E}_s}{\partial \varphi_s^E}. \tag{4.210}$$

Any dry porous solid must always satisfy this condition. *In this section we shall satisfy this condition by replacing the constitutive model, Eq. (4.186), by*

$$\mathcal{E}_s = \mathcal{E}_s(\beta, \eta_s), \tag{4.211}$$

where

$$\varphi_{s0} \equiv 1$$

and

$$\beta \equiv F^s/\varphi_s^E.$$

We see that Eq. (4.210) is satisfied identically by Eq. (4.211). Furthermore, because we desire only plastic pore collapse, we shall assume that $\varphi_s^E = 1$, so that

$$\varphi_s = \varphi_s^P.$$

Following a common practice, we define the *distention ratio* α,

$$\alpha \equiv 1/\varphi_s = 1/\varphi_s^P,$$

and therefore

$$\beta = F^s = F/\alpha.$$

Because the porous solid is dry, we find that the mixture density is [see Eq. (4.180)]

$$\rho = \rho_s$$

and the mixture internal energy is [see Eq. (4.185)]

$$\mathcal{E} = \mathcal{E}_s.$$

Reference Solid

It is important for you to recognize that the internal energy of this porous solid is a known function. To illustrate this point, let us evaluate the true stress $\bar{T}^s$ using Eq. (4.201):

$$\begin{aligned} \mathcal{E} &= \mathcal{E}_s(\beta, \eta_s), \\ \bar{T}^s &= \bar{\rho}_{s0} \frac{\partial \mathcal{E}_s}{\partial F^s} = \bar{\rho}_{s0} \frac{\partial \mathcal{E}_s}{\partial \beta}. \end{aligned}$$

When the solid is fully compacted so that $\alpha = 1$, it is called the *reference solid*. We find that

$$\left.\begin{aligned} \mathcal{E} &= \mathcal{E}_s(F, \eta_s) \\ \bar{T}^s &= \rho_0 \tfrac{\partial \mathcal{E}}{\partial F} \end{aligned}\right\} \quad \text{(reference solid).}$$

Clearly, the expressions for the reference solid are the conventional representations of the internal energy and the equilibrium stress. For example, when $\alpha = 1$, let us assume the true stress is given by the Mie–Grüneisen expression [see Eq. (4.72) in Section 4.2]

$$\bar{T}^s = -p_c(F) - \bar{\rho}\Gamma[\mathcal{E} - \mathcal{E}_c(F)] \quad \text{(reference solid).}$$

When $\alpha \neq 1$, we replace F by β and find that

$$\bar{T}^s = -p_c(F/\alpha) - \bar{\rho}\Gamma[\mathcal{E} - \mathcal{E}_c(F/\alpha)]. \tag{4.212}$$

The functions for the cold pressure p_c and the cold energy $\mathcal{E}_c$ are unaltered by the value of α.

Pore Collapse

Consider the following yield hypothesis for equilibrium states:

$$Y_t(\alpha) \geq \bar{T}^s \geq Y_c(\alpha), \tag{4.213}$$

where Y_t and Y_c are arbitrary functions. We require that

$$Y_c(1) \rightarrow -\infty.$$

We illustrate this condition in Figure 4.62. At equilibrium, the stress $\bar{T}^s$ must fall within the shaded region. We shall implement this yield condition with the following pore-collapse rule:

$$\frac{\partial \hat{\alpha}}{\partial t} = \begin{cases} b(\bar{T}^s - Y_c), & \bar{T}^s < Y_c < 0, \\ 0, & Y_t \geq \bar{T}^s \geq Y_c, \\ b(\bar{T}^s - Y_t), & \bar{T}^s > Y_t > 0, \end{cases} \tag{4.214}$$

where b is an arbitrary positive constant. This collapse rule satisfies the dissipation inequality Eq. (4.204), which in terms of $\bar{T}^s$ and α is

$$\bar{T}^s \frac{\partial \hat{\alpha}}{\partial t} \geq 0,$$

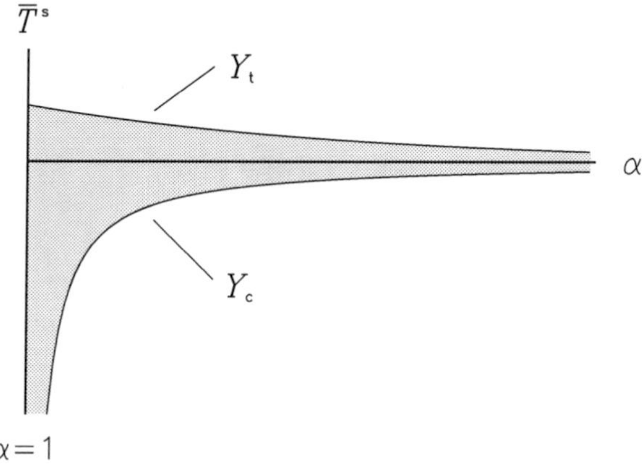

Figure 4.62 The yield condition for pore collapse.

where $H_s^P = -\frac{1}{\alpha^2}\frac{\partial\hat{\alpha}}{\partial t}$, $\Lambda_s = 0$, and $\mathcal{E}_s$ is not an explicit function of φ_s^P [see Eq. (4.211)]. By requiring $Y_c(1) \rightarrow -\infty$, we ensure that the reference solid can support unlimited compressive stress. In contrast, $Y_t(1)$ is finite, which means we only allow the reference solid to support limited tensile stress. Notice that Y_c and Y_t only depend upon α. Thus, if a solid with an initial distention of $\alpha = \alpha_0 > 1$ is first compacted to $\alpha \approx 1$ and then expanded back to $\alpha = \alpha_0$, the final values of Y_c and Y_t will be equal to the initial values. In general, when we subject real solids to this type of cyclic loading, we produce permanent changes in Y_c and Y_t. When these changes are small, the ideal response represented by Figure 4.62 is an appropriate approximation. *Ductile materials* exhibit this type of behavior.

Damage

Cyclic loading of a porous solid often causes damage to the pore structure. In ductile materials such as soft metals, the accumulation of damage may be slow enough to be ignored. However, in *brittle materials* such as glass or rock, the accumulation of damage may be very rapid. We can account for the accumulation of damage by including a *damage variable* $\mathcal{B}$ in the yield condition, Eq. (4.213):

$$Y_t(\alpha)/\mathcal{B} \geq \bar{T}^s \geq Y_c(\alpha)/\mathcal{B},$$

where

$$\frac{\partial\hat{\mathcal{B}}}{\partial t} = K\mathcal{B}\left|\frac{\partial\hat{\alpha}}{\partial t}\right|$$

and K is an arbitrary positive constant. Initially $\mathcal{B} = 1$. As the distention ratio changes, this evolution equation causes $\mathcal{B}$ to grow. We control the rate of growth of damage with K. In Figure 4.63 we illustrate the effect of damage on the yield condition. We can implement this yield condition with damage by using the pore-collapse rule

$$\frac{\partial\hat{\alpha}}{\partial t} = \begin{cases} b(\bar{T}^s - Y_c/\mathcal{B}), & \bar{T}^s < Y_c/\mathcal{B} < 0, \\ 0, & Y_t/\mathcal{B} \geq \bar{T}^s \geq Y_c/\mathcal{B}, \\ b(\bar{T}^s - Y_t/\mathcal{B}), & \bar{T}^s > Y_t/\mathcal{B} > 0. \end{cases} \tag{4.215}$$

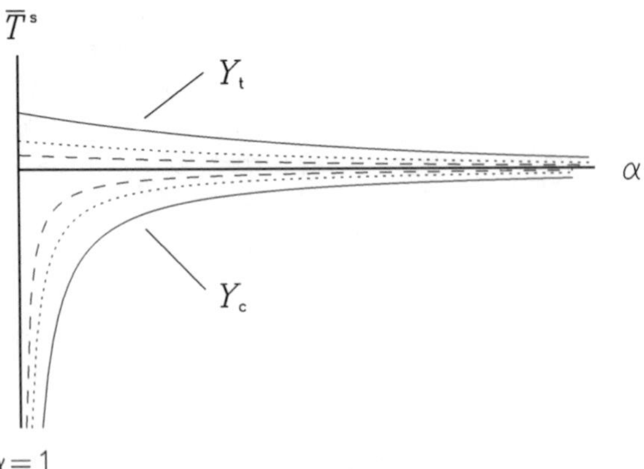

Figure 4.63 The yield condition for several values of damage: $\mathcal{B} = 1$ (—), 2 (· · ·), and 4 (– –).

This is often called a *cumulative-damage model*. Cumulative-damage models can also be used to model a nonporous reference solid. As we subject the solid to a tensile load, the nucleation and accumulation of porosity and damage ultimately results in the failure of the solid.

Governing Equations

This section contains a summary of the equations that govern the response of a dry plastic porous solid. Recall that the mixture stress is

$$T = \sum_{\xi} T_{\xi} = \bar{T}^s/\alpha.$$

By ignoring the specific body forces and external heat supplies, we obtain the following forms for the equations for the conservation of mass, balance of linear momentum, and balance of energy:

$$\frac{\partial \hat{v}}{\partial X} = \frac{\partial \hat{F}}{\partial t},$$

$$\rho_{s0}\frac{\partial \hat{v}}{\partial t} = \frac{\partial}{\partial X}\left(\frac{\hat{\bar{T}}^s}{\hat{\alpha}}\right),$$

$$\rho_{s0}\frac{\partial \hat{\mathcal{E}}}{\partial t} = \left(\frac{\bar{T}^s}{\alpha}\right)\frac{\partial \hat{F}}{\partial t}.$$

These three relationships plus the constitutive equation for $\bar{T}^s$, Eq. (4.212), and the pore-collapse rule for $\partial\hat{\alpha}/\partial t$, Eq. (4.214), are a balanced system of five equations containing the five unknowns, $\bar{T}^s$, F, α, $\mathcal{E}$, and v.

Elastic Waves

We can evaluate the material velocity C of elastic waves in a plastic porous solid from Eq. (3.87) in Section 3.4.2:

$$C = \left(\frac{\partial^2 \mathcal{E}}{\partial F^2}\right)^{1/2}.$$

We apply the chain rule to obtain

$$\frac{\partial \mathcal{E}}{\partial F} = \frac{1}{\alpha}\frac{\partial \mathcal{E}_s}{\partial \beta}, \qquad \frac{\partial^2 \mathcal{E}}{\partial F^2} = \left(\frac{1}{\alpha}\right)^2 \frac{\partial^2 \mathcal{E}_s}{\partial \beta^2},$$

where, by definition, the distention ratio α is constant during an elastic process. The spatial velocity c is

$$c = FC = \beta\left(\frac{\partial^2 \mathcal{E}_s}{\partial \beta^2}\right)^{1/2}.$$

Notice that c depends on β and not separately on α and F. Any combination of the distention ratio α and deformation gradient F that yields the same value of β also yields the same value of c. For example, in a solid where the stress is zero, we find that $\beta = 1$. Thus the value of c in the stress-free porous solid is equal to the value of c in the fully-compacted,

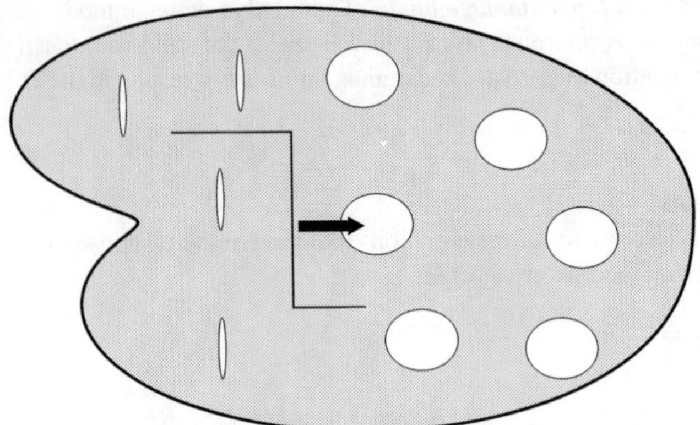

Figure 4.64 A shock wave in a plastic porous solid.

stress-free reference material. *Experimental evidence often contradicts this conclusion that we have derived from the constitutive model for the dry plastic porous solid.*

Shock Waves

Consider a plastic porous solid that is initially at rest and in a stress-free state (see Figure 4.64). The conditions at the initial state are

$$\alpha_- = \alpha_0, \qquad \bar{T}_- = 0,$$

$$F_- = \alpha_0, \qquad \rho_- = \bar{\rho}_{s0}/\alpha_0,$$

$$\beta_- = 1.$$

A shock wave propagates in the positive-X direction. The jump condition for the balance of linear momentum at the shock wave is

$$p_+ = \rho_{s0} U_S^2 (\alpha_0 - F_+),$$

and the jump condition for the balance of energy is

$$\mathcal{E}_+ - \mathcal{E}_- = \frac{p_+}{2\rho_{s0}}(\alpha_0 - F_+).$$

Following the same approach used in Section 4.2.4, we define the functions

$$p_H(F) \equiv \rho_{s0} U_S^2 (\alpha_0 - F)$$

and

$$\mathcal{E}_H(F) \equiv \frac{p_H}{2\rho_{s0}}(\alpha_0 - F) + \mathcal{E}_-.$$

From the Mie–Grüneisen constitutive relation, Eq. (4.212), we obtain the following expression for the true pressure at the Hugoniot state:

$$\bar{p}_+ = p_c(F_+/\alpha_+) + \bar{\rho}_{s+}\Gamma_+[\mathcal{E}_+ - \mathcal{E}_c(F_+/\alpha_+)].$$

We add the quantity $\bar{\rho}_{s+}\Gamma_+\mathcal{E}_-$ to boths sides and rearrange terms to obtain

$$\alpha_+ p_+ - \bar{\rho}_{s+}\Gamma_+(\mathcal{E}_+ - \mathcal{E}_-) = p_c(F_+/\alpha_+) - \bar{\rho}_{s+}\Gamma_+[\mathcal{E}_c(F_+/\alpha_+) - \mathcal{E}_-],$$

where we notice that $\bar{p}_+ = \alpha_+ p_+$. This becomes

$$\left[1 - \frac{\rho_{s+}\Gamma_+}{2\rho_{s0}}(\alpha_0 - F_+)\right]\alpha_+ p_H(F_+) = p_c(F_+/\alpha_+) - \bar{\rho}_{s+}\Gamma_+[\mathcal{E}_c(F_+/\alpha_+) - \mathcal{E}_-], \tag{4.216}$$

which is the desired Hugoniot relationship for the shock wave. If we know the velocity U_S of the shock wave, there are two unknowns in this expression: the deformation gradient F_+ and the distention ratio α_+ at the Hugoniot state. Moreover, the yield hypothesis provides another equation in F_+ and α_+ that we can solve simultaneously with Eq. (4.216) to determine the solution at the Hugoniot state.

***Locking Hugoniot for Constant* $\boldsymbol{\Gamma}$** In general, the Grüneisen parameter $\boldsymbol{\Gamma}$ is a function of the true density of the solid. However, if we assume that Γ is a constant and equal to Γ_0, the Hugoniot relationship Eq. (4.216) becomes

$$\left[1 - \frac{\Gamma_0}{2F_+}(\alpha_0 - F_+)\right]\alpha_+ p_H(F_+) = p_c(F_+/\alpha_+) - \bar{\rho}_{s+}\Gamma_0[\mathcal{E}_c(F_+/\alpha_+) - \mathcal{E}_-].$$

Suppose the shock wave is strong enough to completely collapse the pores in the solid so that $\alpha_+ = 1$. Then

$$\left[1 - \frac{\Gamma_0}{2F_+}(\alpha_0 - F_+)\right] p_H(F_+) = p_c(F_+) - \bar{\rho}_{s+}\Gamma_0[\mathcal{E}_c(F_+) - \mathcal{E}_-].$$

This relationship demonstrates an unusual property of this porous solid. For a particular value of the distention ratio at the initial state, the Hugoniot is vertical and given by $F_+ = 1$. In this case, we see that the value of the right-hand side of the previous relationship is zero. This means either $p_H(1) = 0$ or

$$1 - \frac{\Gamma_0}{2}(\alpha_0 - 1) = 0.$$

This last condition exists, if $\alpha_0 = \alpha_l$, where

$$\alpha_l \equiv \frac{2}{\Gamma_0} + 1. \tag{4.217}$$

Thus, when we select $\alpha_0 = \alpha_l$ and ignore the pore-collapse rule, this plastic porous solid exhibits a vertical Hugoniot and behaves like the porous locking solid. This phenomenon has been observed in numerous experiments on porous metals.

***Locking Hugoniot for Constant* $\boldsymbol{\bar{\rho}_s\Gamma}$** When $\bar{\rho}_s\Gamma$ is a constant and equal to $\bar{\rho}_{s0}\Gamma_0$, we notice that

$$\rho_{s+}\Gamma_+ = \frac{\bar{\rho}_{s+}\Gamma_+}{\alpha_+} = \frac{\bar{\rho}_{s0}\Gamma_0}{\alpha_+}.$$

Substituting this result into the Hugoniot relationship Eq. (4.216), we obtain

$$\left[1 - \frac{\Gamma_0 \alpha_0}{2\alpha_+}(\alpha_0 - F_+)\right]\alpha_+ p_H(F_+) = p_c(F_+/\alpha_+) - \frac{\bar{\rho}_{s0}\Gamma_0}{\alpha_+}[\mathcal{E}_c(F_+/\alpha_+) - \mathcal{E}_-]. \tag{4.218}$$

This porous solid also exhibits a vertical Hugoniot for a particular value of α_0. After substitution of $F_+ = 1$ and $\alpha_+ = 1$, we find that a vertical Hugoniot exists when

$$\alpha_0^2 - \alpha_0 - \frac{2}{\Gamma_0} = 0.$$

This equation has both a positive and negative root. Only the positive root is physically realistic. It is

$$\alpha_l = \frac{1}{2}\left(\sqrt{\frac{8}{\Gamma_0} + 1} + 1\right). \tag{4.219}$$

In Figure 4.65 we illustrate the theoretical Hugoniots for eight samples of stainless steel with different initial distention ratios. We discussed the constitutive parameters for this solid in Section 4.2 where we presented the Mie–Grüneisen solid. The curve on the left is the Hugoniot for the reference solid where $\alpha_0 = 1$. Because we have ignored the yield condition and assumed that $\alpha_+ = 1$, all of the Hugoniots emerge from the same point at $F = 1$ and $p = 0$. Notice that each Hugoniot exhibits a different compression limit, and one is vertical. In this example, the parameters $\Gamma_0 = 1.94$ and $\alpha_l = 1.63$ produce the vertical Hugoniot [see Eq. (4.219)]. When the initial distention ratio is greater than 1.63, the shock wave causes the reference solid to expand, and the true density at the Hugoniot state is *less* than the true density at the initial state.

The source of this expansion is the large heating that the shock wave produces in the porous solid. Figure 4.66 illustrates this point. We show two Hugoniots. The upper solid

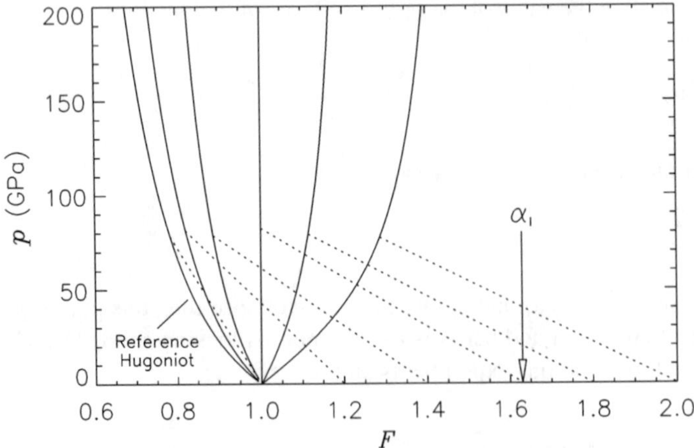

Figure 4.65 Hugoniots (—) for porous stainless steel. The Rayleigh lines (· · ·) connect the Hugoniot to the initial distended state.

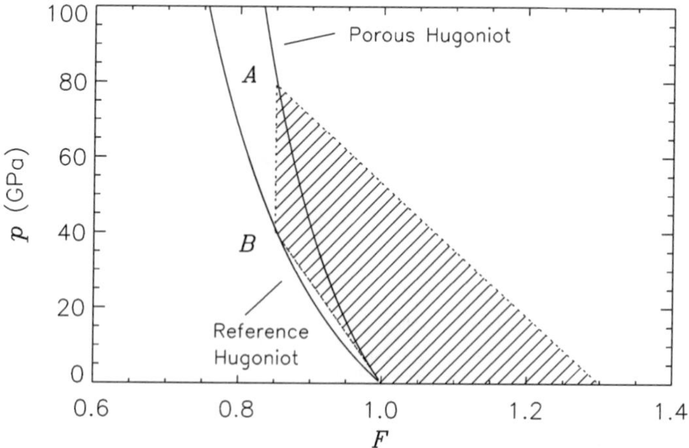

Figure 4.66 The Hugoniots for the reference solid and the porous solid. The cross-hatched area represents the additional energy required to compress the porous solid.

curve is the Hugoniot for a porous sample of stainless steel where $\alpha_0 = 1.3$. We illustrate a particular Hugoniot state A and Rayleigh line. The lower solid curve is the Hugoniot for a fully compacted sample of stainless steel where $\alpha_0 = 1$. We illustrate a Hugoniot state B for this material. Notice that $F_A = F_B$. This means that at the Hugoniot states A and B, both samples have the same cold pressure p_c and cold energy $\mathcal{E}_c$. Thus we can write [see Eq. (4.72)]

$$p_A = p_c + \bar{\rho}_{s0}\Gamma_0(\mathcal{E}_A - \mathcal{E}_c),$$

$$p_B = p_c + \bar{\rho}_{s0}\Gamma_0(\mathcal{E}_B - \mathcal{E}_c).$$

Subtraction yields

$$p_A - p_B = \bar{\rho}_{s0}\Gamma_0(\mathcal{E}_A - \mathcal{E}_B).$$

Indeed, the difference in pressure between the porous and the fully compacted samples is proportional to the difference in internal energy of the two Hugoniot states. Moreover, this difference in internal energy is equal to the cross-hatched area in the figure. Clearly, this area increases when α_0 increases.

In Figure 4.67 we illustrate the same two Hugoniots and Rayleigh lines again. In this figure, we have shaded the respective areas between the Hugoniots and Rayleigh lines. Recall that for weak shocks where the release path is close to the Hugoniot these areas represent the heating caused by each shock wave. Again, it is obvious that the porous solid experiences far greater heating, and as a result, the shock wave in the porous solid dissipates far more rapidly.

Yield Condition In the previous calculations for the Hugoniots of porous stainless steel, we assumed that $\alpha_+ = 1$. Suppose we repeat these calculations using the following

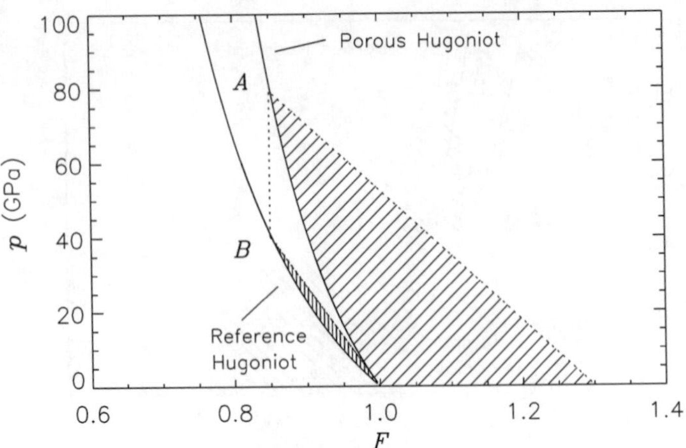

Figure 4.67 The area between the Hugoniot and the Rayleigh line for the reference solid and the porous solid.

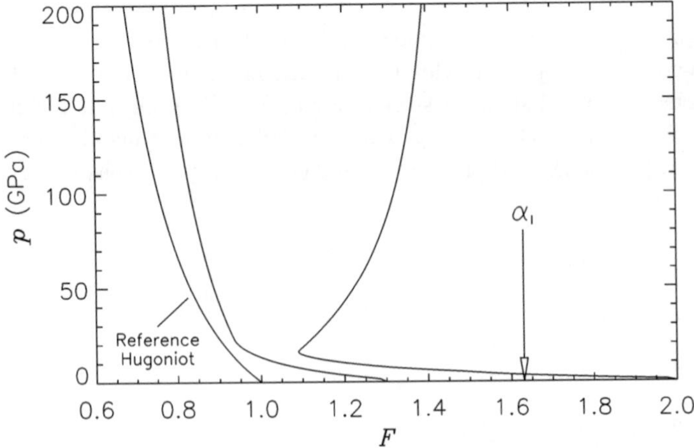

Figure 4.68 The effect of the yield condition on the Hugoniots for porous stainless steel.

empirical yield condition:

$$\begin{aligned}
&Y_c = 0, && \frac{\alpha - 1}{\alpha_0 - 1} \geq \left(\frac{\bar{p}_{\max}}{\bar{p}_{\max} - \bar{p}_{\min}}\right)^2, \\
&\left(\frac{\bar{p}_{\max} + Y_c}{\bar{p}_{\max} - \bar{p}_{\min}}\right)^2 = \frac{\alpha - 1}{\alpha_0 - 1}, && 0 < \frac{\alpha - 1}{\alpha_0 - 1} < \left(\frac{\bar{p}_{\max}}{\bar{p}_{\max} - \bar{p}_{\min}}\right)^2, \\
&Y_c = -\infty, && \alpha = 1,
\end{aligned} \tag{4.220}$$

where $\bar{p}_{\min} = 2$ GPa and $\bar{p}_{\max} = 20$ GPa. We simultaneously solve these expressions with Eq. (4.218) to obtain the Hugoniots in Figure 4.68. We illustrate only three of the original eight Hugoniots. The curve on the left is the Hugoniot for the reference solid. The Hugoniot in the middle is for a solid where $\alpha_0 < \alpha_l$, and the Hugoniot on the right is for a solid where

$\alpha_0 > \alpha_l$. Notice that above $\bar{p}_+ = 20$ GPa, the porous solids are fully compacted by the shock wave and exhibit Hugoniots that are identical to those in Figure 4.65. The curves for the porous samples also exhibit inflection points at $\bar{p}_{\min} = 2$ GPa, but they are difficult to see because of the scale of the plot.

4.4.5 *Elastic-Plastic Porous Solid*

In the previous section, we studied the plastic porous solid. We assumed the internal energy of this solid was

$$\mathcal{E} = \mathcal{E}_s(\beta, \eta_s) \quad \text{(plastic porous solid).}$$

Substitution into Eqs. (4.201) and (4.202) reveals that $\Lambda_s = 0$. Because $\Lambda_s/F = \Lambda_f/F = p_f = 0$, the pore fluid cannot exert a pressure on the solid. Thus we cannot model saturated porous solids with the model for the plastic porous solid. We now turn to a model that can. Let us assume the internal energy of the solid can be partitioned as follows:

$$\mathcal{E}_s\left(F_s, \varphi_s^E, \varphi_s^P, \eta_s\right) = \mathcal{E}_{sr}\left(F^s, \eta_s\right) + \mathcal{E}_{sp}\left(\varphi_s^E, \varphi_s^P\right). \tag{4.221}$$

We assume a special function for $\mathcal{E}_{sp}$:

$$\mathcal{E}_{sp} = \frac{1}{2\omega\left(\varphi_s^P\right)}\left[\ln \varphi_s^E\right]^2, \tag{4.222}$$

where $\omega\left(\varphi_s^P\right)$ is an arbitrary function of φ_s^P. We shall also continue to assume that

$$\varphi_{s0} = 1.$$

Elastic Volume Fraction

When Eq. (4.221) is substituted into Eqs. (4.201) and (4.202), we obtain

$$\bar{T}^s = \bar{\rho}_{s0}\frac{\mathcal{E}_{sr}}{F^s}, \qquad \Omega = -\frac{1}{\omega}\ln \varphi_s^E, \tag{4.223}$$

where

$$\Omega \equiv \frac{\bar{T}^s + \bar{p}_f}{\bar{\rho}_s}.$$

We solve Eq. (4.223) for φ_s^E to obtain

$$\varphi_s^E = e^{-\omega\Omega}, \tag{4.224}$$

and then we substitute the result into Eq. (4.222) to obtain

$$\mathcal{E}_s = \mathcal{E}_{sr}(F^s, \eta_s) + \frac{1}{2}\omega\Omega^2.$$

Notice that

$$\mathcal{E}_{sp} = \frac{1}{2}\omega\Omega^2$$

and $\mathcal{E}_{sp} \to 0$ as $\omega \to 0$. Indeed, we now see that during an elastic process the elastic volume fraction of the porous solid is strictly a function Ω. Therefore, for example, if we heat the dry solid while holding Ω constant, the volume fraction will not change. This phenomenon is commonly observed in dry porous solids.

Reference Solid

Consider a process in which F^s and η_s are constant. This means that $\bar{T}^s$ is constant. We can change Ω by changing $\bar{p}_f$. Clearly, if we increase the pore pressure, the internal energy of the solid should increase. Therefore, we require that

$$\omega \geq 0.$$

When $\varphi_s^P = 1$, we shall also require that

$$\omega(1) = 0. \tag{4.225}$$

Then Eq. (4.224) shows that $\varphi_s^E = 1$, and thus $\varphi_s = 1$. Therefore, Eq. (4.221) yields

$$\mathcal{E}_s(F, 1, 1, \eta_s) = \mathcal{E}_{sr}(F^s, \eta_s).$$

Thus $\mathcal{E}_{sr}$ is the internal energy expression for the fully compacted reference solid, and by Eq. (4.223), we can obtain the true stress $\bar{T}^s$ in the porous material from the internal energy of the reference solid. This connection between the stress in the porous material and the internal energy in the fully compacted material was employed by Herrmann[†] to derive a similar theory for a dry porous solid. However, Herrmann equated the partial stress T^s to the derivative of $\mathcal{E}_{sr}$. Here the dissipation inequality requires us to use the true stress $\bar{T}^s$ instead of the partial stress T^s.

Plastic Volume Fraction

We shall control the plastic volume fraction with a yield condition and a pore-collapse rule for H_s^P. The pore-collapse rule must satisfy the dissipation inequality Eq. (4.204):

$$-\left(\bar{T}^s + \frac{\Lambda_s}{F} + \varphi_s^P \bar{\rho}_s \frac{\partial \mathcal{E}_s}{\partial \varphi_s^P}\right) H_s^P \geq 0.$$

We substitute Eq. (4.222) and divide the result by $\bar{\rho}_s$ to obtain

$$-\left[1 - \frac{\varphi_s^P \Omega}{2} \frac{d\omega}{d\varphi_s^P}\right] \Omega H_s^P \geq 0.$$

We construct a yield hypothesis for this material by modifying Eq. (4.213) as follows:

$$Y_t(\alpha) \geq \bar{\rho}_s \Omega \geq Y_c(\alpha).$$

[†] See "Constitutive equation for the dynamic compaction of ductile porous materials" by Herrmann, W. in *Journal of Applied Physics* **40**, pp. 2490–2499 (1969).

We implement this yield condition by modifying the pore-collapse rule to satisfy the new dissipation inequality [see Eq. (4.214)]:

$$H_s^P = \begin{cases} -b(\bar{\rho}_s\Omega - Y_c)\left[1 - \frac{\varphi_s^P\Omega}{2}\frac{d\omega}{d\varphi_s^P}\right], & \bar{\rho}_s\Omega < Y_c < 0, \\ 0, & Y_t \geq \bar{\rho}_s\Omega \geq Y_c, \\ -b(\bar{\rho}_s\Omega - Y_t)\left[1 - \frac{\varphi_s^P\Omega}{2}\frac{d\omega}{d\varphi_s^P}\right], & \bar{\rho}_s\Omega > Y_t > 0, \end{cases}$$

where b is an arbitrary positive constant. Examination of Eq. (4.222) tells us that the internal energy of the solid depends upon the value of ω. When $H_s^P \neq 0$, the value of ω changes. The terms contained in the square brackets of the flow rule compensate for this change in internal energy.

Elastic Wave Velocity

Recall that the wave velocity c of the dry plastic porous solid is not a function of the distention ratio. The elastic-plastic porous solid has an elastic wave velocity that does depend on the distention of the solid and on the properties of the pore fluid. We obtain this wave velocity from the derivative of the stress,

$$\frac{\partial T}{\partial F} = \frac{\partial}{\partial F}(\varphi_s \bar{T}^s - \varphi_f \bar{p}_f) = \frac{\partial}{\partial F}[\varphi_s(\bar{T}^s + \bar{p}_f) - \bar{p}_f],$$

where, because we are considering an elastic process, we hold φ_s^P constant so that $H_s^P = 0$. To evaluate this expression we first evaluate the isentropic stiffnesses of the solid C_s^η and the fluid C_f^η:

$$C_s^\eta = F^s \frac{\partial \bar{T}^s}{\partial F^s},$$

$$C_f^\eta = \bar{\rho}_f \frac{\partial \bar{p}_f}{\partial \bar{\rho}_f}.$$

We also notice that

$$\frac{1}{\varphi_s^E}\frac{\partial \varphi_s^E}{\partial \Omega} = -\omega.$$

Using these relationships, we obtain

$$\frac{\partial T}{\partial F} = \frac{\varphi_s C_s^\eta}{F^s}\frac{\partial F^s}{\partial F} - \frac{\varphi_f C_f^\eta}{\bar{\rho}_f}\frac{\partial \bar{\rho}_f}{\partial F} - \varphi_s \bar{\rho}_s \omega \Omega \frac{\partial \Omega}{\partial F}. \tag{4.226}$$

Our objective now is to evaluate the partial derivatives on the right-hand side of this expression. We start by differentiating the logarithm of

$$F^s = \varphi_s^E \varphi_s^P F$$

to obtain

$$\frac{1}{F^s}\frac{\partial F^s}{\partial F} = \frac{1}{F} - \omega \frac{\partial \Omega}{\partial F}. \tag{4.227}$$

We obtain the next expression by differentiating the logarithm of

$$\bar{\rho}_s \varphi_s^E \varphi_s^P F = \rho_{s0},$$

which gives

$$\frac{1}{\bar{\rho}_s}\frac{\partial \bar{\rho}_s}{\partial F} - \omega\frac{\partial \Omega}{\partial F} + \frac{1}{F} = 0. \tag{4.228}$$

Next we differentiate

$$\bar{\rho}_f(1 - \varphi_s^E \varphi_s^P)F = \rho_{f0}$$

to obtain

$$\frac{1}{\bar{\rho}_f}\frac{\partial \bar{\rho}_f}{\partial F} + \frac{\omega\varphi_s}{\varphi_f}\frac{\partial \Omega}{\partial F} + \frac{1}{F} = 0. \tag{4.229}$$

Then we differentiate

$$\bar{\rho}_s\Omega = \bar{T}^s + \bar{p}_f$$

to obtain

$$\frac{\partial \Omega}{\partial F} = \frac{1}{\bar{\rho}_s}\left[\frac{C_s^\eta}{F^s}\frac{\partial F^s}{\partial F} + \frac{C_f^\eta}{\bar{\rho}_f}\frac{\partial \bar{\rho}_f}{\partial F}\right] - \frac{\Omega}{\bar{\rho}_s}\frac{\partial \bar{\rho}_s}{\partial F}.$$

Substituting Eqs. (4.227)–(4.229) into this result gives

$$\frac{\partial \Omega}{\partial F} = \frac{\epsilon}{F}\left(1 + \epsilon\omega + \frac{\omega C_f^\eta}{\bar{\rho}_s\varphi_f}\right)^{-1},$$

where

$$\epsilon = \frac{C_s^\eta - C_f^\eta}{\bar{\rho}_s} + \Omega.$$

We substitute this expression into Eqs. (4.227) and (4.229), and then we use the results in Eq. (4.226) to show that

$$\frac{\partial T}{\partial F} = \frac{1}{F}(\varphi_s C_s^\eta + \varphi_f C_f^\eta) - \frac{\omega\epsilon^2\rho_{s0}}{F^2}\left(1 + \epsilon\omega + \frac{\omega C_f^\eta}{\bar{\rho}_s\varphi_f}\right)^{-1}. \tag{4.230}$$

This is the desired result that we can substitute into the expression for the material wave velocity to obtain [see Eq. (3.86) in Section 3.4.2]

$$C = \left[\frac{1}{\rho_0 F}(\varphi_s C_s^\eta + \varphi_f C_f^\eta) - \frac{\omega\epsilon^2\rho_{s0}}{\rho_0 F^2}\left(1 + \epsilon\omega + \frac{\omega C_f^\eta}{\bar{\rho}_s\varphi_f}\right)^{-1}\right]^{1/2}.$$

The spatial wave velocity is

$$c = \left[\frac{1}{\rho}(\varphi_s C_s^\eta + \varphi_f C_f^\eta) - \frac{\omega\epsilon^2\rho_{s0}}{\rho_0}\left(1 + \epsilon\omega + \frac{\omega C_f^\eta}{\bar{\rho}_s\varphi_f}\right)^{-1}\right]^{1/2}. \tag{4.231}$$

These are the expressions for the velocity of an elastic wave in a saturated elastic-plastic porous solid. They are valid as long as $H_s^P = 0$. They do not represent the velocity of structured waves where $H_s^P \neq 0$. We now examine several special cases of these results.

Reference Solid We can obtain the velocity of an elastic wave propagating in the reference solid by letting $\varphi_s = 1$ and $\varphi_f = 0$. Under these conditions, Eq. (4.225) requires that $\omega = 0$. From Eq. (4.231), we find that the velocity is $c = c_r$, where

$$c_r = \left(\frac{C_s^\eta}{\bar{\rho}_{s0}}\right)^{1/2}. \tag{4.232}$$

Dry Solid If fluid is not present in the pores, then $C_f^\eta = 0$. Thus

$$c = \left[\frac{C_s^\eta}{\bar{\rho}_s} - \omega\epsilon^2(1+\epsilon\omega)^{-1}\right]^{1/2},$$

where

$$\epsilon = \frac{C_s^\eta + \bar{T}^s}{\bar{\rho}_s}.$$

When the dry solid is in a stress-free state, we know that $F = 1$ and $\bar{T}^s = 0$. Thus

$$c^2 = c_r^2\left(1 + \omega c_r^2\right)^{-1}.$$

Solving for ω, we find that

$$\omega = \frac{1}{c^2} - \frac{1}{c_r^2}.$$

This is an extremely useful result because it demonstrates how the constitutive parameter ω can be prescribed by using measurements of the velocity of the elastic wave. Because we have required that $\omega \geq 0$, we find

$$c \leq c_r.$$

For the special case when $c = c_r$, we find that $\omega = 0$. Hence the elastic volume fraction φ_s^E is always equal to one. This is similar to the result for the plastic porous solid, where we found that c is independent of the distention ratio and $c = c_r$.

The stiffness of the reference solid is related to the velocity c_r by Eq. (4.232). Let us denote the stiffness of the reference solid at the stress-free state as

$$K_r = \bar{\rho}_{s0} c_r^2.$$

The stiffness of the dry porous solid at the stress-free state is

$$K = \rho_{s0} c^2.$$

Using the result $c \leq c_r$ yields

$$K \leq \varphi_{s0} K_r,$$

which means the porous solid is less stiff than the reference solid.

Partially Saturated Solid Suppose the solid is partially filled with fluid. The sum of the volume fractions of the solid and the fluid is less than one. Due to the presence of the pore fluid, the mixture density is greater than the density of the dry solid ($\rho > \rho_s$). However, the fluid cannot develop an interface pressure. Consequently,

$$\Lambda_\xi / F = \bar{p}_f = 0.$$

An expedient method of introducing these conditions into our results for the elastic-plastic porous solid is to let $C_f^\eta = 0$. Then Eq. (4.231) gives

$$c = \left[\frac{\varphi_s C_s^\eta}{\rho} - \frac{\omega \epsilon^2 \rho_{s0}}{\rho_0} (1 + \epsilon \omega)^{-1} \right]^{1/2},$$

where

$$\epsilon = \frac{C_s^\eta + \bar{T}^s}{\bar{p}_s}.$$

Notice that, as we add fluid to a dry solid, the velocity initially decreases because ρ increases. As we saturate the solid with fluid, the velocity reaches a minimum value and then increases. The velocity in the saturated solid is given by Eq. (4.231).

4.4.6 *Comparison of Models for Porous Solids*

We have considered three different constitutive relations for a porous solid. They are the porous locking solid, the dry plastic porous solid, and the elastic-plastic porous solid. We began our discussion of mixture theories for porous solids by recognizing that formulating a useful mixture theory is an art in which we must balance the needs of our analysis against the complexity of the resulting formulation. The comparison of these three formulations provides an excellent illustration of how this works.

Suppose we are given the task of analyzing a particular porous solid. Then consider that in each portion of this section, we have constructed a different constitutive model for the same porous solid. The porous locking solid is the least complex of the three models. In this model, the reference solid is rigid, and because of this assumption, to use this model we only need information about the initial porosity of the material. For applications in which we have a highly distended dry porous solid, the actual compression of the reference solid may produce an effect that is negligible in comparison to the collapse of the pores, and this model may be totally adequate to describe the physical behavior of the porous solid.

However, suppose that we must consider situations in which the solid has only a small initial porosity. We might then consider introducing the next step in complexity into our analysis. Here we allow the reference solid to deform, but we still only allow plastic changes in the volume fraction. As we have seen, this leads to the model for the dry plastic porous solid. This model requires additional constitutive information about both the reference solid and the collapse rule for the plastic change in the volume fraction.

Although the dry plastic porous solid may be satisfactory for many applications, it still cannot accommodate a situation in which pore fluid is present. This requires the most complex model we have discussed, the elastic-plastic porous solid. This model requires even more constitutive information. In particular, we must determine exactly how the propagation velocities of elastic waves are influenced by the elastic change of the volume fraction.

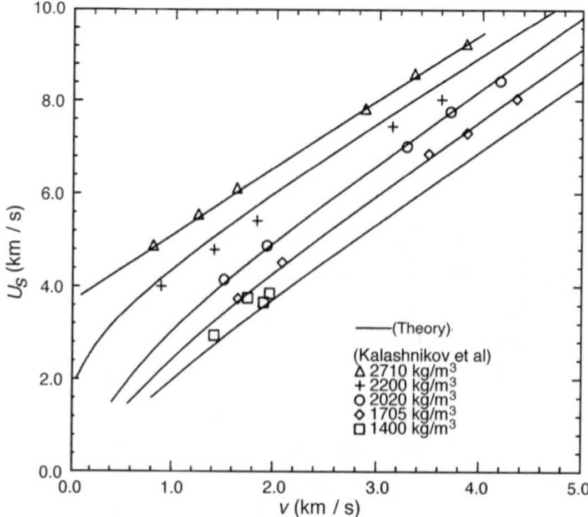

Figure 4.69 Comparison of measured and calculated Hugoniots for five samples of porous calcite. The 2200-kg/m^3 sample is saturated with water. The rest are dry. Reprinted from Drumheller, D. S. and Grady, D. E., "The dynamic response of porous calcium carbonate minerals." In *Shock Waves in Condensed Matter, Proceedings of the Fourth American Physical Society Topical Conference on Shock Waves in Condensed Matter, held July 22–25, 1985, in Spokane, Washington*, ed. Gupta, Y. M., Copyright 1985, p. 311, with kind permission from Plenum Publishing Corporation, New York.

Clearly, more complex models can also be formulated that will require even more constitutive information. When faced with this type of analysis, you are the one who must determine how much constitutive information is required and how complex a mathematical formulation can reasonably be solved.

4.4.7 *Shock and Rarefaction Waves*

Mixture theories offer a convenient method for the analysis of both dry and saturated porous solids. We shall briefly consider an example of this utility. The mixture we shall consider is composed of powered calcite saturated with water. We use the tabular constitutive equation of Kerley† to represent water, and we use a Mie–Grüneisen model to represent powdered calcite. We show the measured Hugoniots reported by Kalashnikov, Pvlovskiy, Simakov, and Trunin‡ for four dry calcite samples with different initial porosities (see Figure 4.69). We have used these data to determine that the Grüneisen parameter Γ obeys

$$\bar{\rho}_s \Gamma F^{2.25} = \bar{\rho}_{s0} \Gamma_0,$$

† See "Theoretical equations of state for the detonation products of explosives" by Kerley, G. I. in *Proceedings of the Eighth Symposium on Detonation*, ed. Short, J. M., pp. 540–547 (Naval Surface Weapon Center, 1986).

‡ See "Dynamic compressibility of calcite-group minerals" by Kalashnikov, N. G., Pvlovskiy, M. N., Simakov, G. V., and Trunin, R. F. in *Izv., Earth Physics* **2**, pp. 23–29 (1973).

where we have used the following constitutive parameters for calcite:

$$\bar{\rho}_{s0} = 2.71\,\text{Mg/m}^3, \qquad \Gamma_0 = 0.8,$$

$$c_0 = 3.65\,\text{km/s}, \qquad \bar{p}_{\text{max}} = 4.0\,\text{GPa},$$

$$s = 1.46, \qquad \bar{p}_{\text{min}} = 0.1\,\text{GPa}.$$

Here we also use the yield condition Eq. (4.220). The illustrated results are from Drumheller and Grady. Their calculated Hugoniots are the solid lines. The 2,710-kg/m^3 material is the fully-compacted reference solid that exhibits a linear Hugoniot. We also show Hugoniot data for a saturated sample with a mixture density of 2,200 kg/m^3. Here we notice some disagreement between the data and the calculation. Because the properties of calcite are determined by the comparison to the dry samples, and the properties of water are determined by tabulated values, the parameters in the theory are completely determined and cannot be further adjusted to fit the measured data for the saturated sample. However, Drumheller and Grady noticed that the values of ρ_0, $\bar{\rho}_{s0}$, and φ_ξ reported by Kalashnikov et. al. are inconsistent with respect to the known density of water. Slight errors in the measurements of volume fractions as well as partial saturation of the samples may account for the observed difference.

In a later work, Drumheller reported calculations of shock and rarefaction waves in water-saturated porous calcite (see Figure 4.70). He used a model that closely approximates our theory for the elastic-plastic porous solid. Solutions of the governing equations based on the material parameters listed above were obtained by numerical integration. We compare these calculations to the data of Grady in Figure 4.70. These data are histories of velocity v measured in a flyer-plate experiment. We observe that the wave interactions within the thin flyer plate produce two sets of shock waves and rarefaction waves within the saturated-

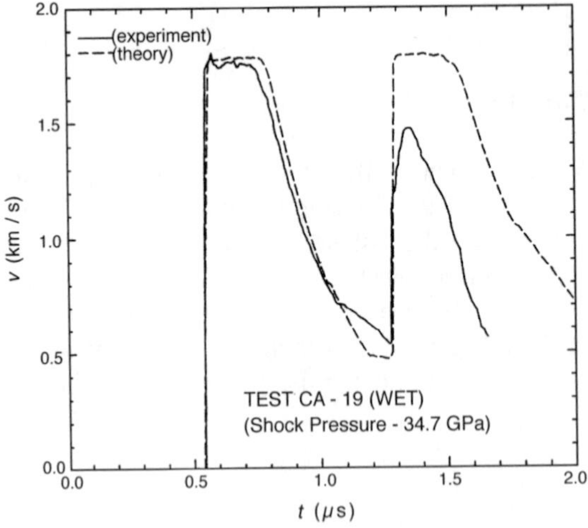

Figure 4.70 Comparison of measured and calculated profiles in a 1750-kg/m^3 sample of saturated porous calcite. Reprinted from Drumheller, D. S., "Hypervelocity impact of mixtures." In *Hypervelocity Impact, Proceeding of the 1986 Symposium, San Antonio, Texas, 21–24 October 1986*, ed. Anderson Jr., C. E., Copyright 1986, p. 267, with kind permission from Elsevier Science Ltd. The Boulevard, Langford Lane, Kidlington 0X5 1GB, UK.

calcite target plate. The comparison between theory and experiment is quite good for the first shock wave and most of the first rarefaction wave. Substantial error appears in the comparisons for the second set of waves. This may be a result of damage to the electrical leads of the gage assembly.

Although these comparisons are quite encouraging, they are far from completely satisfying. We are left wondering if something is missing from our theory or if some experimental artifact is present in the data. Indeed, developing new constitutive theories is a difficult task because we must basically infer the constitutive model and then test it against experiment. Often this gets harder as more data become available. This is clearly evident in Figure 4.70, where, if the experimental apparatus had been shut off at 1 μs, we would have been quite happy with the comparison and declared the theory a total success. Moreover, it is altogether possible that the same data can be modeled by two theories that are fundamentally quite different. Finally, we point out that it is easy to be mislead by experimental data. When we see a graph like Figure 4.70, we are inclined to interpret these profiles as a progression of simple waves, but often this is not true. The calculated profiles for the porous locking solid contained in Figure 4.60 are a classic illustration of how this can lead to serious error in our interpretation of data. Each wave profile in this figure looks like a shock wave followed by a rarefaction wave. We might think that the rarefaction wave is a simple wave, until we remember that the reference material for the porous locking solid is rigid. This means the velocity of a simple wave is infinite. It is this instantaneous communication between the shock wave and the boundary of the material that has produced a profile that looks like a simple rarefaction wave with a finite velocity but is not.

4.4.8 *Exercises*

4.32 We present a solution for a shock wave propagating in the porous locking solid in Section 4.4.1. Recall that the stress on the boundary of this solid is zero.

(a) Use the laws of balance of momentum and balance of energy to demonstrate that the momentum and the total energy in this material must be constant.

(b) Compute the total momentum from our solution and demonstrate that it is indeed constant.

(c) Discuss why this material can only store energy as heat or kinetic energy.

(d) Compute the kinetic energy for our solution and demonstrate that it decreases with time.

4.33 In Figure 4.60, we plot the profiles of the solution for the porous locking solid for $\varphi_0 = 0.50$. We plot these profiles in the material description.

(a) Derive an expression for the motion $x = \hat{x}(X, t)$ of this solid.

(b) For values of $t/t_0 = 0.1$, 0.4, and 0.9, replot these profiles in both the material and spatial descriptions.

4.34 Verify the multiplicative decomposition of the volume fraction, Eq. (4.177).

4.35 In a mixture where all constituents have the same motion, show that

$$\rho \frac{\partial \hat{\mathcal{E}}}{\partial t} = \sum_{\xi} \rho_{\xi} \frac{\partial \hat{\mathcal{E}}_{\xi}}{\partial t}.$$

4.36 Starting with the equation for the balance of energy for a constituent of a mixture, Eq. (4.183), verify our result for the balance of energy of the pore fluid, Eq. (4.197).

4.37 We present the dissipation inequality for a fluid-saturated porous solid in Eq. (4.199). We use this inequality to derive the constitutive constraints in Eqs. (4.200)–(4.203). The remaining inequalities are given in Eqs. (4.204)–(4.205). Determine the conditions that allow us to conclude that

$$\bar{T}^s = -\frac{\Lambda_s}{F} - \varphi_s^P \bar{\rho}_s \frac{\partial \mathcal{E}_s}{\partial \varphi_s^P},$$

$$\vartheta_f = \vartheta_s.$$

4.38 Verify Eq. (4.209) for the speed of sound in a fluid mixture.

4.39 In Figure 4.66, we compare two Rayleigh-line processes. Both processes occur in samples of stainless steel. One sample has no initial porosity, whereas in the other sample, $\alpha_0 = 1.3$. Both samples are compressed by a shock wave to the same Hugoniot density. Use the figure to determine the pressure difference of the two Hugoniot states. Use the Mie–Grüneisen theory to describe the reference solid. Let

$$\bar{\rho}_s \Gamma_s = \bar{\rho}_{s0} \Gamma_{s0}, \qquad \bar{\rho}_{s0} = 7{,}890 \text{ kg/m}^3,$$

$$c_0 = 4.58 \text{ km/s}, \qquad s = 1.49,$$

$$\Gamma_{s0} = 1.94, \qquad c_v = 447 \text{ J/kg/K},$$

$$\vartheta_- = 295 \text{ K}.$$

(a) Determine the temperature difference between the two Hugoniot states.

(b) Repeat the calculation for the porous sample using $\alpha_0 = 1.03$, and determine the temperature difference between the reference sample and the porous sample when $\alpha_+ = 1$ and $F_+ = 0.85$.

4.40 Start by verifying Eq. (4.226). Then repeat the derivation of the spatial wave velocity, Eq. (4.231).

4.5 Detonation and Phase Transformation

Waves are capable of inducing chemical reactions and phase transformations. Indeed, small blasting caps are used every day to generate waves that initiate chemical reactions in high explosives. The subsequent release of chemical energy in high explosives can produce and sustain a steady detonation wave. We can also see vapor trails streaming off the wings of airplanes. This is a good example of a phase transformation induced by a wave. Moreover, the most violent events on earth are thought to be caused by sudden phase transformations. The eruption of the Mount Saint Helens volcano in the state of Washington on May 18, 1980 and the eruption of the volcanic island of Krakatoa in the South Pacific in August 1883 are both attributed to the sudden transformation of liquid to vapor. It is estimated that Mount Saint Helens exhibited an energy flux of 25 GW/m^2 from its vent and ejected 0.12 km^3 of material. The total energy released was equivalent to the detonation of 48 megatons of dynamite.† This event is dwarfed by the Krakatoa eruption that ejected 4 km^3 of material to an altitude of 30 km. This eruption created tidal waves that killed more than 30,000 people and dust that dramatically altered weather patterns around the globe for the next year.

The study of chemical reactions and phase changes is a highly advanced science. Numerous books have been devoted to both fields. We do not attempt a detailed study of either subject here. Rather, we select two topics to provide an introduction to this subject. Our topics are the detonation of a combustible gas and a polymorphic phase transformation induced by a shock wave.

† See "Blast dynamics at Mount St Helens on 18 May 1980" by Kieffer, S. W. in *Nature* **291**, pp. 568–570 (1981).

4.5.1 *Detonation*

Consider a half space of combustible gas. Suppose that we ignite the gas at the boundary. Often this initiates a slow combustion process called a *flame* or *deflagration*. Under normal circumstances this flame propagates through the half space at a few meters per second. However, under certain conditions we find that this flame transforms into a fast combustion process, which propagates as a steady wave at several thousand meters per second. This is called a *detonation*. Detonations occur not only in combustible gases but also in combustible solids and liquids. A typical example is the common fertilizer, ammonium nitrate. When saturated with fuel oil, the resulting mixture will support detonation.

The original theory of detonation of a combustible gas was independently developed by Chapman† (1899) and Jouguet (1917)‡. They assumed that detonation is produced by an instantaneous chemical reaction. Their theory leads to a characterization of detonation as a shock wave. Later work by Zel'dovich, von Neumann, and Doering in the middle of the twentieth century led to a description of detonation as a steady wave with a finite rate of chemical reaction. In this section we present this steady-wave analysis for an ideal gas with constant specific heat.

In our discussion of the saturated porous solid, we discovered that a new conjugate pair of variables $(\Lambda_\xi, \varphi_\xi)$ was required to describe the special internal features of the solid. This pair represented the volume fraction φ_ξ and the interface pressure Λ_ξ. The combustible gas also has special features that require a new conjugate pair (Q, λ). The flux term in this pair is the *reaction variable* λ. By definition, λ does not have physical dimensions, and

$$0 \leq \lambda \leq 1.$$

In the unburned gas $\lambda = 0$; in the fully burned gas $\lambda = 1$. The force term in this pair is called the *reaction energy* Q. It has physical dimensions of energy per unit mass, and it represents the energy that is released by burning the gas. The constitutive equations for a combustible gas are assumed to be

$$\begin{aligned} \mathcal{E} &= \mathcal{E}(\rho, \eta, \lambda), & Q &= Q(\rho, \eta, \lambda), \\ \vartheta &= \vartheta(\rho, \eta, \lambda), & \partial\lambda/\partial t &= \mathcal{L}(\rho, \eta, \lambda), \\ p &= p(\rho, \eta, \lambda). \end{aligned}$$

In this section we consider only a very special type of combustible gas. For the constitutive equation for $\mathcal{E}$, we assume

$$c_v \ln\left(\frac{\mathcal{E} + \lambda Q}{\mathcal{E}_0}\right) = NR\ln(\rho/\rho_0) + \eta - \eta_0, \tag{4.233}$$

and for the constitutive equation for Q, we assume

$$Q = \text{positive constant.}$$

For any process where λ is constant, the product λQ merely adds a constant to the energy of the material, and Eq. (4.233) reduces to the fundamental statement for an ideal gas with a constant c_v [see Eq. (4.4)].

† See "On the rate of explosions in gases" by D. L. Chapman in *Philosophical Magazine* **44**, pp. 90–104 (1899).

‡ See "Méchanique des explosifs. Etude de dynamique chinmique," O. Doin et fils, Paris, pp. 516 (1917).

Because p is not a function of the rate of change of density, we can derive both the temperature and pressure from the partial derivatives of the internal energy, Eq. (4.233). We determine the temperature from the partial derivative with respect to η,

$$\left(\frac{c_v}{\mathcal{E}+\lambda Q}\right)\frac{\partial \mathcal{E}}{\partial \eta} = 1,$$

which yields

$$c_v\vartheta = c_v\frac{\partial \mathcal{E}}{\partial \eta} = \mathcal{E}+\lambda Q.$$

Notice that as λ increases, the temperature of the gas increases.

We obtain the pressure from the derivative of the internal energy with respect to F,

$$\left(\frac{c_v}{\mathcal{E}+\lambda Q}\right)\frac{\partial \mathcal{E}}{\partial F} = -\frac{NR}{F},$$

which yields

$$p = -\rho_0\frac{\partial \mathcal{E}}{\partial F} = \frac{NR}{c_v}\rho(\mathcal{E}+\lambda Q).$$

Because $NR/c_v = \Gamma$, we obtain

$$p = \rho\Gamma(\mathcal{E}+\lambda Q) = NR\rho\vartheta. \tag{4.234}$$

With the exception of the chemical-reaction rate, we now have explicit relationships for all of the constitutive equations.

Dissipation Inequality

Recall that we must constrain the constitutive equations to ensure that any admissible thermodynamic process does not violate the second law of thermodynamics. We determine these constraints by combining the balance of energy with the dissipation inequality. We begin by applying the chain rule to the balance of energy to obtain

$$\rho_0\vartheta\frac{\partial \hat{\eta}}{\partial t} + \rho_0\frac{\partial \mathcal{E}}{\partial F}\frac{\partial \hat{F}}{\partial t} + \rho_0\frac{\partial \mathcal{E}}{\partial \lambda}\frac{\partial \hat{\lambda}}{\partial t} = -p\frac{\partial \hat{F}}{\partial t} + \rho_0 r.$$

From Eq. (4.233), we find that $\partial\mathcal{E}/\partial\lambda = -Q$. Thus the energy equation reduces to

$$\vartheta\frac{\partial \hat{\eta}}{\partial t} - r = Q\frac{\partial \hat{\lambda}}{\partial t}.$$

When we combine this equation with the second law of thermodynamics [see Eq. (3.31) in Section 3.2.3],

$$\vartheta\frac{\partial \hat{\eta}}{\partial t} - r \geq 0,$$

we obtain

$$Q\frac{\partial \hat{\lambda}}{\partial t} \geq 0.$$

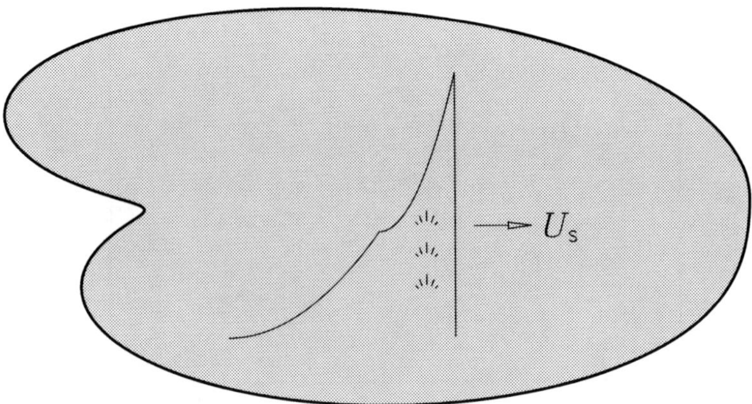

Figure 4.71 A detonation propagating as a steady wave in a combustible gas.

Because $Q > 0$, then

$$\frac{\partial \hat{\lambda}}{\partial t} \geq 0,$$

which means that the constitutive relationship for the chemical-reaction rate $\partial \hat{\lambda}/\partial t$ must ensure that λ only increases to release increasing amounts of energy. This is called an *exothermic* chemical reaction.

Hugoniot Relationship

Suppose a detonation propagates through a combustible gas as a steady wave with velocity U_S (see Figure 4.71). For the initial state in front of the detonation, we prescribe

$$\lambda = 0, \qquad \rho = \rho_0, \qquad \vartheta = \vartheta_0.$$

The state behind the detonation is

$$\lambda = \lambda_+ = 1, \qquad \rho = \rho_+, \qquad \vartheta = \vartheta_+.$$

We can derive the Hugoniot relationship for the fully burned gas by using the jump condition for the balance of linear momentum [see Eq. (2.147) in Section 2.7.2],

$$-NR(\rho_+\vartheta_+ - \rho_0\vartheta_0) = \rho_0 U_S^2 (F_+ - 1),$$

and the jump condition for the balance of energy [see Eq. (3.111) in Section 3.4.5],

$$\rho_0 c_v \vartheta_+ - \rho_0 \lambda_+ Q - \rho_0 c_v \vartheta_0 = -NR\rho_0\vartheta_0(F_+ - 1) + \frac{1}{2}\rho_0 U_S^2 (F_+ - 1)^2,$$

where $F_+ = \rho_0/\rho_+$. Following the procedure used in Section 4.1.5, we solve these two relationships to obtain

$$\gamma(1 - F_+)\left\{\left[F_+ + \frac{1}{2}(\gamma - 1)(F_+ - 1)\right]M^2 - 1\right\} = \frac{\lambda_+ Q}{c_v \vartheta_0}, \tag{4.235}$$

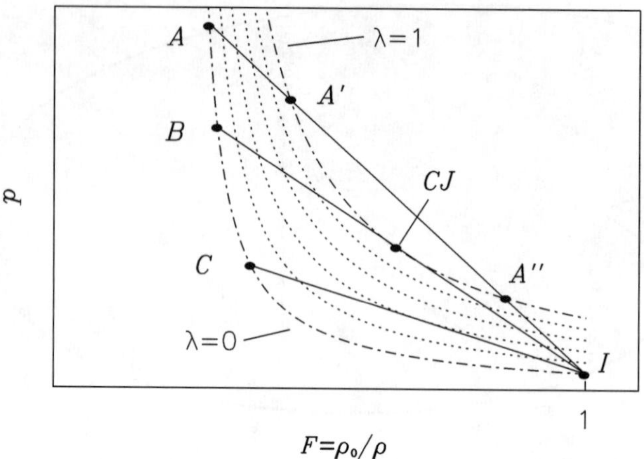

Figure 4.72 The Hugoniots (— · —) for the unburned gas, $\lambda = 0$, and fully burned gas, $\lambda = 1$. Dotted lines are for constant $\lambda = 0.2$, 0.4, 0.6, and 0.8.

where the Grüneisen ratio is $\Gamma = \gamma - 1$ and the Mach number is $M = U_S/c_0$. This is the desired relationship for the Hugoniot curve of a detonation. It is illustrated in Figure 4.72. We show the Hugoniot curves for the unburned and fully burned gas as dot-dashed lines. They are obtained by respectively setting $\lambda_+ = 0$ and $\lambda_+ = 1$. We require that the chemical reaction always results in $\lambda_+ = 1$ behind the detonation; however, it is instructive to plot additional curves of constant λ. We show four such curves as dotted lines. Obviously, these are not Hugoniot curves.

Compression Limit As $M \to \infty$, the Hugoniot relationship reduces to

$$F_+ + \frac{1}{2}(\gamma - 1)(F_+ - 1) = 0.$$

The solution to this expression is

$$F_+ = \frac{\gamma - 1}{\gamma + 1},$$

which shows that the combustible gas has the same compression limit as the nonreacting ideal gas.

Deflagration The Hugoniot states that we show in Figure 4.72 are associated with compression of the gas ($F_+ < 1$). Processes that end at these states are called *detonations*. We notice that Hugoniot states also exist for $F_+ > 1$ and $p_+ < p_0$. Processes that end at these states are called *deflagrations* or *flames*. Deflagrations can also result in steady waves. For example, the study of deflagration is important to the design of internal combustion engines, where deflagration is desirable and detonation is not. We do not study deflagration here.

Chapman–Jouguet Process

We show three Rayleigh lines in Figure 4.72. They connect the initial state I to states A, B, and C on the Hugoniot curve for the unburned gas. The steepest Rayleigh line, which intersects the Hugoniot for the unburned gas at state A, also intersects the Hugoniot for the fully burned gas at both states A' and A''. The Rayleigh line that intersects state B is tangent to the Hugoniot for the fully burned gas at state *CJ*, and the Rayleigh line that intersects state C never intersects the Hugoniot for the fully burned gas. We shall discuss each of these Rayleigh-line processes. Let us begin with the Rayleigh line that is tangent to the Hugoniot for the fully burned gas at state *CJ*. This is called a *Chapman–Jouguet process* or simply a *CJ* process. The point where this Rayleigh line is tangent to the Hugoniot is called the *Chapman–Jouguet state* or simply the *CJ* state. We shall show that this state yields the minimum propagation velocity for a steady detonation.

Consider the Rayleigh line that extends from the state I through the *CJ* state to the Hugoniot state B. Because we show two Hugoniots, one for the fully burned gas and one for the unburned gas, it is useful to think of two separate state surfaces. Thus the Hugoniot for $\lambda = 0$ lies on the state surface for the unburned gas, and the Hugoniot for $\lambda = 1$ lies on the state surface for the fully burned gas. Recall that the gas is not viscous. Therefore, the Rayleigh line connecting state I to state B represents a shock wave. Indeed, the state jumps between states I and B on the state surface for the unburned gas without any change in the reaction variable λ. It is this shock wave that initiates the chemical reaction. Then the chemical reaction causes λ to increase. Provided that the chemical reaction has produced a steady wave, this increase in λ can only occur along the Rayleigh line. The illustration shows that as λ increases from 0 to 1, the process must move back down the Rayleigh line from state B to the *CJ* state. However, although this process appears to retrace its path in the illustration, the chemical reaction actually carries the process off the state surface for the unburned gas and onto the state surface for the fully burned gas. We show the wave profile resulting from these processes in Figure 4.73. The head of the wave is the jump between state I and state B. This is followed by a rarefaction wave that moves the gas from state B to the *CJ* state. This portion of the wave is also called the *reaction zone*. We can determine

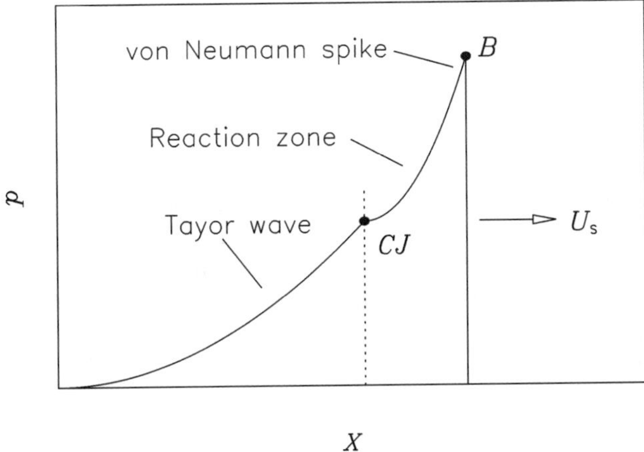

Figure 4.73 The wave profile of a *CJ* process.

the profile of the reaction zone solely with the constitutive equation for the rate of chemical reaction

$$\partial\hat{\lambda}/\partial t = \mathcal{L}(\rho, \eta, \lambda).$$

Because of the shape of the wave profile in the reaction zone, it is often called the *von Neumann spike*. When we substitute the Rayleigh-line solutions in Eqs. (3.104) in Section 3.4.4 and the steady wave variable $\hat{\varphi} = X - U_S t$ into the previous expression, we obtain an ordinary differential equation in λ. We can integrate this equation to obtain the profile of the spike. Behind the reaction zone is another wave called the *Taylor wave*. The Taylor wave is not necessarily a steady wave.

Two properties of the von Neumann spike are quite unusual. First, the wave in the reaction zone is a steady rarefaction wave. Until now, we have found that steady rarefaction waves cannot exist in an ideal gas because they violate the dissipation inequality. However, in a combustible gas, the dissipation inequality requires λ to increase and drives the rarefaction process to a steady wave. The second unusual property of the von Neumann spike is that it is not influenced by the trailing Taylor wave. This is because the velocity U_S of the detonation is equal to the material wave velocity C at the *CJ* state. We can demonstrate this fact by showing that the Hugoniot curve for the fully burned gas is tangent to an isentrope of the fully burned gas at the *CJ* state. Consider the jump condition for the balance of energy of the fully burned gas [see Eq. (3.112)]:

$$\mathcal{E}_+ - \mathcal{E}_0 + \frac{1}{2\rho_0}(p_+ + p_0)(F_+ - 1) = 0. \tag{4.236}$$

Suppose we describe this curve with a single parameter F_+. Let us take the derivative of this expression with respect to F_+. Expanding the derivative of $\mathcal{E}_+$ with the chain rule yields

$$\vartheta_+ \frac{d\eta_+}{dF_+} + \frac{1}{2\rho_0}\left[\frac{dp_+}{dF_+}(F_+ - 1) - (p_+ - p_0)\right] = 0. \tag{4.237}$$

At the *CJ* state, we know that the slope of the Hugoniot curve is equal to the slope of the Rayleigh line:

$$\frac{dp_+}{dF_+} = \frac{p_+ - p_0}{F_+ - 1}. \tag{4.238}$$

Combining these two results, we obtain

$$\frac{d\eta_+}{dF_+} = 0.$$

Thus the entropy is constant along the Hugoniot at the *CJ* state, and therefore the Hugoniot is tangent to an isentrope. This is clearly evident in Figure 4.74. Here we illustrate the Hugoniot for the fully burned gas along with the isentropes. Entropy increases up and to the right. The Hugoniot for the fully burned gas is tangent to an isentrope at the *CJ* state. Notice that the entropy η_+ on the Hugoniot is a minimum at this point. It is left as an exercise to derive this minimum condition from Eq. (4.237).

Because the *CJ* state is special, the pressure and the deformation gradient at the *CJ* state are denoted by p_{CJ} and F_{CJ}. The material wave velocity at the *CJ* state is

$$C_{CJ}^2 = -\frac{1}{\rho_0}\frac{\partial p}{\partial F} = -\frac{1}{\rho_0}\frac{dp_+}{dF_+},$$

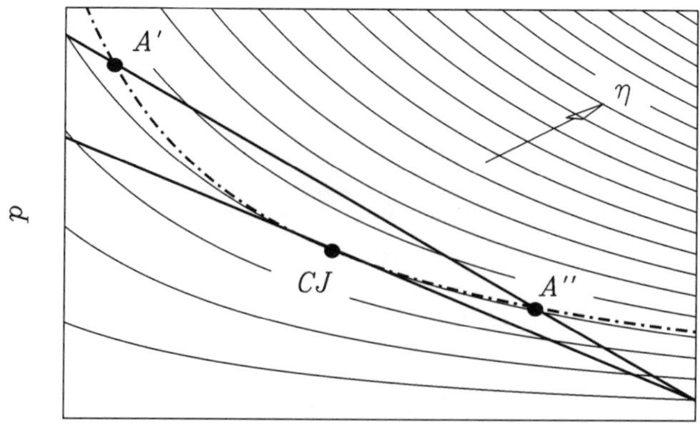

Figure 4.74 Comparison of the Hugoniot for the fully burned gas (— · —) to the isentropes of the fully burned gas (—).

and the detonation velocity is

$$U_{CJ}^2 = -\frac{1}{\rho_0}\left(\frac{p_+ - p_0}{F_+ - 1}\right). \qquad (4.239)$$

Substituting these results into Eq. (4.238), we find that

$$C_{CJ} = U_{CJ}.$$

Now let us suppose the Taylor wave is a rarefaction wave. The previous result demonstrates that the portion of Taylor wave in contact with the *CJ* state propagates at the same velocity as the detonation, and the remainder of the Taylor wave propagates at a slower velocity. Thus the von Neumann spike propagates as a steady wave even as the rarefaction Taylor wave continues to spread out. However, if the Taylor wave is a compression wave, it will eventually overtake the von Neumann spike.

Minimum Detonation Velocity We have assumed that once a chemical reaction is initiated by the leading shock wave, it will go to completion. Therefore, behind the reaction zone, $\lambda = 1$. Consider the Rayleigh line that connects states *I* and *C* (see Figure 4.72). This Rayleigh line does not cross a state where $\lambda = 1$. A detonation of this type cannot be realized as a steady-wave process. Consequently, for a detonation to exist as a steady-wave process,

$$U_S \geq U_{CJ}.$$

You should realize that a detonation initiated at state *C* might evolve to a steady wave. However, to do so, the energy in the chemical reaction must be sufficient to eventually drive the leading shock wave to state *B*.

Chapman–Jouguet State

Suppose we assume that $p_{CJ} \gg p_0$. This leads to a simplified description of the *CJ* state. We derive these expressions by first noticing that

$$p_+ = \rho_+ \Gamma(\mathcal{E}_+ + Q).$$

Then we substitute this expression into the jump condition for the balance of energy Eq. (4.236),

$$\frac{p_+}{\rho_+ \Gamma} - Q + \frac{p_+}{2\rho_0}(F_+ - 1) = 0,$$

where we have neglected the terms containing p_0. This can be rewritten as

$$[(\gamma + 1)F_+ - (\gamma - 1)]p_+ = 2\rho_0 \Gamma Q. \tag{4.240}$$

We differentiate this result with respect to F_+ to obtain

$$[(\gamma + 1)F_+ - (\gamma - 1)]\frac{dp_+}{dF_+} + (\gamma + 1)p_+ = 0.$$

Having derived this result, we now use Eq. (4.238), which also holds at the *CJ* state. Because $p_{CJ} \gg p_0$, we approximate this equation as

$$\frac{dp_+}{dF_+} = \frac{p_+}{F_+ - 1}.$$

When the previous two expressions are combined, we obtain the deformation gradient at the *CJ* state,

$$F_{CJ} = \frac{\gamma}{\gamma + 1}.$$

This is an interesting result: It shows that the compression at the *CJ* state lies halfway between the initial state at $F = 1$ and the compression limit at $F = (\gamma - 1)/(\gamma + 1)$.

The velocity v at the *CJ* state is

$$\frac{v_{CJ}}{U_{CJ}} = 1 - F_{CJ} = \frac{1}{\gamma + 1}.$$

We obtain the pressure at the *CJ* state from Eq. (4.240),

$$p_{CJ} = 2\rho_0(\gamma - 1)Q,$$

and we obtain the propagation velocity of the detonation from Eq. (4.239),

$$U_{CJ}^2 = 2(\gamma^2 - 1)Q. \tag{4.241}$$

Notice that the square of the propagation velocity of the *CJ* process is related directly to the reaction energy Q. Often the chemical energy of an explosive is specified by the value of U_{CJ}.

Strong and Weak Detonations

Let us turn our attention to the detonation represented by the Rayleigh line that connects state I to state A (see Figure 4.72). This detonation is composed of a shock wave followed by a steady rarefaction wave. The shock wave causes the state of the gas to jump from state I to state A. Both of these states lay on the Hugoniot curve of the unburned gas, and $\lambda = 0$ at both states. The shock wave initiates the chemical reaction. It is followed by a rarefaction wave that moves the gas from state A across the reaction zone to state A' where $\lambda = 1$. This is called an *overdriven detonation* because the state A' lies above the *CJ* state. Notice that the Rayleigh line crosses the Hugoniot for the burned gas twice at states A' and A''. This suggests the existence of a second type of steady detonation that moves the gas directly from state I to state A''. We shall show that such a process cannot exist as a steady detonation.

The existence of the overdriven detonation depends upon the relative values of the material wave velocity C and the detonation velocity U_S. If $C > U_S$ at state A', the Taylor wave can overrun the reaction zone, and if $C < U_S$ at state A', the Taylor wave cannot keep up with the reaction zone. Either situation will alter the profile of the wave in the reaction zone. To determine the relationship between C and U_S, let us recall that the jump condition for the balance of linear momentum yields the velocity of the detonation,

$$-\rho_0 U_S^2 = \frac{p_+ - p_0}{F_+ - 1}.$$

We notice that the right-hand side of this relationship is the slope of the Rayleigh line through state A' in Figure 4.74. Likewise, the material wave velocity C is given by

$$-\rho_0 C^2 = \frac{\partial p}{\partial F}.$$

This is the slope of the isentrope. Indeed, Figure 4.74 shows that, for all Hugoniot states above the *CJ* state including state A', the isentropes are steeper than the Rayleigh line and

$$C > U_S \quad \text{for } p_+ > p_{CJ}.$$

Similarly,

$$C < U_S \quad \text{for } p_+ < p_{CJ}.$$

At state A', the Taylor wave overtakes the reaction zone and changes the profile of the von Neumann spike. A steady wave exists only when the pressure in the Taylor wave is constant and equal to the pressure at state A'. This is called a *strong detonation*. For a reaction zone that connects state I directly to state A'', the Taylor wave cannot keep up with the reaction zone. The profile of the wave in the reaction zone will change regardless of the profile of the Taylor wave. Thus state A'' cannot be reached by a steady-wave process. State A'' is called a *weak detonation*. Even though our simple combustible gas does not allow a weak detonation to exist as a steady wave, weak detonations do exist in more complex systems.

Example of a CJ Process

Consider a combustible gas where the velocity U_{CJ} is known. We assume the chemical-reaction rate of this gas is

$$\frac{\partial \hat{\lambda}}{\partial t} = k(1-\lambda)^n, \tag{4.242}$$

where n and k are prescribed constants. Given that we also know the values of the gas constant γ and the specific heat c_v, we have sufficient information to calculate the profile of the von Neumann spike for a steady detonation. We shall now demonstrate how to do this. To simplify the calculations and highlight the salient features of this example, we shall assume that the final state behind the von Neumann spike is the *CJ* state and that $p_{CJ} \gg p_0$. We start by analyzing the balance of energy and balance of linear momentum for a Rayleigh-line process with a steady-wave velocity of U_{CJ}. This yields not only the jump conditions for the leading shock wave, it also yields relationships for F and p on the Rayleigh line in terms of λ. Then we integrate Eq. (4.242) and substitute the solution for $\lambda(t)$ into these results to determine the profile of the von Neumann spike.

We illustrate the Rayleigh-line process for the initial jump and the trailing rarefaction wave in the reaction zone in Figure 4.75. The balance of energy for the Rayleigh-line process is

$$c_v\vartheta - \lambda Q = \frac{1}{2}U_{CJ}^2(1-F)^2,$$

where, because $p_{CJ} \gg p_0$, we have neglected the terms for the initial state. The reaction energy is known. From Eq. (4.241), it is

$$Q = \frac{U_{CJ}^2}{2(\gamma^2-1)}.$$

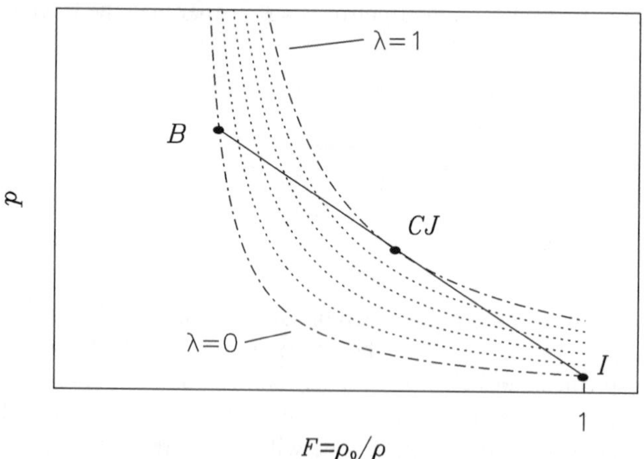

Figure 4.75 The Hugoniots (— · —) for the unburned gas, $\lambda = 0$, and fully burned gas, $\lambda = 1$. Dotted lines are for constant $\lambda = 0.2, 0.4, 0.6$, and 0.8.

The balance of linear momentum for the Rayleigh-line process is

$$p = \rho_0 U_{CJ}^2 (1 - F), \tag{4.243}$$

where we have neglected p_0. Initially, the gas moves from state I to state B. Because $\lambda = 0$ and the gas is not viscous, this process is a jump. Thus we can only use these relationships to evaluate the conditions at state B. Then the gas moves from state B to state CJ. This is a smooth process, and the previous relationships hold everywhere along this path. Along this process, the pressure in the gas is [see Eq. (4.234)]

$$pF = \rho_0(\gamma - 1)c_v \vartheta.$$

We simultaneously solve the previous four relationships to obtain

$$\left[\left(\frac{\gamma + 1}{\gamma - 1}\right)F - 1\right](1 - F) = \frac{\lambda}{\gamma^2 - 1}.$$

When $\lambda = 0$ this is the Hugoniot relationship for the unburned gas, and when $\lambda = 1$ this is the Hugoniot relationship for the fully burned gas. When $0 \le \lambda \le 1$, this relationship yields the solution for F in the reaction zone. The solution is

$$F = \frac{\gamma \pm \sqrt{1 - \lambda}}{\gamma + 1}.$$

There are two branches to this solution. If we use the $+$ branch, then $F = 1$ at state B where $\lambda = 0$. This is a weak detonation that does not yield a steady-wave solution. Thus we must use

$$F = \frac{\gamma - \sqrt{1 - \lambda}}{\gamma + 1}, \tag{4.244}$$

which is the strong detonation. Notice that

$$F = \frac{\gamma - 1}{\gamma + 1} \quad \text{at } \lambda = 0,$$

which is the compression limit of the gas. This result is an artifact of the assumption $p_{CJ} \gg p_0$. Substituting Eq. (4.244) into Eq. (4.243), we obtain

$$p = \rho_0 U_{CJ}^2 \left(\frac{1 + \sqrt{1 - \lambda}}{\gamma + 1}\right).$$

These are the desired solutions for F and p in terms of λ. Notice that the pressure at state B, where $\lambda = 0$, is twice that at the CJ state, where $\lambda = 1$.

Now we can integrate Eq. (4.242) and substitute the results into these solutions. We consider three possible chemical reactions in which $n = 0$, $1/2$, and 1. Integration yields

$$1 - \lambda = \begin{cases} 1 - kt, & n = 0, \\ \left(1 - \frac{1}{2}kt\right)^2, & n = \frac{1}{2}, \\ e^{-kt}, & n = 1. \end{cases}$$

We show the profiles of the von Neumann spike for each of these three choices in Figure 4.76. When $n = 0$ and $n = 1/2$, we illustrate a Taylor wave with a constant pressure following the reaction zone. For $n = 1$, the reaction zone is infinitely wide.

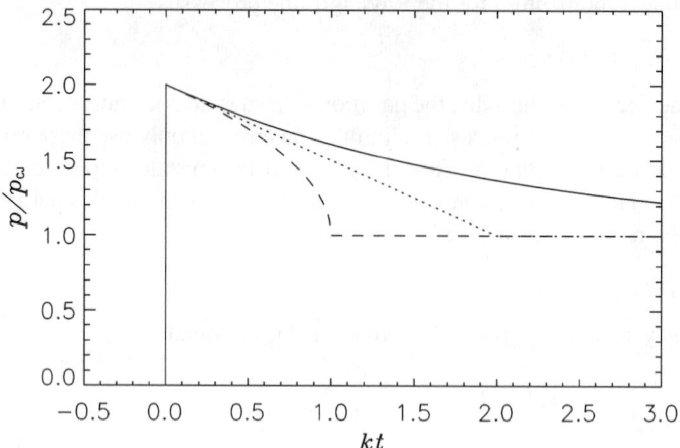

Figure 4.76 The wave profile of a *CJ* process for $n = 0$ (— —), $n = 1/2$ (· · ·), and $n = 1$ (—).

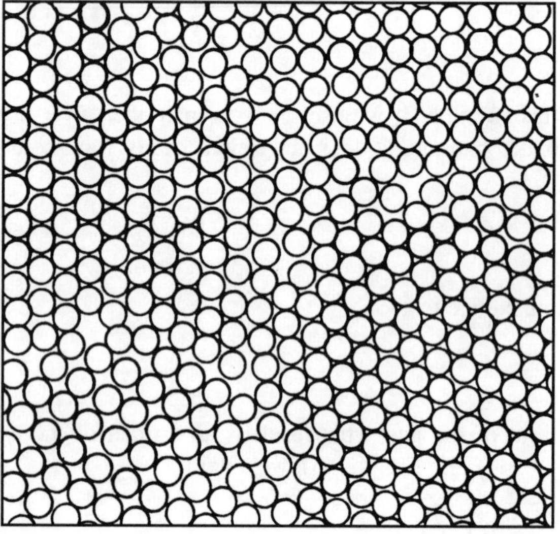

Figure 4.77 Grain boundaries during a polymorphic phase transition. Reprinted from Mason, C., *Introductory Physical Metallurgy*, 1947, p. 33, with kind permission from American Society for Metals, Cleveland.

4.5.2 *Polymorphic Phase Transformations*

The arrangement of atoms in the crystalline structure of many solids can change as the solid moves across the state surface. This type of phase transformation is called a *polymorphic phase transformation.* We show an example of a solid with two polymorphic phases in Figure 4.77. Notice the four regions in this illustration. Each is called a *grain*. In the upper-right and lower-left grains of the illustration, the atoms are arranged in a cubic pattern. In the remaining grains, the atoms are arranged in a hexagonal pattern. The cubic grains represent one phase of the material, and the hexagonal grains represent another. In

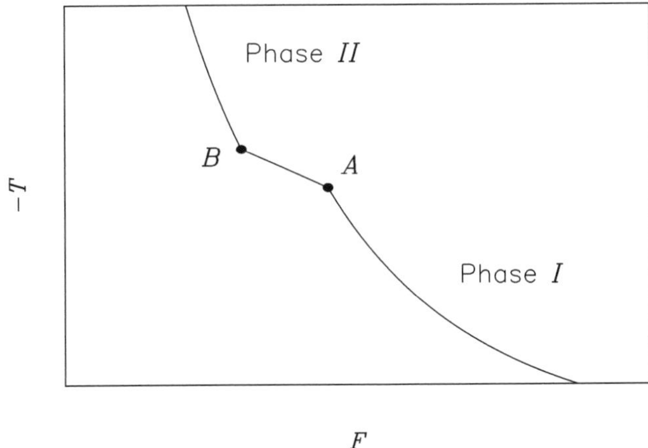

Figure 4.78 The isentrope of a polymorphic phase transition.

the cubic grains, the layers of atoms in the illustrated grains are covered by other layers whose atoms fit over the gaps between these atoms. In the hexagonal grains, the next layers of atoms lay directly on top of the atoms in the illustration. Thus the atoms in the cubic grains are more densely packed than those in the hexagonal grains. As an example, both diamond and graphite are crystalline forms of carbon. Diamond is a cubic crystal, which is the hardest substance known. Graphite is a hexagonal crystal and one of the softest solids.

Polymorphic phase transformations can occur under conditions of constant pressure and temperature. Indeed, polymorphic phase transitions are quite complex and beyond our scope of study. Their effect on shock waves can be dramatic. In this section we show how these phase transitions cause shock waves to split and how they also give rise to rarefaction shock waves. We illustrate the isentrope of a solid with a polymorphic phase transition in Figure 4.78. States A and B are the boundary states of the phase transition. Below state A, the solid exists in phase I, and above state B, the solid exists in phase II. In phase II, the atoms are more closely packed than in phase I. Thus the solid exhibits an accelerated change in volume between states A and B as the atoms rearrange themselves.

Consider a solid with this polymorphic phase transformation. It is initially in phase I, at rest, and free of stress. For solids, we have found that it is often appropriate to use the weak-shock assumption. Recall that a weak shock wave causes a negligible change in entropy, and therefore, we can approximate the Hugoniot curve of a weak shock by the isentrope of the initial state. We illustrate such an isentrope in Figure 4.79. It now represents the Hugoniot curve for the shock wave, and we show it with several Rayleigh lines. Suppose the stress behind the shock wave is T_D. This shock wave causes the solid to move from the initial state, which is in phase I, to the final state D, which is in phase II. We see that this Rayleigh line only intersects the Hugoniot at the initial state and the final state. All Rayleigh lines with Hugoniot states above state C cause the solid to transform from phase I to phase II in a single jump.

Now suppose the stress behind the shock wave is at state B. We cannot draw a straight Rayleigh line from the initial state to state B, because it would intersect the Hugoniot three times. Instead we must draw two Rayleigh lines, one from the initial state to state A, and

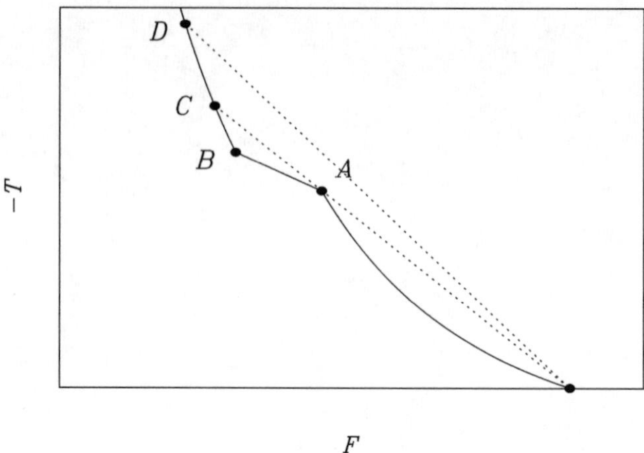

Figure 4.79 The Hugoniot curve (—) of the weak shock with several Rayleigh lines (· · ·).

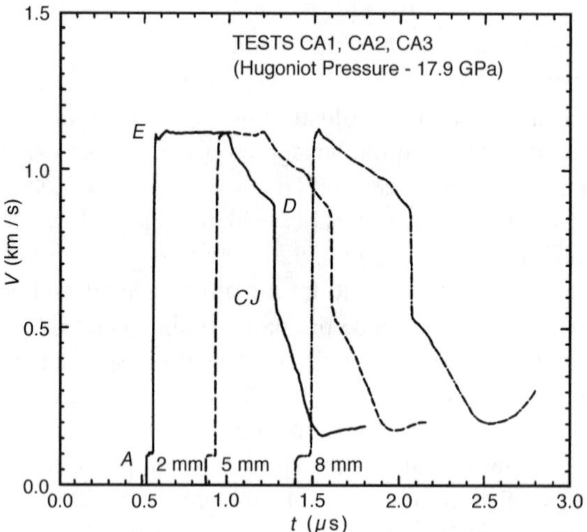

Figure 4.80 Shock wave profiles in crystalline calcite. We show flyer-plate data from experiments on three samples with thicknesses of 2, 5, and 8 mm. Reprinted from Grady, D. E., "High-pressure release-wave measurements and phase transformation in $CACO_3$." In *Shock Waves in Condensed Matter, Proceedings of the Fourth American Physical Society Topical Conference on Shock Waves in Condensed Matter, held July 22–25, 1985, in Spokane, Washington*, ed. Gupta, Y. M., Copyright 1985, p. 311, with kind permission from Plenum Publishing Corporation, New York.

another one from state A to state B. Because the slope of the Rayleigh line is directly related to the velocity of the shock wave, we see that this wave must split into two shock waves. We previously observed this phenomena for shock waves propagating through elastic-plastic solids. In that case, the elastic precursor wave propagated ahead of the slower plastic wave. In the present case, a phase-I precursor propagates ahead of the slower mixed-phase wave. In Figure 4.80, we show an example of a splitting shock wave in solid calcite that exhibits

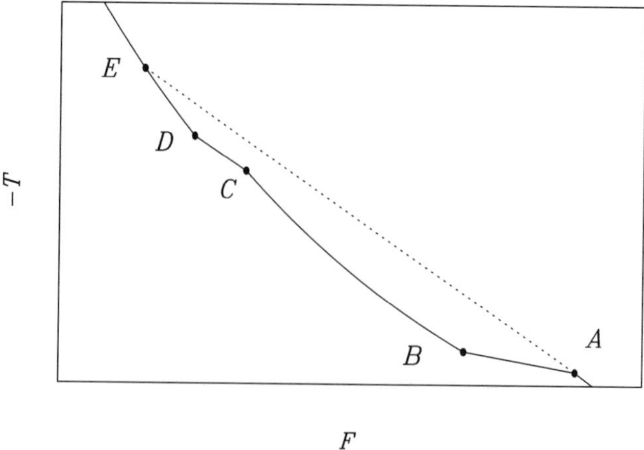

Figure 4.81 A solid with two polymorphic phase transitions. The Rayleigh line intersects the first phase transition at state A and misses the second phase transition.

several polymorphic phase transitions. Three wave histories are shown. We have labeled certain states on the leading history in order to facilitate our discussion. For a moment, we shall concentrate on the leading compression waves in each history. We shall return to the trailing rarefaction waves later. The data show a leading wave that has split at state A into two waves. The splitting results from the calcite *I-II-III* phase transition that produces a precursor with the low amplitude of $v = 0.1$ m/s. (In this material there are actually two closely spaced phase transitions at this point that separate three distinct phases. To simplify the discussion we shall treat them as a single transition.) Each history represents a wave that has propagated a different distance. Notice that the separation between the two waves increases with propagation distance. Calcite is a complex solid, and another phase transition occurs across the second shock wave. This additional phase transition does not cause the second shock wave to split, because the Rayleigh line of the second shock wave does not intersect the Hugoniot curve. This is schematically illustrated in Figure 4.81. States A and B are the boundary states for the *I-II-III* transition, and states C and D are the boundary states for the next transition. Here the precursor wave causes the solid to move from the initial state to state A. The Rayleigh line for the precursor is coincident with, and hidden by, the isentrope. The second wave moves the solid from state A to state E and misses the second phase transition.

While the phase transition between states C and D does not influence the histories of the compression wave in Figure 4.80, it does have a striking effect on the rarefaction wave. In Figure 4.80, the history of this rarefaction wave is comprised of three portions, a structured wave between states E and D, a rarefaction shock wave between states D and CJ, and another structured wave below state CJ. The leading portion of the rarefaction wave is a structured wave that causes the velocity to decrease from the Hugoniot state E down the isentrope to state D. The next portion of the rarefaction wave is a rarefaction shock wave. Indeed, when we compare the three measured histories, we see that it is a steady wave propagating at a constant velocity. Examination of the isentrope in the region of the higher-pressure phase transition illustrates why this wave exists (see Figure 4.82).

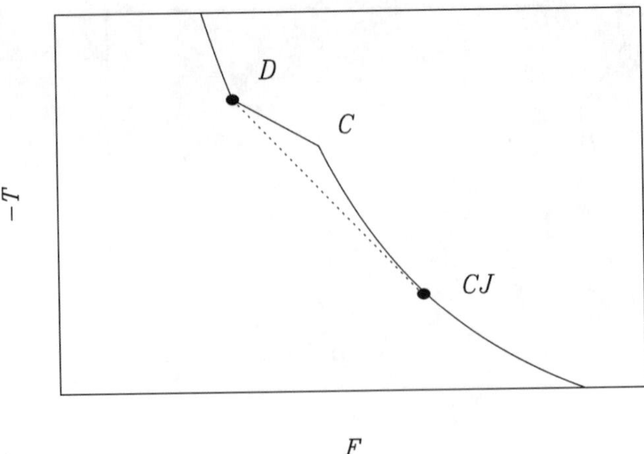

Figure 4.82 The isentrope (—) in the region of the second phase transition and the Rayleigh line (– –) of the rarefaction shock wave.

Recall that compression shock waves exist when

$$\frac{\partial^2 T}{\partial F^2} \leq 0.$$

This condition causes the characteristic contours of a structured compression wave to converge and form a shock discontinuity. Now notice that the isentrope violates this condition in the neighborhood of state C. Thus compression shock waves cannot exist in this region. Moreover, we see that the characteristic contours of a structured rarefaction wave now converge to form a shock discontinuity. The Rayleigh line of the rarefaction shock wave intersects state D and the Chapman–Jouguet state CJ. Recall from our discussion of detonation that at the CJ state, the velocity of a structured wave is equal to the velocity of a shock wave. Indeed, the present rarefaction shock wave is analogous to the steady rarefaction wave that forms the trailing portion of von Neumann spike in a detonation.

The description of the release process is completed when we recognize that another structured rarefaction wave forms below state CJ. The only anomaly between our description and the data in Figure 4.80 occurs at state D. In Figure 4.82, we show a discontinuity between the slopes of the isentrope and the Rayleigh line. This means the structured wave between states E and D should propagate faster than the rarefaction shock wave between states D and CJ. A plateau of constant velocity should separate the leading structured rarefaction wave and the rarefaction shock wave. These data do not exhibit such a plateau. This suggests that a smooth transition, perhaps like another CJ state, exists between the slopes of the isentrope and the Rayleigh line at state D. However, this conjecture can only be verified by future investigation into the nature of nonlinear wave phenomena.

4.5.3 *Exercises*

4.41 Verify Eq. (4.235), which is the expression for the Hugoniot of an explosive gas.

4.42 Consider a combustible gas. The jump condition for the balance of energy is given by Eq. (4.236). Use this expression to show that the entropy on the Hugoniot for the fully burned gas is minimum at the CJ state.

4.43 In a certain explosive gas we observe a detonation with a Mach number of $M = 10$. We also find that the speed of sound and density of the unburned gas are $c_0 = 375$ m/s and $\rho_0 = 1.1$ kg/m^3. Assume that this is a *CJ* detonation, $\gamma = 7/5$, and the detonation pressure is large with respect to the initial pressure of the unburned gas. Assume that $p_{CJ} \gg p_-$.

(a) Determine the reaction energy Q of the unburned gas.

(b) Determine the peak pressure of the von Neumann spike and the pressure at the *CJ* state.

4.44 Consider our example of a *CJ* process in the last subsection of Section 4.5.1.

(a) Verify that along the Rayleigh line

$$\frac{\vartheta}{\vartheta_{CJ}} = (1 + \sqrt{1-\lambda})(\gamma - \sqrt{1-\lambda})/\gamma.$$

(b) Plot ϑ/ϑ_{CJ} as a function of kt for the three reaction equations that we consider in this example.

APPENDIX A

Numerical Methods

People routinely employ a variety of numerical methods to analyze waves in materials with complex constitutive equations and boundary conditions. Indeed, it may appear to you that the number of numerical solution methods is equal to or greater than the number of numerical analysts. However, most of these algorithms are based on a few simple concepts. Here we shall discuss these concepts by examining several algorithms. We shall solve the linear wave equation with a numerical algorithm derived from the method of characteristics, and we shall solve the equations describing nonlinear waves with a numerical algorithm derived from the method of finite differences. Through these examples, you will become acquainted with the fundamental issues in the numerical analysis of waves. You will need to understand these issues to successfully select and apply any good commercial software package to analyze your wave-propagation problems.

We can only obtain analytic solutions to the equations describing nonlinear waves for a very limited set of geometries, constitutive equations, and boundary conditions. For example, the Riemann integrals yield a solution for a one-dimensional simple wave in an infinite volume of nonlinear-elastic material; however, obtaining analytic solutions for the interaction of two simple waves ranges from difficult to impossible depending upon the constitutive equations of the medium. Without the advent of the digital computer and sophisticated methods of numerical analysis, advances in the field of nonlinear wave motion and the development of constitutive theories for dynamic loading of materials would have stagnated over the past 40 years. In this appendix, we shall *not* attempt to present a complete discussion of these numerical methods. Indeed, this would unnecessarily double the length of this book, because in most applications it is reasonable to view the numerical algorithm as a tool, which we can use without necessarily understanding exactly how it works. However, like other tools, some numerical algorithms are not suited to the job, and even the right algorithm can be used incorrectly. Thus, instead of attempting to develop a detailed understanding of these methods, we shall instead study several basic one-dimensional algorithms and use them to illustrate their proper use.

A.1 Some General Comments

Recall that in Chapter 2, we made extensive use of the X–t and the x–t diagrams. We drew the characteristic contours of propagating waves in these diagrams. These contours can represent either waves with smooth solutions obtained from differential equations or waves with discontinuous solutions obtained from jump conditions. By now it is obvious that both time and position in these diagrams are smooth continuous quantities. The fundamental idea behind numerical analysis is to turn the coordinates in either the X–t diagram

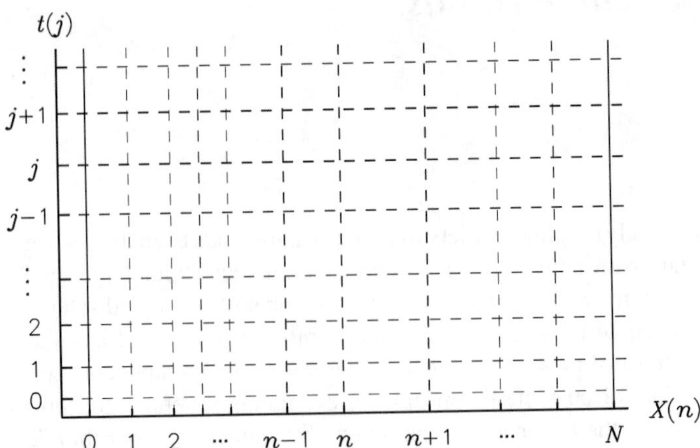

Figure A.1 The X–t diagram after we have discretized position and time. The left and right boundaries are at $X(0)$ and $X(N)$, respectively.

or the x–t diagram into discrete quantities. We illustrate this process with the X–t diagram in Figure A.1. Here we have replaced the continuous functions X and t with the discrete sets of quantities $X(n)$ and $t(j)$, where n and j are positive integers. From the equation for the motion, Eq. (1.37), we see that

$$x = \hat{x}[X(n), t(j)] \equiv x(n, j).$$

The discrete intervals in time are called *time cycles* or simply *cycles*, and the discrete intervals in position are called *cells*. Notice that we have selected nonuniform cycles and cells. It is the numerical analyst's task to write *algebraic* expressions to evaluate the solutions for motion, velocity, stress, etc. at the points $[X(n), t(j)]$. When the analyst works with material coordinates and evaluates the solutions at $[X(n), t(j)]$, we call the result a *Lagrangian algorithm*. When instead the analyst discretizes the spatial coordinates to obtain solutions at $[x(n), t(j)]$, we call the result an *Eulerian algorithm*. For one-dimensional motion, Lagrangian algorithms have a unique advantage over Eulerian algorithms because we can always place a discrete position $X(n)$ at each material interface and boundary. Thus Lagrangian cells only contain one material. When we use an Eulerian algorithm the positions of the material interfaces and boundaries move with respect to the discrete points $x(n)$. Thus Eulerian cells can contain multiple materials. Keeping track of the exact positions of the interfaces and boundaries places an additional computational burden on the Eulerian algorithm. We illustrate one of these difficulties in Figure A.2. Here we use a bold line to represent the actual position of the left boundary of a solid. Suppose this boundary separates the solid to the right from a vacuum to the left. During the first time cycle from $t(0)$ to $t(1)$, the boundary moves partially across the first cell of the solid. Thus at $t(1)$ most of the material in the cell between $x(0)$ and $x(1)$ is a solid and the rest is a vacuum. To implement a numerical algorithm we must necessarily assume that the material properties in this cell are uniform. Therefore, in the numerical computation for this cell we are forced to replace the actual properties of the solid by properties that approximate a mixture of solid and vacuum. During the next time cycle when we advance the computation from $t(1)$ to $t(2)$, a portion of the mixture in the cell between $x(0)$ and $x(1)$ will move into the cell between $x(1)$ and $x(2)$. Thus after this cycle of the computation, two cells will contain materials with mixed properties. In fact,

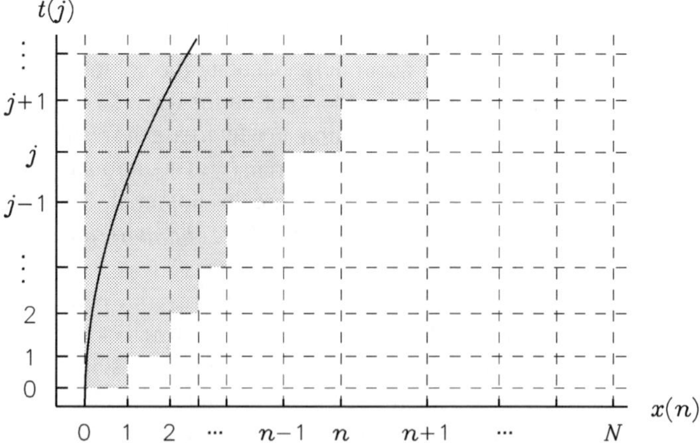

Figure A.2 An x–t diagram that has been discretized. The bold line illustrates the actual position of the left boundary. The shaded region represents the material dispersion in the discretized computation.

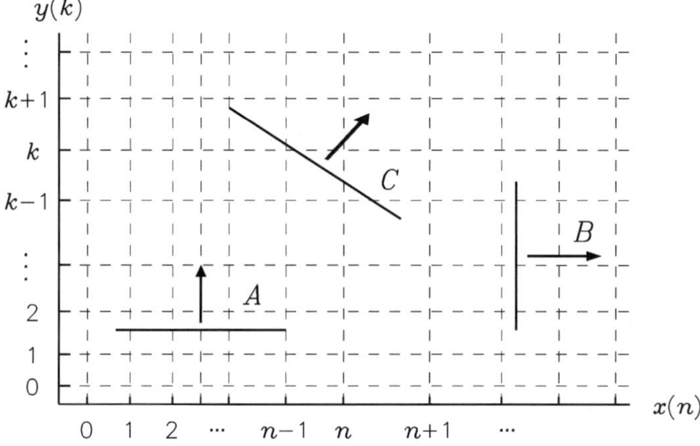

Figure A.3 Three separate one-dimensional shock waves traveling in different directions.

each cycle of the computation will cause the mixture to *diffuse* into a neighboring cell. Unless the Eulerian algorithm is designed carefully, the material can be "vaporized" by the numerical computation. Clearly, designers of Lagrangian algorithms do not face this problem.

In computations in two and three spatial dimensions, this advantage of the Lagrangian algorithm outweighs its disadvantages until we encounter large deformations. To understand the difficulties of computing the motion of bodies with large deformations, consider the computation of the motion of a piece of graph paper. Suppose we have written an algorithm that evaluates the motion at the grid points on the paper. Now wad up the paper. When we distort the computational grid this way, we find that Lagrangian algorithms lose accuracy, take longer to compute the solutions, and often become unstable. In these cases, the Eulerian algorithms have significant advantages over the Lagrangian algorithms.

In addition, both types of algorithms suffer from other limitations. For example, we illustrate one good test case for both types of algorithms in Figure A.3. Here we illustrate three one-dimensional shock waves propagating in a discretized *two-dimensional space*. Except for the direction of propagation, all three waves have identical steady-wave solutions. Typically, most numerical algorithms can accurately calculate the solutions for waves A and B, but they often have difficulty calculating the solution for wave C.

Computational grids can also have their own motion, in which case the resulting algorithm is neither Lagrangian (grid moves with material) or Eulerian (grid is stationary). These types of algorithms are currently at the leading edge of software development. *However, here we shall limit our discussion to one-dimensional motion where the material description is preferred.*

The selection of the size of the cells between each of the material points $X(n)$ in the material description is another important issue. In general, if we select smaller cells, we will obtain more accurate numerical approximations to the analytical solutions. However, by decreasing this interval, we increase the total number of points $X(n)$ in the calculation. This increases the amount of computation required to generate a solution. In particular, by doubling the number of points $X(n)$, we generally expect to increase the amount of computation by a factor of *four*. To see why the number of computations increases by four instead of two, let us examine how these computations are executed. Suppose we prescribe the initial conditions of the solution at the time $t = t(0)$. The left boundary of the material is at $X(0)$ and the right boundary is at $X(N)$, where $n = 0, 1, 2, \ldots, N$ (see Figure A.1). Now we prescribe the initial motion, velocity, and temperature as

$$x(n, 0) = x_n^0, \qquad v(n, 0) = v_n^0, \qquad \vartheta(n, 0) = \vartheta_n^0,$$

where x_n^0, v_n^0, and ϑ_n^0 are three sets of numbers. Suppose a stress $T_l[t]$ is applied to the left boundary. Then

$$T[X(0), t(j)] \equiv T(0, j) = T_l[t(j)].$$

Then suppose we fix the right boundary,

$$v[X(N), t(j)] \equiv v(N, j) = 0.$$

For the purpose of discussion, let us assume that these initial conditions are sufficient to completely define the numerical solution at $t = t(0)$. Our objective now is to select a value of $t = t(1)$ and then use the numerical algorithm, which is a set of algebraic equations, to compute the numerical solution for each $X(n)$ at this new time. After we have done this, we select a value of $t = t(2)$ and then compute the numerical solution at each $X(n)$ for this time. Thus cycle by cycle we march the solution forward in time. A critical part of this process is the determination of the time interval for each cycle. Later in this appendix, we shall derive an expression for the maximum time interval for each cycle. However, here we appeal to the concept of causality to estimate this quantity.

Let us suppose that at time $t = t(j)$ the material wave velocity between $X(n-1)$ and $X(n)$ is approximately constant and denoted by $C(n - \frac{1}{2}, j)$. The time $\delta t(n - \frac{1}{2}, j)$ required for a structured wave to travel from $X(n-1)$ to $X(n)$ is

$$\delta t\left(n - \frac{1}{2}, j\right) \equiv [X(n) - X(n-1)]/C\left(n - \frac{1}{2}, j\right), \tag{A.1}$$

where each value of n yields a different value of $\delta t(n - \frac{1}{2}, j)$. Now we argue that the accuracy of our numerical solution will be enhanced by resolving the propagation of individual waves between each of the points $X(n)$. Thus we require that

$$t(j+1) - t(j) \leq \delta t_{cr}(j), \tag{A.2}$$

where $\delta t_{cr}(j)$ is called the *critical time step*,

$$t_{cr}(j) \equiv \text{minimum}\left[\delta t\left(n - \frac{1}{2}, j\right)\right], \quad \text{where } 0 < n \leq N.$$

Now if we halve the spatial interval $[X(n) - X(n-1)]$, we will halve the first time interval $[t(1) - t(0)]$ as well as every future time interval in the solution. This results in four times as many solution points in the X–t diagram.

In some algorithms, the solution at $[X(n), t(j+1)]$ depends only on the solutions at $t(j)$ and earlier. It does not depend on the solution at any other position at the time $t(j+1)$. These types of algorithms are called *explicit algorithms* because the *unknown* solution at $[X(n), t(j+1)]$ is expressed explicitly in terms of *known* solutions at $t(j)$ and earlier. We shall discuss several explicit algorithms in this appendix. In other types of algorithms the solution at $[X(n), t(j+1)]$ does depend upon the solution at other spatial points at the time $t(j+1)$; these types of algorithms are called *implicit algorithms*. For example, the finite-element method yields implicit algorithms. Implicit algorithms lose accuracy when we allow the time interval to exceed $t_{cr}(j)$, and explicit algorithms become unstable when we violate this condition. *Thus, regardless of the type of algorithm you might use, violating this condition will seriously compromise your wave solution.* In general, implicit algorithms are *forgiving*. When these algorithms are misused by increasing the time interval beyond the critical time step, they often yield *well-behaved* wave solutions that are totally meaningless. Explicit algorithms are *unforgiving*. When misused in this way, they usually become unstable.

Some numerical algorithms only solve jump conditions for discontinuous solutions, some only solve differential equations for smooth solutions, and some solve both. The algorithms that employ jump conditions alone are based upon the concept that a smooth structured wave can be approximated as a sequence of weak shock waves (see Exercise 2.43 in Section 2.7.8). In contrast, the algorithms that only employ approximate solutions to the differential equations are based upon the concept that a smooth steady wave propagates at the same velocity and satisfies the same jump conditions as a shock wave. In this type of algorithm, artificially large viscous stresses are introduced into the calculation. This increases the thickness of the steady wave profile until it spreads across a number of cells, producing a solution that satisfies the smoothness requirements of the particular algorithm that we are using. Provided the thickness of the steady wave is small in comparison to the curvature of the wave front, the artificially large viscous stresses do not alter either the propagation velocity or the jump conditions for the wave.

We shall now turn to the discussion of a set of algorithms based upon the method of characteristics for the linear wave equation. These algorithms are very simple and efficient and actually yield analytic solutions under certain conditions. After this we discuss an explicit, finite-difference Lagrangian algorithm. This one-dimensional algorithm employs an artificial viscosity term to smooth the solutions over several cells. It is a classical method developed in the 1940s by J. von Neumann and R. D. Richtmyer. By comparing these two methods we shall illustrate the relationship between the method of characteristics and the method of finite differences. We shall also show that the methods, which are suitable for nonlinear problems, are not suitable for linear problems.

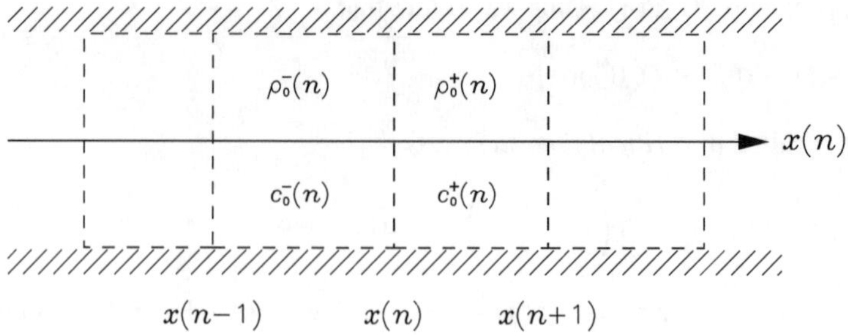

Figure A.4 The discrete layers for a linear-elastic body.

A.2 Analysis of the Linear Wave Equation

In this section, we shall discuss two numerical algorithms for solving the linear wave equation. The linear wave equation describes waves in linear-elastic materials. Thus, there is no distinction between the *X–t* and *x–t* diagrams. We shall develop these algorithms by discretizing the *x–t* diagram in a very special way. Because we use time and position, the algorithm is classified as a *time-domain* approach. Other popular methods for analyzing the linear wave equation are classified as *frequency-domain* approaches. These methods are based upon the Fourier transform of the wave equation† and are outside the scope of this work.

We discretize the position x with the set of quantities $x(n)$, where $n = 0, 1, 2, \ldots, N$. Between each set of adjacent points, say $x(n)$ and $x(n-1)$, we assume that the speed of sound c_0 and the density ρ_0 are constant. Thus we can picture the body to be composed of a series of imaginary layers (see Figure A.4). The density and sound speed in the layer to the right of $x(n)$ are denoted by $\rho_0^+(n)$ and $c_0^+(n)$, and to the left, they are denoted by $\rho_0^-(n)$ and $c_0^-(n)$. Thus, for example, $c_0^-(n) = c_0^+(n-1)$. We shall require that the discrete values of $x(n)$ and $t(j)$ obey the following special relationships:

$$x(n+1) - x(n) = c_0^+(n)\Delta, \qquad t(j) = j\Delta, \tag{A.3}$$

where Δ is an arbitrary positive constant. By comparing this expression to Eq. (A.1), we see that Δ is equal to the critical time step. The trick in selecting Δ is to choose it so that each physical interface and boundary of the body is either exactly at or negligibly close to an imaginary interface that we created by the discretization of x. We must also be careful to select a value of Δ that is small enough to accurately approximate the initial conditions and boundary conditions.

A.2.1 *First Algorithm*

As a consequence of Eq. (A.3), all of the points $[x(n), t(j)]$ are connected together by characteristic contours (see Figure A.5). This means we can use the characteristic

† For example, see *Introduction to Elastic Wave Propagation* by Bedford, A. and Drumheller, D. S. (Wiley, 1994).

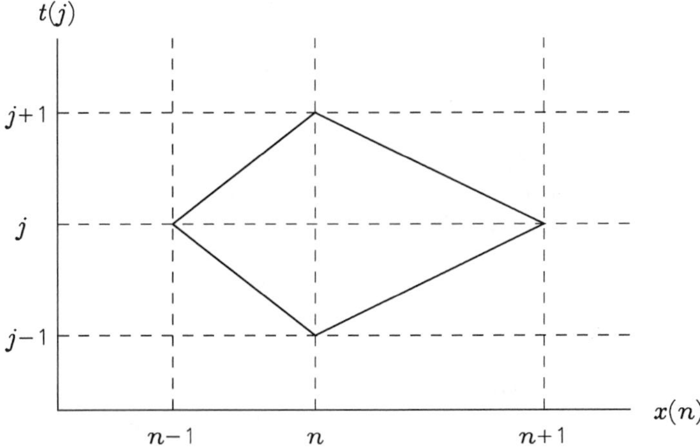

Figure A.5 Selected characteristic contours in the discretized x–t diagram.

relationships of Eqs. (2.68)–(2.69),

$$
\begin{aligned}
v - \frac{T}{z_0} &= \text{constant}, \qquad \text{positive-sloping characteristics,} \\
v + \frac{T}{z_0} &= \text{constant}, \qquad \text{negative-sloping characteristics,}
\end{aligned}
\tag{A.4}
$$

to derive algebraic relationships between the solutions at each of these points in the x–t diagram. We can then use the resulting algorithm to evaluate the solution to the wave equation at the boundaries and all of the physical as well as imaginary interfaces of the body.

For a given position n, the acoustic impedances to the left and right of $x(n)$ are $z(n)^- = \rho_0^- c_0^-$ and $z(n)^+ = \rho_0^+ c_0^+$, respectively. Next we define

$$
z(n) \equiv \frac{1}{2}[z(n)^+ + z(n)^-]. \tag{A.5}
$$

With Eq. (A.4), we can establish relationships between the values of v and T at the points that are connected by the characteristic contours in Figure A.5. From the two characteristic contours at the top of the "diamond," we obtain

$$
z(n)^- v(n, j+1) - T(n, j+1) = z(n)^- v(n-1, j) - T(n-1, j) \tag{A.6}
$$

and

$$
z(n)^+ v(n, j+1) + T(n, j+1) = z(n)^+ v(n+1, j) + T(n+1, j). \tag{A.7}
$$

We add these two equations to obtain

$$
\begin{aligned}
2z(n)v(n, j+1) = {} & z(n)^+ v(n+1, j) + z(n)^- v(n-1, j) \\
& + T(n+1, j) - T(n-1, j).
\end{aligned}
\tag{A.8}
$$

We can use Eqs. (A.6) and (A.8) to analyze initial-boundary-value problems in the body. First we prescribe the initial conditions $T(n, 0)$ and $v(n, 0)$. Then we use Eq. (A.8) to determine $v(n, 1)$, and we use Eq. (A.6) to determine $T(n, 1)$. We can carry out these calculations in any order for all values of n except $n = 0$ and $n = N$. At the left boundary, $n = 0$, we must use Eq. (A.7), and we must prescribe either $v(0, 1)$ or $T(0, 1)$. Similarly, at the right boundary, $n = N$, we must use Eq. (A.6), and we must prescribe either $v(N, 1)$ or $T(N, 1)$. This process is repeated for successive values of j until a chosen final time is reached.

Each application of Eqs. (A.6) and (A.8) requires nine arithmetic operations. In the next section we describe an alternative algorithm that determines only the velocity. It is simpler and more efficient both for computing simple examples by hand and for solving complicated problems using a computer.

A.2.2 *Second Algorithm*

In this method we employ four relations, Eqs. (A.6) and (A.7) together with the two characteristic relations we obtain from the bottom of the diamond in Figure A.5:

$$z(n)^- v(n, j-1) + T(n, j-1) = z(n)^- v(n-1, j) + T(n-1, j) \tag{A.9}$$

and

$$z(n)^+ v(n, j-1) - T(n, j-1) = z(n)^+ v(n+1, j) - T(n+1, j). \tag{A.10}$$

We add these four relations to obtain the new algorithm:

$$z(n)[v(n, j+1) + v(n, j-1)] = z(n)^+ v(n+1, j) + z(n)^- v(n-1, j). \tag{A.11}$$

This equation requires only four arithmetic operations to determine $v(n, j+1)$. When the acoustic impedance is constant at n, then $z(n) = z(n)^+ = z(n)^-$, and the algorithm simplifies to

$$v(n, j+1) + v(n, j-1) = v(n+1, j) + v(n-1, j). \tag{A.12}$$

Our sequence of calculations using Eq. (A.11) is roughly the same as for the first algorithm, but because T does not appear explicitly, we handle initial and boundary conditions somewhat differently. We prescribe the initial conditions $T(n, 0)$ and $v(n, 0)$ as before. Then we use Eq. (A.8) to determine $v(n, 1)$, and from that time on, we use Eq. (A.11) to determine v. When we prescribe T at the left boundary, we evaluate Eqs. (A.7) and (A.10) for $n = 0$ and then sum the results to obtain

$$v(0, j+1) + v(0, j-1) = 2v(1, j) - \frac{1}{z(0)^+}[T(0, j+1) - T(0, j-1)], \tag{A.13}$$

and when T is specified at the right boundary, we get

$$v(N, j+1) + v(N, j-1) = 2v(N-1, j) + \frac{1}{z(N)^-}[T(N, j+1) - T(N, j-1)]. \tag{A.14}$$

These two algorithms have an interesting property: The values of $T(n, j)$ and $v(n, j)$ at points where $j+n$ is even are determined by the boundary and initial conditions at points where $j+n$ is even, and the values of $T(n, j)$ and $v(n, j)$ at points where $j+n$ is odd are determined by the boundary and initial conditions at points where $j+n$ is odd. This is most apparent in Eq. (A.12). You can compute either the even or odd solution independently, reducing the number of computations by a half.

A.2.3 *Examples*

1. Reflection at an Interface Consider two bonded layers of linear-elastic material that have acoustic impedances of $z = 2$ in the left layer and $z = 1$ in the right layer. Suppose that the left boundary is subjected to a step in velocity $v(0, t) = 3$ for $t > 0$.We select a value for Δ that divides each layer into five imaginary layers. In Figure A.6, we illustrate the odd solution to the wave equation for v as a matrix of numbers. Inspection of this matrix reveals that the left boundary condition results in a wave of amplitude 3 that propagates in the positive-x direction at the speed of sound. When it intersects the interface at $n = 5$, it produces a transmitted wave of amplitude 4 and a reflected wave of amplitude 1. In this matrix, the transmitted wave appears to have the same velocity as the incident wave. However, you should notice that when the speeds of sound in the two layers are different, $x(6) - x(5) \neq x(5) - x(4)$. Thus while each wave in this solution travels one index n for every increment in time Δ, the transmitted wave travels a different distance.

Originally, we started with an empty table. Let us see how we calculated the numbers in this table. The values for $v(0, j)$ and $v(10, j)$ are obtained from the boundary conditions.

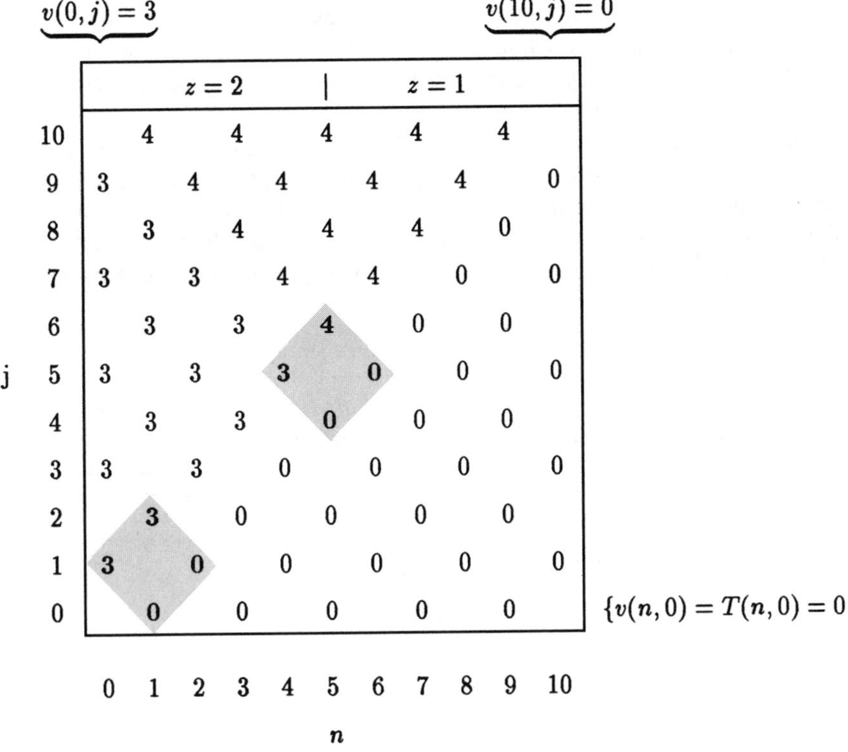

Figure A.6 The numerical solution for $v(n, j)$ for two layers with acoustic impedances $z = 2$ and $z = 1$.

The values for $v(n,0)$ and $T(n,0)$ are obtained from the initial conditions. Next we use Eq. (A.8) to evaluate $v(n,1)$ at $n = 2, 4, 6, 8$. Then we use Eq. (A.11) to compute $v(n,2)$ at $n = 1, 3, 5, 7, 9$. We continue to apply Eq. (A.11) for each value of j until the table is complete. The speed of these calculations can be increased by noticing that we can use Eq. (A.12) for all values of n except at the physical interface between layers where $n = 5$. For example, we see that

$$v(1,2) = v(0,1) + v(2,1) - v(1,0) = 3 + 0 - 0 = 3.$$

This particular calculation is denoted by the bold numbers in the lower left corner of the table. These numbers form a diamond pattern, and the algorithm requires that the sum of the numbers on the left and right corners of the diamond must be equal to the sum of the numbers on the top and bottom corners. This *diamond rule* holds at every interior point in the table except when the center of the diamond is located at the physical interface $n = 5$. Here $z(5)^+ \neq z(5)^-$, so we must use Eq. (A.11). For example, at $j = 5$ and $n = 5$, this algorithm yields

$$v(5,6) = \frac{2}{3}v(6,5) + \frac{4}{3}v(4,5) - v(5,4) = \frac{2}{3} \times 0 + \frac{4}{3} \times 3 - 0 = 4.$$

This is illustrated with a second set of bold numbers at the center of the table.

2. *Multilayered Medium* Our previous example illustrated wave behavior using an example that is simple enough to compute by hand, but the real utility of these methods becomes evident when they are applied to layered media with large numbers of layers. As an example, we consider a medium consisting of 300 layers. Each layer has the same thickness and the same speed of sound c_0. Thus we choose Δ such that $c_0\Delta$ is equal to the thickness of each layer. The acoustic impedance of each layer is prescribed to be $z^-(n) = n$. Thus, the acoustic impedance of the first layer is 1, that of the second layer is 2, and so on. We subject the left boundary of this medium to a unit step in velocity, $v(0,t) = 1$, for $t > 0$. We evaluate the odd solution from Eq. (A.11) by using the following program:

- Initialize storage:
- *z is a vector with* 301 *elements initialized to* $z(n) = n$. *v is a matrix with* $(301, 301)$ *elements initialized to* 0. *Subscripts start at* 0.
- Using Eq. (A.5), compute $\frac{z(n)^+}{z(n)}$ and $\frac{z(n)^-}{z(n)}$:
- *for* $(n = 1$ *to* 299 *by increments of* $1)$ *begin*

  $$zp(n) = 2 * z(n+1)/(z(n) + z(n+1))$$
  $$zm(n) = 2 * z(n)/(z(n) + z(n+1))$$

- *end for*
- Prescribe left boundary condition:
- *for* $(j = 1$ *to* 299 *by increments of* $2)$ $v(0,j) = 1$

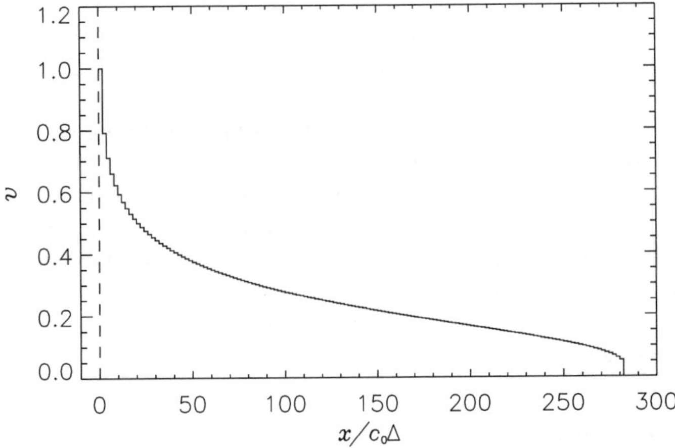

Figure A.7 The wave profile in a multilayered medium at $j = 281$.

- Start time loop to compute solution:
- *for* $(j = 1$ *to* 299 *by increments of* 1) *begin*

  if j *is even* $ns = 2$ *else* $ns = 1$

  Using Eq. (A.11), compute odd solution – exclude right boundary:

  for $(n = ns$ *to* 299 *by increments of* 2) $v(n, j+1) =$
  $zp(n) * v(n+1, j) + zm(n) * v(n-1, j) - v(n, j-1)$
- *end for*

The computer memory must be large enough to hold the values of v for all values of n and j simultaneously, including the values of the even solution, which remain as zeros.

We show the computed profile of the velocity of the material at time $t = 281\Delta$ in Figure A.7. The large number of layers results in a relatively smooth transition in the acoustic impedance from $z = 1$ at the left boundary to $z = 300$ at the right boundary. The boundary condition at $n = 0$ causes a jump in velocity to propagate into the first layer; however, the large number of layers produce an enormous number of reflections that cause the wave to evolve into a structured ramp wave. You should notice that this profile represents an analytical solution to this example.

A.3 The von Neumann–Richtmyer Algorithm

There are many numerical algorithms for analyzing nonlinear waves, but perhaps the oldest and still most widely used method was developed by J. von Neumann and R. D. Richtmyer. We shall now discuss this method. To successfully employ this algorithm we must understand two important concepts: the critical time step and artificial viscosity.

In a linear-elastic material, the velocity of a wave is always equal to c_0. In the preceding section, we used this fact to great advantage. We selected a time step and spatial increments that satisfied Eq. (A.3). Thus every application of the algorithm was executed at the critical time step. This means that the characteristics from one grid location will always exactly intersect the neighboring grid locations. However, in nonlinear materials, there are multiple wave speeds that change with time, and we cannot use this procedure. Thus although we may execute portions of the calculation at or near the critical time step, we must execute other portions well below the critical time step.

Rather than analyze the nonlinear elastic equations, we shall now derive a simple finite-difference approximation to the linear wave equation and use it to illustrate two points about nonlinear computations. First we analyze the finite-difference algorithm to demonstrate that it is unstable when $\delta t(n, j)$ exceeds the critical time step. Then we show the effects that are introduced into the solution when we operate the algorithm with time steps that are smaller than the critical time step.

A.3.1 *Critical Time Step*

Consider the wave equation

$$\frac{\partial^2 v}{\partial t^2} = c^2 \frac{\partial^2 v}{\partial x^2}.$$

Let us derive a finite-difference algorithm to solve this equation in the x–t diagram. To simplify our analysis we shall assume that c is a constant. We discretize time and position as follows:

$$t(j) = j\delta t, \qquad x(n) = n\delta x,$$

where δt and δx are prescribed constants. In the method of finite differences, derivatives are approximated with algebraic ratios. For example,

$$\frac{\partial v}{\partial t}\left(n, j + \frac{1}{2}\right) \approx \frac{v(n, j+1) - v(n, j)}{\delta t},$$

where $\frac{\partial v}{\partial t}(n, j + \frac{1}{2})$ denotes that we have evaluated the time derivative half way between $t(j+1)$ and $t(j)$. Similarly,

$$\frac{\partial^2 v}{\partial t^2}(n, j) \approx \frac{1}{\delta t}\left[\frac{\partial v}{\partial t}\left(n, j + \frac{1}{2}\right) - \frac{\partial v}{\partial t}\left(n, j - \frac{1}{2}\right)\right].$$

We can combine the previous two results to obtain

$$\frac{\partial^2 v}{\partial t^2}(n, j) \approx \frac{1}{(\delta t)^2}[v(n, j+1) - 2v(n, j) + v(n, j-1)].$$

Similarly, we can obtain the following finite-difference approximation for the spatial derivative in the wave equation:

$$\frac{\partial^2 v}{\partial x^2}(n, j) \approx \frac{1}{(\delta x)^2}[v(n+1, j) - 2v(n, j) + v(n-1, j)].$$

Thus the finite-difference approximation of the wave equation is

$$v(n, j+1) - 2v(n, j) + v(n, j-1) = \alpha^2[v(n+1, j) - 2v(n, j) + v(n-1, j)], \qquad \text{(A.15)}$$

where

$$\alpha \equiv \frac{c\delta t}{\delta x}.$$

Notice that this is an explicit algorithm. The unknown value of velocity at the new time $v(n, j+1)$ depends only on known values of velocity at earlier times.

For the special case when $\alpha = 1$, the time and spatial increments satisfy Eq. (A.1), and the finite-difference algorithm Eq. (A.15) reduces to Eq. (A.12). *Thus we find that this method of finite differences and the method of characteristics yield identical algorithms when we operate at the critical time step.*

However, in nonlinear computations we shall be forced to select $\alpha \neq 1$. Let us determine how Eq. (A.15) behaves under these conditions. In particular, let us examine the following solution:

$$v(j,n) = G^j \mathrm{e}^{ikn}, \tag{A.16}$$

where $i = \sqrt{-1}$ and G and k are arbitrary constants. By substituting this solution into Eq. (A.15) and noticing that $2\cos k = \mathrm{e}^{ik} + \mathrm{e}^{-ik}$, we obtain

$$G^2 - 2\beta G + 1 = 0,$$

where

$$\beta = 1 - \alpha^2(1 - \cos k).$$

The solution to this quadratic equation is

$$G = \beta \pm \sqrt{\beta^2 - 1}.$$

From Eq. (A.16), we see that v will grow without bound if $|G| > 1$. If $\beta^2 > 1$, we see that one of the roots for G yields $|G| > 1$ and the corresponding solution is unstable. However, if $\beta^2 \leq 1$, then both roots obey $|G| = 1$, and the solution is stable. We require that the solution for v must be stable for all real values of k. Thus $\beta^2 \leq 1$ must hold for all real values of k. This is only possible if $\alpha \leq 1$. Thus the requirement for stability of these solutions is

$$\delta t \leq \delta x / c.$$

This verifies our original statement of the critical time step in Eq. (A.2).

Although we have required α to be a constant here, in nonlinear applications of the method of finite differences, α is a function of position as well as time. We shall now show that the value of α has a strong influence on the numerical solution. To do this, we shall use the algorithm Eq. (A.15) to calculate a wave in a thick, homogeneous, linear-elastic material. We discretize x with 41 points. The boundaries of the material are located at $x(0) = 0$ and $x(40) = 40\delta x$. We apply a step in velocity at the left boundary, $v(0, j) = 1$, and we hold the right boundary fixed, $v(40, j) = 0$. The initial conditions are $v(n, 0) = v(n, 1) = 0$, which hold for every value of n except $n = 0$. We calculate the solution to this problem from $t = 0$ to $t = 25\delta x/c$. This is done three times using $\alpha = 0.2$, 0.8, and 1.0. The respective maximum values of j are $j = 125$, 31, and 25. We illustrate the results of these three computations in Figure A.8. The dotted line for $\alpha = 1$ matches the analytic solution given by the method of characteristics. The two remaining solutions for $\alpha = 0.2$ and 0.8 exhibit large oscillations in the neighborhood of the front of the wave. These oscillations are distributed over approximately 20 discrete spatial points. They propagate with the wave

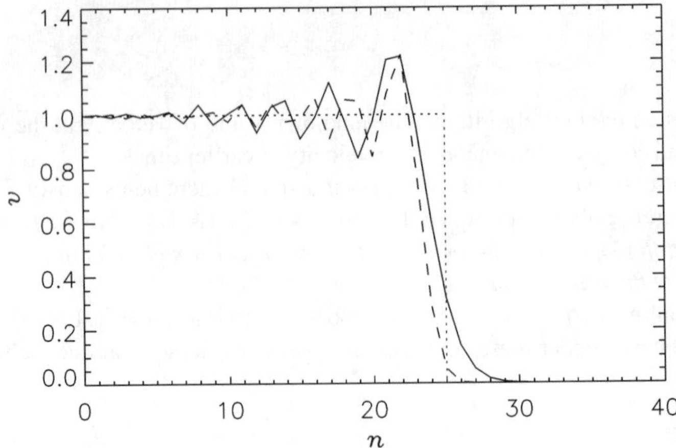

Figure A.8 Wave profiles from the finite-difference algorithm at $t = 25\delta x/c$ for $\alpha = 0.2$ (—), 0.8 (– –), and 1.0 (· · ·).

and do not diminish in amplitude. Thus we observe a curious behavior. If we operate at the critical time step, we can generate an analytical solution, but otherwise the algorithm spreads the wave over many spatial points until the solution satisfies the smoothness conditions for the partial derivatives in the wave equation. Moreover, if we select $\alpha > 1$, the oscillations will become unstable and grow without bound.

A.3.2 *A Nonlinear Algorithm*

The classic algorithm for solving the Lagrangian form of the one-dimensional equations for nonlinear waves was developed by J. von Neumann and R. D. Richtmyer. They discretized the X–t diagram to obtain $X(n)$ and $t(j)$. The increments in both time and space were not uniform. They evaluated certain discretized quantities, such as the current position $x(n, j)$, at the intersections of $X(n)$ and $t(j)$. However, they evaluated other quantities, such a stress $T(n - \frac{1}{2}, j)$, between these intersections. The locations or *centering* of various quantities will become clear as we proceed with the development of the algorithm.

This algorithm employs the following approximations of the partial derivatives with respect to time and space:

$$\frac{\partial \hat{\Psi}}{\partial t}\left(n, j + \frac{1}{2}\right) \approx \frac{\Psi(n, j+1) - \Psi(n, j)}{t(j+1) - t(j)},$$

$$\frac{\partial \hat{\Psi}}{\partial X}\left(n + \frac{1}{2}, j\right) \approx \frac{\Psi(n+1, j) - \Psi(n, j)}{X(n+1) - X(n)},$$

where Ψ is an arbitrary variable. We shall also use the averaging rules

$$\Psi\left(n, j + \frac{1}{2}\right) \approx \frac{1}{2}[\Psi(n, j+1) + \Psi(n, j)],$$

$$\Psi\left(n + \frac{1}{2}, j\right) \approx \frac{1}{2}[\Psi(n+1, j) + \Psi(n, j)].$$

Our objective now is to derive a set of algebraic expressions that will allow us to cycle the numerical solution forward in time. To do this, we define the time step $\delta t(j+\frac{1}{2})$ to be

$$\delta t\left(j+\frac{1}{2}\right) \equiv t(j+1)-t(j).$$

The mass $m(n-\frac{1}{2})$ between two adjacent points is

$$m\left(n-\frac{1}{2}\right) = \rho_0\left(n-\frac{1}{2}\right)[X(n)-X(n-1)],$$

where $\rho_0(n-\frac{1}{2})$ is the reference density of the discretized material. We use the definition of the deformation gradient to obtain the following finite-difference approximation:

$$F\left(n-\frac{1}{2}, j+1\right) = \frac{x(n, j+1)-x(n-1, j+1)}{X(n)-X(n-1)}. \tag{A.17}$$

The velocity of the material is equal to the material derivative of the current position. The finite-difference approximation of this relationship is

$$v\left(n, j+\frac{1}{2}\right) = \frac{x(n, j+1)-x(n, j)}{t(j+1)-t(j)},$$

which becomes

$$x(n, j+1) = x(n, j)+\delta t\left(j+\frac{1}{2}\right)v\left(n, j+\frac{1}{2}\right). \tag{A.18}$$

Similarly, we can obtain the velocity from the acceleration by using

$$v\left(n, j+\frac{1}{2}\right) = v\left(n, j-\frac{1}{2}\right)+\frac{1}{2}\left[\delta t\left(j+\frac{1}{2}\right)+\delta t\left(j-\frac{1}{2}\right)\right]a(n, j). \tag{A.19}$$

We obtain a finite-difference expression for the acceleration $a(n, j)$ from the equation for the balance of linear momentum

$$\rho_0 a = \frac{\partial \hat{T}}{\partial X}.$$

We approximate the reference density as

$$\rho_0(n) = \frac{m\left(n+\frac{1}{2}\right)+m\left(n-\frac{1}{2}\right)}{X(n+1)-X(n-1)},$$

so that the balance of linear momentum yields

$$a(n, j) = 2\left[\frac{T\left(n+\frac{1}{2}, j\right)-T\left(n-\frac{1}{2}, j\right)}{m\left(n+\frac{1}{2}\right)+m\left(n-\frac{1}{2}\right)}\right]. \tag{A.20}$$

Nonlinear-Elastic Material
Notice that the acceleration depends upon the stress. For the moment, let us limit our discussion to a nonlinear-elastic material. The constitutive relationship for the equilibrium stress of this material $T = T_e[F]$ yields

$$T\left(n - \frac{1}{2}, j\right) = T_e\left[F\left(n - \frac{1}{2}, j\right)\right]. \tag{A.21}$$

With this constitutive equation, we now have a complete numerical algorithm. To illustrate this, suppose we execute a calculation where the values of position, velocity, and stress are known at $t = t(j)$ and earlier. At the new time $t(j+1)$, we advance this calculation by first determining the critical time step and selecting $\delta t(j + \frac{1}{2})$ to be some prescribed fraction of the critical time step. Then we start at the left boundary $x(0)$ and proceed to the right. At any interior point $N > n > 0$, the solution at $t = t(j + 1)$ is determined as follows:

- Solve Eq. (A.20) for $a(n, j)$.
- Solve Eq. (A.19) for $v(n, j + \frac{1}{2})$.
- Solve Eq. (A.18) for $x(n, j + 1)$.
- Because $x(n - 1, j + 1)$ is known from the calculation for the preceding value of n, solve Eq. (A.17) for $F(n - \frac{1}{2}, j + 1)$.
- Solve Eq. (A.21) for $T(n - \frac{1}{2}, j + 1)$.

We use a slightly modified sequence at the boundaries where either the stresses $T(0, j)$ and $T(N, j)$ or the motions $v(0, j)$ and $v(N, j)$ are prescribed. After we have determined all of the positions $x(n, j+1)$, velocities $v(n, j+\frac{1}{2})$, and stresses $T(n-\frac{1}{2}, j+1)$, we can repeat this process for $t = t(j + 2)$. However, as we carry out this procedure, our solution will misbehave and either exhibit oscillations of the type illustrated in Figure A.8 or the solution will become unstable. The answer to this difficulty was first proposed by von Neumann and Richtmyer. They introduced an artificial viscous stress into the constitutive equation, Eq. (A.21).

A.3.3 *Artificial Viscosity*

We have already studied the viscous stress term [see Eq. (4.56)]

$$T = \frac{\rho_0 b_n}{F}\frac{\partial \hat{F}}{\partial t}\left|\frac{\partial \hat{F}}{\partial t}\right|^{n-1}.$$

When $n = 2$, we obtained an expression that we called quadratic viscosity. Let us consider a slightly different form of quadratic viscosity:

$$Q = \begin{cases} -\rho_0(b_2 \delta x D)^2, & D < 0, \\ 0, & D \geq 0, \end{cases} \tag{A.22}$$

where we recall that

$$D = \frac{1}{F}\frac{\partial \hat{F}}{\partial t}.$$

Here we denote the viscous stress by the symbol Q, which is the traditional notation used in numerical analysis for an *artificial viscosity. The only purpose of the term Q is to spread* *a steady wave or a shock wave over an artificially large distance to accommodate the smoothness requirements for our numerical algorithm.* In keeping with this purpose, we only allow artificial viscosity during compression, and we normalize the viscosity coefficient b_2 to the length increment δx. Indeed, we shall present an example where this definition of Q results in a steady wave with a thickness of $b_2\delta x$. The finite-difference representation of the *nonzero component* of Q is

$$Q\left(n-\frac{1}{2}, j-\frac{1}{2}\right) = -\rho_0\left(n-\frac{1}{2}, j-\frac{1}{2}\right) \times \left[b_2\delta x\left(n-\frac{1}{2}, j-\frac{1}{2}\right) D\left(n-\frac{1}{2}, j-\frac{1}{2}\right)\right]^2, \tag{A.23}$$

where

$$D\left(n-\frac{1}{2}, j-\frac{1}{2}\right) = \frac{2}{\delta t\left(j-\frac{1}{2}\right)}\left[\frac{F\left(n-\frac{1}{2}, j\right) - F\left(n-\frac{1}{2}, j-1\right)}{F\left(n-\frac{1}{2}, j\right) + F\left(n-\frac{1}{2}, j-1\right)}\right] \tag{A.24}$$

and

$$\delta x\left(n-\frac{1}{2}, j-\frac{1}{2}\right) = \frac{1}{2}[x(n, j) - x(n-1, j) + x(n, j-1) - x(n-1, j-1)]. \tag{A.25}$$

We emphasize that this is *our* choice for Q and many other forms are often used. However, in general, most people select $Q = 0$ during expansion of the material, unless the material supports rarefaction shock waves.

Recall from the previous paragraph that the first step in our algorithm is to solve Eq. (A.20) for $a(n, j)$. This means that $T(n - \frac{1}{2}, j)$ and $T(n + \frac{1}{2}, j)$ must be known. Therefore, we use the finite-difference approximation

$$T\left(n-\frac{1}{2}, j\right) = T_e\left[F\left(n-\frac{1}{2}, j\right)\right] + Q\left(n-\frac{1}{2}, j-\frac{1}{2}\right). \tag{A.26}$$

Notice that the stress $T(n - \frac{1}{2}, j)$ is equal to the properly centered value of the equilibrium stress $T_e[F(n - \frac{1}{2}, j)]$ plus the value of the viscous stress $Q(n - \frac{1}{2}, j - \frac{1}{2})$ that is centered a half a time step behind the time $t(j)$. At this point in the algorithm, we can evaluate $Q(n - \frac{1}{2}, j - \frac{1}{2})$ but not $Q(n - \frac{1}{2}, j)$. We are now in a position to illustrate the von Neumann–Richtmyer algorithm by solving a simple nonlinear wave problem. Consider a nonlinear-elastic material with a linear Hugoniot, which has the following properties:

$$\rho_0 = 7820 \text{ kg/m}^3, \quad c_0 = 4.4 \text{ km/s}, \quad s = 1.44.$$

The thickness of the material is 20 mm, and initially it is at rest. At $t = 0$, we suddenly change the velocity of the left boundary to a constant value of 700 m/s. The right boundary is stress free. We discretize the thickness into 20 increments and use the following computer program to solve this problem:

- Initialize storage:
- *rho is a scalar equal to the initial density; rho* = 7820.
- *m is a scalar equal to mass of each spatial increment;* $m = 7.82$.
- dX *is a scalar equal to initial spatial increment;* $dX = 0.001$.
- *c0 is a scalar equal to the sound speed;* $c0 = 4400$.
- *s is a scalar equal to the Hugoniot slope;* $s = 1.44$.
- *x is a vector with 21 elements equal to the initial positions;* $x(n) = 0.001 * n$, $0 \leq n \leq 20$.
- *v is a vector with 21 elements equal to the initial velocities;* $v(n) = 0$.
- *T is a vector with 21 elements equal to the initial stresses. The stress at* $n - \frac{1}{2}$ *is stored in element n;* $T(n) = 0.0$.
- *F is a vector with 21 elements equal to the initial deformation gradient. The deformation gradient at* $n - \frac{1}{2}$ *is stored in element n;* $F(n) = 1.0$.
- *b2 is the artificial viscosity coefficient;* $b2 = 7$.
- *alpha is a time-step factor; alpha* = 0.4
- *dt is the initial time step;* $dt = 0.0000001/c_0$.
- *xn is a vector with 21 elements.*
- Prescribe left boundary condition and initial time:
- $v(0) = 700$
- *time* = 0
- Start time loop to compute solution:
- *for* ($j = 1$ *to* 180 *by increments of* 1) *begin*

Use Eq. (2.180) to compute critical time step, $dtcr = \delta x / C$:

$dtcr = 1$
for ($n = 1$ *to* 20 *by increments of* 1) *begin*

$C = c0 * sqrt((1 + s * (1 - F(n)))/(1 - s * (1 - F(n)))\hat{}3)$

$dtcr = \min(dtcr, dX/C)$

end for

Determine $dtn = \delta t(j + \frac{1}{2})$:

$dtn = \min(1.05 * dt, alpha * dtcr)$
$time = time + dtn$

Use Eq. (A.18) to evaluate $x(0, j + 1)$ at left boundary:

$xn(0) = x(0) + dtn * v(0)$

Prescribe the right boundary condition:

$T(20) = 0$

Start spatial loop to compute solution at $j+1$:

for $(n = 1$ *to* 19 *by increments of* $1)$ *begin*

Use Eq. (A.20) to evaluate $a(n, j)$:

$a = (T(n+1) - T(n))/m$

Use Eq. (A.19) to evaluate $v(n, j+\frac{1}{2})$:

$v(n) = v(n) + 0.5 * (dt + dtn) * a$

Use Eq. (A.18) to evaluate $x(n, j+1)$:

$xn(n) = x(n) + dtn * v(n)$

Evaluate $F(n-\frac{1}{2}, j+1)$:

$Fnew = (xn(n) - xn(n-1))/dX$

Use Eqs. (A.23)–(A.25) to evaluate $Q(n-\frac{1}{2}, j+\frac{1}{2})$:

if $D < 0$ *then begin*

$\cdot\, D = 2/dtn * (Fnew - F(n))/(Fnew + F(n))$

$\cdot\, delx = 0.5 * (xn(n) - xn(n-1) + x(n) - x(n-1))$

$\cdot\, Q = -rho * (b2 * delx)\hat{}2 * D\hat{}2$

end if else $Q = 0$

Use Eqs. (2.179) and (A.26) to evaluate $T(n-\frac{1}{2}, j+1)$:

$Te = -rho * c0\hat{}2 * (1 - Fnew)/(1 - s * (1 - Fnew))\hat{}2$

$T(n) = Te + Q$

Update deformation gradient:

$F(n) = Fnew$

end for

Update positions and time step:

for $(n = 0$ *to* 19 *by increments of* $1)$ *do* $x(n) = xn(n)$

$dt = dtn$

if $(j$ *is integer multiple of* $20)$ *then plot* $v(n)$ *and print time*

- *end for*

In this computation we use

$$\delta t_{cr} = \text{minimum}(\delta X/C)$$

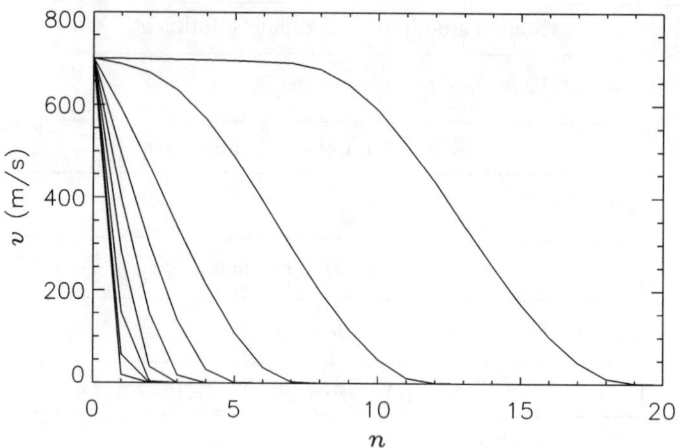

Figure A.9 Calculation of a steady wave in a nonlinear-elastic material with a linear-Hugoniot. Profiles are shown at every 20th time step when $t = [0.00079, 0.0029, 0.0084, 0.023, 0.062, 0.166, 0.441, 1.17, 2.40]$ μs.

to estimate the critical time step. We start the calculation with a very small time step $\delta t(\frac{1}{2}) = \delta t_{cr}/1000$. Then we allow the time step to grow by a factor of 1.05 per time step until it equals $0.4 \times \delta t_{cr}$. Thus we execute the computation at 40% of this estimate of the critical time step. Many other codes use a more sophisticated estimate of the critical time step, which allows execution of the algorithm as close as 95% of the critical time step.

We show the results of this calculation in Figure A.9. We have graphed the material velocity v against the discrete index n. Because the cell size is uniform, this is a plot of velocity in the material description. Profiles are plotted after every 20 time steps. They are *not* evenly spaced in time. Notice that the viscosity coefficient $b_2 = 7$ causes the discontinuous solution to evolve to a steady wave that is spread over approximately 7 discrete spatial points. The linear-Hugoniot model yields a steady-wave velocity of $U_S = 4400 + 1.44 \times 700 = 5.41$ km/s for this solution. This numerical solution yields the same wave propagation velocity.

The spreading of the solution in Figure A.9 and its evolution into a steady wave is the result of a balance between the artificial viscosity and the nonlinear material response. The nonlinear response causes the wave to steepen whereas the artificial viscosity causes the wave to disperse. In linear-elastic materials the nonlinear effect is not present. Consequently, the artificial viscosity used in this algorithm causes continued spreading of linear-elastic waves as time is advanced. Because steady waves are not attained, codes with artificial viscosity are not particularly useful for solving linear-elastic wave problems.

A.3.4 *Energy Equation and the Ideal Gas*

In this section we shall illustrate how the equation for the balance of energy can be included in the von Neumann–Richtmyer algorithm. We shall calculate the response of a volume of ideal gas to a sudden jump in velocity at the left boundary. This numerical solution exhibits an anomaly called the *wall-heating effect*.

For one-dimensional motion in the absence of heat conduction and external heating, the equation for the balance of energy is [see Eq. (3.10)]

$$\rho_0 \frac{\partial \hat{\mathcal{E}}}{\partial t} = T \frac{\partial \hat{F}}{\partial t}.$$

After substitution of the equation for the equilibrium pressure of an ideal gas [see Eq. (4.14)], we obtain

$$\frac{\partial \hat{\mathcal{E}}}{\partial t} = \left[(1-\gamma)\mathcal{E} + \frac{Q}{\rho} \right] D,$$

where Q is the artificial viscosity given by Eq. (A.22). The finite-difference approximation of this expression is

$$\frac{\mathcal{E}(n-\frac{1}{2}, j+1) - \mathcal{E}(n-\frac{1}{2}, j)}{\delta t(j+\frac{1}{2})} = \left\{ \frac{1-\gamma}{2} \left[\mathcal{E}\left(n-\frac{1}{2}, j+1\right) + \mathcal{E}\left(n-\frac{1}{2}, j\right) \right] + \frac{Q(n-\frac{1}{2}, j+\frac{1}{2})}{\rho(n-\frac{1}{2}, j+\frac{1}{2})} \right\} D\left(n-\frac{1}{2}, j+\frac{1}{2}\right).$$

This expression yields

$$\mathcal{E}\left(n-\frac{1}{2}, j+1\right) = \frac{\mathcal{E}\left(n-\frac{1}{2}, j\right) + \left[\left(\frac{1-\gamma}{2}\right)\mathcal{E}\left(n-\frac{1}{2}, j\right) + \dfrac{Q\left(n-\frac{1}{2}, j+\frac{1}{2}\right)}{\rho\left(n-\frac{1}{2}, j+\frac{1}{2}\right)} \right] \delta D\left(n-\frac{1}{2}, j+\frac{1}{2}\right)}{\left[1 - \left(\frac{1-\gamma}{2}\right)\delta D\left(n-\frac{1}{2}, j+\frac{1}{2}\right) \right]},$$

where

$$\rho\left(n-\frac{1}{2}, j+\frac{1}{2}\right) = \frac{1}{2}\rho_0\left(n-\frac{1}{2}\right) \left[\frac{1}{F\left(n-\frac{1}{2}, j+1\right)} + \frac{1}{F\left(n-\frac{1}{2}, j\right)} \right]$$

and

$$\delta D\left(n-\frac{1}{2}, j+\frac{1}{2}\right) = 2\left[\frac{F\left(n-\frac{1}{2}, j+1\right) - F\left(n-\frac{1}{2}, j\right)}{F\left(n-\frac{1}{2}, j+1\right) + F\left(n-\frac{1}{2}, j\right)} \right].$$

This result can be used in the algorithm listed at the end of Section A.3.2 as the fifth step directly after we evaluate $F(n-\frac{1}{2}, j+1)$. Once we determine $\mathcal{E}(n-\frac{1}{2}, j+1)$, the stress in the ideal gas is

$$T\left(n-\frac{1}{2}, j+1\right) = (1-\gamma)\rho_0\left(n-\frac{1}{2}\right) \frac{\mathcal{E}\left(n-\frac{1}{2}, j+1\right)}{F\left(n-\frac{1}{2}, j+1\right)} + Q\left(n-\frac{1}{2}, j+\frac{1}{2}\right).$$

To illustrate this algorithm, we repeat the calculation of the previous section after we replace the nonlinear-elastic material with an ideal gas. The following properties for air are

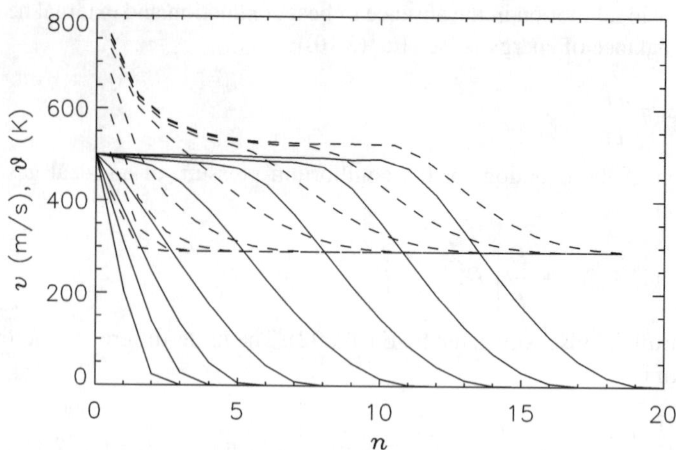

Figure A.10 Calculation of a steady wave in a ideal gas. Profiles for v (—) and ϑ (– –) are shown at every 20th time step when $t = [0.1, 0.367, 1.07, 2.96, 7.18, 11.2, 14.9, 18.6]$ μs.

used for the initial state:

$$\begin{aligned} \rho_0 &= 1.22 \text{ kg/m}^3, & \gamma &= 1.4, \\ c_v &= 733 \text{ J/kg-K}, & \vartheta_0 &= 290 \text{ K}, \\ b_2 &= 4, & alpha &= 0.25. \end{aligned}$$

We apply a constant velocity of 500 m/s to the left boundary. And the critical time step is determined by the material wave velocity for an ideal gas [see Eq. (4.27)]. The initial time step is $10^{-6}/c_0$.

We illustrate the solutions for both the velocity and the temperature in Figure A.10. The sharp temperature rise at the left boundary is called the *wall-heating effect.* Wall heating is not an artifact of the numerical procedure. Rather, it is caused by the heating due to the large viscous stresses at the early stages of the calculation. It will occur regardless of whether we use artificial viscosity or a physical viscosity term in the numerical procedure. Because heat conduction is absent, these anomalous temperatures remain in the numerical solution as time is advanced. One method of suppressing this anomaly is to include heat conduction in the numerical calculation†.

A.3.5 *Subcycling and the Elastic-Plastic Material*

In the coding structure of most wave-propagation computer algorithms, the equation for the balance of energy and the constitutive equations are usually solved simultaneously in a separate subroutine or procedure. A procedure is designed for each constitutive description. Often the procedure is quite complex. In some cases, while the main portion of the algorithm advances the solution from $t(j)$ to $t(j+1)$ in a single step or *cycle*, the numerical algorithm in the procedure for the constitutive description advances this portion of the solution from $t(j)$ to $t(j+1)$ in much smaller time steps. This process is called *subcycling* and it can greatly reduce the speed of the calculation. However, subcycling is an

† See "Errors for calculations of strong shocks using an artificial viscosity and an artificial heat flux" by Noh, W. F. in *Journal of Computational Physics* **72**, pp. 78–120 (1987).

extremely versatile tool. Often we use it to test a new constitutive formulation against experimental data. If the constitutive formulation proves to be of significant value in predicting these data, we can develop faster and more specialized solution algorithms later.

The constitutive model for the elastic-plastic material is widely employed in the numerical analysis of waves. It is usually solved with a very efficient numerical algorithm. However, to illustrate the method of subcycling, we shall analyze an elastic-plastic material with this type of algorithm. Consider an elastic-plastic material where the mean normal stress is given by

$$\sigma = -\frac{\rho_0 c_0^2 (1 - F)}{[1 - s(1 - F)]^2}$$

and the deviatoric stress is

$$T - \tilde{T} = 2\mu(F^E - 1).$$

Here the elastic and plastic components of the deformation gradient obey [see Eq. (4.138)]

$$F = F^E F^P$$

and the stress is

$$\begin{aligned} T &= \sigma + \frac{2}{3}(T - \tilde{T}) \\ &= -\frac{\rho_0 c_0^2 (1 - F)}{[1 - s(1 - F)]^2} + \frac{4}{3}\mu(F^E - 1). \end{aligned} \tag{A.27}$$

The flow rule for the plastic component of the deformation gradient is [see Eq. (4.164)]

$$\frac{\partial \hat{F}^P}{\partial t} = \begin{cases} bF^P \left(\frac{2}{3}|T - \tilde{T}| - \frac{2}{3}Y\right)^2 \operatorname{sgn}(T - \tilde{T}), & |T - \tilde{T}| \geq Y, \\ 0, & |T - \tilde{T}| < Y. \end{cases}$$

To apply the method of subcycling to these equations, we employ the first four steps of the von Neumann–Richtmyer algorithm listed at the end of Section A.3.2. With these four steps we solve for $a(n, j)$, $v(n, j - \frac{1}{2})$, $x(n, j + 1)$, and $F(n - \frac{1}{2}, j + 1)$. For the elastic-plastic material, we must next calculate $F^P(n - \frac{1}{2}, j + 1)$ before we can calculate $T(n - \frac{1}{2}, j + 1)$. To do this we shall change the flow rule into an ordinary differential equation of the following form:

$$\frac{d\mathcal{F}^P}{dt} = f(\mathcal{F}^P, t),$$

and integrate it with a standard numerical integration procedure such as a Runga-Kutta method‡.

To convert the flow rule into this form, we assume that both the deformation gradient and the plastic portion of the deformation gradient at $n - \frac{1}{2}$ are continuous functions of t, which we denote by $\mathcal{F}(t)$ and $\mathcal{F}^P(t)$, respectively. Between $t(j)$ and $t(j + 1)$ the deformation gradient is given by the following linear interpolation:

$$\mathcal{F}(t) = F\left(n - \frac{1}{2}, j\right) + \frac{F\left(n - \frac{1}{2}, j + 1\right) - F\left(n - \frac{1}{2}, j\right)}{t(j + 1) - t(j)}[t - t(j)].$$

‡ See *Numerical Recipes: The Art of Scientific Computing* by Press, W. H., Flannery, B. P., Teukolsky, S. A., and Vetterling, W. T. (Cambridge University Press, 1987).

Therefore,

$$T - \tilde{T} = 2\mu[\mathcal{F}(t)/\mathcal{F}^P(t) - 1],$$

and the flow rule becomes

$$\frac{d\mathcal{F}^P}{dt} = \begin{cases} \frac{16}{9}\mu^2 b\mathcal{F}^P\left(|\mathcal{B}| - \frac{Y}{2\mu}\right)^2 \operatorname{sgn}(\mathcal{B}), & |\mathcal{B}| \geq \frac{Y}{2\mu}, \\ 0, & |\mathcal{B}| < \frac{Y}{2\mu}, \end{cases}$$

where

$$\mathcal{B} = \mathcal{F}(t)/\mathcal{F}^P(t) - 1.$$

This is the required form for our ordinary differential equation. To integrate this equation for $\mathcal{F}^P(t)$ from $t(j)$ to $t(j+1)$, we use the initial condition $\mathcal{F}^P[t(j)] = F^P(n-\frac{1}{2}, j)$. After the integration procedure reaches $t = t(j+1)$, we let $F^P(n-\frac{1}{2}, j+1) = \mathcal{F}^P[t(j+1)]$.

We complete the algorithm by using Eq. (A.27) to calculate the stress $T(n-\frac{1}{2}, j+1)$. We do not include artificial viscosity in the stress expression because the flow rule is dissipative and we shall discretize the material to the scale of the thickness of the physical steady wave. Of course, it is also possible to use artificially large values of the coefficient b to accommodate coarser discretization of the spatial coordinate.

We illustrate this procedure by calculating the solution for a body with 200 spatial meshes where $\delta X = 1\ \mu$m. The body is initially free of stress and at rest. At $t = 0$ the velocity of the left boundary jumps to a constant value of 280 m/s while the right boundary is held fixed. The parameters for the computations are

$$\begin{aligned} \rho_0 &= 7.89\ \text{Mg/m}^3, & c_0 &= 4.403\ \text{km/s}, \\ s &= 1.441, & \mu &= 78.6\ \text{GPa}, \\ b &= 1.26 \times 10^{-11}\ (\text{Pa-s}^2)^{-1}, & Y &= 0.9\ \text{GPa}, \\ alpha &= 0.4. \end{aligned}$$

The initial time step is $10^{-8}/c_0 = 10^{-8}/4403$, and the critical time step is computed using Eq. (2.180). In Figure A.11, we use seven profiles to illustrate the evolution of the wave. The development of the structured plastic wave and the elastic precursor is evident. Because the plastic flow rule does not directly influence the elastic precursor, a small oscillation is present in this wave. Artificial viscosity can be used to suppress this oscillation.

A.3.6 *Early Use of the von Neumann–Richtmyer Algorithm*

The first applications of the von Neumann–Richtmyer algorithm predate the use of digital computers. In the 1940s during the Second World War, this algorithm was used to design the first atomic bombs in Los Alamos, New Mexico. A young physicist, named Richard Feynman, collected a group of people into one large room. He arranged them at desks with mechanical calculators such that each person represented a discretized spatial mesh point in the material. Each person's job was to advance their mesh point one time step after another time step until the calculation was completed. The waves in the calculation propagated from desk to desk, reflected off the "boundary desks," and returned again desk by desk.

This paradigm for calculating shock waves seems antiquated by todays standards. We have replaced all of those people with a high-speed digital computer that advances the

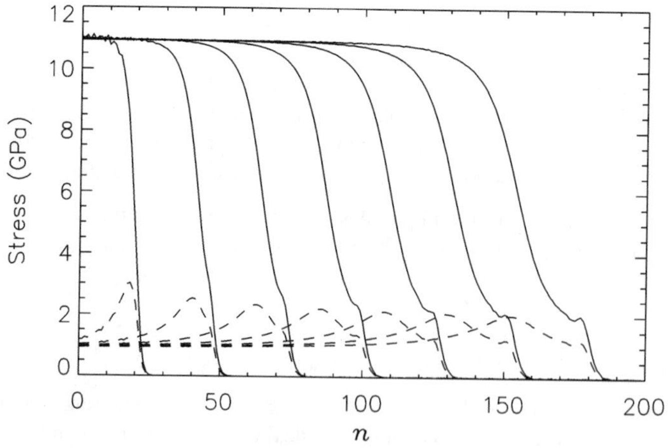

Figure A.11 The axial component of stress $-T$ (—) and the deviatoric stress $\tilde{T} - T$ (– –) in an elastic-plastic material. Profiles are shown at times $t = [3.7, 8.3, 12.9, 17.5, 22.2, 26.8, 31.4]$ ns

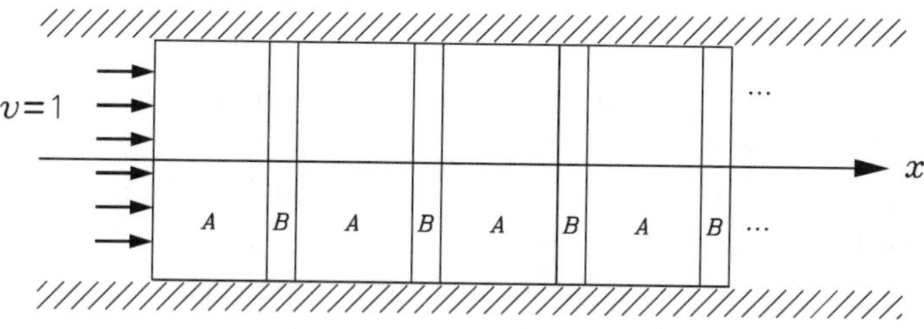

Figure A.12 Exercise A.1

calculation many millions of times faster. We have also found new uses for these calculations such as modeling the operation of a lithotripter, which is a noninvasive medical technique for fragmenting human kidney stones with shock waves. However, some old ideas persist, and the most advanced applications of this algorithm use the method of parallel processing, which involves collecting a large number of microcomputers into a single network. Each computer's job is to advance its mesh point one time step after another time step until the calculation is completed.

A.3.7 *Exercises*

A.1 Consider the layered material that we have illustrated in Figure A.12. It is constructed of 50 layers of two types of linear-elastic material. The odd-numbered layers are constructed of material A in which the speed of sound is c_A and the acoustic impedance is z_A. All odd-numbered layers have the same thickness $9c_A\Delta$, where Δ is a prescribed time increment. The even-numbered layers are constructed of material B. The speed of sound, acoustic impedance, and thickness of the even-numbered layers are c_B, $z_B = 300z_A$, and $c_B\Delta$, respectively. This periodically layered medium is initially at rest and free of stress. At $t = 0$, the velocity of the left boundary jumps to $v = 1$ and remains constant. The right boundary is fixed. Use our linear algorithm to calculate the resulting wave. At the interface between the 4th and 5th layers, plot the wave history for v from $t = 0$ to $t = 500\Delta$.

The resulting wave profile is called an *undulating bore* because it resembles a tidal bore or tidal surge up an estuary. For the tidal bore, the structure of the undulation is determined by the depth of the estuary, whereas here it is determined by the thicknesses of the layers. The shape of this profile is not an artifact of the numerical algorithm. In fact, our algorithm for the analysis of the linear-elastic wave equation yields an analytical solution to this problem.

A.2 Repeat the calculation shown in Figure A.9, but this time assume the material is twice as thick so that $0 \le n \le 40$. Determine the velocity profiles for both the spatial and material descriptions from $t = 0$ to $t = 11\ \mu s$ for increments of $1\ \mu s$. It may be necessary for you to adjust the maximum time step.

(a) Plot these results when the right boundary condition is $T(40) = 0$.

(b) Plot these results when the right boundary condition is $v(40) = 0$.

A.3 In Section 2.6.2 we examined the analytic solution for a centered rarefaction wave in an adiabatic gas. Consider the calculation for the ideal gas, which is shown in Figure A.10. If we change the boundary conditions for this calculation to $v(0) = 0$ and $T(20) = -10$ kPa, a centered rarefaction wave will be generated at the right boundary.

(a) By using the analytical solution for a centered rarefaction wave, compute the velocity behind the tail contour of the wave.

(b) Set *alpha* $= 0.2$ and compute the numerical solution to $t = 0.5$ ms.

(c) Graph the solutions for v and ϑ at every 20th time step in the material description.

(d) Graph the solution for v at every 20th time step in the spatial description.

(e) Compare your numerical solution for the velocity behind the rarefaction wave to your analytical solution from part (a).

A.4 In Section A.3.4 we derive an algorithm for the ideal gas. Use this method to derive an algorithm for the Mie–Grüneisen material.

A.5 Estimate the velocity of the steady wave in Figure A.10 and compare your result to the analytical solution for the Hugoniot of an ideal gas.

A.6 In Figure A.10, we show a calculation that uses a quadratic artificial viscosity to compute a steady wave in an ideal gas.

(a) Derive an algorithm using the linear artificial viscosity,

$$Q = \begin{cases} \rho_0 b_1 \delta x D, & D < 0, \\ 0, & D \ge 0. \end{cases}$$

(b) Use this Q to repeat the calculations in Figure A.10.

(c) Select a value of b_1 to produce steady wave profiles that you can compare to those in Figure A.10.

Material Properties

Many of the properties for linear-elastic materials in this section are compiled from results reported by Selfridge[†]. The linear-Hugoniot data are largely complied from Marsh[‡]. The plastic flow-rule data are from Swegle and Grady[#].

Table B.1 Relationships between Young's modulus E^0, Poisson's ratio ν, and the Lamé constants λ and μ.

	E^0	ν	μ	λ
E^0, ν			$\dfrac{E^0}{2(1+\nu)}$	$\dfrac{E^0\nu}{(1+\nu)(1-2\nu)}$
E^0, μ		$\dfrac{E^0-2\mu}{2\mu}$		$\dfrac{\mu(E^0-2\mu)}{3\mu-E^0}$
E^0, λ		$\dfrac{2\lambda}{E^0+\lambda+R}$	$\dfrac{E^0-3\lambda+R}{4}$	
μ, ν	$2\mu(1+\nu)$			$\dfrac{2\mu\nu}{1-2\nu}$
ν, λ	$\dfrac{\lambda(1+\nu)(1-2\nu)}{\nu}$		$\dfrac{\lambda(1-2\nu)}{2\nu}$	
μ, λ	$\dfrac{\mu(3\lambda+2\mu)}{\lambda+\mu}$	$\dfrac{\lambda}{2(\lambda+\mu)}$		

Note: $R = \sqrt{(E^0)^2 + 9\lambda^2 + 2E^0\lambda} > 0$

Table B.2 Elastic constants of various solids.

Material	ρ_0 (Mg/m^3)	c_0 (km/s)	c_{0T} (km/s)	E^0 (GPa)	ν	λ (GPa)	μ (GPa)
Aluminum (rolled)	2.70	6.42	3.04	67.6	0.35	61.3	24.9
Beryllium	1.87	12.8	8.88	309.0	0.04	15.7	147.0
Bismuth	9.8	2.2	1.1	31.6	0.33	23.7	11.8
Brass (yellow)	8.64	4.70	2.10	104.0	0.37	114.0	38.1
Cadmium	8.6	2.8	1.5	50.2	0.29	28.7	19.3
Copper (rolled)	8.93	5.01	2.27	126.0	0.37	132.0	46.0
Epon 828	1.21	2.83	1.23	5.06	0.38	6.02	1.83
Fused silica	2.20	5.70	3.75	69.2	0.11	9.60	30.9

[†] See "Approximate material properties in isotropic materials" by Selfridge, A. R. in *IEEE Transactions on Sonics and Ultrasonics* **SU-32**, pp. 381–394 (1985).

[‡] See *LASL Shock Hugoniot Data*, ed. Marsh, S. P. (University of California Press, 1980).

[#] See "Shock viscosity and the prediction of shock wave rise times" by Swegle, J. W. and Grady, D. E. in *Journal of Applied Physics* **58**, pp. 692–701 (1985).

Table B.2 (*Cont.*)

Material	ρ_0 (Mg/m^3)	c_0 (km/s)	c_{0T} (km/s)	E^0 (GPa)	ν	λ (GPa)	μ (GPa)
Glass (pyrex)	2.24	5.64	3.28	59.9	0.24	23.0	24.0
Gold (hard drawn)	19.7	3.24	1.20	80.5	0.42	150.0	28.3
Ice	0.91	3.99	1.98	9.61	0.33	7.40	3.59
Iron	7.69	5.9	3.2	203.0	0.29	110.0	78.7
Iron (cast)	7.22	4.6	2.6	123.0	0.26	55.1	48.8
Lead	11.2	2.2	0.7	15.8	0.44	43.2	5.48
Magnesium	1.73	5.8	3.0	41.2	0.31	27.1	15.6
Molybdenum	10.0	6.3	3.4	299.0	0.29	165.0	115.0
Nickel	8.84	5.6	3.0	206.0	0.29	118.0	79.5
Platinum	21.4	3.26	1.73	167.0	0.30	99.3	64.0
Silver	10.6	3.6	1.6	74.7	0.37	83.1	27.1
Steel (mild)	7.8	5.9	3.2	206.0	0.29	111.0	79.8
Steel (stainless)	7.89	5.79	3.1	196.0	0.29	112.0	75.8
Tantalum	16.6	4.1	2.00	178.0	0.34	146.0	66.4
Titanium carbide	5.15	8.27	5.16	323.0	0.18	77.9	137.0
Tungsten	19.4	5.2	2.9	415.0	0.27	198.0	163.0
Zircaloy	9.36	4.72	2.36	139.0	0.33	104.0	52.1
Nylon (6/6)	1.12	2.6	1.1	3.77	0.39	4.86	1.35
Polyethylene	0.90	1.95	0.54	0.76	0.45	2.89	0.26
Polystyrene	1.05	2.4	1.15	3.75	0.35	3.27	1.38

Table B.3 Elastic constants of various liquids.

Material	ρ_0 (Mg/m^3)	c_0 (km/s)	λ (GPa)
Acetonyl acetone	0.72	1.40	1.42
Alcohol (ethanol 25°C)	0.79	1.20	1.15
Alcohol (methanol)	0.79	1.10	0.96
Argon (87 K)	1.43	0.84	1.00
Carbon tetrachloride (25°C)	1.59	0.92	1.36
Chloroform (25°C)	1.49	0.98	1.45
Fluorinert (FC-40)	1.86	0.64	0.76
Gasoline	0.80	1.25	1.25
Glycol (polyethylene 200)	1.08	1.62	2.85
Helium-4 (2 K)	0.14	0.22	0.01
Honey (Sue Bee orange)	1.42	2.03	5.85
Kerosene	0.81	1.32	1.41
Nitromethane	1.13	1.33	1.99
Oil (baby)	0.82	1.43	1.67
Oil (jojoba)	1.17	1.45	2.47
Oil (olive)	0.91	1.44	1.91
Oil (peanut)	0.91	1.43	1.88
Oil (SAE 30)	0.87	1.7	2.51
Oil (sperm)	0.88	1.44	1.82
Turpentine (25°C)	0.88	1.25	1.38
Water (20°C)	1.10	1.4	2.16

Table B.4 Elastic constants of various gases.

Material	ρ_0 (kg/m^3)	c_0 (km/s)	λ (MPa)
Air (dry at 273 K)	1.29	0.33	0.14
Carbon dioxide (273 K)	1.97	0.25	0.13
Helium (273 K)	0.17	0.96	0.16
Nitrogen (273 K)	1.25	0.33	0.13
Oxygen (273 K)	1.42	0.31	0.14
Oxygen (293 K)	1.32	0.32	0.14

Table B.5 Linear-Hugoniot constants of various elements. The Hugoniot is $U_s = c_0 + sv$. The speed of sound is c_L, and the velocity of a transverse wave is c_{0T}. Notice that the constant c_0 in the Hugoniot relation is not necessarily equal to the speed of sound (See Section 2.7.7).

Material	ρ_0 (Mg/m^3)	c_0 (km/s)	s	Γ	c_L (km/s)	c_{0T} (km/s)
Beryllium, sintered	1.85	7.99	1.13		13.2	8.97
Cadmium	8.64	2.48	1.64		3.20	1.65
Calcium	1.55	3.63	0.94		4.39	2.49
Carbon, diamond	3.19	7.81	1.43			
Carbon, foamed	0.56	0.36	1.22			
Carbon, foamed	0.48	0.26	1.18			
Carbon, foamed	0.32	0.63	0.99			
Carbon, powdered graphite	0.47	0.44	1.44			
Carbon, graphite	1.01	0.79	1.30			
Cobalt	8.82	4.77	1.28			
Copper	8.92	3.91	1.51	1.96	4.76	2.33
Copper, sintered	5.7	0.71	1.97		2.68	1.52
Gold	19.2	3.07	1.54	2.97	3.25	1.19
Iron	7.85	3.57	1.92	1.69		
Lead	11.3	2.03	1.47		2.25	0.89
Lithium	0.53	4.58	1.15			
Magnesium, AZ31B	1.78	4.52	1.26	1.43	5.74	3.15
Mercury	13.5	1.75	1.72		1.45	0.00
Nickel	8.88	4.59	1.44		5.79	3.13
Platinum	21.4	3.63	1.47	2.40	4.08	1.76
Silver	10.5	3.27	1.55		3.71	1.66
Sodium	0.97	2.58	1.24			
Tantalum	16.7	3.43	1.19		4.16	2.09
Tin	7.29	2.59	1.49		3.43	1.77
Tungsten	19.2	4.04	1.23		5.22	2.89
Zinc	7.14	3.03	1.55			

Table B.6 Linear-Hugoniot constants of various solids. The Hugoniot is $U_s = c_0 + sv$. The speed of sound is c_L, and the velocity of a transverse wave is c_{0T}. The constant c_0 is not necessarily equal to the speed of sound (See Section 2.7.7).

Material	ρ_0 (Mg/m^3)	c_0 (km/s)	s	Γ	c_L (km/s)	c_{0T} (km/s)
Aluminum, 1100	2.71	5.38	1.34	2.25	6.38	3.16
Aluminum, 2024	2.78	5.37	1.29	2.00	6.36	3.16
Aluminum, 921T	2.83	5.15	1.37	2.10	6.29	3.11
Brass	8.45	3.47	1.43		4.41	2.13
Steel 304	7.89	4.58	1.49		5.77	3.12
Uranium-8.3 wt% Mo	17.3	2.66	1.51		3.08	1.32
Sodium chloride, powdered-unpressed	0.87	1.06	1.25			
Tantalum carbide	12.6	3.32	1.49			
Uranium dioxide	6.3	0.43	1.70			
Oil shale, Green River Rifle, Colorado	2.19	3.78	1.15			
Eclogite Sunnmore, Norway	3.55	6.26	1.02		7.35	4.44
Tuff, Nevada nuclear test site	1.7	1.32	1.41			
Tuff, Nevada water-saturated	1.9	1.92	1.54			
Adiprene	1.09	2.33	1.54			
Melmac	1.45	3.51	1.21			
Micarta	1.40	3.05	1.42		2.67	1.50
Neoprene	1.44	2.79	1.42			
Paraffin	0.92	3.12	1.47		2.18	0.83
Phenolic furfural-filled	1.38	2.85	1.40			
Polychloro -trifluoroethylene	2.12	2.03	1.64		1.74	0.77
Polyurethane, foamed	0.28	0.87	1.04			
Polyurethane, foamed	0.16	0.32	1.15			
Polyurethane, foamed	0.09	1.11	0.78			
Silastic, RTV-521	1.37	1.84	1.44			
Epoxy 40 vol% forsterite	2.21	2.90	1.81			
Silicon carbide 80 wt% carbon	1.32	0.89	1.42			
Birch wood	0.69	0.65	1.44			
Cherry wood	0.60	0.64	1.37			
Douglas fir	0.54	0.45	1.38			
Sugar pine	0.45	0.45	1.33			

Table B.7 Linear-Hugoniot constants of various liquids. The Hugoniot is $U_s = c_0 + sv$. The speed of sound is c_L. The constant c_0 is not necessarily equal to the speed of sound (See Section 2.7.7).

Material	ρ_0 (Mg/m^3)	c_0 (km/s)	s	Γ	c_L (km/s)
Acetone	0.785	1.94	1.38		
Ethyl alcohol	0.786	1.73	1.57		
Methyl alcohol	0.792	1.78	1.49		
Ammonia, liquid	0.726	2.00	1.51		
Ethylene glycol	1.112	2.15	1.55		1.60

Table B.8 Values of the plastic flow-rule coefficient for selected materials [see Eq. (4.164)].

Material	b (Pa^2-s)$^{-1}$
Al	1×10^{-10}
Be	5×10^{-11}
Bi	6×10^{-10}
Cu	3×10^{-10}
Fe	3×10^{-11}
MgO	7×10^{-12}
Fused silica	3×10^{-12}
Stainless steel	1×10^{-11}
U	2×10^{-12}

References

Achenbach, J. D., *Wave Propagation in Elastic Solids*, North-Holland Publishing Company, Amsterdam, 1975.

Asay, J. R. and Shahinpoor, M. (editors), *High-Pressure Shock Compression of Solids*, Springer-Verlag, New York, 1992.

Bedford, A. and Drumheller, D. S., *Introduction to Elastic Wave Propagation*, Wiley, Chichester, 1994.

Brekhovskikh, L. M., *Waves in Layered Media*, Academic Press, New York, 1980.

Brillouin, L., *Wave Propagation in Periodic Structures*, Dover Publications, Inc., New York, 1953.

Brillouin, L., *Wave Propagation and Group Velocity*, Academic Press, New York, 1960.

Burke, J., *The Kinetics of Phase Transformation in Metals*, Pergamon Press, Oxford, 1965.

Burke, J. J. and Weiss, V. (editors), *Shock Waves and the Mechanical Properties of Solids*, Syracuse University Press, Syracuse, 1971.

Callen, H. B., *Thermodynamics*, John Wiley, New York, 1960.

Courant, R. and Friedrichs, K. O., *Supersonic Flow and Shock Waves*, Springer-Verlag, Heidelberg, 1985.

Ewing, W. M., Jardetzky, W. S., and Press, F., *Elastic Waves in Layered Media*, McGraw-Hill Book Company, New York, 1957.

Fickett, W., *Introduction to Detonation Theory*, University of California Press, Berkeley, 1985.

Graff, K. F., *Wave Motion in Elastic Solids*, Ohio State University Press, Columbus, Ohio, 1975.

Graham, R. A., *Solids Under High-Pressure Shock Compression*, Springer-Verlag, New York, 1993.

Guggenheim, M. A., *Thermodynamics*, North-Holland Publishing Company, Amsterdam, 1967.

Kinslow, R. (editor), *High-Velocity Impact Phenomena*, Academic Press, New York 1970.

Kolsky, H., *Stress Waves in Solids*, Dover Publications, Inc., New York, 1963.

Lighthill, J., *Waves in Fluids*, Cambridge University Press, Cambridge, 1980.

Malvern, L. E., *Introduction to the Mechanics of a Continuous Medium*, Prentice-Hall, Englewood Cliffs, 1969.

Marsh, S. P. (editor), *LASL Shock Hugoniot Data*, University of California Press, Berkeley, 1980.

Miklowitz, J., *The Theory of Elastic Waves and Waveguides*, North-Holland Publishing Company, Amsterdam, 1978.

Morse, P. M. and Ingard, K. U., *Theoretical Acoustics*, McGraw-Hill Book Company, New York, 1968.

Press, W. H., Flannery, B. P., Teukolsky, S. A., and Vetterling, W. T., *Numerical Recipes*, Cambridge University Press, Cambridge, 1986.

Roache, P. J., *Computational Fluid Dynamics*, Hermosa Publishers, Albuquerque, New Mexico, 1972.

Sears, F. W., *An Introduction to Thermodynamics, the Kinetic Theory of Gases, and Statistical Mechanics*, Addison-Wesley Publishing Company, Reading, Massachusetts, 1953.

Shapiro, A. H., *The Dynamics and Thermodynamics of Compressible Fluid Flow*, vols. I and II, The Ronald Press Company, New York, 1953.

Sommerfeld, A., *Thermodynamics and Statistical Mechanics*, Academic Press, New York, 1964.

ıll, C., *Rational Thermodynamics*, McGraw-Hill Book Company, New York, 1969.
ıam, G. B., *Linear and Nonlinear Waves*, Wiley, New York, 1974.
ʼdovich, Y. B. and Raizer, Y. P., *Physics of Shock Waves and High-Temperature Hydrodynamic Phenomena*, vols. I and II, Academic Press, New York, 1966.
Zukas, J. A. (editor), *High Velocity Impact Dynamics*, Wiley, New York, 1990.

Index